HESSIAN POLYHEDRA, INVARIANT THEORY AND APPELL HYPERGEOMETRIC FUNCTIONS

HESSIAN POLYHEDRA, INVARIANT THEORY AND APPELL HYPERGEOMETRIC FUNCTIONS

Lei Yang

Peking University, China

World Scientific

NEW JERSEY • LONDON • SINGAPORE • BEIJING • SHANGHAI • HONG KONG • TAIPEI • CHENNAI • TOKYO

Published by

World Scientific Publishing Co. Pte. Ltd.
5 Toh Tuck Link, Singapore 596224
USA office: 27 Warren Street, Suite 401-402, Hackensack, NJ 07601
UK office: 57 Shelton Street, Covent Garden, London WC2H 9HE

Library of Congress Cataloging-in-Publication Data
Names: Yang, Lei (Mathematics professor), author.
Title: Hessian polyhedra, invariant theory, and Appell hypergeometric functions /
by Lei Yang (Peking University, China).
Description: New Jersey : World Scientific, 2017. | Includes bibliographical references.
Identifiers: LCCN 2017042755 | ISBN 9789813209473 (hardcover : alk. paper)
Subjects: LCSH: Affine algebraic groups. | Group schemes (Mathematics) |
Transformations (Mathematics) | Polyhedra. | Hypergeometric functions.
Classification: LCC QA179 .Y36 2017 | DDC 516/.156--dc23
LC record available at https://lccn.loc.gov/2017042755

British Library Cataloguing-in-Publication Data
A catalogue record for this book is available from the British Library.

For any available supplementary material, please visit
http://www.worldscientific.com/worldscibooks/10.1142/10434#t=suppl

Preface

In this monograph, we give the complex counterpart of Klein's classic book on the icosahedron, i.e., a story about complex regular polyhedra. We show that the following four apparently disjoint theories: the symmetries of the Hessian polyhedra (geometry), the resolution of some system of algebraic equations (algebra), the system of partial differential equations of Appell hypergeometric functions (analysis) and the modular equation of Picard modular functions (arithmetic) are in fact dominated by the structure of a single object, the Hessian group $\mathfrak{G}'_{216}$. There are two finite unitary groups generated by reflections corresponding to this collineation group. One, of order 648, is the symmetry group of the complex regular polyhedron $3\{3\}3\{3\}3$, and the other, of order 1296, is the symmetry group of the regular complex polyhedron $2\{4\}3\{3\}3$ or its reciprocal $3\{3\}3\{4\}2$.

Contents

Part I. Geometry and arithmetic associated with Appell hypergeometric functions

Part II. Hessian polyhedra, invariant theory and Appell hypergeometric functions

Part III. Galois representations arising from twenty-seven lines on a cubic surface and the arithmetic associated with Hessian polyhedra

Part I. Geometry and arithmetic associated with Appell hypergeometric functions

Contents

1. Introduction

Mathematicians often speak of the unity of their subject. Such a unity from the 19th century origins in the history of linear ordinary differential equations, especially those closely connected with elliptic and modular functions. It is concerned with the impact of group theoretical and geometrical ideas upon the problem of understanding the nature of the solutions to a differential equation (see [21], [22] and [13] as excellent surveys).

The theory of hypergeometric functions has a long history (see [23], [1], [2], [17], [59], [61] and [62]). The first occurrence of hypergeometric series goes back to the 17th century. Euler was the first to consider the series as a function of one variable x:

$$F(a,b;c;x)=1+\frac{a\cdot b}{1\cdot 2}x+\frac{a(a+1)\cdot b(b+1)}{1\cdot 2\cdot c(c+1)}x^2+\cdots.$$

Euler found that F is a solution of the hypergeometric differential equation

$$x(1-x)y''+[c-(a+b+1)x]y'-aby=0$$

and up to a constant factor F can be expressed as a hypergeometric integral

$$\int_0^1 t^{a-1}(1-t)^{c-a-1}(1-xt)^{-b}dt.$$

The next developments are closely connected with the name of Gauss, who considered the hypergeometric function as a function of a complex variable. In a letter to Schumacher (September 17, 1808), Gauss outlined his new modern point of view (Gauss' program) as follows:

(1) Proof of the existence of functions which satisfy our differential equation.

(2) Investigation of the properties of these functions. We look upon every differential equation as defining a specific (possibly new) class of transcendental functions.

For this project the work of many outstanding mathematicians was needed: Abel, Cauchy, Dedekind, Fuchs, Jacobi, Klein, Kummer, Poincaré, Riemann, Schwarz, Weierstrass and others. The theory of hypergeometric functions was a central theme in the function theory of one complex variable. Now, we will give a survey as follows:

In 1812, Gauss studied the hypergeometric equation

$$x(1-x)\frac{d^2y}{dx^2} + [c - (a+b+1)x]\frac{dy}{dx} - aby = 0.$$

This equation is important in its own right as a linear ordinary differential equation for which explicit power-series solutions can be given. Gauss' work in this direction was extended by Riemann in 1857, who introduced the crucial idea of analytically continuing the solutions around their singularities in the complex domain. Riemann's monodromy approach gave a thorough global understanding of the solutions of the hypergeometric equation. In 1872, Schwarz studied the hypergeometric equations all of whose solutions were algebraic functions (see [39]). It led to Schwarz's discovery of a new class of transcendental functions associated with the hypergeometric equation which turned out to be of vital importance in the theory of automorphic functions. The concept of Schwarzian derivatives played a central role in his study.

It is known that the Schwarzian derivative

$$\{w; x\} := \frac{w'''}{w'} - \frac{3}{2}\left(\frac{w''}{w'}\right)^2$$

satisfies that

$$\left\{\frac{aw+b}{cw+d}; x\right\} = \{w; x\} \quad \text{for} \quad \begin{pmatrix} a & b \\ c & d \end{pmatrix} \in GL(2, \mathbb{C}),$$

where x is a complex number and w is a complex-valued function of x.

The Schwarzian derivative enters into several branches of complex analysis. It plays an important role in the theory of linear second order differential equations, in conformal mapping, and in the study of univalence and of uniformization (see [16], [30], [31], [59], [60], [63] and [64]).

The Schwarzian derivative also appears in the conformal field theory (see [37]), which plays an important role in mathematics, statistical mechanics and string theory. Some infinite dimensional algebras are related to the conformal field theory, such as Virasoro algebra, Kac-Moody Lie algebra and Hopf algebra.

Let us recall some basic facts about the Virasoro algebra (see [4] and [58]). For $A = \mathbb{C}[t, t^{-1}]$, $\partial = \frac{d}{dt}$, the classical residue

$$\text{res} : \Omega \to \mathbb{C}, \quad \text{res}\left(\sum_i c_i t^i dt\right) = c_{-1}$$

defines the isomorphism $\Omega/dA = \mathbb{C}$. The Virasoro algebra V is a central extension of the Lie algebra $A\frac{d}{dt}$ by $\mathbb{C}z_0$ with the commutation rule

$$\left[f\frac{d}{dt}, g\frac{d}{dt}\right]_{z_0} = \left(f\frac{dg}{dt} - g\frac{df}{dt}\right)\frac{d}{dt} + \frac{1}{12}\mathrm{res}\, g\frac{d^3 f}{dt^3}dt \cdot z_0.$$

The element z_0 or its proper value in a representation space of V is called the central charge.

The list of fascinating aspects of the Korteweg-de Vries equation is long. The well-known Miura transformations, Schwarzian derivatives and the Hill operator are three examples from this list. It is known that (see [24], [50], [10], [55], [56] and [57]) the evolution equation

$$u_t = u_{xxx} - \frac{3}{2}u_{xx}^2 u_x^{-1}$$

for a function $u(x,t)$ is invariant with respect to the group $PSL(2)$ of linear fractional transformations

$$u \mapsto \frac{au+b}{cu+d},$$

and implies that the Korteweg-de Vries equation

$$v_t = v_{xxx} + 6vv_x$$

for the variable

$$v = \frac{1}{2}\left(u_{xxx}u_x^{-1} - \frac{3}{2}u_{xx}^2 u_x^{-2}\right) = \frac{1}{2}\{u, x\}.$$

The above evolution equation is equivalent to

$$\frac{u_t}{u_x} = \{u, x\}.$$

The hypergeometric equation is also important because it contains many interesting equations as special cases. We are interested in the Legendre's equation:

$$(1-\kappa^2)\frac{d^2y}{d\kappa^2} + \frac{1-3\kappa^2}{\kappa}\frac{dy}{d\kappa} - y = 0.$$

There are two linearly independent solutions:

$$K = \int_0^1 \frac{dx}{\sqrt{(1-x^2)(1-\kappa^2x^2)}}, \quad K' = \int_1^{\frac{1}{\kappa}} \frac{dx}{\sqrt{(1-x^2)(1-\kappa^2x^2)}},$$

which are the periods of elliptic integrals considered as functions of the modulus κ^2. This equation was studied by Legendre, Gauss, Abel, Jacobi and Kummer. It is important in the study of the arithmetical theory of modular functions.

The arithmetical theory of modular functions is a result of the synthesis of numerous ideas of many outstanding mathematicians (see [53] for an exciting survey). The complex multiplication of elliptic functions is one of the main sources of this theory. Complex multiplication theory is usually associated with the name of Kronecker, who contributed much to it. Some of the ideas of complex multiplication theory can be found in earlier papers by Gauss, Abel and Eisenstein (see [53] and [27]).

The transform problem for elliptic functions, which is related to the study of isogenies of elliptic curves, appeared on the historic stage in the following way. In 1750's, Fagnano and Euler discovered a transform of $\frac{dx}{\sqrt{F(x)}}$ to $\frac{dy}{\sqrt{G(y)}}$, where F and G are polynomials of degree four. This transform is given by the equation $f(x,y)=0$, where f is a polynomial of second degree in each variable. In 1775, Landen found the Landen transform which is of order two. In 1824, Legendre discovered a transform of order three. However, transforms of orders 3, 5 and 7 had been known to Gauss since 1808.

The general transform problem for an arbitrary order was stated by Jacobi [19] (see also [54] for the details). He proved the existence of a transform of any odd order n, namely $y = U(x)/V(x)$, where U and V are mutually prime polynomials, U is an odd polynomial of degree n, and V is an even polynomial of degree $n-1$, and

$$\frac{dy}{\sqrt{(1-y^2)(1-\lambda^2 y^2)}} = \frac{1}{M}\frac{dx}{\sqrt{(1-x^2)(1-\kappa^2 x^2)}},$$

where M is the so-called multiplicator and λ is the transformed modulus (κ being given). This leads to the study of modular equations which are of great importance in the theory of elliptic functions (see [53]). Let n be an odd prime. The roots of the modular equation are the $(n+1)$ values of moduli λ which are images of the modulus κ via all the maps of order n. For $n=3$, Legendre obtained (1824) that the modular equation has the form: $\sqrt{\kappa\lambda}+\sqrt{\kappa'\lambda'}=1$, where $\kappa^2+\kappa'^2=1$ and $\lambda^2+\lambda'^2=1$. Jacobi derived the modular equation for $n=3,5$ (see [54]). In 1832, Galois claimed that the Galois group of the modular equation is $PGL(2,\mathbb{F}_n)$, containing the subgroup $PSL(2,\mathbb{F}_n)$ of index 2 and of order $\frac{1}{2}n(n^2-1)$. For and only for $n=5,7,11$, $PSL(2,\mathbb{F}_n)$ contains a subgroup of index n. Hence in these cases the modular equation can be reduced to an equation of degree n. In 1858, Hermite (see [15], pp.5-12), Kronecker and Brioschi used the modular equation to solve algebraic equations of degree four and five. A general equation of degree four is reduced to the modular equation of order three, a general equation of degree five is reduced to the modular equation of degree five. Klein (1879) studied the equation with the group $PGL(2,\mathbb{F}_7)$, the Galois group of the modular equation of degree seven.

Klein's work grew out of an attempt to reformulate the theory of modular functions and modular transformations without using the theory of elliptic functions. This had been almost completely achieved by Dedekind in his celebrated paper of 1877 (see [7]). Klein (see [20]) enriched Dedekind's approach with an explicit use of group theoretic ideas and an essentially Galois theoretic approach to the fields of rational and modular functions. Both Dedekind (see [7]) and Klein (see [20]) relied on the theory of the hypergeometric equation at technical points in their arguments. In particular, for $0, 1, \lambda, \infty \in \mathbb{P}^1(\mathbb{C})$, $\lambda \neq 0, 1$, Klein (see [20]) set

$$J = J(\lambda) = \frac{4}{27} \frac{(\lambda^2 - \lambda + 1)^3}{\lambda^2(\lambda - 1)^2}. \tag{A.0}$$

It is well-known that the Gauss hypergeometric function

$$F(a, b; c; x) = \frac{\Gamma(c)}{\Gamma(a)\Gamma(c-a)} \int_0^1 t^{a-1}(1-t)^{c-a-1}(1-tx)^{-b} dt$$

satisfies the following equation:

$$x(1-x)\frac{d^2y}{dx^2} + [c - (a+b+1)x]\frac{dy}{dx} - aby = 0.$$

For the group $GL(2)$, the elliptic integral

$$y = \int_0^1 \frac{dt}{\sqrt{t(t-1)(t-x)}} \tag{A.1}$$

satisfies the following equation:

$$x(x-1)\frac{d^2y}{dx^2} + (2x-1)\frac{dy}{dx} + \frac{1}{4}y = 0. \tag{A.2}$$

Thus, a solution is given by

$$y = F\left(\frac{1}{2}, \frac{1}{2}; 1; x\right), \quad (a, b, c) = \left(\frac{1}{2}, \frac{1}{2}, 1\right). \tag{A.3}$$

The elliptic modular function is associated to the following equation (see [7] and [20]):

$$x(1-x)\frac{d^2y}{dx^2} + \left(\frac{2}{3} - \frac{7}{6}x\right)\frac{dy}{dx} - \frac{1}{144}y = 0. \tag{A.4}$$

A solution is given by

$$y = F\left(\frac{1}{12}, \frac{1}{12}; \frac{2}{3}; x\right), \quad (a, b, c) = \left(\frac{1}{12}, \frac{1}{12}, \frac{2}{3}\right). \tag{A.5}$$

Poincaré's papers (see [36]) in the 1880's on Fuchsian and Kleinian groups are of great interest. The papers are the climax of the geometric theory of functions initiated by Riemann, and the ideal representatives of the unity between analysis, geometry, topology and algebra. In particular, Poincaré proved two theorems:

Theorem 1. *Between two Fuchsian functions with the same group and without any essential singular point which is not a consequence of their definition, there is an algebraic relation.*

Theorem 2. *Any Fuchsian function $F(z)$ permits the integration of a linear equation with algebraic coefficients in the following way. If we put*

$$x = F(z), \quad y_1 = \sqrt{\frac{dF}{dz}}, \quad y_2 = z\sqrt{\frac{dF}{dz}},$$

then y_1 and y_2 satisfy the differential equation

$$\frac{d^2y}{dx^2} = y\phi(x),$$

where $\phi(x)$ is algebraic in x.

In fact, the hypergeometric functions are heavily involved in special functions, representation theory, theory of automorphic forms, elliptic curves and integrals and more algebraic geometry, and number theory.

In 1880 the two variable story began when Appell gave a generalization of the hypergeometric equation as an equation in two variables, and found four double series generalizing the hypergeometric series which he called F_1, F_2, F_3 and F_4. Appell showed that each of these new functions satisfies two simultaneous partial differential equations (see [3]). In particular, the function $F_1 = F_1(a; b, b'; c; x, y)$ satisfies the following partial differential equations:

$$\begin{cases} x(1-x)\dfrac{\partial^2 z}{\partial x^2} + y(1-x)\dfrac{\partial^2 z}{\partial x \partial y} + [c-(a+b+1)x]\dfrac{\partial z}{\partial x} - by\dfrac{\partial z}{\partial y} - abz = 0, \\ y(1-y)\dfrac{\partial^2 z}{\partial y^2} + x(1-y)\dfrac{\partial^2 z}{\partial x \partial y} + [c-(a+b'+1)y]\dfrac{\partial z}{\partial y} - b'x\dfrac{\partial z}{\partial x} - ab'z = 0. \end{cases}$$

Picard (see [32], [33], [34] and [35]) studied the Appell hypergeometric partial differential equations and the hyperfuchsian functions, which led him to investigate the theory of algebraic surfaces. In particular, in his paper [33], Picard generalized Poincaré's Theorem 1 and Theorem 2 as follows:

Theorem 3. *Suppose that three hyperfuchsian functions*

$$x = F_1(\xi, \eta), \quad y = F_2(\xi, \eta), \quad z = F(\xi, \eta)$$

satisfy the algebraic relation

$$f(x, y, z) = 0.$$

One can form a system of three partial differential equations

$$\begin{cases} \dfrac{\partial^2 z}{\partial x^2} = a\dfrac{\partial z}{\partial x} + b\dfrac{\partial z}{\partial y} + cz, \\ \dfrac{\partial^2 z}{\partial x \partial y} = a_1\dfrac{\partial z}{\partial x} + b_1\dfrac{\partial z}{\partial y} + c_1 z, \\ \dfrac{\partial^2 z}{\partial y^2} = a_2\dfrac{\partial z}{\partial x} + b_2\dfrac{\partial z}{\partial y} + c_2 z, \end{cases}$$

where a, b, c are rational functions of x, y and z (z is an algebraic function of x, y defined by $f(x, y, z) = 0$.) These three equations have three linearly independent solutions z_1, z_2, z_3. If one sets

$$\frac{z_2}{z_1} = \xi, \quad \frac{z_3}{z_1} = \eta,$$

these equations can be resolved by x, y as follows:

$$x = F_1(\xi, \eta), \quad y = F_2(\xi, \eta).$$

In 1921, Giraud [12] studied the two variable automorphic functions and the complex hyperbolic geometry. According to [12] and [33], set $\xi = \frac{z_2}{z_1}$ and $\eta = \frac{z_3}{z_1}$, we have

$$\begin{vmatrix} \frac{\partial z_1}{\partial x} & \frac{\partial z_1}{\partial y} & z_1 \\ \frac{\partial z_2}{\partial x} & \frac{\partial z_2}{\partial y} & z_2 \\ \frac{\partial z_3}{\partial x} & \frac{\partial z_3}{\partial y} & z_3 \end{vmatrix} = z_1^2 \begin{vmatrix} 0 & 0 & z_1 \\ \frac{\partial\left(\frac{z_2}{z_1}\right)}{\partial x} & \frac{\partial\left(\frac{z_2}{z_1}\right)}{\partial y} & z_2 \\ \frac{\partial\left(\frac{z_3}{z_1}\right)}{\partial x} & \frac{\partial\left(\frac{z_3}{z_1}\right)}{\partial y} & z_3 \end{vmatrix} = z_1^3 \frac{\partial(\xi, \eta)}{\partial(x, y)} = 1,$$

when z_1, z_2 and z_3 satisfy the following relations:

$$z_1 = \sqrt[3]{\frac{\partial(x, y)}{\partial(\xi, \eta)}}, \quad z_2 = \xi \sqrt[3]{\frac{\partial(x, y)}{\partial(\xi, \eta)}}, \quad z_3 = \eta \sqrt[3]{\frac{\partial(x, y)}{\partial(\xi, \eta)}}.$$

In [51] and [52], Terada considered the multidimensional analogue of the hypergeometric function F_1 (Lauricella function) and studied arithmetically defined discontinuous groups.

In 1964, Shimura (see [45]) considered the classifying spaces for the isomorphy classes of plane curves of the following types:

(1) $y^3 = p_4(x)$ (non-hyperelliptic),

(2) $y^4 = p_2(x) \cdot p_3(x)^2$ (hyperelliptic),

where $p_k(x)$ is a polynomial of degree k and without multiple root. These are curves of genus 3. By a result of Shimura (see [45]) these curves are essentially parametrized by

(1) $\mathbb{B}^2/U(2, 1; \mathbb{Z}[\omega])$, $\omega = \frac{-1+\sqrt{-3}}{2}$,

(2) $\mathbb{B}^2/U(2, 1; \mathbb{Z}[i])$, $i = \sqrt{-1}$,

where $\mathbb{B}^2 = \{(z_1, z_2) \in \mathbb{C}^2 : |z_1|^2 + |z_2|^2 < 1\}$ is the complex ball, $U(2, 1) = \{g \in GL(3, \mathbb{C}) : g^* I_{2,1} g = I_{2,1}\}$ with $I_{2,1} = \text{diag}(1, 1, -1)$. $U(2, 1; \mathbb{Z}[\omega]) = \{g = (a_{ij}) \in U(2, 1) : a_{ij} \in \mathbb{Z}[\omega]\}$ and $U(2, 1; \mathbb{Z}[i]) = \{g = (a_{ij}) \in U(2, 1) : a_{ij} \in \mathbb{Z}[i]\}$.

Moreover, Shimura (see [45]) proved that in these two cases, the field of all automorphic functions on $\mathbb{B}^2$ with respect to $\Gamma = U(2,1;\mathbb{Z}[\omega])$ or $\Gamma = U(2,1;\mathbb{Z}[i])$ is a purely transcendental extension of $\mathbb{C}$ of dimension 2.

In fact, Picard (see [32]) investigated the curve $y^3 = p_4(x)$ (Picard curve) and observed that moduli of such curves give automorphic functions on $\mathbb{B}^2$. Heegner (see [14]) studied the abelian integrals associated to these curves and the transformations of automorphic functions. Moreover, Picard curves and Picard modular functions were extensively studied by Shiga (see [40], [41], [42], [43] and [44]).

In his monographs [17], [18], Holzapfel studied a concrete hypergeometric object of complex dimension two and the complete solution of the associated system of Euler partial differential equations. Holzapfel's results and methods involve several parts of mathematics, notably algebraic geometry (theory of surfaces, Gauss-Manin connection), theory of automorphic forms, monodromy groups of differential equations and number theory.

In their papers [5] and [6], Cohen and Wolfart classified the algebraic Appell-Lauricella functions F_1 in several variables, which extended the famous 1872 list of Schwarz of the algebraic Gauss hypergeometric functions (see [39]). The monodromy groups of the Appell-Lauricella functions F_1 in several variables can be discontinuous, but this does not in general entail that they be arithmetically defined (see [8], [28] and [9]). Nonetheless, Cohen and Wolfart associated to them families of abelian varieties for which they play a role similar to that of modular groups.

In fact, the geometric objects associated to Appell hypergeometric partial differential equations are significant in automorphic representation theory and arithmetic algebraic geometry (see the monograph [26] as an excellent survey, Rogawski's work [38], [11] and Shimura's work [46], [48]). According to [26], the Picard surfaces have an immediate geometric appeal, their connected components being quotients of two-dimensional balls by discrete groups were introduced by Picard during his study of the monodromy of Appell hypergeometric partial differential equations. We are interested in the groups which are defined by discrete groups associated to unitary groups in three variables over imaginary quadratic extensions of $\mathbb{Q}$. The quotients are non-compact. The corresponding varieties are investigated in [26]. As for many Shimura varieties the basic arithmetic properties of the Picard surfaces are most easily presented as aspects of the solution of a moduli problem.

In fact, after Kronecker, the history of the development of complex multiplication theory at the end of the 19th century is mostly connected with the name of Weber. Takagi, Fueter and Hasse devoted to the proof of Kronecker's Jugendtraum. Deuring introduced the algebraic approach to complex multiplication (see [53]). The complex multiplication theory which realizes the Jugendtraum is one-dimensional in the following sense: it considers modular functions in one variable and one-dimensional abelian varieties, that is, elliptic curves.

In 1911, Hecke studied complex multiplications for Hilbert modular functions in two variables. In 1955, some principal ideas of multidimensional complex multiplication theory were introduced by Weil and by Shimura and Taniyama. A thorough development of these ideas was accomplished by Shimura and Taniyama (see [47] and [49]). In a sense, the theory of Shimura varieties can be considered as a far-reaching generalization of complex multiplication theory.

However, Ovseevič (see [29]) showed that if $\dim A \neq 1$, where A is an abelian variety, that is, if $[K : \mathbb{Q}] \neq 2$, then the analogue of the completeness theorem is false: the maximal Shimura-Taniyama field does not coincide with the maximal abelian extension of K. Thus if a CM-field is not complex quadratic, abelian varieties give only a partial solution of Hilbert's 12th problem.

According to Langlands (see [25]), the 12th problem of Hilbert reminds us of the relationship of three subjects. The first is the theory of class fields or abelian extensions of number fields. The second is the algebraic theory of elliptic curves and more generally, abelian varieties. The third is the theory of automorphic functions which is connected with the study of abelian varieties, especially of their moduli.

In this paper, we will study the monodromy of Appell hypergeometric partial differential equations, which lead us to find four derivatives which are associated to the group $GL(3)$. Our four derivatives have the remarkable properties, which play a role similar to the Schwarzian derivative in the case of $GL(2)$ and Gauss hypergeometric equation. We find that Appell hypergeometric partial differential equations can be reduced to four nonlinear partial differential equations by the use of our four derivatives. We will investigate these four nonlinear partial differential equations. We find that the Jacobian of the map which is defined by the above nonlinear system satisfies a system of linear partial differential equations.

We are interested in some particular cases which grow out of the maps on the complex hyperbolic space and the unitary group in three variables $U(2,1) = \{g \in GL(3, \mathbb{C}) : g^* J g = J\}$ where $J = \begin{pmatrix} & & 1 \\ & -1 & \\ 1 & & \end{pmatrix}$. The corresponding complex domain $\mathfrak{S}_2 = \{(z_1, z_2) \in \mathbb{C}^2 : z_1 + \overline{z_1} - z_2\overline{z_2} > 0\}$ is the simplest example of a bounded symmetric domain which is not a tube domain. In these cases, the system of linear partial differential equations has the remarkable properties. Moreover, it can be solved by Appell hypergeometric partial differential equations. This lead us to study the properties of solutions. In particular, we find the η-functions associated to the unitary group $U(2,1)$, which has the interesting properties. Unlike the Dedekind η-functions in the case of $GL(2)$, our η-functions are involved in the non-commutative translations.

Our η-functions lead us to study the arithmetic of Picard curves, especially the transform problems (moduli problems) and the corresponding modular equations, which can be considered as a natural generalization of complex multiplication. Hence, they are related to Hilbert's 12th problem. Our approach is as follows: instead of the Jacobian variety of Picard curves, which is an abelian variety, we will investigate the Picard curves themselves. We will give a rational transformation between two irrational integrals which are associated to some algebraic surfaces of degree seven. We find a modular equation which gives the relation between the moduli of two Picard curves. On the other hand, by the investigation of the solutions, we find some interesting transformations. The solutions have some remarkable properties under these transformations. This leads us to obtain the J-invariants associated to the Picard curves.

Finally, we will study a system of nonlinear evolution equations which are associated to our four derivatives. We find that they are $GL(3)$-invariant. Moreover, our derivatives satisfy two first order nonlinear partial differential equations, which play a role similar to the Korteweg-de Vries equation in the case of $GL(2)$.

Now, we will give a survey of our main results.

Main Results.

Let us define the following four derivatives:

Definition 1.1. Four derivatives associated to the group $GL(3)$ are given by:

$$\{u_1, u_2; x, y\}_x := \frac{\begin{vmatrix} \frac{\partial u_1}{\partial x} & \frac{\partial u_2}{\partial x} \\ \frac{\partial^2 u_1}{\partial x^2} & \frac{\partial^2 u_2}{\partial x^2} \end{vmatrix}}{\begin{vmatrix} \frac{\partial u_1}{\partial x} & \frac{\partial u_2}{\partial x} \\ \frac{\partial u_1}{\partial y} & \frac{\partial u_2}{\partial y} \end{vmatrix}}, \qquad \{u_1, u_2; x, y\}_y := \frac{\begin{vmatrix} \frac{\partial u_1}{\partial y} & \frac{\partial u_2}{\partial y} \\ \frac{\partial^2 u_1}{\partial y^2} & \frac{\partial^2 u_2}{\partial y^2} \end{vmatrix}}{\begin{vmatrix} \frac{\partial u_1}{\partial y} & \frac{\partial u_2}{\partial y} \\ \frac{\partial u_1}{\partial x} & \frac{\partial u_2}{\partial x} \end{vmatrix}},$$

$$[u_1, u_2; x, y]_x := \frac{\begin{vmatrix} \frac{\partial u_1}{\partial y} & \frac{\partial u_2}{\partial y} \\ \frac{\partial^2 u_1}{\partial x^2} & \frac{\partial^2 u_2}{\partial x^2} \end{vmatrix} + 2\begin{vmatrix} \frac{\partial u_1}{\partial x} & \frac{\partial u_2}{\partial x} \\ \frac{\partial^2 u_1}{\partial x \partial y} & \frac{\partial^2 u_2}{\partial x \partial y} \end{vmatrix}}{\begin{vmatrix} \frac{\partial u_1}{\partial x} & \frac{\partial u_2}{\partial x} \\ \frac{\partial u_1}{\partial y} & \frac{\partial u_2}{\partial y} \end{vmatrix}}, \tag{1.1}$$

$$[u_1, u_2; x, y]_y := \frac{\begin{vmatrix} \frac{\partial u_1}{\partial x} & \frac{\partial u_2}{\partial x} \\ \frac{\partial^2 u_1}{\partial y^2} & \frac{\partial^2 u_2}{\partial y^2} \end{vmatrix} + 2\begin{vmatrix} \frac{\partial u_1}{\partial y} & \frac{\partial u_2}{\partial y} \\ \frac{\partial^2 u_1}{\partial x \partial y} & \frac{\partial^2 u_2}{\partial x \partial y} \end{vmatrix}}{\begin{vmatrix} \frac{\partial u_1}{\partial y} & \frac{\partial u_2}{\partial y} \\ \frac{\partial u_1}{\partial x} & \frac{\partial u_2}{\partial x} \end{vmatrix}},$$

where x and y are complex numbers, u_1 and u_2 are complex-valued functions of x and y.

In fact, the derivation of our four derivatives comes from the resolution of Appell hypergeometric partial differential equations by the elimination method. The three linearly independent solutions z_1, z_2 and z_3 are reduced to the two variables $u_1 = \frac{z_1}{z_3}$ and $u_2 = \frac{z_2}{z_3}$, which satisfy four nonlinear partial differential equations (see section 2, in particular, the formulas (2.6)). This is different from that in Yoshida's work [59], [60], [63] and [64].

Proposition 1.2 ($GL(3)$-invariance). *Our derivatives are invariant under linear fractional transformation acting on the first arguments:*

$$\begin{cases} \{v_1, v_2; x, y\}_x = \{u_1, u_2; x, y\}_x, \\ \{v_1, v_2; x, y\}_y = \{u_1, u_2; x, y\}_y, \\ [v_1, v_2; x, y]_x = [u_1, u_2; x, y]_x, \\ [v_1, v_2; x, y]_y = [u_1, u_2; x, y]_y, \end{cases} \tag{1.2}$$

where

$$(v_1, v_2) = \gamma(u_1, u_2) = \left(\frac{a_1 u_1 + a_2 u_2 + a_3}{c_1 u_1 + c_2 u_2 + c_3}, \frac{b_1 u_1 + b_2 u_2 + b_3}{c_1 u_1 + c_2 u_2 + c_3} \right) \tag{1.3}$$

for $\gamma = \begin{pmatrix} a_1 & a_2 & a_3 \\ b_1 & b_2 & b_3 \\ c_1 & c_2 & c_3 \end{pmatrix} \in GL(3, \mathbb{C})$.

Thus our derivatives are differential invariants of the general linear group acting as a linear fractional transformation. In particular, we have the following:

Corollary 1.3 (Vanishing property). *Our derivatives vanish when* (w_1, w_2) *are linear fractional functions of* (z_1, z_2)*:*

$$\begin{cases} \{w_1, w_2; z_1, z_2\}_{z_1} = 0, \\ \{w_1, w_2; z_1, z_2\}_{z_2} = 0, \\ [w_1, w_2; z_1, z_2]_{z_1} = 0, \\ [w_1, w_2; z_1, z_2]_{z_2} = 0, \end{cases} \tag{1.4}$$

where

$$(w_1, w_2) = \gamma(z_1, z_2) = \left(\frac{a_1 z_1 + a_2 z_2 + a_3}{c_1 z_1 + c_2 z_2 + c_3}, \frac{b_1 z_1 + b_2 z_2 + b_3}{c_1 z_1 + c_2 z_2 + c_3} \right),$$

for $\gamma = \begin{pmatrix} a_1 & a_2 & a_3 \\ b_1 & b_2 & b_3 \\ c_1 & c_2 & c_3 \end{pmatrix} \in GL(3, \mathbb{C})$.

Proposition 1.4 (Main Proposition 1) (Change of variables). *The change of independent variables obeys the laws:*

$$\begin{aligned} &\{u_1, u_2; x, y\}_x \\ &= \frac{\left(\frac{\partial w_1}{\partial x}\right)^3}{\frac{\partial(w_1, w_2)}{\partial(x,y)}} \{u_1, u_2; w_1, w_2\}_{w_1} + \frac{\left(\frac{\partial w_1}{\partial x}\right)^2 \frac{\partial w_2}{\partial x}}{\frac{\partial(w_1, w_2)}{\partial(x,y)}} [u_1, u_2; w_1, w_2]_{w_1} \\ &\quad + \frac{\left(\frac{\partial w_2}{\partial x}\right)^3}{\frac{\partial(w_2, w_1)}{\partial(x,y)}} \{u_1, u_2; w_1, w_2\}_{w_2} + \frac{\frac{\partial w_1}{\partial x} \left(\frac{\partial w_2}{\partial x}\right)^2}{\frac{\partial(w_2, w_1)}{\partial(x,y)}} [u_1, u_2; w_1, w_2]_{w_2} + \{w_1, w_2; x, y\}_x. \end{aligned} \tag{1.5}$$

$$\begin{aligned} &\{u_1, u_2; x, y\}_y \\ &= \frac{\left(\frac{\partial w_1}{\partial y}\right)^3}{\frac{\partial(w_2, w_1)}{\partial(x,y)}} \{u_1, u_2; w_1, w_2\}_{w_1} + \frac{\left(\frac{\partial w_1}{\partial y}\right)^2 \frac{\partial w_2}{\partial y}}{\frac{\partial(w_2, w_1)}{\partial(x,y)}} [u_1, u_2; w_1, w_2]_{w_1} \\ &\quad + \frac{\left(\frac{\partial w_2}{\partial y}\right)^3}{\frac{\partial(w_1, w_2)}{\partial(x,y)}} \{u_1, u_2; w_1, w_2\}_{w_2} + \frac{\frac{\partial w_1}{\partial y} \left(\frac{\partial w_2}{\partial y}\right)^2}{\frac{\partial(w_1, w_2)}{\partial(x,y)}} [u_1, u_2; w_1, w_2]_{w_2} + \{w_1, w_2; x, y\}_y. \end{aligned} \tag{1.6}$$

$$
\begin{aligned}
&[u_1, u_2; x, y]_x \\
&= \frac{3(\frac{\partial w_1}{\partial x})^2 \frac{\partial w_1}{\partial y}}{\frac{\partial(w_1,w_2)}{\partial(x,y)}} \{u_1, u_2; w_1, w_2\}_{w_1} \\
&\quad + \frac{(\frac{\partial w_1}{\partial x})^2 \frac{\partial w_2}{\partial y} + 2\frac{\partial w_1}{\partial x}\frac{\partial w_2}{\partial x}\frac{\partial w_1}{\partial y}}{\frac{\partial(w_1,w_2)}{\partial(x,y)}} [u_1, u_2; w_1, w_2]_{w_1} \\
&\quad + \frac{3(\frac{\partial w_2}{\partial x})^2 \frac{\partial w_2}{\partial y}}{\frac{\partial(w_2,w_1)}{\partial(x,y)}} \{u_1, u_2; w_1, w_2\}_{w_2} \\
&\quad + \frac{(\frac{\partial w_2}{\partial x})^2 \frac{\partial w_1}{\partial y} + 2\frac{\partial w_1}{\partial x}\frac{\partial w_2}{\partial x}\frac{\partial w_2}{\partial y}}{\frac{\partial(w_2,w_1)}{\partial(x,y)}} [u_1, u_2; w_1, w_2]_{w_2} \\
&\quad + [w_1, w_2; x, y]_x .
\end{aligned} \tag{1.7}
$$

$$
\begin{aligned}
&[u_1, u_2; x, y]_y \\
&= \frac{3\frac{\partial w_1}{\partial x}(\frac{\partial w_1}{\partial y})^2}{\frac{\partial(w_2,w_1)}{\partial(x,y)}} \{u_1, u_2; w_1, w_2\}_{w_1} \\
&\quad + \frac{\frac{\partial w_2}{\partial x}(\frac{\partial w_1}{\partial y})^2 + 2\frac{\partial w_1}{\partial x}\frac{\partial w_1}{\partial y}\frac{\partial w_2}{\partial y}}{\frac{\partial(w_2,w_1)}{\partial(x,y)}} [u_1, u_2; w_1, w_2]_{w_1} \\
&\quad + \frac{3\frac{\partial w_2}{\partial x}(\frac{\partial w_2}{\partial y})^2}{\frac{\partial(w_1,w_2)}{\partial(x,y)}} \{u_1, u_2; w_1, w_2\}_{w_2} \\
&\quad + \frac{\frac{\partial w_1}{\partial x}(\frac{\partial w_2}{\partial y})^2 + 2\frac{\partial w_2}{\partial x}\frac{\partial w_1}{\partial y}\frac{\partial w_2}{\partial y}}{\frac{\partial(w_1,w_2)}{\partial(x,y)}} [u_1, u_2; w_1, w_2]_{w_2} \\
&\quad + [w_1, w_2; x, y]_y .
\end{aligned} \tag{1.8}
$$

Corollary 1.5 (Exchange of variables). *The exchange of the first and the second arguments obeys the laws:*

$$
\begin{aligned}
&\{w_1, w_2; x, y\}_x \\
&= -\frac{(\frac{\partial w_1}{\partial x})^3}{\frac{\partial(w_1,w_2)}{\partial(x,y)}} \{x, y; w_1, w_2\}_{w_1} - \frac{(\frac{\partial w_1}{\partial x})^2 \frac{\partial w_2}{\partial x}}{\frac{\partial(w_1,w_2)}{\partial(x,y)}} [x, y; w_1, w_2]_{w_1} \\
&\quad + \frac{(\frac{\partial w_2}{\partial x})^3}{\frac{\partial(w_1,w_2)}{\partial(x,y)}} \{x, y; w_1, w_2\}_{w_2} + \frac{\frac{\partial w_1}{\partial x}(\frac{\partial w_2}{\partial x})^2}{\frac{\partial(w_1,w_2)}{\partial(x,y)}} [x, y; w_1, w_2]_{w_2},
\end{aligned} \tag{1.9}
$$

$$
\begin{aligned}
&\{w_1, w_2; x, y\}_y \\
&= \frac{(\frac{\partial w_1}{\partial y})^3}{\frac{\partial(w_1,w_2)}{\partial(x,y)}} \{x, y; w_1, w_2\}_{w_1} + \frac{(\frac{\partial w_1}{\partial y})^2 \frac{\partial w_2}{\partial y}}{\frac{\partial(w_1,w_2)}{\partial(x,y)}} [x, y; w_1, w_2]_{w_1} \\
&\quad - \frac{(\frac{\partial w_2}{\partial y})^3}{\frac{\partial(w_1,w_2)}{\partial(x,y)}} \{x, y; w_1, w_2\}_{w_2} - \frac{\frac{\partial w_1}{\partial y}(\frac{\partial w_2}{\partial y})^2}{\frac{\partial(w_1,w_2)}{\partial(x,y)}} [x, y; w_1, w_2]_{w_2},
\end{aligned} \tag{1.10}
$$

$$
\begin{aligned}
&[w_1, w_2; x, y]_x \\
&= -\frac{3(\frac{\partial w_1}{\partial x})^2 \frac{\partial w_1}{\partial y}}{\frac{\partial(w_1,w_2)}{\partial(x,y)}}\{x, y; w_1, w_2\}_{w_1} - \frac{(\frac{\partial w_1}{\partial x})^2 \frac{\partial w_2}{\partial y} + 2\frac{\partial w_1}{\partial x}\frac{\partial w_2}{\partial x}\frac{\partial w_1}{\partial y}}{\frac{\partial(w_1,w_2)}{\partial(x,y)}}[x, y; w_1, w_2]_{w_1} \\
&\quad + \frac{3(\frac{\partial w_2}{\partial x})^2 \frac{\partial w_2}{\partial y}}{\frac{\partial(w_1,w_2)}{\partial(x,y)}}\{x, y; w_1, w_2\}_{w_2} + \frac{(\frac{\partial w_2}{\partial x})^2 \frac{\partial w_1}{\partial y} + 2\frac{\partial w_1}{\partial x}\frac{\partial w_2}{\partial x}\frac{\partial w_2}{\partial y}}{\frac{\partial(w_1,w_2)}{\partial(x,y)}}[x, y; w_1, w_2]_{w_2},
\end{aligned} \tag{1.11}
$$

$$
\begin{aligned}
&[w_1, w_2; x, y]_y \\
&= \frac{3\frac{\partial w_1}{\partial x}(\frac{\partial w_1}{\partial y})^2}{\frac{\partial(w_1,w_2)}{\partial(x,y)}}\{x, y; w_1, w_2\}_{w_1} + \frac{\frac{\partial w_2}{\partial x}(\frac{\partial w_1}{\partial y})^2 + 2\frac{\partial w_1}{\partial x}\frac{\partial w_1}{\partial y}\frac{\partial w_2}{\partial y}}{\frac{\partial(w_1,w_2)}{\partial(x,y)}}[x, y; w_1, w_2]_{w_1} \\
&\quad - \frac{3\frac{\partial w_2}{\partial x}(\frac{\partial w_2}{\partial y})^2}{\frac{\partial(w_1,w_2)}{\partial(x,y)}}\{x, y; w_1, w_2\}_{w_2} - \frac{\frac{\partial w_1}{\partial x}(\frac{\partial w_2}{\partial y})^2 + 2\frac{\partial w_2}{\partial x}\frac{\partial w_1}{\partial y}\frac{\partial w_2}{\partial y}}{\frac{\partial(w_1,w_2)}{\partial(x,y)}}[x, y; w_1, w_2]_{w_2}.
\end{aligned} \tag{1.12}
$$

We may apply a linear fractional transformation to the second arguments instead of the first. The result is as follows:

Corollary 1.6 ($GL(3)$-covariance). *For*

$$
(w_1, w_2) = \gamma(x, y) = \left(\frac{a_1x + a_2y + a_3}{c_1x + c_2y + c_3}, \frac{b_1x + b_2y + b_3}{c_1x + c_2y + c_3}\right), \tag{1.13}
$$

where $\gamma = \begin{pmatrix} a_1 & a_2 & a_3 \\ b_1 & b_2 & b_3 \\ c_1 & c_2 & c_3 \end{pmatrix} \in GL(3, \mathbb{C})$, *the following four identities hold:*

$$
\begin{aligned}
&\{u_1, u_2; x, y\}_x \\
&= \frac{A_1^3}{\Delta(c_1x + c_2y + c_3)^3}\left\{u_1, u_2; \frac{a_1x + a_2y + a_3}{c_1x + c_2y + c_3}, \frac{b_1x + b_2y + b_3}{c_1x + c_2y + c_3}\right\}_{\frac{a_1x+a_2y+a_3}{c_1x+c_2y+c_3}} \\
&\quad + \frac{A_1^2B_1}{\Delta(c_1x + c_2y + c_3)^3}\left[u_1, u_2; \frac{a_1x + a_2y + a_3}{c_1x + c_2y + c_3}, \frac{b_1x + b_2y + b_3}{c_1x + c_2y + c_3}\right]_{\frac{a_1x+a_2y+a_3}{c_1x+c_2y+c_3}} \\
&\quad - \frac{B_1^3}{\Delta(c_1x + c_2y + c_3)^3}\left\{u_1, u_2; \frac{a_1x + a_2y + a_3}{c_1x + c_2y + c_3}, \frac{b_1x + b_2y + b_3}{c_1x + c_2y + c_3}\right\}_{\frac{b_1x+b_2y+b_3}{c_1x+c_2y+c_3}} \\
&\quad - \frac{A_1B_1^2}{\Delta(c_1x + c_2y + c_3)^3}\left[u_1, u_2; \frac{a_1x + a_2y + a_3}{c_1x + c_2y + c_3}, \frac{b_1x + b_2y + b_3}{c_1x + c_2y + c_3}\right]_{\frac{b_1x+b_2y+b_3}{c_1x+c_2y+c_3}},
\end{aligned} \tag{1.14}
$$

$$
\begin{aligned}
&\{u_1,u_2;x,y\}_y\\
&=-\frac{A_2^3}{\Delta(c_1x+c_2y+c_3)^3}\left\{u_1,u_2;\frac{a_1x+a_2y+a_3}{c_1x+c_2y+c_3},\frac{b_1x+b_2y+b_3}{c_1x+c_2y+c_3}\right\}_{\frac{a_1x+a_2y+a_3}{c_1x+c_2y+c_3}}\\
&\quad-\frac{A_2^2B_2}{\Delta(c_1x+c_2y+c_3)^3}\left[u_1,u_2;\frac{a_1x+a_2y+a_3}{c_1x+c_2y+c_3},\frac{b_1x+b_2y+b_3}{c_1x+c_2y+c_3}\right]_{\frac{a_1x+a_2y+a_3}{c_1x+c_2y+c_3}}\\
&\quad+\frac{B_2^3}{\Delta(c_1x+c_2y+c_3)^3}\left\{u_1,u_2;\frac{a_1x+a_2y+a_3}{c_1x+c_2y+c_3},\frac{b_1x+b_2y+b_3}{c_1x+c_2y+c_3}\right\}_{\frac{b_1x+b_2y+b_3}{c_1x+c_2y+c_3}}\\
&\quad+\frac{A_2B_2^2}{\Delta(c_1x+c_2y+c_3)^3}\left[u_1,u_2;\frac{a_1x+a_2y+a_3}{c_1x+c_2y+c_3},\frac{b_1x+b_2y+b_3}{c_1x+c_2y+c_3}\right]_{\frac{b_1x+b_2y+b_3}{c_1x+c_2y+c_3}},
\end{aligned}
\tag{1.15}
$$

$$
\begin{aligned}
&[u_1,u_2;x,y]_x\\
&=\frac{3A_1^2A_2}{\Delta(c_1x+c_2y+c_3)^3}\left\{u_1,u_2;\frac{a_1x+a_2y+a_3}{c_1x+c_2y+c_3},\frac{b_1x+b_2y+b_3}{c_1x+c_2y+c_3}\right\}_{\frac{a_1x+a_2y+a_3}{c_1x+c_2y+c_3}}\\
&\quad+\frac{A_1^2B_2+2A_1B_1A_2}{\Delta(c_1x+c_2y+c_3)^3}\left[u_1,u_2;\frac{a_1x+a_2y+a_3}{c_1x+c_2y+c_3},\frac{b_1x+b_2y+b_3}{c_1x+c_2y+c_3}\right]_{\frac{a_1x+a_2y+a_3}{c_1x+c_2y+c_3}}\\
&\quad-\frac{3B_1^2B_2}{\Delta(c_1x+c_2y+c_3)^3}\left\{u_1,u_2;\frac{a_1x+a_2y+a_3}{c_1x+c_2y+c_3},\frac{b_1x+b_2y+b_3}{c_1x+c_2y+c_3}\right\}_{\frac{b_1x+b_2y+b_3}{c_1x+c_2y+c_3}}\\
&\quad-\frac{B_1^2A_2+2A_1B_1B_2}{\Delta(c_1x+c_2y+c_3)^3}\left[u_1,u_2;\frac{a_1x+a_2y+a_3}{c_1x+c_2y+c_3},\frac{b_1x+b_2y+b_3}{c_1x+c_2y+c_3}\right]_{\frac{b_1x+b_2y+b_3}{c_1x+c_2y+c_3}},
\end{aligned}
\tag{1.16}
$$

$$
\begin{aligned}
&[u_1,u_2;x,y]_y\\
&=-\frac{3A_1A_2^2}{\Delta(c_1x+c_2y+c_3)^3}\left\{u_1,u_2;\frac{a_1x+a_2y+a_3}{c_1x+c_2y+c_3},\frac{b_1x+b_2y+b_3}{c_1x+c_2y+c_3}\right\}_{\frac{a_1x+a_2y+a_3}{c_1x+c_2y+c_3}}\\
&\quad-\frac{B_1A_2^2+2A_1A_2B_2}{\Delta(c_1x+c_2y+c_3)^3}\left[u_1,u_2;\frac{a_1x+a_2y+a_3}{c_1x+c_2y+c_3},\frac{b_1x+b_2y+b_3}{c_1x+c_2y+c_3}\right]_{\frac{a_1x+a_2y+a_3}{c_1x+c_2y+c_3}}\\
&\quad+\frac{3B_1B_2^2}{\Delta(c_1x+c_2y+c_3)^3}\left\{u_1,u_2;\frac{a_1x+a_2y+a_3}{c_1x+c_2y+c_3},\frac{b_1x+b_2y+b_3}{c_1x+c_2y+c_3}\right\}_{\frac{b_1x+b_2y+b_3}{c_1x+c_2y+c_3}}\\
&\quad+\frac{A_1B_2^2+2B_1A_2B_2}{\Delta(c_1x+c_2y+c_3)^3}\left[u_1,u_2;\frac{a_1x+a_2y+a_3}{c_1x+c_2y+c_3},\frac{b_1x+b_2y+b_3}{c_1x+c_2y+c_3}\right]_{\frac{b_1x+b_2y+b_3}{c_1x+c_2y+c_3}}.
\end{aligned}
\tag{1.17}
$$

Here, $\Delta=\det(\gamma)$ *and*

$$
\begin{cases}
A_1=(a_1c_2-a_2c_1)y+(a_1c_3-a_3c_1),\\
A_2=(a_2c_1-a_1c_2)x+(a_2c_3-a_3c_2),\\
B_1=(b_1c_2-b_2c_1)y+(b_1c_3-b_3c_1),\\
B_2=(b_2c_1-b_1c_2)x+(b_2c_3-b_3c_2).
\end{cases}
$$

The Proposition 1.4 implies that

$$\begin{pmatrix} \{u_1, u_2; x, y\}_x \\ \{u_1, u_2; x, y\}_y \\ [u_1, u_2; x, y]_x \\ [u_1, u_2; x, y]_y \end{pmatrix} = \frac{M(w_1, w_2; x, y)}{\frac{\partial(w_1,w_2)}{\partial(x,y)}} \begin{pmatrix} \{u_1, u_2; w_1, w_2\}_{w_1} \\ \{u_1, u_2; w_1, w_2\}_{w_2} \\ [u_1, u_2; w_1, w_2]_{w_1} \\ [u_1, u_2; w_1, w_2]_{w_2} \end{pmatrix} + \begin{pmatrix} \{w_1, w_2; x, y\}_x \\ \{w_1, w_2; x, y\}_y \\ [w_1, w_2; x, y]_x \\ [w_1, w_2; x, y]_y \end{pmatrix}, \tag{1.18}$$

where

$$M(w_1, w_2; x, y) :=$$
$$\begin{pmatrix} \begin{array}{c} (\frac{\partial w_1}{\partial x})^3 \\ -(\frac{\partial w_1}{\partial y})^3 \end{array} & \begin{array}{c} -(\frac{\partial w_2}{\partial x})^3 \\ (\frac{\partial w_2}{\partial y})^3 \end{array} & \begin{array}{c} (\frac{\partial w_1}{\partial x})^2 \frac{\partial w_2}{\partial x} \\ -(\frac{\partial w_1}{\partial y})^2 \frac{\partial w_2}{\partial y} \end{array} & \begin{array}{c} -\frac{\partial w_1}{\partial x}(\frac{\partial w_2}{\partial x})^2 \\ \frac{\partial w_1}{\partial y}(\frac{\partial w_2}{\partial y})^2 \end{array} \\ 3(\frac{\partial w_1}{\partial x})^2 \frac{\partial w_1}{\partial y} & -3(\frac{\partial w_2}{\partial x})^2 \frac{\partial w_2}{\partial y} & \begin{array}{c} (\frac{\partial w_1}{\partial x})^2 \frac{\partial w_2}{\partial y} \\ +2\frac{\partial w_1}{\partial x}\frac{\partial w_2}{\partial x}\frac{\partial w_1}{\partial y} \end{array} & \begin{array}{c} -(\frac{\partial w_2}{\partial x})^2 \frac{\partial w_1}{\partial y} \\ -2\frac{\partial w_1}{\partial x}\frac{\partial w_2}{\partial x}\frac{\partial w_2}{\partial y} \end{array} \\ -3\frac{\partial w_1}{\partial x}(\frac{\partial w_1}{\partial y})^2 & 3\frac{\partial w_2}{\partial x}(\frac{\partial w_2}{\partial y})^2 & \begin{array}{c} -\frac{\partial w_2}{\partial x}(\frac{\partial w_1}{\partial y})^2 \\ -2\frac{\partial w_1}{\partial x}\frac{\partial w_1}{\partial y}\frac{\partial w_2}{\partial y} \end{array} & \begin{array}{c} \frac{\partial w_1}{\partial x}(\frac{\partial w_2}{\partial y})^2 \\ +2\frac{\partial w_2}{\partial x}\frac{\partial w_1}{\partial y}\frac{\partial w_2}{\partial y} \end{array} \end{pmatrix}. \tag{1.19}$$

Proposition 1.7 (Cocycle condition).

$$M(w_1, w_2; x, y) \cdot M(u_1, u_2; w_1, w_2) = M(u_1, u_2; x, y). \tag{1.20}$$

Set

$$U_{(w_1,w_2;x,y)} := \begin{pmatrix} M(w_1, w_2; x, y) & c\frac{\partial(w_1,w_2)}{\partial(x,y)} \begin{pmatrix} \{w_1, w_2; x, y\}_x \\ \{w_1, w_2; x, y\}_y \\ [w_1, w_2; x, y]_x \\ [w_1, w_2; x, y]_y \end{pmatrix} \\ 0 & \frac{\partial(w_1,w_2)}{\partial(x,y)} \end{pmatrix}. \tag{1.21}$$

Then the above proposition implies that

$$U_{(w_1,w_2;x,y)} \cdot U_{(u_1,u_2;w_1,w_2)} = U_{(u_1,u_2;x,y)}. \tag{1.22}$$

The equation

$$U_{W,Z}U_{U,W} = U_{U,Z}$$

with $W = (w_1, w_2)$, $Z = (z_1, z_2)$ and $U = (u_1, u_2)$ is a cocycle condition, which means that the U's can be interpreted as transition functions of a vector bundle. More precisely, when we choose an open cover $\Sigma = \bigcup_i \Sigma_i$, with a local coordinate Z_i on Σ_i, then the above equation permits us to interpret the U_{Z_i,Z_j} as transition functions on the intersection $\Sigma_i \cap \Sigma_j$. Let $\mathcal{V}_0$ be this vector bundle. It is a five dimensional sub-bundle of the infinite dimensional vector bundle $\mathcal{V}$.

The next Main Proposition 2 gives the relation between the deformation of the Jacobian and our four derivatives:

Proposition 1.8 (Main Proposition 2) (Deformation property). *The following identity holds:*

$$
\begin{aligned}
&\frac{\partial(f_1,f_2)}{\partial(z_1,z_2)}\\
=&\frac{\partial(\widehat{f_1},\widehat{f_2})}{\partial(w_1,w_2)}-\widehat{f_1}\frac{\partial\widehat{f_1}}{\partial w_2}\{z_1,z_2;w_1,w_2\}_{w_1}-\widehat{f_2}\frac{\partial\widehat{f_2}}{\partial w_1}\{z_1,z_2;w_1,w_2\}_{w_2}\\
&-\widehat{f_1}\frac{\partial\widehat{f_2}}{\partial w_2}\frac{\begin{vmatrix}\frac{\partial z_1}{\partial w_2} & \frac{\partial z_2}{\partial w_2}\\ \frac{\partial^2 z_1}{\partial w_1^2} & \frac{\partial^2 z_2}{\partial w_1^2}\end{vmatrix}}{\begin{vmatrix}\frac{\partial z_1}{\partial w_1} & \frac{\partial z_2}{\partial w_1}\\ \frac{\partial z_1}{\partial w_2} & \frac{\partial z_2}{\partial w_2}\end{vmatrix}}-\left(\widehat{f_2}\frac{\partial\widehat{f_1}}{\partial w_2}-\widehat{f_1}\frac{\partial\widehat{f_1}}{\partial w_1}\right)\frac{\begin{vmatrix}\frac{\partial z_1}{\partial w_1} & \frac{\partial z_2}{\partial w_1}\\ \frac{\partial^2 z_1}{\partial w_1\partial w_2} & \frac{\partial^2 z_2}{\partial w_1\partial w_2}\end{vmatrix}}{\begin{vmatrix}\frac{\partial z_1}{\partial w_1} & \frac{\partial z_2}{\partial w_1}\\ \frac{\partial z_1}{\partial w_2} & \frac{\partial z_2}{\partial w_2}\end{vmatrix}}\\
&-\widehat{f_2}\frac{\partial\widehat{f_1}}{\partial w_1}\frac{\begin{vmatrix}\frac{\partial z_1}{\partial w_1} & \frac{\partial z_2}{\partial w_1}\\ \frac{\partial^2 z_1}{\partial w_2^2} & \frac{\partial^2 z_2}{\partial w_2^2}\end{vmatrix}}{\begin{vmatrix}\frac{\partial z_1}{\partial w_2} & \frac{\partial z_2}{\partial w_2}\\ \frac{\partial z_1}{\partial w_1} & \frac{\partial z_2}{\partial w_1}\end{vmatrix}}-\left(\widehat{f_1}\frac{\partial\widehat{f_2}}{\partial w_1}-\widehat{f_2}\frac{\partial\widehat{f_2}}{\partial w_2}\right)\frac{\begin{vmatrix}\frac{\partial z_1}{\partial w_2} & \frac{\partial z_2}{\partial w_2}\\ \frac{\partial^2 z_1}{\partial w_1\partial w_2} & \frac{\partial^2 z_2}{\partial w_1\partial w_2}\end{vmatrix}}{\begin{vmatrix}\frac{\partial z_1}{\partial w_2} & \frac{\partial z_2}{\partial w_2}\\ \frac{\partial z_1}{\partial w_1} & \frac{\partial z_2}{\partial w_1}\end{vmatrix}}\\
&+\widehat{f_1}^2\frac{\begin{vmatrix}\frac{\partial^2 z_1}{\partial w_1^2} & \frac{\partial^2 z_2}{\partial w_1^2}\\ \frac{\partial^2 z_1}{\partial w_1\partial w_2} & \frac{\partial^2 z_2}{\partial w_1\partial w_2}\end{vmatrix}}{\begin{vmatrix}\frac{\partial z_1}{\partial w_1} & \frac{\partial z_2}{\partial w_1}\\ \frac{\partial z_1}{\partial w_2} & \frac{\partial z_2}{\partial w_2}\end{vmatrix}}+\widehat{f_2}^2\frac{\begin{vmatrix}\frac{\partial^2 z_1}{\partial w_2^2} & \frac{\partial^2 z_2}{\partial w_2^2}\\ \frac{\partial^2 z_1}{\partial w_1\partial w_2} & \frac{\partial^2 z_2}{\partial w_1\partial w_2}\end{vmatrix}}{\begin{vmatrix}\frac{\partial z_1}{\partial w_2} & \frac{\partial z_2}{\partial w_2}\\ \frac{\partial z_1}{\partial w_1} & \frac{\partial z_2}{\partial w_1}\end{vmatrix}}+\widehat{f_1}\widehat{f_2}\frac{\begin{vmatrix}\frac{\partial^2 z_1}{\partial w_1^2} & \frac{\partial^2 z_2}{\partial w_1^2}\\ \frac{\partial^2 z_1}{\partial w_2^2} & \frac{\partial^2 z_2}{\partial w_2^2}\end{vmatrix}}{\begin{vmatrix}\frac{\partial z_1}{\partial w_1} & \frac{\partial z_2}{\partial w_1}\\ \frac{\partial z_1}{\partial w_2} & \frac{\partial z_2}{\partial w_2}\end{vmatrix}},
\end{aligned}
\tag{1.23}
$$

where

$$
(f_1,f_2)=(\widehat{f_1},\widehat{f_2})\begin{pmatrix}\frac{\partial z_1}{\partial w_1} & \frac{\partial z_2}{\partial w_1}\\ \frac{\partial z_1}{\partial w_2} & \frac{\partial z_2}{\partial w_2}\end{pmatrix}.
$$

Now, put

$$
\phi(f_1,f_2):=\iint\frac{\partial(f_1,f_2)}{\partial(t_1,t_2)}dt_1\wedge dt_2.
$$

We define an operation

$$
\begin{aligned}
&\left[f_1\frac{\partial}{\partial t_1}\otimes\frac{\partial}{\partial t_2},f_2\frac{\partial}{\partial t_1}\otimes\frac{\partial}{\partial t_2}\right]^*_{Z_0}\\
:=&c\cdot\phi(f_1,f_2)-f_1\frac{\partial f_1}{\partial t_2}\frac{\partial}{\partial t_1}-f_1\frac{\partial f_2}{\partial t_2}\frac{\partial}{\partial t_1}-\left(f_2\frac{\partial f_1}{\partial t_2}-f_1\frac{\partial f_1}{\partial t_1}\right)\frac{\partial}{\partial t_1}\\
&-f_2\frac{\partial f_2}{\partial t_1}\frac{\partial}{\partial t_2}-f_2\frac{\partial f_1}{\partial t_1}\frac{\partial}{\partial t_2}-\left(f_1\frac{\partial f_2}{\partial t_1}-f_2\frac{\partial f_2}{\partial t_2}\right)\frac{\partial}{\partial t_2}\\
&+f_1^2\frac{\partial^2}{\partial t_1^2}+f_2^2\frac{\partial^2}{\partial t_2^2}+f_1f_2\frac{\partial^2}{\partial t_1\partial t_2}.
\end{aligned}
$$

Note that

$$f_1\frac{\partial f_1}{\partial t_2}+f_1\frac{\partial f_2}{\partial t_2}+\left(f_2\frac{\partial f_1}{\partial t_2}-f_1\frac{\partial f_1}{\partial t_1}\right)=(f_1+f_2)\frac{\partial f_1}{\partial t_2}+f_1\left(\frac{\partial f_2}{\partial t_2}-\frac{\partial f_1}{\partial t_1}\right)$$

and

$$f_2\frac{\partial f_2}{\partial t_1}+f_2\frac{\partial f_1}{\partial t_1}+\left(f_1\frac{\partial f_2}{\partial t_1}-f_2\frac{\partial f_2}{\partial t_2}\right)=(f_1+f_2)\frac{\partial f_2}{\partial t_1}+f_2\left(\frac{\partial f_1}{\partial t_1}-\frac{\partial f_2}{\partial t_2}\right).$$

Modulo the terms of second order partial derivatives, we define the following operation:

$$\begin{aligned}&\left[f_1\frac{\partial}{\partial t_1}\otimes\frac{\partial}{\partial t_2},f_2\frac{\partial}{\partial t_1}\otimes\frac{\partial}{\partial t_2}\right]_{Z_0}\\ :=&\left[(f_1+f_2)\frac{\partial f_1}{\partial t_2}+f_1\left(\frac{\partial f_2}{\partial t_2}-\frac{\partial f_1}{\partial t_1}\right)\right]\frac{\partial}{\partial t_1}\\ &+\left[(f_1+f_2)\frac{\partial f_2}{\partial t_1}+f_2\left(\frac{\partial f_1}{\partial t_1}-\frac{\partial f_2}{\partial t_2}\right)\right]\frac{\partial}{\partial t_2}+c\cdot\phi(f_1,f_2).\end{aligned}$$

The significance of our four derivatives come from the following proposition: with the help of four derivatives, a system of linear partial differential equations can be transformed into a system of nonlinear partial differential equations.

Proposition 1.9 (Transformation property). *For the system of partial differential equations:*

$$\begin{cases}r=a_1p+a_2q+a_3z,\\ s=b_1p+b_2q+b_3z,\\ t=c_1p+c_2q+c_3z,\end{cases}$$

where $a_i=a_i(x,y)$, $b_i=b_i(x,y)$, $c_i=c_i(x,y)$ $(i=1,2,3)$ *and* $r=\frac{\partial^2 z}{\partial x^2}$, $s=\frac{\partial^2 z}{\partial x\partial y}$, $t=\frac{\partial^2 z}{\partial y^2}$, $p=\frac{\partial z}{\partial x}$, $q=\frac{\partial z}{\partial y}$, *assume that there exists three linearly independent solutions:* z_1, z_2 *and* z_3. *Put* $u_1=z_1/z_3$ *and* $u_2=z_2/z_3$. *Then*

$$\begin{cases}\{u_1,u_2;x,y\}_x=a_2,\\ \{u_1,u_2;x,y\}_y=c_1,\\ \ [u_1,u_2;x,y]_x=2b_2-a_1,\\ \ [u_1,u_2;x,y]_y=2b_1-c_2.\end{cases}\tag{1.24}$$

The next step is to solve this system of nonlinear partial differential equations. It is well known that nonlinear partial differential equations are extremely difficult to solve. Nevertheless, in our case, we can not only solve them, but also in the following explicit form:

Theorem 1.10 (Main Theorem 1) (The explicit solutions of nonlinear partial differential system of equations). *For the over-determined system of second* *order, nonlinear (quasilinear) complex partial differential equations:*

$$\begin{cases} \{w_1, w_2; v_1, v_2\}_{v_1} = F_1(v_1, v_2), \\ \{w_1, w_2; v_1, v_2\}_{v_2} = F_2(v_1, v_2), \\ [w_1, w_2; v_1, v_2]_{v_1} = P_1(v_1, v_2), \\ [w_1, w_2; v_1, v_2]_{v_2} = P_2(v_1, v_2), \end{cases} \tag{1.25}$$

set the Jacobian

$$\Delta := \frac{\partial(v_1, v_2)}{\partial(w_1, w_2)} = \begin{vmatrix} \frac{\partial v_1}{\partial w_1} & \frac{\partial v_2}{\partial w_1} \\ \frac{\partial v_1}{\partial w_2} & \frac{\partial v_2}{\partial w_2} \end{vmatrix}. \tag{1.26}$$

Then $z = \sqrt[3]{\Delta}$ satisfies the following linear differential equations:

$$\begin{cases} \dfrac{\partial^2 z}{\partial v_1^2} + \dfrac{1}{3}P_1\dfrac{\partial z}{\partial v_1} - F_1\dfrac{\partial z}{\partial v_2} + \left(\dfrac{\partial F_1}{\partial v_2} - \dfrac{1}{3}\dfrac{\partial P_1}{\partial v_1} - \dfrac{2}{9}P_1^2 - \dfrac{2}{3}F_1P_2\right)z = 0, \\ \dfrac{\partial^2 z}{\partial v_1 \partial v_2} - \dfrac{1}{3}P_2\dfrac{\partial z}{\partial v_1} - \dfrac{1}{3}P_1\dfrac{\partial z}{\partial v_2} + \left(\dfrac{1}{3}\dfrac{\partial P_2}{\partial v_1} + \dfrac{1}{3}\dfrac{\partial P_1}{\partial v_2} + \dfrac{1}{9}P_1P_2 - F_1F_2\right)z = 0, \\ \dfrac{\partial^2 z}{\partial v_2^2} + \dfrac{1}{3}P_2\dfrac{\partial z}{\partial v_2} - F_2\dfrac{\partial z}{\partial v_1} + \left(\dfrac{\partial F_2}{\partial v_1} - \dfrac{1}{3}\dfrac{\partial P_2}{\partial v_2} - \dfrac{2}{9}P_2^2 - \dfrac{2}{3}F_2P_1\right)z = 0. \end{cases} \tag{1.27}$$

Moreover, the above system (1.27) of linear partial differential equations are also solvable in the following:

Theorem 1.11 (Main Theorem 2) (The explicit solutions of linear partial differential system of equations). *For the system of linear partial differential equations:*

$$\begin{cases} \dfrac{\partial^2 z}{\partial v_1^2} + \dfrac{1}{3}P_1\dfrac{\partial z}{\partial v_1} - F_1\dfrac{\partial z}{\partial v_2} + \left(\dfrac{\partial F_1}{\partial v_2} - \dfrac{1}{3}\dfrac{\partial P_1}{\partial v_1} - \dfrac{2}{9}P_1^2 - \dfrac{2}{3}F_1P_2\right)z = 0, \\ \dfrac{\partial^2 z}{\partial v_1 \partial v_2} - \dfrac{1}{3}P_2\dfrac{\partial z}{\partial v_1} - \dfrac{1}{3}P_1\dfrac{\partial z}{\partial v_2} + \left(\dfrac{1}{3}\dfrac{\partial P_2}{\partial v_1} + \dfrac{1}{3}\dfrac{\partial P_1}{\partial v_2} + \dfrac{1}{9}P_1P_2 - F_1F_2\right)z = 0, \\ \dfrac{\partial^2 z}{\partial v_2^2} + \dfrac{1}{3}P_2\dfrac{\partial z}{\partial v_2} - F_2\dfrac{\partial z}{\partial v_1} + \left(\dfrac{\partial F_2}{\partial v_1} - \dfrac{1}{3}\dfrac{\partial P_2}{\partial v_2} - \dfrac{2}{9}P_2^2 - \dfrac{2}{3}F_2P_1\right)z = 0, \end{cases} \tag{1.28}$$

put

$$P_1 = \frac{\alpha}{v_1} + \frac{\beta}{v_1 - 1} + \frac{\gamma}{v_1 - v_2}, \quad P_2 = \frac{\alpha}{v_2} + \frac{\beta}{v_2 - 1} + \frac{\gamma}{v_2 - v_1}, \tag{1.29}$$

$$F_1 = \frac{-\gamma v_2(v_2 - 1)}{v_1(v_1 - 1)(v_1 - v_2)}, \quad F_2 = \frac{-\gamma v_1(v_1 - 1)}{v_2(v_2 - 1)(v_2 - v_1)}. \tag{1.30}$$

Set

$$w = v_1^{-\frac{1}{3}\alpha} v_2^{-\frac{1}{3}\alpha}(v_1 - 1)^{-\frac{1}{3}\beta}(v_2 - 1)^{-\frac{1}{3}\beta}(v_1 - v_2)^{\frac{2}{3}\gamma} z, \tag{1.31}$$

then the function $w = w(v_1, v_2)$ satisfies the following equations:

$$\begin{cases} \dfrac{\partial^2 w}{\partial v_1^2} + \left(\dfrac{\alpha}{v_1} + \dfrac{\beta}{v_1 - 1} + \dfrac{-\gamma}{v_1 - v_2}\right)\dfrac{\partial w}{\partial v_1} + \dfrac{\gamma v_2(v_2-1)}{v_1(v_1-1)(v_1-v_2)}\dfrac{\partial w}{\partial v_2} \\ + \dfrac{(1-\alpha-\beta)\gamma}{v_1(v_1-1)} w = 0, \\ \dfrac{\partial^2 w}{\partial v_2^2} + \left(\dfrac{\alpha}{v_2} + \dfrac{\beta}{v_2 - 1} + \dfrac{-\gamma}{v_2 - v_1}\right)\dfrac{\partial w}{\partial v_2} + \dfrac{\gamma v_1(v_1-1)}{v_2(v_2-1)(v_2-v_1)}\dfrac{\partial w}{\partial v_1} \\ + \dfrac{(1-\alpha-\beta)\gamma}{v_2(v_2-1)} w = 0, \\ \dfrac{\partial^2 w}{\partial v_1 \partial v_2} = \dfrac{-\gamma}{v_1 - v_2}\dfrac{\partial w}{\partial v_1} + \dfrac{\gamma}{v_1 - v_2}\dfrac{\partial w}{\partial v_2}. \end{cases} \tag{1.32}$$

A solution is given by

$$\begin{aligned} w = &\, const \cdot F_1(\alpha+\beta-1; -\gamma, -\gamma; \alpha-\gamma; v_1, v_2) \\ &+ const \cdot F_1(\alpha+\beta-1; -\gamma, -\gamma; \beta-\gamma; 1-v_1, 1-v_2), \end{aligned} \tag{1.33}$$

where $F_1 = F_1(a; b, b'; c; x, y)$ is the two variable hypergeometric function (see [3]). Consequently,

$$\begin{aligned} z = &\, v_1^{\frac{1}{3}\alpha} v_2^{\frac{1}{3}\alpha} (v_1-1)^{\frac{1}{3}\beta}(v_2-1)^{\frac{1}{3}\beta}(v_1-v_2)^{-\frac{2}{3}\gamma} \\ &\times [const \cdot F_1(\alpha+\beta-1; -\gamma, -\gamma; \alpha-\gamma; v_1, v_2) \\ &+ const \cdot F_1(\alpha+\beta-1; -\gamma, -\gamma; \beta-\gamma; 1-v_1, 1-v_2)]. \end{aligned} \tag{1.34}$$

In fact, there exists even some structure of $GL(3)$-invariance as follows: when (w_1, w_2) is a solution of the system of nonlinear partial differential equations:

$$\begin{cases} \{w_1, w_2; v_1, v_2\}_{v_1} = F_1(v_1, v_2), \\ \{w_1, w_2; v_1, v_2\}_{v_2} = F_2(v_1, v_2), \\ [w_1, w_2; v_1, v_2]_{v_1} = P_1(v_1, v_2), \\ [w_1, w_2; v_1, v_2]_{v_2} = P_2(v_1, v_2), \end{cases}$$

the general solution has the form

$$\left(\frac{a_1 w_1 + a_2 w_2 + a_3}{c_1 w_1 + c_2 w_2 + c_3}, \frac{b_1 w_1 + b_2 w_2 + b_3}{c_1 w_1 + c_2 w_2 + c_3}\right),$$

where $\begin{pmatrix} a_1 & a_2 & a_3 \\ b_1 & b_2 & b_3 \\ c_1 & c_2 & c_3 \end{pmatrix} \in GL(3, \mathbb{C})$.

It is known that the two variable hypergeometric function

$$F_1(a; b, b'; c; x, y) = \frac{\Gamma(c)}{\Gamma(a)\Gamma(c-a)} \int_0^1 t^{a-1}(1-t)^{c-a-1}(1-tx)^{-b}(1-ty)^{-b'} dt$$

satisfies the following equations:

$$\begin{cases} x(1-x)\dfrac{\partial^2 z}{\partial x^2} + y(1-x)\dfrac{\partial^2 z}{\partial x \partial y} + [c-(a+b+1)x]\dfrac{\partial z}{\partial x} - by\dfrac{\partial z}{\partial y} - abz = 0, \\ y(1-y)\dfrac{\partial^2 z}{\partial y^2} + x(1-y)\dfrac{\partial^2 z}{\partial x \partial y} + [c-(a+b'+1)y]\dfrac{\partial z}{\partial y} - b'x\dfrac{\partial z}{\partial x} - ab'z = 0. \end{cases}$$

For the group $GL(3)$, the Picard integral

$$z = \int_0^1 \frac{dt}{\sqrt[3]{t(t-1)(t-x)(t-y)}}$$

satisfies the following equations:

$$\begin{cases} x(1-x)\dfrac{\partial^2 z}{\partial x^2} + y(1-x)\dfrac{\partial^2 z}{\partial x \partial y} + \left(1-\dfrac{5}{3}x\right)\dfrac{\partial z}{\partial x} - \dfrac{1}{3}y\dfrac{\partial z}{\partial y} - \dfrac{1}{9}z = 0, \\ y(1-y)\dfrac{\partial^2 z}{\partial y^2} + x(1-y)\dfrac{\partial^2 z}{\partial x \partial y} + \left(1-\dfrac{5}{3}y\right)\dfrac{\partial z}{\partial y} - \dfrac{1}{3}x\dfrac{\partial z}{\partial x} - \dfrac{1}{9}z = 0. \end{cases} \tag{1.35}$$

Thus, a solution is given by

$$z = F_1\left(\frac{1}{3}; \frac{1}{3}, \frac{1}{3}; 1; x, y\right), \quad (a, b, b', c) = \left(\frac{1}{3}, \frac{1}{3}, \frac{1}{3}, 1\right). \tag{1.36}$$

It is a higher dimensional analogue of the hypergeometric differential equation (A.2) (see also (A.3)) satisfied by the elliptic integral (A.1).

The Picard modular function is associated to the following equations:

$$\begin{cases} x(1-x)\dfrac{\partial^2 z}{\partial x^2} + y(1-x)\dfrac{\partial^2 z}{\partial x \partial y} + \left(1-\dfrac{3}{2}x\right)\dfrac{\partial z}{\partial x} - \dfrac{1}{4}y\dfrac{\partial z}{\partial y} - \dfrac{1}{16}z = 0, \\ y(1-y)\dfrac{\partial^2 z}{\partial y^2} + x(1-y)\dfrac{\partial^2 z}{\partial x \partial y} + \left(1-\dfrac{3}{2}y\right)\dfrac{\partial z}{\partial y} - \dfrac{1}{4}x\dfrac{\partial z}{\partial x} - \dfrac{1}{16}z = 0. \end{cases} \tag{1.37}$$

A solution is given by

$$z = F_1\left(\frac{1}{4}; \frac{1}{4}, \frac{1}{4}; 1; x, y\right), \quad (a, b, b', c) = \left(\frac{1}{4}, \frac{1}{4}, \frac{1}{4}, 1\right). \tag{1.38}$$

It is a higher dimensional analogue of the hypergeometric differential equation (A.4) (see also (A.5)) satisfied by the elliptic modular function.

The η-function associated to the unitary group $U(2,1)$ is given as follows:

$$\eta(w_1, w_2) := v_1^{-\frac{1}{12}} v_2^{-\frac{1}{12}} (1-v_1)^{-\frac{1}{18}} (1-v_2)^{-\frac{1}{18}} (v_1-v_2)^{-\frac{1}{18}} \left[\frac{\partial(v_1, v_2)}{\partial(w_1, w_2)}\right]^{\frac{1}{9}}, \tag{1.39}$$

which is a higher dimensional analogue of the Dedekind η-function (see [7]). It satisfies the following transformation:

$$\eta\left(\frac{a_1 w_1 + a_2 w_2 + a_3}{c_1 w_1 + c_2 w_2 + c_3}, \frac{b_1 w_1 + b_2 w_2 + b_3}{c_1 w_1 + c_2 w_2 + c_3}\right) = \Delta^{-\frac{1}{9}} (c_1 w_1 + c_2 w_2 + c_3)^{\frac{1}{3}} \eta(w_1, w_2), \tag{1.40}$$

where $\begin{pmatrix} a_1 & a_2 & a_3 \\ b_1 & b_2 & b_3 \\ c_1 & c_2 & c_3 \end{pmatrix} \in GL(3,\mathbb{C})$ and $\Delta = \begin{vmatrix} a_1 & a_2 & a_3 \\ b_1 & b_2 & b_3 \\ c_1 & c_2 & c_3 \end{vmatrix}$.

Let $U(2,1) = \{g \in GL(3,\mathbb{C}) : g^* J g = J\}$ where $J = \begin{pmatrix} & & 1 \\ & -1 & \\ 1 & & \end{pmatrix}$. The corresponding complex domain $\mathfrak{S}_2 = \{(z_1, z_2) \in \mathbb{C}^2 : z_1 + \overline{z_1} - z_2\overline{z_2} > 0\}$ is the simplest example of a bounded symmetric domain which is not a tube domain. Put $\rho(Z) := z_1 + \overline{z_1} - z_2\overline{z_2}$ for $Z = (z_1, z_2) \in \mathfrak{S}_2$.

The imaginary quadratic field $K = \mathbb{Q}(\sqrt{-3})$ is called the field of Eisenstein numbers. Its ring of integers

$$\mathcal{O}_K = \{a/2 + b\sqrt{-3}/2 : a, b \in \mathbb{Z}, a \equiv b (\mathrm{mod} 2)\} = \mathbb{Z} + \mathbb{Z}[\omega]$$

with $\omega = \frac{-1+\sqrt{-3}}{2}$ is called the ring of Eisenstein integers.

Let

$$T_1 = \begin{pmatrix} 1 & 1 & -\omega \\ & 1 & 1 \\ & & 1 \end{pmatrix}, \quad T_2 = \begin{pmatrix} 1 & \omega & -\omega \\ & 1 & \overline{\omega} \\ & & 1 \end{pmatrix}, \quad S = -\overline{\omega} J = \begin{pmatrix} & & -\overline{\omega} \\ & \overline{\omega} & \\ -\overline{\omega} & & \end{pmatrix}, \tag{1.41}$$

$$U_1 = \begin{pmatrix} 1 & & \\ & -\omega & \\ & & 1 \end{pmatrix}, \quad U_2 = \begin{pmatrix} -1 & & \\ & -\omega & \\ & & -1 \end{pmatrix}. \tag{1.42}$$

The actions of the above five generators on $(w_1, w_2) \in \mathfrak{S}_2$ are given as follows:

$$\begin{cases} T_1(w_1, w_2) = (w_1 + w_2 - \omega, w_2 + 1), \\ T_2(w_1, w_2) = (w_1 + \omega w_2 - \omega, w_2 + \overline{\omega}), \\ S(w_1, w_2) = \left(\dfrac{1}{w_1}, -\dfrac{w_2}{w_1}\right), \\ U_1(w_1, w_2) = (w_1, -\omega w_2), \\ U_2(w_1, w_2) = (w_1, \omega w_2), \end{cases} \tag{1.43}$$

where T_1 and T_2 can be considered as two non-commutative translations.

Put

$$\eta_1(w_1, w_2) := \eta\left(\begin{pmatrix} 3 & & \\ & 1-\omega & \\ & & 1 \end{pmatrix}(w_1, w_2)\right) = \eta(3w_1, (1-\omega)w_2), \tag{1.44}$$

$$\eta_2(w_1, w_2) := \eta\left(\begin{pmatrix} 1 & & \\ & 1-\omega & \\ & & 3 \end{pmatrix}(w_1, w_2)\right) = \eta\left(\frac{w_1}{3}, \frac{w_2}{1-\overline{\omega}}\right). \tag{1.45}$$

$$\eta_3(w_1, w_2) := \eta_2([T_1, T_2]^{-1}(w_1, w_2)). \tag{1.46}$$

$$\eta_4(w_1, w_2) := \eta_1(S^3[T_1, T_2]^{-1} S^3(w_1, w_2)). \tag{1.47}$$

$$\eta_5(w_1,w_2) := \eta_2([T_1,T_2]^{-1}S^3[T_1,T_2]^{-1}S^3(w_1,w_2)). \tag{1.48}$$

Here,

$$[T_1,T_2] := T_1T_2T_1^{-1}T_2^{-1} = \begin{pmatrix} 1 & 0 & \overline{\omega}-\omega \\ & 1 & 0 \\ & & 1 \end{pmatrix}. \tag{1.49}$$

In section 4, we will study the transformation property of these η-functions under some non-commutative translations.

Put

$$\begin{cases} \varphi_0(w_1,w_2) = \dfrac{\eta_1(w_1,w_2)}{\eta_2(w_1,w_2)} =: \kappa_0^{\frac{1}{9}} =: k_0^{\frac{1}{27}}, \\ \varphi_1(w_1,w_2) = \dfrac{\eta_1(w_1,w_2)}{\eta_3(w_1,w_2)} =: \kappa_1^{\frac{1}{9}} =: k_1^{\frac{1}{27}}, \\ \varphi_2(w_1,w_2) = \dfrac{\eta_4(w_1,w_2)}{\eta_2(w_1,w_2)} =: \kappa_2^{\frac{1}{9}} =: k_2^{\frac{1}{27}}, \\ \varphi_3(w_1,w_2) = \dfrac{\eta_4(w_1,w_2)}{\eta_5(w_1,w_2)} =: \kappa_3^{\frac{1}{9}} =: k_3^{\frac{1}{27}}. \end{cases} \tag{1.50}$$

Here, there are four functions φ_0, φ_1, φ_2 and φ_3, $C_4^2 = 6$. In fact, the dimension of the moduli space of algebraic curves of genus three is: $\dim\mathcal{M}_3 = 3\times 3-3 = 6$. Namely, we define

$$\begin{cases} K_1 := \displaystyle\int_0^1 \frac{dx}{\sqrt[3]{x^2(1-x)(1-k_2x)(1-k_3x)}}, \\ K_2 := \displaystyle\int_0^1 \frac{dx}{\sqrt[3]{x^2(1-x)(1-k_3x)(1-k_1x)}}, \\ K_3 := \displaystyle\int_0^1 \frac{dx}{\sqrt[3]{x^2(1-x)(1-k_1x)(1-k_2x)}}. \end{cases} \tag{1.51}$$

$$\begin{cases} K_4 := \displaystyle\int_0^1 \frac{dx}{\sqrt[3]{x^2(1-x)(1-k_0x)(1-k_1x)}}, \\ K_5 := \displaystyle\int_0^1 \frac{dx}{\sqrt[3]{x^2(1-x)(1-k_0x)(1-k_2x)}}, \\ K_6 := \displaystyle\int_0^1 \frac{dx}{\sqrt[3]{x^2(1-x)(1-k_0x)(1-k_3x)}}. \end{cases} \tag{1.52}$$

From the viewpoint of group representation theory, the classical trigonometric integrals and elliptic integrals can be interpreted as follows:

$GL(1)$: trigonometric functions, cyclotomic units.

$$\frac{dy}{\sqrt{1-y^2}} = \frac{1}{M}\frac{dx}{\sqrt{1-x^2}}.$$

$GL(2)$: elliptic functions, elliptic units, complex multiplication.

$$\frac{dy}{\sqrt{(1-y^2)(1-\lambda^2 y^2)}} = \frac{1}{M} \frac{dx}{\sqrt{(1-x^2)(1-\kappa^2 x^2)}}.$$

This viewpoint, i.e. the transformation property of elliptic integrals, especially the higher order transformation with modular equation of higher degree, plays a central role in the work of Jacobi (see [19]), which leads to the arithmetical theory of complex multiplication for elliptic modular functions. In particular, Jacobi (see [19], p. 28) obtained the following celebrated formula for the transformation of the fifth order:

$$\frac{v(1-uv^3)dy}{\sqrt{1-y^2}\sqrt{1-v^8 y^2}} = \frac{(v-u^5)dx}{\sqrt{1-x^2}\sqrt{1-u^8 x^2}}, \tag{A.6}$$

where

$$y = \frac{v(v-u^5)x + u^3(u^2+v^2)(v-u^5)x^3 + u^{10}(1-uv^3)x^5}{v^2(1-uv^3) + uv^2(u^2+v^2)(v-u^5)x^2 + u^6 v^3(v-u^5)x^4}, \tag{A.7}$$

the moduli u and v satisfy the modular equation

$$u^6 - v^6 + 5u^2 v^2(u^2 - v^2) + 4uv(1 - u^4 v^4) = 0. \tag{A.8}$$

For the group $GL(3)$, we have the following theorem which can be considered as a natural generalization of the above story from $GL(1)$, $GL(2)$ to $GL(3)$. In particular, it extends the arithmetical theory of complex multiplication from elliptic modular functions to Picard modular functions.

Let us consider the following two homogeneous polynomials of degree 9:

$$\begin{cases} f_1(x,y,z) := x^2 y^2 z^2 (z-x)(z-\lambda_1^3 x)(z-\lambda_2^3 x), \\ f_2(x,y,z) := x^2 y^2 z^2 (z-y)(z-\lambda_0^3 y)(z-\lambda_3^3 y). \end{cases} \tag{1.53}$$

We will study the following two differential forms:

$$\begin{cases} \omega_1 := \dfrac{xdy\wedge dz + ydz\wedge dx + zdx\wedge dy}{\sqrt[3]{f_1(x,y,z)}}, \\ \omega_2 := \dfrac{xdy\wedge dz + ydz\wedge dx + zdx\wedge dy}{\sqrt[3]{f_2(x,y,z)}}. \end{cases} \tag{1.54}$$

The following theorem can be considered as a higher dimensional analogue of Jacobi's celebrated work [19], §§1–34 on the transformation property of elliptic integrals.

Theorem 1.12 (Main Theorem 3) (Higher order transformation of double irrational integrals with modular equation of higher degree). *The following identity of differential forms holds:*

$$\begin{aligned} &\frac{dw_1 \wedge dw_2}{\sqrt[3]{w_1^2 w_2^2 (1-w_1)(1-\lambda_1^3 w_1)(1-\lambda_2^3 w_1)}} \\ &= -\frac{5t_1}{t_2^2} \frac{dt_1 \wedge dt_2}{\sqrt[3]{t_1^2 t_2^2 (1-t_1)(1-\kappa_1^3 t_1)(1-\kappa_2^3 t_1)}}, \end{aligned} \tag{1.55}$$

where

$$w_1 = \frac{(\beta+\gamma)t_1+\alpha}{\beta t_1+(\alpha+\gamma)}, \quad w_2 = \frac{t_1[(\beta+\gamma)t_1+\alpha]^2[\beta t_1+(\alpha+\gamma)]}{t_2^5}, \tag{1.56}$$

which is a rational transformation of order five. Here,

$$\begin{cases} \alpha = \dfrac{(v_1-1)(v_2-1)(v_1+v_2-2)-(u_1-1)(u_2-1)(u_1+u_2-2)}{2(u_1-1)(u_2-1)(v_1-1)(v_2-1)}, \\ \beta = \dfrac{(u_1-1)(u_2-1)(2u_1u_2-u_1-u_2)-(v_1-1)(v_2-1)(2v_1v_2-v_1-v_2)}{2(u_1-1)(u_2-1)(v_1-1)(v_2-1)}, \\ \gamma = \dfrac{(v_1-1)(v_2-1)}{(u_1-1)(u_2-1)}, \end{cases} \tag{1.57}$$

with

$$u_1 = \kappa_1^3, \quad u_2 = \kappa_2^3, \quad v_1 = \lambda_1^3, \quad v_2 = \lambda_2^3.$$

Moreover, the moduli (κ_1, κ_2) *and* (λ_1, λ_2) *satisfy the modular equation:*

$$(\kappa_1^3-1)(\kappa_2^3-1)(\kappa_1^3-\kappa_2^3) = (\lambda_1^3-1)(\lambda_2^3-1)(\lambda_1^3-\lambda_2^3), \tag{1.58}$$

which is an algebraic variety of dimension three. The corresponding two algebraic surfaces are:

$$\begin{cases} w_3^3 = w_1^2w_2^2(1-w_1)(1-\lambda_1^3w_1)(1-\lambda_2^3w_1), \\ t_3^3 = t_1^2t_2^2(1-t_1)(1-\kappa_1^3t_1)(1-\kappa_2^3t_1), \end{cases} \tag{1.59}$$

which give the homogeneous algebraic equations of degree seven:

$$\begin{cases} w_3^3w_4^4 = w_1^2w_2^2(w_4-w_1)(w_4-\lambda_1^3w_1)(w_4-\lambda_2^3w_1), \\ t_3^3t_4^4 = t_1^2t_2^2(t_4-t_1)(t_4-\kappa_1^3t_1)(t_4-\kappa_2^3t_1). \end{cases} \tag{1.60}$$

Corollary 1.13 (The other form of higher order transformation of double irrational integrals with modular equation of higher degree). *The following identity holds:*

$$\frac{dy_1\wedge dy_2}{\sqrt[3]{(1-y_1^3)(1-\lambda_1^3y_1^3)(1-\lambda_2^3y_1^3)}} = -\frac{5x_1^3}{x_2^6}\frac{dx_1\wedge dx_2}{\sqrt[3]{(1-x_1^3)(1-\kappa_1^3x_1^3)(1-\kappa_2^3x_1^3)}}, \tag{1.61}$$

where

$$y_1^3 = \frac{(\beta+\gamma)x_1^3+\alpha}{\beta x_1^3+(\alpha+\gamma)}, \quad y_2^3 = \frac{x_1^3[(\beta+\gamma)x_1^3+\alpha]^2[\beta x_1^3+(\alpha+\gamma)]}{x_2^{15}}, \tag{1.62}$$

and

$$\begin{cases} \alpha = \dfrac{(v_1-1)(v_2-1)(v_1+v_2-2)-(u_1-1)(u_2-1)(u_1+u_2-2)}{2(u_1-1)(u_2-1)(v_1-1)(v_2-1)}, \\ \beta = \dfrac{(u_1-1)(u_2-1)(2u_1u_2-u_1-u_2)-(v_1-1)(v_2-1)(2v_1v_2-v_1-v_2)}{2(u_1-1)(u_2-1)(v_1-1)(v_2-1)}, \\ \gamma = \dfrac{(v_1-1)(v_2-1)}{(u_1-1)(u_2-1)}, \end{cases}$$

with

$$u_1 = \kappa_1^3, \quad u_2 = \kappa_2^3, \quad v_1 = \lambda_1^3, \quad v_2 = \lambda_2^3.$$

Moreover, the moduli (κ_1, κ_2) *and* (λ_1, λ_2) *satisfy the modular equation:*

$$(\kappa_1^3 - 1)(\kappa_2^3 - 1)(\kappa_1^3 - \kappa_2^3) = (\lambda_1^3 - 1)(\lambda_2^3 - 1)(\lambda_1^3 - \lambda_2^3).$$

The following results can be considered as a higher dimensional analogue of Schwarz's celebrated triangular s-functions [39] and their transformation properties. We will study the system of partial differential equations:

$$\begin{cases} \{w_1, w_2; v_1, v_2\}_{v_1} = F_1(\gamma; v_1, v_2), \\ \{w_1, w_2; v_1, v_2\}_{v_2} = F_2(\gamma; v_1, v_2), \\ [w_1, w_2; v_1, v_2]_{v_1} = P_1(\alpha, \beta, \gamma; v_1, v_2), \\ [w_1, w_2; v_1, v_2]_{v_2} = P_2(\alpha, \beta, \gamma; v_1, v_2), \end{cases} \tag{1.63}$$

where

$$\begin{cases} F_1(\gamma; v_1, v_2) = \dfrac{-\gamma v_2(v_2 - 1)}{v_1(v_1 - 1)(v_1 - v_2)}, \\ F_2(\gamma; v_1, v_2) = \dfrac{-\gamma v_1(v_1 - 1)}{v_2(v_2 - 1)(v_2 - v_1)}, \\ P_1(\alpha, \beta, \gamma; v_1, v_2) = \dfrac{\alpha}{v_1} + \dfrac{\beta}{v_1 - 1} + \dfrac{\gamma}{v_1 - v_2}, \\ P_2(\alpha, \beta, \gamma; v_1, v_2) = \dfrac{\alpha}{v_2} + \dfrac{\beta}{v_2 - 1} + \dfrac{\gamma}{v_2 - v_1}. \end{cases} \tag{1.64}$$

The solutions are denoted by

$$w_1 = s_1(\alpha, \beta, \gamma; v_1, v_2), \quad w_2 = s_2(\alpha, \beta, \gamma; v_1, v_2). \tag{1.65}$$

Consequently,

$$\begin{cases} \{s_1(\alpha, \beta, \gamma; v_1, v_2), s_2(\alpha, \beta, \gamma; v_1, v_2); v_1, v_2\}_{v_1} = F_1(\gamma; v_1, v_2), \\ \{s_1(\alpha, \beta, \gamma; v_1, v_2), s_2(\alpha, \beta, \gamma; v_1, v_2); v_1, v_2\}_{v_2} = F_2(\gamma; v_1, v_2), \\ [s_1(\alpha, \beta, \gamma; v_1, v_2), s_2(\alpha, \beta, \gamma; v_1, v_2); v_1, v_2]_{v_1} = P_1(\alpha, \beta, \gamma; v_1, v_2), \\ [s_1(\alpha, \beta, \gamma; v_1, v_2), s_2(\alpha, \beta, \gamma; v_1, v_2); v_1, v_2]_{v_2} = P_2(\alpha, \beta, \gamma; v_1, v_2), \end{cases} \tag{1.66}$$

Let

$$T = \begin{pmatrix} -1 & 0 & 1 \\ & -1 & 1 \\ & & 1 \end{pmatrix}, \quad S_1 = \begin{pmatrix} 1 & 0 & 0 \\ 0 & 0 & 1 \\ 0 & 1 & 0 \end{pmatrix}, \quad S_2 = \begin{pmatrix} & & 1 \\ & 1 & \\ 1 & & \end{pmatrix}. \tag{1.67}$$

Then

$$T(x, y) = (1 - x, 1 - y). \tag{1.68}$$

$$\begin{cases} S_1(x,y) = \left(\frac{x}{y}, \frac{1}{y}\right), \\ S_1T(x,y) = \left(\frac{1-x}{1-y}, \frac{1}{1-y}\right), \\ TS_1(x,y) = \left(\frac{y-x}{y}, \frac{y-1}{y}\right), \\ S_1TS_1(x,y) = \left(\frac{y-x}{y-1}, \frac{y}{y-1}\right). \end{cases} \tag{1.69}$$

$$\begin{cases} S_2(x,y) = \left(\frac{1}{x}, \frac{y}{x}\right), \\ S_2T(x,y) = \left(\frac{1}{1-x}, \frac{1-y}{1-x}\right), \\ TS_2(x,y) = \left(\frac{x-1}{x}, \frac{x-y}{x}\right), \\ S_2TS_2(x,y) = \left(\frac{x}{x-1}, \frac{x-y}{x-1}\right). \end{cases} \tag{1.70}$$

We have

$$\begin{cases} \{s_1(\beta,\alpha,\gamma;v_1,v_2), s_2(\beta,\alpha,\gamma;v_1,v_2);v_1,v_2\}_{v_1} = F_1(-\gamma;T(v_1,v_2)), \\ \{s_1(\beta,\alpha,\gamma;v_1,v_2), s_2(\beta,\alpha,\gamma;v_1,v_2);v_1,v_2\}_{v_2} = F_2(-\gamma;T(v_1,v_2)), \\ [s_1(\beta,\alpha,\gamma;v_1,v_2), s_2(\beta,\alpha,\gamma;v_1,v_2);v_1,v_2]_{v_1} = -P_1(\alpha,\beta,\gamma;T(v_1,v_2)), \\ [s_1(\beta,\alpha,\gamma;v_1,v_2), s_2(\beta,\alpha,\gamma;v_1,v_2);v_1,v_2]_{v_2} = -P_2(\alpha,\beta,\gamma;T(v_1,v_2)). \end{cases} \tag{1.71}$$

$$\begin{cases} \{s_1(\alpha,\gamma,\beta;v_1,v_2), s_2(\alpha,\gamma,\beta;v_1,v_2);v_1,v_2\}_{v_1} = F_1(-\beta;S_1(v_1,v_2)), \\ \{s_1(\alpha,\gamma,\beta;v_1,v_2), s_2(\alpha,\gamma,\beta;v_1,v_2);v_1,v_2\}_{v_2} = F_2(-\beta;S_2(v_1,v_2)), \\ [s_1(\alpha,\gamma,\beta;v_1,v_2), s_2(\alpha,\gamma,\beta;v_1,v_2);v_1,v_2]_{v_1} = \frac{P_1(\alpha,\beta,\gamma;S_1(v_1,v_2))}{v_2}, \\ [s_1(\alpha,\gamma,\beta;v_1,v_2), s_2(\alpha,\gamma,\beta;v_1,v_2);v_1,v_2]_{v_2} = \frac{P_2(\alpha,\beta,\gamma;S_2(v_1,v_2))}{v_1}. \end{cases} \tag{1.72}$$

$$\begin{cases} \{s_1(\gamma,\alpha,\beta;v_1,v_2), s_2(\gamma,\alpha,\beta;v_1,v_2);v_1,v_2\}_{v_1} = F_1(\beta;S_1T(v_1,v_2)), \\ \{s_1(\gamma,\alpha,\beta;v_1,v_2), s_2(\gamma,\alpha,\beta;v_1,v_2);v_1,v_2\}_{v_2} = F_2(\beta;S_2T(v_1,v_2)), \\ [s_1(\gamma,\alpha,\beta;v_1,v_2), s_2(\gamma,\alpha,\beta;v_1,v_2);v_1,v_2]_{v_1} = \frac{P_1(\alpha,\beta,\gamma;S_1T(v_1,v_2))}{v_2-1}, \\ [s_1(\gamma,\alpha,\beta;v_1,v_2), s_2(\gamma,\alpha,\beta;v_1,v_2);v_1,v_2]_{v_2} = \frac{P_2(\alpha,\beta,\gamma;S_2T(v_1,v_2))}{v_1-1}. \end{cases} \tag{1.73}$$

$$\begin{cases} \{s_1(\beta,\gamma,\alpha;v_1,v_2), s_2(\beta,\gamma,\alpha;v_1,v_2);v_1,v_2\}_{v_1} = F_1(\alpha;TS_1(v_1,v_2)), \\ \{s_1(\beta,\gamma,\alpha;v_1,v_2), s_2(\beta,\gamma,\alpha;v_1,v_2);v_1,v_2\}_{v_2} = F_2(\alpha;TS_2(v_1,v_2)), \\ [s_1(\beta,\gamma,\alpha;v_1,v_2), s_2(\beta,\gamma,\alpha;v_1,v_2);v_1,v_2]_{v_1} = -\dfrac{P_1(\alpha,\beta,\gamma;TS_1(v_1,v_2))}{v_2}, \\ [s_1(\beta,\gamma,\alpha;v_1,v_2), s_2(\beta,\gamma,\alpha;v_1,v_2);v_1,v_2]_{v_2} = -\dfrac{P_2(\alpha,\beta,\gamma;TS_2(v_1,v_2))}{v_1}. \end{cases} \tag{1.74}$$

$$\begin{cases} \{s_1(\gamma,\beta,\alpha;v_1,v_2), s_2(\gamma,\beta,\alpha;v_1,v_2);v_1,v_2\}_{v_1} = F_1(-\alpha;S_1TS_1(v_1,v_2)), \\ \{s_1(\gamma,\beta,\alpha;v_1,v_2), s_2(\gamma,\beta,\alpha;v_1,v_2);v_1,v_2\}_{v_2} = F_2(-\alpha;S_2TS_2(v_1,v_2)), \\ [s_1(\gamma,\beta,\alpha;v_1,v_2), s_2(\gamma,\beta,\alpha;v_1,v_2);v_1,v_2]_{v_1} = \dfrac{P_1(\alpha,\beta,\gamma;S_1TS_1(v_1,v_2))}{1-v_2}, \\ [s_1(\gamma,\beta,\alpha;v_1,v_2), s_2(\gamma,\beta,\alpha;v_1,v_2);v_1,v_2]_{v_2} = \dfrac{P_2(\alpha,\beta,\gamma;S_2TS_2(v_1,v_2))}{1-v_1}. \end{cases} \tag{1.75}$$

Proposition 1.14 (Main Proposition 3) (Transformation properties). *The triangular s-functions have the following properties of transformations:*
(1)

$$\begin{cases} \{s_1(\alpha,\beta,\gamma;T(v_1,v_2)), s_2(\alpha,\beta,\gamma;T(v_1,v_2));v_1,v_2\}_{v_1} = F_1(\gamma;v_1,v_2), \\ \{s_1(\alpha,\beta,\gamma;T(v_1,v_2)), s_2(\alpha,\beta,\gamma;T(v_1,v_2));v_1,v_2\}_{v_2} = F_2(\gamma;v_1,v_2), \\ [s_1(\alpha,\beta,\gamma;T(v_1,v_2)), s_2(\alpha,\beta,\gamma;T(v_1,v_2));v_1,v_2]_{v_1} = P_1(\beta,\alpha,\gamma;v_1,v_2), \\ [s_1(\alpha,\beta,\gamma;T(v_1,v_2)), s_2(\alpha,\beta,\gamma;T(v_1,v_2));v_1,v_2]_{v_2} = P_2(\beta,\alpha,\gamma;v_1,v_2). \end{cases} \tag{1.76}$$

(2)

$$\begin{cases} \quad \{s_1(\alpha,\beta,\gamma;S_1(v_1,v_2)), s_2(\alpha,\beta,\gamma;S_1(v_1,v_2));v_1,v_2\}_{v_1} \\ =F_1(\gamma;v_1,v_2), \\ \quad \{s_1(\alpha,\beta,\gamma;S_2(v_1,v_2)), s_2(\alpha,\beta,\gamma;S_2(v_1,v_2));v_1,v_2\}_{v_2} \\ =F_2(\gamma;v_1,v_2), \\ \quad [s_1(\alpha,\beta,\gamma;S_1(v_1,v_2)), s_2(\alpha,\beta,\gamma;S_1(v_1,v_2));v_1,v_2]_{v_1} \\ =P_1(\alpha,-2\gamma,\beta+3\gamma;v_1,v_2), \\ \quad [s_1(\alpha,\beta,\gamma;S_2(v_1,v_2)), s_2(\alpha,\beta,\gamma;S_2(v_1,v_2));v_1,v_2]_{v_2} \\ =P_2(\alpha,-2\gamma,\beta+3\gamma;v_1,v_2). \end{cases} \tag{1.77}$$

(3)

$$\begin{cases} \{s_1(\alpha,\beta,\gamma;S_1T(v_1,v_2)),s_2(\alpha,\beta,\gamma;S_1T(v_1,v_2));v_1,v_2\}_{v_1} \\ =F_1(\gamma;v_1,v_2), \\ \{s_1(\alpha,\beta,\gamma;S_2T(v_1,v_2)),s_2(\alpha,\beta,\gamma;S_2T(v_1,v_2));v_1,v_2\}_{v_2} \\ =F_2(\gamma;v_1,v_2), \\ [s_1(\alpha,\beta,\gamma;S_1T(v_1,v_2)),s_2(\alpha,\beta,\gamma;S_1T(v_1,v_2));v_1,v_2]_{v_1} \\ =P_1(-2\gamma,\alpha,\beta+3\gamma;v_1,v_2), \\ [s_1(\alpha,\beta,\gamma;S_2T(v_1,v_2)),s_2(\alpha,\beta,\gamma;S_2T(v_1,v_2));v_1,v_2]_{v_2} \\ =P_2(-2\gamma,\alpha,\beta+3\gamma;v_1,v_2). \end{cases} \tag{1.78}$$

(4)

$$\begin{cases} \{s_1(\alpha,\beta,\gamma;TS_1(v_1,v_2)),s_2(\alpha,\beta,\gamma;TS_1(v_1,v_2));v_1,v_2\}_{v_1} \\ =F_1(\gamma;v_1,v_2), \\ \{s_1(\alpha,\beta,\gamma;TS_2(v_1,v_2)),s_2(\alpha,\beta,\gamma;TS_2(v_1,v_2));v_1,v_2\}_{v_2} \\ =F_2(\gamma;v_1,v_2), \\ [s_1(\alpha,\beta,\gamma;TS_1(v_1,v_2)),s_2(\alpha,\beta,\gamma;TS_1(v_1,v_2));v_1,v_2]_{v_1} \\ =P_1(\beta,-2\gamma,\alpha+3\gamma;v_1,v_2), \\ [s_1(\alpha,\beta,\gamma;TS_2(v_1,v_2)),s_2(\alpha,\beta,\gamma;TS_2(v_1,v_2));v_1,v_2]_{v_2} \\ =P_2(\beta,-2\gamma,\alpha+3\gamma;v_1,v_2). \end{cases} \tag{1.79}$$

(5)

$$\begin{cases} \{s_1(\alpha,\beta,\gamma;S_1TS_1(v_1,v_2)),s_2(\alpha,\beta,\gamma;S_1TS_1(v_1,v_2));v_1,v_2\}_{v_1} \\ =F_1(\gamma;v_1,v_2), \\ \{s_1(\alpha,\beta,\gamma;S_2TS_2(v_1,v_2)),s_2(\alpha,\beta,\gamma;S_2TS_2(v_1,v_2));v_1,v_2\}_{v_2} \\ =F_2(\gamma;v_1,v_2), \\ [s_1(\alpha,\beta,\gamma;S_1TS_1(v_1,v_2)),s_2(\alpha,\beta,\gamma;S_1TS_1(v_1,v_2));v_1,v_2]_{v_1} \\ =P_1(-2\gamma,\beta,\alpha+3\gamma;v_1,v_2), \\ [s_1(\alpha,\beta,\gamma;S_2TS_2(v_1,v_2)),s_2(\alpha,\beta,\gamma;S_2TS_2(v_1,v_2));v_1,v_2]_{v_2} \\ =P_2(-2\gamma,\beta,\alpha+3\gamma;v_1,v_2). \end{cases} \tag{1.80}$$

Corollary 1.15. *The triangular s-functions have the following properties of transformations:*

(1)

$$\begin{cases} \{s_1(\alpha,\beta,\gamma;T(v_1,v_2)),s_2(\alpha,\beta,\gamma;T(v_1,v_2));v_1,v_2\}_{v_1} \\ =\{s_1(\alpha,\beta,\gamma;v_1,v_2),s_2(\alpha,\beta,\gamma;v_1,v_2);v_1,v_2\}_{v_1}, \\ \{s_1(\alpha,\beta,\gamma;T(v_1,v_2)),s_2(\alpha,\beta,\gamma;T(v_1,v_2));v_1,v_2\}_{v_2} \\ =\{s_1(\alpha,\beta,\gamma;v_1,v_2),s_2(\alpha,\beta,\gamma;v_1,v_2);v_1,v_2\}_{v_2}, \\ [s_1(\alpha,\beta,\gamma;T(v_1,v_2)),s_2(\alpha,\beta,\gamma;T(v_1,v_2));v_1,v_2]_{v_1} \\ =[s_1(\beta,\alpha,\gamma;v_1,v_2),s_2(\beta,\alpha,\gamma;v_1,v_2);v_1,v_2]_{v_1}, \\ [s_1(\alpha,\beta,\gamma;T(v_1,v_2)),s_2(\alpha,\beta,\gamma;T(v_1,v_2));v_1,v_2]_{v_2} \\ =[s_1(\beta,\alpha,\gamma;v_1,v_2),s_2(\beta,\alpha,\gamma;v_1,v_2);v_1,v_2]_{v_2}. \end{cases} \tag{1.81}$$

(2)

$$\begin{cases} \{s_1(\alpha,\beta,\gamma;S_1(v_1,v_2)),s_2(\alpha,\beta,\gamma;S_1(v_1,v_2));v_1,v_2\}_{v_1} \\ =\{s_1(\alpha,\beta,\gamma;v_1,v_2),s_2(\alpha,\beta,\gamma;v_1,v_2);v_1,v_2\}_{v_1}, \\ \{s_1(\alpha,\beta,\gamma;S_2(v_1,v_2)),s_2(\alpha,\beta,\gamma;S_2(v_1,v_2));v_1,v_2\}_{v_2} \\ =\{s_1(\alpha,\beta,\gamma;v_1,v_2),s_2(\alpha,\beta,\gamma;v_1,v_2);v_1,v_2\}_{v_2}, \\ [s_1(\alpha,\beta,\gamma;S_1(v_1,v_2)),s_2(\alpha,\beta,\gamma;S_1(v_1,v_2));v_1,v_2]_{v_1} \\ =[s_1(\alpha,-2\gamma,\beta+3\gamma;v_1,v_2),s_2(\alpha,-2\gamma,\beta+3\gamma;v_1,v_2);v_1,v_2]_{v_1}, \\ [s_1(\alpha,\beta,\gamma;S_2(v_1,v_2)),s_2(\alpha,\beta,\gamma;S_2(v_1,v_2));v_1,v_2]_{v_2} \\ =[s_1(\alpha,-2\gamma,\beta+3\gamma;v_1,v_2),s_2(\alpha,-2\gamma,\beta+3\gamma;v_1,v_2);v_1,v_2]_{v_2}. \end{cases} \tag{1.82}$$

(3)

$$\begin{cases} \{s_1(\alpha,\beta,\gamma;S_1T(v_1,v_2)),s_2(\alpha,\beta,\gamma;S_1T(v_1,v_2));v_1,v_2\}_{v_1} \\ =\{s_1(\alpha,\beta,\gamma;v_1,v_2),s_2(\alpha,\beta,\gamma;v_1,v_2);v_1,v_2\}_{v_1}, \\ \{s_1(\alpha,\beta,\gamma;S_2T(v_1,v_2)),s_2(\alpha,\beta,\gamma;S_2T(v_1,v_2));v_1,v_2\}_{v_2} \\ =\{s_1(\alpha,\beta,\gamma;v_1,v_2),s_2(\alpha,\beta,\gamma;v_1,v_2);v_1,v_2\}_{v_2}, \\ [s_1(\alpha,\beta,\gamma;S_1T(v_1,v_2)),s_2(\alpha,\beta,\gamma;S_1T(v_1,v_2));v_1,v_2]_{v_1} \\ =[s_1(-2\gamma,\alpha,\beta+3\gamma;v_1,v_2),s_2(-2\gamma,\alpha,\beta+3\gamma;v_1,v_2);v_1,v_2]_{v_1}, \\ [s_1(\alpha,\beta,\gamma;S_2T(v_1,v_2)),s_2(\alpha,\beta,\gamma;S_2T(v_1,v_2));v_1,v_2]_{v_2} \\ =[s_1(-2\gamma,\alpha,\beta+3\gamma;v_1,v_2),s_2(-2\gamma,\alpha,\beta+3\gamma;v_1,v_2);v_1,v_2]_{v_2}. \end{cases} \tag{1.83}$$

(4)

$$\begin{cases} \{s_1(\alpha,\beta,\gamma;TS_1(v_1,v_2)),s_2(\alpha,\beta,\gamma;TS_1(v_1,v_2));v_1,v_2\}_{v_1} \\ =\{s_1(\alpha,\beta,\gamma;v_1,v_2),s_2(\alpha,\beta,\gamma;v_1,v_2);v_1,v_2\}_{v_1}, \\ \{s_1(\alpha,\beta,\gamma;TS_2(v_1,v_2)),s_2(\alpha,\beta,\gamma;TS_2(v_1,v_2));v_1,v_2\}_{v_2} \\ =\{s_1(\alpha,\beta,\gamma;v_1,v_2),s_2(\alpha,\beta,\gamma;v_1,v_2);v_1,v_2\}_{v_2}, \\ [s_1(\alpha,\beta,\gamma;TS_1(v_1,v_2)),s_2(\alpha,\beta,\gamma;TS_1(v_1,v_2));v_1,v_2]_{v_1} \\ =[s_1(\beta,-2\gamma,\alpha+3\gamma;v_1,v_2),s_2(\beta,-2\gamma,\alpha+3\gamma;v_1,v_2);v_1,v_2]_{v_1}, \\ [s_1(\alpha,\beta,\gamma;TS_2(v_1,v_2)),s_2(\alpha,\beta,\gamma;TS_2(v_1,v_2));v_1,v_2]_{v_2} \\ =[s_1(\beta,-2\gamma,\alpha+3\gamma;v_1,v_2),s_2(\beta,-2\gamma,\alpha+3\gamma;v_1,v_2);v_1,v_2]_{v_2}. \end{cases} \tag{1.84}$$

(5)

$$\begin{cases} \{s_1(\alpha,\beta,\gamma;S_1TS_1(v_1,v_2)),s_2(\alpha,\beta,\gamma;S_1TS_1(v_1,v_2));v_1,v_2\}_{v_1} \\ =\{s_1(\alpha,\beta,\gamma;v_1,v_2),s_2(\alpha,\beta,\gamma;v_1,v_2);v_1,v_2\}_{v_1}, \\ \{s_1(\alpha,\beta,\gamma;S_2TS_2(v_1,v_2)),s_2(\alpha,\beta,\gamma;S_2TS_2(v_1,v_2));v_1,v_2\}_{v_2} \\ =\{s_1(\alpha,\beta,\gamma;v_1,v_2),s_2(\alpha,\beta,\gamma;v_1,v_2);v_1,v_2\}_{v_2}, \\ [s_1(\alpha,\beta,\gamma;S_1TS_1(v_1,v_2)),s_2(\alpha,\beta,\gamma;S_1TS_1(v_1,v_2));v_1,v_2]_{v_1} \\ =[s_1(-2\gamma,\beta,\alpha+3\gamma;v_1,v_2),s_2(-2\gamma,\beta,\alpha+3\gamma;v_1,v_2);v_1,v_2]_{v_1}, \\ [s_1(\alpha,\beta,\gamma;S_2TS_2(v_1,v_2)),s_2(\alpha,\beta,\gamma;S_2TS_2(v_1,v_2));v_1,v_2]_{v_2} \\ =[s_1(-2\gamma,\beta,\alpha+3\gamma;v_1,v_2),s_2(-2\gamma,\beta,\alpha+3\gamma;v_1,v_2);v_1,v_2]_{v_2}. \end{cases} \tag{1.85}$$

We will give some invariants for Picard curves, which play a similar role as Klein's J-invariant (A.0) for elliptic curves. The significant properties of Picard curves come from the following facts:

Definition. Suppose that C is an algebraic curve of genus $g \geq 2$. Let $\{\omega_1, \cdots, \omega_g\}$ be a basis of $\Omega^1(C)$. Then the map

$$\varphi_K : C \to \mathbb{P}^{g-1}$$
$$p \mapsto [\omega_1(p), \cdots, \omega_g(p)]$$

is called a canonical map over C.

In fact, Suppose that z is a local coordinate near the point p and $\omega_\alpha = f_\alpha(z)dz$ $(\alpha = 1, 2, \cdots, g)$. Then

$$\varphi_K(p) = [f_1(z(p)), \cdots, f_g(z(p))].$$

Theorem. *Any algebraic curve of genus $g = 2$ is a hyperelliptic curve.*

Definition. Suppose that C is a non-hyperelliptic algebraic curve of genus $g \geq 2$. Let

$$\varphi_K : C \to \mathbb{P}^{g-1}$$

be a canonical map. Then

$$\varphi_K(C) \subset \mathbb{P}^{g-1}$$

is called a canonical curve.

Let C be a canonical curve of genus g. Then $\deg(C) = 2g - 2$.

Theorem. *Any canonical curve of genus $g = 3$ is a smooth quartic plane algebraic curve.*

Theorem. *Any canonical curve of genus $g = 4$ is the intersection of a quadratic surface and a cubic surface in $\mathbb{P}^3$.*

For $0, 1, \lambda_1, \lambda_2, \infty \in \mathbb{P}^1(\mathbb{C})$, $\lambda_1, \lambda_2 \notin \{0, 1\}$ and $\lambda_1 \neq \lambda_2$, set

$$\begin{cases} J_1 = J_1(\lambda_1, \lambda_2) := \dfrac{\lambda_2^2(\lambda_2 - 1)^2}{\lambda_1^2(\lambda_1 - 1)^2(\lambda_1 - \lambda_2)^2}, \\ J_2 = J_2(\lambda_1, \lambda_2) := \dfrac{\lambda_1^2(\lambda_1 - 1)^2}{\lambda_2^2(\lambda_2 - 1)^2(\lambda_2 - \lambda_1)^2}. \end{cases} \tag{1.86}$$

The corresponding Picard curve is

$$y^3 = x(x-1)(x-\lambda_1)(x-\lambda_2). \tag{1.87}$$

The functions J_1 and J_2 satisfy the following relations:

$$\begin{aligned} &J_1(\lambda_1,\lambda_2) = J_1(T(\lambda_1,\lambda_2)) = J_1(S_1(\lambda_1,\lambda_2)) = J_1(S_1T(\lambda_1,\lambda_2)) \\ =&J_1(TS_1(\lambda_1,\lambda_2)) = J_1(S_1TS_1(\lambda_1,\lambda_2)). \end{aligned} \tag{1.88}$$

$$\begin{aligned} &J_2(\lambda_1,\lambda_2) = J_2(T(\lambda_1,\lambda_2)) = J_2(S_2(\lambda_1,\lambda_2)) = J_2(S_2T(\lambda_1,\lambda_2)) \\ =&J_2(TS_2(\lambda_1,\lambda_2)) = J_2(S_2TS_2(\lambda_1,\lambda_2)). \end{aligned} \tag{1.89}$$

In fact, J_1 and J_2 can be considered as a higher dimensional analogue of Klein's celebrated J-invariant (A.0).

We will study the following four evolution equations for $GL(3)$:

$$\frac{\begin{vmatrix} \frac{\partial u_1}{\partial t_1} & \frac{\partial u_2}{\partial t_1} \\ \frac{\partial u_1}{\partial x} & \frac{\partial u_2}{\partial x} \end{vmatrix}}{\begin{vmatrix} \frac{\partial u_1}{\partial y} & \frac{\partial u_2}{\partial y} \\ \frac{\partial u_1}{\partial x} & \frac{\partial u_2}{\partial x} \end{vmatrix}} = \{u_1, u_2; x, y\}_x, \quad \frac{\begin{vmatrix} \frac{\partial u_1}{\partial t_1} & \frac{\partial u_2}{\partial t_1} \\ \frac{\partial u_1}{\partial y} & \frac{\partial u_2}{\partial y} \end{vmatrix}}{\begin{vmatrix} \frac{\partial u_1}{\partial y} & \frac{\partial u_2}{\partial y} \\ \frac{\partial u_1}{\partial x} & \frac{\partial u_2}{\partial x} \end{vmatrix}} = [u_1, u_2; x, y]_x, \tag{1.90}$$

$$\frac{\begin{vmatrix} \frac{\partial u_1}{\partial t_2} & \frac{\partial u_2}{\partial t_2} \\ \frac{\partial u_1}{\partial x} & \frac{\partial u_2}{\partial x} \end{vmatrix}}{\begin{vmatrix} \frac{\partial u_1}{\partial x} & \frac{\partial u_2}{\partial x} \\ \frac{\partial u_1}{\partial y} & \frac{\partial u_2}{\partial y} \end{vmatrix}} = [u_1, u_2; x, y]_y, \quad \frac{\begin{vmatrix} \frac{\partial u_1}{\partial t_2} & \frac{\partial u_2}{\partial t_2} \\ \frac{\partial u_1}{\partial y} & \frac{\partial u_2}{\partial y} \end{vmatrix}}{\begin{vmatrix} \frac{\partial u_1}{\partial x} & \frac{\partial u_2}{\partial x} \\ \frac{\partial u_1}{\partial y} & \frac{\partial u_2}{\partial y} \end{vmatrix}} = \{u_1, u_2; x, y\}_y, \tag{1.91}$$

where the functions $u_1 = u_1(x, y; t_1, t_2)$ and $u_2 = u_2(x, y; t_1, t_2)$. The most important property of the above equations is that they are invariant with respect to the group $GL(3, \mathbb{C})$ of linear fractional transformations:

$$(u_1, u_2) \mapsto \left(\frac{a_1u_1 + a_2u_2 + a_3}{c_1u_1 + c_2u_2 + c_3}, \frac{b_1u_1 + b_2u_2 + b_3}{c_1u_1 + c_2u_2 + c_3}\right)$$

with $\gamma = \begin{pmatrix} a_1 & a_2 & a_3 \\ b_1 & b_2 & b_3 \\ c_1 & c_2 & c_3 \end{pmatrix} \in GL(3, \mathbb{C})$.

Theorem 1.16 (Main Theorem 4) (Reduction property of evolution equations). *For the system of partial differential equations:*

$$\frac{\begin{vmatrix} \frac{\partial u_1}{\partial t_1} & \frac{\partial u_2}{\partial t_1} \\ \frac{\partial u_1}{\partial x} & \frac{\partial u_2}{\partial x} \end{vmatrix}}{\begin{vmatrix} \frac{\partial u_1}{\partial y} & \frac{\partial u_2}{\partial y} \\ \frac{\partial u_1}{\partial x} & \frac{\partial u_2}{\partial x} \end{vmatrix}} = \{u_1, u_2; x, y\}_x, \quad \frac{\begin{vmatrix} \frac{\partial u_1}{\partial t_1} & \frac{\partial u_2}{\partial t_1} \\ \frac{\partial u_1}{\partial y} & \frac{\partial u_2}{\partial y} \end{vmatrix}}{\begin{vmatrix} \frac{\partial u_1}{\partial y} & \frac{\partial u_2}{\partial y} \\ \frac{\partial u_1}{\partial x} & \frac{\partial u_2}{\partial x} \end{vmatrix}} = [u_1, u_2; x, y]_x, \tag{1.92}$$

$$\frac{\begin{vmatrix} \frac{\partial u_1}{\partial t_2} & \frac{\partial u_2}{\partial t_2} \\ \frac{\partial u_1}{\partial x} & \frac{\partial u_2}{\partial x} \end{vmatrix}}{\begin{vmatrix} \frac{\partial u_1}{\partial x} & \frac{\partial u_2}{\partial x} \\ \frac{\partial u_1}{\partial y} & \frac{\partial u_2}{\partial y} \end{vmatrix}} = [u_1, u_2; x, y]_y, \quad \frac{\begin{vmatrix} \frac{\partial u_1}{\partial t_2} & \frac{\partial u_2}{\partial t_2} \\ \frac{\partial u_1}{\partial y} & \frac{\partial u_2}{\partial y} \end{vmatrix}}{\begin{vmatrix} \frac{\partial u_1}{\partial x} & \frac{\partial u_2}{\partial x} \\ \frac{\partial u_1}{\partial y} & \frac{\partial u_2}{\partial y} \end{vmatrix}} = \{u_1, u_2; x, y\}_y, \tag{1.93}$$

set

$$\begin{cases} v_1 := \{u_1, u_2; x, y\}_x, \\ w_1 := [u_1, u_2; x, y]_x, \\ v_2 := \{u_1, u_2; x, y\}_y, \\ w_2 := [u_1, u_2; x, y]_y. \end{cases} \tag{1.94}$$

Then v_1, v_2, w_1 and w_2 satisfy the following two equations:

$$\begin{cases} \dfrac{\partial w_1}{\partial t_2} + \dfrac{\partial v_2}{\partial t_1} - v_1 \dfrac{\partial v_2}{\partial y} - v_2 \dfrac{\partial w_1}{\partial x} + w_1 \dfrac{\partial v_2}{\partial x} + w_2 \dfrac{\partial w_1}{\partial y} = 0, \\ \dfrac{\partial w_2}{\partial t_1} + \dfrac{\partial v_1}{\partial t_2} - v_1 \dfrac{\partial w_2}{\partial y} - v_2 \dfrac{\partial v_1}{\partial x} + w_1 \dfrac{\partial w_2}{\partial x} + w_2 \dfrac{\partial v_1}{\partial y} = 0. \end{cases} \tag{1.95}$$

In fact, The above two equations are invariant under the action of

$$\begin{cases} v_1(x, y; t_1, t_2) \mapsto v_1(x - a_1 t_1, y - b_1 t_2; t_1, t_2), \\ v_2(x, y; t_1, t_2) \mapsto v_2(x - a_2 t_1, y - b_2 t_2; t_1, t_2), \\ w_1(x, y; t_1, t_2) \mapsto w_1(x - a_1 t_1, y - b_1 t_2; t_1, t_2) + a_2, \\ w_2(x, y; t_1, t_2) \mapsto w_2(x - a_2 t_1, y - b_2 t_2; t_1, t_2) + b_1. \end{cases} \tag{1.96}$$

This paper consists of seven sections. In section two, we will study the monodromy of Appell hypergeometric partial differential equations and give the construction of our four derivatives. We will investigate the properties of our four derivatives. In section three, we will study the system of nonlinear partial differential equations associated to our four derivatives. We will give a solution which is associated to the Picard modular function. In section four, we will introduce the η-functions associated to the unitary group $U(2,1)$ and study their properties, especially the relation with Picard curves. In section five, we will study the transform problems (moduli problems) and the associated modular equations which give the connection between the moduli of two Picard curves. In section six, we will study the properties of solutions of the above system of nonlinear partial differential equations, which we call the triangular s-functions. We obtain the J-invariants associated to the Picard curves. In section seven, we will investigate a system of nonlinear evolution equations associated to our four derivatives.

Acknowledgements. The author thanks the referees very much for their helpful comments.

2. Four derivatives and their properties

Let us recall some basic facts about algebraic curves, which show the significant properties of Picard curves.

Definition. Suppose that C is an algebraic curve of genus $g \geq 2$. Let $\{\omega_1, \cdots, \omega_g\}$ be a basis of $\Omega^1(C)$. Then the map

$$\varphi_K : C \to \mathbb{P}^{g-1}$$
$$p \mapsto [\omega_1(p), \cdots, \omega_g(p)]$$

is called a canonical map over C.

In fact, Suppose that z is a local coordinate near the point p and $\omega_\alpha = f_\alpha(z)dz$ $(\alpha = 1, 2, \cdots, g)$. Then

$$\varphi_K(p) = [f_1(z(p)), \cdots, f_g(z(p))].$$

Theorem. *Any algebraic curve of genus $g = 2$ is a hyperelliptic curve.*

Definition. Suppose that C is a non-hyperelliptic algebraic curve of genus $g \geq 2$. Let

$$\varphi_K : C \to \mathbb{P}^{g-1}$$

be a canonical map. Then

$$\varphi_K(C) \subset \mathbb{P}^{g-1}$$

is called a canonical curve.

Let C be a canonical curve of genus g. Then $\deg(C) = 2g - 2$.

Theorem. *Any canonical curve of genus $g = 3$ is a smooth quartic plane algebraic curve.*

Theorem. *Any canonical curve of genus $g = 4$ is the intersection of a quadratic surface and a cubic surface in $\mathbb{P}^3$.*

Let us consider the Picard integral

$$z(x, y) = \int_0^1 \frac{dt}{\sqrt[3]{t(t-1)(t-x)(t-y)}}.$$

The corresponding Picard curve is given by

$$w^3 = t(t-1)(t-x)(t-y).$$

In fact,

$$z(x, y) = -x^{-\frac{1}{3}} y^{-\frac{1}{3}} \int_0^1 t^{-\frac{1}{3}}(1-t)^{-\frac{1}{3}}(1-tx^{-1})^{-\frac{1}{3}}(1-ty^{-1})^{-\frac{1}{3}} dt.$$

It is known that the two variable hypergeometric function (Appell hypergeometric function)

$$F_1(a; b, b'; c; x, y) = \frac{\Gamma(c)}{\Gamma(a)\Gamma(c-a)} \int_0^1 t^{a-1}(1-t)^{c-a-1}(1-tx)^{-b}(1-ty)^{-b'} dt$$

satisfies the following equations:

$$\begin{cases} x(1-x)r + y(1-x)s + [c - (a+b+1)x]p - byq - abz = 0, \\ y(1-y)t + x(1-y)s + [c - (a+b'+1)y]q - b'xp - ab'z = 0, \end{cases}$$

where

$$r = \frac{\partial^2 z}{\partial x^2}, \quad s = \frac{\partial^2 z}{\partial x \partial y}, \quad t = \frac{\partial^2 z}{\partial y^2}, \quad p = \frac{\partial z}{\partial x}, \quad q = \frac{\partial z}{\partial y}.$$

Thus,

$$z(x,y) = -\frac{\Gamma\left(\frac{2}{3}\right)^2}{\Gamma\left(\frac{4}{3}\right)} x^{-\frac{1}{3}} y^{-\frac{1}{3}} F_1\left(\frac{2}{3}; \frac{1}{3}, \frac{1}{3}; \frac{4}{3}; \frac{1}{x}, \frac{1}{y}\right).$$

Set $u = \frac{1}{x}$ and $v = \frac{1}{y}$. $f(u,v) := F_1\left(\frac{2}{3}; \frac{1}{3}, \frac{1}{3}; \frac{4}{3}; u, v\right)$ satisfies the following equations:

$$\begin{cases} u(1-u)\dfrac{\partial^2 f}{\partial u^2} + v(1-u)\dfrac{\partial^2 f}{\partial u \partial v} + \left(\dfrac{4}{3} - 2u\right)\dfrac{\partial f}{\partial u} - \dfrac{1}{3}v\dfrac{\partial f}{\partial v} - \dfrac{2}{9}f = 0, \\ v(1-v)\dfrac{\partial^2 f}{\partial v^2} + u(1-v)\dfrac{\partial^2 f}{\partial u \partial v} + \left(\dfrac{4}{3} - 2v\right)\dfrac{\partial f}{\partial v} - \dfrac{1}{3}u\dfrac{\partial f}{\partial u} - \dfrac{2}{9}f = 0. \end{cases}$$

By a straightforward calculation, we find that $z = z(x,y)$ satisfies the following equations:

$$\begin{cases} x(1-x)r + y(1-x)s + \left(1 - \dfrac{5}{3}x\right)p - \dfrac{1}{3}yq - \dfrac{1}{9}z = 0, \\ y(1-y)t + x(1-y)s + \left(1 - \dfrac{5}{3}y\right)q - \dfrac{1}{3}xp - \dfrac{1}{9}z = 0. \end{cases}$$

Let ω_1, ω_2 and ω_3 be three linearly independent solutions of the above partial differential equations. Set

$$\frac{\omega_2}{\omega_1} = u(x,y), \quad \frac{\omega_3}{\omega_1} = v(x,y).$$

In his paper [32], Picard proved that

$$2\mathrm{Re}(v) + |u|^2 < 0.$$

In general, according to [3], p.54, Théorème b and p.81, we have the following Theorem:

Theorem. *Suppose that there are four functions z_1, z_2, z_3 and z_4 which satisfy the equations:*

$$\begin{cases} r = a_1 p + a_2 q + a_3 z, \\ s = b_1 p + b_2 q + b_3 z, \\ t = c_1 p + c_2 q + c_3 z, \end{cases}$$

where $a_i = a_i(x,y)$, $b_i = b_i(x,y)$ and $c_i = c_i(x,y)$ $(i = 1,2,3)$ are functions of x and y. There exists a linear relation of constant coefficients for all these functions such that

$$C_1 z_1 + C_2 z_2 + C_3 z_3 + C_4 z_4 = 0.$$

The above partial differential equations can admit three linearly independent solutions.

Now, let us consider the following partial differential equations:

$$\begin{cases} x(1-x)\dfrac{\partial^2 z}{\partial x^2}+y(1-x)\dfrac{\partial^2 z}{\partial x\partial y}+[c-(a+b+1)x]\dfrac{\partial z}{\partial x}-by\dfrac{\partial z}{\partial y}-abz=0,\\ y(1-y)\dfrac{\partial^2 z}{\partial y^2}+x(1-y)\dfrac{\partial^2 z}{\partial x\partial y}+[c-(a+b'+1)y]\dfrac{\partial z}{\partial y}-b'x\dfrac{\partial z}{\partial x}-ab'z=0, \end{cases} \tag{2.1}$$

which are equivalent to the following three equations:

$$\begin{cases} \dfrac{\partial^2 z}{\partial x^2}+\left(\dfrac{c-b'}{x}+\dfrac{a+b-c+1}{x-1}+\dfrac{b'}{x-y}\right)\dfrac{\partial z}{\partial x}\\ \quad -\dfrac{by(y-1)}{x(x-1)(x-y)}\dfrac{\partial z}{\partial y}+\dfrac{abz}{x(x-1)}=0,\\ \dfrac{\partial^2 z}{\partial y^2}+\left(\dfrac{c-b}{y}+\dfrac{a+b'-c+1}{y-1}+\dfrac{b}{y-x}\right)\dfrac{\partial z}{\partial y}\\ \quad -\dfrac{b'x(x-1)}{y(y-1)(y-x)}\dfrac{\partial z}{\partial x}+\dfrac{ab'z}{y(y-1)}=0,\\ \dfrac{\partial^2 z}{\partial x\partial y}=\dfrac{b'}{x-y}\dfrac{\partial z}{\partial x}-\dfrac{b}{x-y}\dfrac{\partial z}{\partial y}. \end{cases} \tag{2.2}$$

There are three linearly independent solutions z_1, z_2 and z_3. Consider the monodromy:

$$\begin{cases} w_1=a_1z_1+a_2z_2+a_3z_3,\\ w_2=b_1z_1+b_2z_2+b_3z_3,\\ w_3=c_1z_1+c_2z_2+c_3z_3. \end{cases}$$

We have

$$\begin{cases} \dfrac{w_1}{w_3}=\dfrac{a_1z_1+a_2z_2+a_3z_3}{c_1z_1+c_2z_2+c_3z_3}=\dfrac{a_1\frac{z_1}{z_3}+a_2\frac{z_2}{z_3}+a_3}{c_1\frac{z_1}{z_3}+c_2\frac{z_2}{z_3}+c_3},\\ \dfrac{w_2}{w_3}=\dfrac{b_1z_1+b_2z_2+b_3z_3}{c_1z_1+c_2z_2+c_3z_3}=\dfrac{b_1\frac{z_1}{z_3}+b_2\frac{z_2}{z_3}+b_3}{c_1\frac{z_1}{z_3}+c_2\frac{z_2}{z_3}+c_3}. \end{cases}$$

Set $u_1=\frac{z_1}{z_3}$, $u_2=\frac{z_2}{z_3}$ and $v_1=\frac{w_1}{w_3}$, $v_2=\frac{w_2}{w_3}$, then

$$v_1=\frac{a_1u_1+a_2u_2+a_3}{c_1u_1+c_2u_2+c_3},\qquad v_2=\frac{b_1u_1+b_2u_2+b_3}{c_1u_1+c_2u_2+c_3}.$$

Note that

$$\begin{cases} \dfrac{\partial z_1}{\partial x}z_3-z_1\dfrac{\partial z_3}{\partial x}=z_3^2\dfrac{\partial u_1}{\partial x},\\ \dfrac{\partial z_1}{\partial y}z_3-z_1\dfrac{\partial z_3}{\partial y}=z_3^2\dfrac{\partial u_1}{\partial y},\\ \dfrac{\partial z_2}{\partial x}z_3-z_2\dfrac{\partial z_3}{\partial x}=z_3^2\dfrac{\partial u_2}{\partial x},\\ \dfrac{\partial z_2}{\partial y}z_3-z_2\dfrac{\partial z_3}{\partial y}=z_3^2\dfrac{\partial u_2}{\partial y}. \end{cases}$$

We find that

$$\begin{cases} \dfrac{\partial^2 z_1}{\partial x^2} z_3 - z_1 \dfrac{\partial^2 z_3}{\partial x^2} = z_3^2 \dfrac{\partial^2 u_1}{\partial x^2} + 2z_3 \dfrac{\partial z_3}{\partial x}\dfrac{\partial u_1}{\partial x}, \\ \dfrac{\partial^2 z_1}{\partial y^2} z_3 - z_1 \dfrac{\partial^2 z_3}{\partial y^2} = z_3^2 \dfrac{\partial^2 u_1}{\partial y^2} + 2z_3 \dfrac{\partial z_3}{\partial y}\dfrac{\partial u_1}{\partial y}, \\ \dfrac{\partial^2 z_2}{\partial x^2} z_3 - z_2 \dfrac{\partial^2 z_3}{\partial x^2} = z_3^2 \dfrac{\partial^2 u_2}{\partial x^2} + 2z_3 \dfrac{\partial z_3}{\partial x}\dfrac{\partial u_2}{\partial x}, \\ \dfrac{\partial^2 z_2}{\partial y^2} z_3 - z_2 \dfrac{\partial^2 z_3}{\partial y^2} = z_3^2 \dfrac{\partial^2 u_2}{\partial y^2} + 2z_3 \dfrac{\partial z_3}{\partial y}\dfrac{\partial u_2}{\partial y}. \end{cases}$$

$$\begin{cases} \dfrac{\partial^2 z_1}{\partial x \partial y} z_3 - z_1 \dfrac{\partial^2 z_3}{\partial x \partial y} = z_3^2 \dfrac{\partial^2 u_1}{\partial x \partial y} + z_3 \dfrac{\partial z_3}{\partial x}\dfrac{\partial u_1}{\partial y} + z_3 \dfrac{\partial z_3}{\partial y}\dfrac{\partial u_1}{\partial x}, \\ \dfrac{\partial^2 z_2}{\partial x \partial y} z_3 - z_2 \dfrac{\partial^2 z_3}{\partial x \partial y} = z_3^2 \dfrac{\partial^2 u_2}{\partial x \partial y} + z_3 \dfrac{\partial z_3}{\partial x}\dfrac{\partial u_2}{\partial y} + z_3 \dfrac{\partial z_3}{\partial y}\dfrac{\partial u_2}{\partial x}. \end{cases}$$

Substituting these expressions in the above three partial differential equations, we get the next six equations:

$$\begin{cases} \dfrac{\partial^2 u_1}{\partial x^2} + \left(2\dfrac{\frac{\partial z_3}{\partial x}}{z_3} + \dfrac{c-b'}{x} + \dfrac{a+b-c+1}{x-1} + \dfrac{b'}{x-y} \right) \dfrac{\partial u_1}{\partial x} \\ \quad - \dfrac{by(y-1)}{x(x-1)(x-y)} \dfrac{\partial u_1}{\partial y} = 0, \\ \dfrac{\partial^2 u_2}{\partial x^2} + \left(2\dfrac{\frac{\partial z_3}{\partial x}}{z_3} + \dfrac{c-b'}{x} + \dfrac{a+b-c+1}{x-1} + \dfrac{b'}{x-y} \right) \dfrac{\partial u_2}{\partial x} \\ \quad - \dfrac{by(y-1)}{x(x-1)(x-y)} \dfrac{\partial u_2}{\partial y} = 0, \\ \dfrac{\partial^2 u_1}{\partial y^2} + \left(2\dfrac{\frac{\partial z_3}{\partial y}}{z_3} + \dfrac{c-b}{y} + \dfrac{a+b'-c+1}{y-1} + \dfrac{b}{y-x} \right) \dfrac{\partial u_1}{\partial y} \\ \quad - \dfrac{b'x(x-1)}{y(y-1)(y-x)} \dfrac{\partial u_1}{\partial x} = 0, \\ \dfrac{\partial^2 u_2}{\partial y^2} + \left(2\dfrac{\frac{\partial z_3}{\partial y}}{z_3} + \dfrac{c-b}{y} + \dfrac{a+b'-c+1}{y-1} + \dfrac{b}{y-x} \right) \dfrac{\partial u_2}{\partial y} \\ \quad - \dfrac{b'x(x-1)}{y(y-1)(y-x)} \dfrac{\partial u_2}{\partial x} = 0. \end{cases}$$

$$\begin{cases} \dfrac{\partial^2 u_1}{\partial x \partial y} + \left(\dfrac{\frac{\partial z_3}{\partial y}}{z_3} - \dfrac{b'}{x-y} \right) \dfrac{\partial u_1}{\partial x} + \left(\dfrac{\frac{\partial z_3}{\partial x}}{z_3} + \dfrac{b}{x-y} \right) \dfrac{\partial u_1}{\partial y} = 0, \\ \dfrac{\partial^2 u_2}{\partial x \partial y} + \left(\dfrac{\frac{\partial z_3}{\partial y}}{z_3} - \dfrac{b'}{x-y} \right) \dfrac{\partial u_2}{\partial x} + \left(\dfrac{\frac{\partial z_3}{\partial x}}{z_3} + \dfrac{b}{x-y} \right) \dfrac{\partial u_2}{\partial y} = 0. \end{cases}$$

It is known that the Schwarzian derivative

$$\{w;x\} := \frac{w'''}{w'} - \frac{3}{2}\left(\frac{w''}{w'}\right)^2$$

satisfies

$$\left\{\frac{aw+b}{cw+d};x\right\} = \{w;x\} \quad \text{for} \quad \begin{pmatrix} a & b \\ c & d \end{pmatrix} \in GL(2,\mathbb{C}).$$

Now, we define the following four derivatives:

Definition 2.1. Four derivatives associated to the group $GL(3)$ are given by:

$$\{u_1,u_2;x,y\}_x := \frac{\begin{vmatrix} \frac{\partial u_1}{\partial x} & \frac{\partial u_2}{\partial x} \\ \frac{\partial^2 u_1}{\partial x^2} & \frac{\partial^2 u_2}{\partial x^2} \end{vmatrix}}{\begin{vmatrix} \frac{\partial u_1}{\partial x} & \frac{\partial u_2}{\partial x} \\ \frac{\partial u_1}{\partial y} & \frac{\partial u_2}{\partial y} \end{vmatrix}}, \quad \{u_1,u_2;x,y\}_y := \frac{\begin{vmatrix} \frac{\partial u_1}{\partial y} & \frac{\partial u_2}{\partial y} \\ \frac{\partial^2 u_1}{\partial y^2} & \frac{\partial^2 u_2}{\partial y^2} \end{vmatrix}}{\begin{vmatrix} \frac{\partial u_1}{\partial y} & \frac{\partial u_2}{\partial y} \\ \frac{\partial u_1}{\partial x} & \frac{\partial u_2}{\partial x} \end{vmatrix}}, \tag{2.3}$$

$$[u_1,u_2;x,y]_x := \frac{\begin{vmatrix} \frac{\partial u_1}{\partial y} & \frac{\partial u_2}{\partial y} \\ \frac{\partial^2 u_1}{\partial x^2} & \frac{\partial^2 u_2}{\partial x^2} \end{vmatrix} + 2\begin{vmatrix} \frac{\partial u_1}{\partial x} & \frac{\partial u_2}{\partial x} \\ \frac{\partial^2 u_1}{\partial x \partial y} & \frac{\partial^2 u_2}{\partial x \partial y} \end{vmatrix}}{\begin{vmatrix} \frac{\partial u_1}{\partial x} & \frac{\partial u_2}{\partial x} \\ \frac{\partial u_1}{\partial y} & \frac{\partial u_2}{\partial y} \end{vmatrix}}, \tag{2.4}$$

$$[u_1,u_2;x,y]_y := \frac{\begin{vmatrix} \frac{\partial u_1}{\partial x} & \frac{\partial u_2}{\partial x} \\ \frac{\partial^2 u_1}{\partial y^2} & \frac{\partial^2 u_2}{\partial y^2} \end{vmatrix} + 2\begin{vmatrix} \frac{\partial u_1}{\partial y} & \frac{\partial u_2}{\partial y} \\ \frac{\partial^2 u_1}{\partial x \partial y} & \frac{\partial^2 u_2}{\partial x \partial y} \end{vmatrix}}{\begin{vmatrix} \frac{\partial u_1}{\partial y} & \frac{\partial u_2}{\partial y} \\ \frac{\partial u_1}{\partial x} & \frac{\partial u_2}{\partial x} \end{vmatrix}}, \tag{2.5}$$

where x and y are complex numbers, u_1 and u_2 are complex-valued functions of x and y.

The above six equations give that

$$\begin{cases} \{u_1,u_2;x,y\}_x = \dfrac{by(y-1)}{x(x-1)(x-y)}, \\ \{u_1,u_2;x,y\}_y = \dfrac{b'x(x-1)}{y(y-1)(y-x)}, \\ [u_1,u_2;x,y]_x = \dfrac{c-b'}{x} + \dfrac{a+b-c+1}{x-1} + \dfrac{b'-2b}{x-y}, \\ [u_1,u_2;x,y]_y = \dfrac{c-b}{y} + \dfrac{a+b'-c+1}{y-1} + \dfrac{b-2b'}{y-x}. \end{cases} \tag{2.6}$$

In general, we have the following result:

Proposition 2.2. *For the system of partial differential equations:*

$$\begin{cases} r = a_1p + a_2q + a_3z, \\ s = b_1p + b_2q + b_3z, \\ t = c_1p + c_2q + c_3z, \end{cases} \tag{2.7}$$

where $a_i = a_i(x,y)$, $b_i = b_i(x,y)$, $c_i = c_i(x,y)$ $(i = 1,2,3)$ *and* $r = \frac{\partial^2 z}{\partial x^2}$, $s = \frac{\partial^2 z}{\partial x \partial y}$, $t = \frac{\partial^2 z}{\partial y^2}$, $p = \frac{\partial z}{\partial x}$, $q = \frac{\partial z}{\partial y}$, *there exists three linearly independent solutions:* z_1, z_2 *and* z_3. *Put* $u_1 = z_1/z_3$ *and* $u_2 = z_2/z_3$. *Then*

$$\begin{cases} \{u_1, u_2; x, y\}_x = a_2, \\ \{u_1, u_2; x, y\}_y = c_1, \\ [u_1, u_2; x, y]_x = 2b_2 - a_1, \\ [u_1, u_2; x, y]_y = 2b_1 - c_2. \end{cases} \tag{2.8}$$

Proposition 2.3. *Our derivatives are invariant under linear fractional transformation acting on the first arguments*

$$\begin{cases} \{v_1, v_2; x, y\}_x = \{u_1, u_2; x, y\}_x, \\ \{v_1, v_2; x, y\}_y = \{u_1, u_2; x, y\}_y, \\ [v_1, v_2; x, y]_x = [u_1, u_2; x, y]_x, \\ [v_1, v_2; x, y]_y = [u_1, u_2; x, y]_y, \end{cases} \tag{2.9}$$

where

$$(v_1, v_2) = \gamma(u_1, u_2) = \left(\frac{a_1 u_1 + a_2 u_2 + a_3}{c_1 u_1 + c_2 u_2 + c_3}, \frac{b_1 u_1 + b_2 u_2 + b_3}{c_1 u_1 + c_2 u_2 + c_3} \right) \tag{2.10}$$

for $\gamma = \begin{pmatrix} a_1 & a_2 & a_3 \\ b_1 & b_2 & b_3 \\ c_1 & c_2 & c_3 \end{pmatrix} \in GL(3, \mathbb{C})$.

Proof. We have

$$\begin{vmatrix} \frac{\partial v_1}{\partial x} & \frac{\partial v_2}{\partial x} \\ \frac{\partial v_1}{\partial y} & \frac{\partial v_2}{\partial y} \end{vmatrix} = \Delta (c_1 u_1 + c_2 u_2 + c_3)^{-3} \begin{vmatrix} \frac{\partial u_1}{\partial x} & \frac{\partial u_2}{\partial x} \\ \frac{\partial u_1}{\partial y} & \frac{\partial u_2}{\partial y} \end{vmatrix},$$

where

$$\Delta = \begin{vmatrix} a_1 & a_2 & a_3 \\ b_1 & b_2 & b_3 \\ c_1 & c_2 & c_3 \end{vmatrix}.$$

Note that

$$\begin{vmatrix} a & b \\ c + c' & d + d' \end{vmatrix} = \begin{vmatrix} a & b \\ c & d \end{vmatrix} + \begin{vmatrix} a & b \\ c' & d' \end{vmatrix}.$$

A straightforward calculation gives that

$$\begin{vmatrix} \frac{\partial v_1}{\partial x} & \frac{\partial v_2}{\partial x} \\ \frac{\partial^2 v_1}{\partial x^2} & \frac{\partial^2 v_2}{\partial x^2} \end{vmatrix} = \Delta (c_1 u_1 + c_2 u_2 + c_3)^{-3} \begin{vmatrix} \frac{\partial u_1}{\partial x} & \frac{\partial u_2}{\partial x} \\ \frac{\partial^2 u_1}{\partial x^2} & \frac{\partial^2 u_2}{\partial x^2} \end{vmatrix},$$

$$\begin{aligned} \begin{vmatrix} \frac{\partial v_1}{\partial y} & \frac{\partial v_2}{\partial y} \\ \frac{\partial^2 v_1}{\partial x^2} & \frac{\partial^2 v_2}{\partial x^2} \end{vmatrix} =& 2\Delta (c_1 u_1 + c_2 u_2 + c_3)^{-4} \left(c_1 \frac{\partial u_1}{\partial x} + c_2 \frac{\partial u_2}{\partial x}\right) \begin{vmatrix} \frac{\partial u_1}{\partial x} & \frac{\partial u_2}{\partial x} \\ \frac{\partial u_1}{\partial y} & \frac{\partial u_2}{\partial y} \end{vmatrix} \\ &+ \Delta (c_1 u_1 + c_2 u_2 + c_3)^{-3} \begin{vmatrix} \frac{\partial u_1}{\partial y} & \frac{\partial u_2}{\partial y} \\ \frac{\partial^2 u_1}{\partial x^2} & \frac{\partial^2 u_2}{\partial x^2} \end{vmatrix}. \end{aligned}$$

$$\begin{vmatrix} \frac{\partial v_1}{\partial x} & \frac{\partial v_2}{\partial x} \\ \frac{\partial^2 v_1}{\partial x \partial y} & \frac{\partial^2 v_2}{\partial x \partial y} \end{vmatrix} = -\Delta(c_1u_1 + c_2u_2 + c_3)^{-4}(c_1\frac{\partial u_1}{\partial x} + c_2\frac{\partial u_2}{\partial x}) \begin{vmatrix} \frac{\partial u_1}{\partial x} & \frac{\partial u_2}{\partial x} \\ \frac{\partial u_1}{\partial y} & \frac{\partial u_2}{\partial y} \end{vmatrix}$$
$$+ \Delta(c_1u_1 + c_2u_2 + c_3)^{-3} \begin{vmatrix} \frac{\partial u_1}{\partial x} & \frac{\partial u_2}{\partial x} \\ \frac{\partial^2 u_1}{\partial x \partial y} & \frac{\partial^2 u_2}{\partial x \partial y} \end{vmatrix}.$$

Hence,

$$\{v_1, v_2; x, y\}_x = \{u_1, u_2; x, y\}_x, \quad [v_1, v_2; x, y]_x = [u_1, u_2; x, y]_x.$$

Exchange the position of x and y, we get the other identities.

□

Thus our derivatives are differential invariants of the general linear group acting as a linear fractional transformation.

Corollary 2.4. *Our derivatives vanish when (w_1, w_2) are linear fractional functions of (z_1, z_2):*

$$\begin{cases} \{w_1, w_2; z_1, z_2\}_{z_1} = 0, \\ \{w_1, w_2; z_1, z_2\}_{z_2} = 0, \\ [w_1, w_2; z_1, z_2]_{z_1} = 0, \\ [w_1, w_2; z_1, z_2]_{z_2} = 0, \end{cases} \tag{2.11}$$

where

$$(w_1, w_2) = \gamma(z_1, z_2) = \left(\frac{a_1z_1 + a_2z_2 + a_3}{c_1z_1 + c_2z_2 + c_3}, \frac{b_1z_1 + b_2z_2 + b_3}{c_1z_1 + c_2z_2 + c_3} \right), \tag{2.12}$$

for $\gamma = \begin{pmatrix} a_1 & a_2 & a_3 \\ b_1 & b_2 & b_3 \\ c_1 & c_2 & c_3 \end{pmatrix} \in GL(3, \mathbb{C})$.

Proof. Note that

$$\begin{cases} \{z_1, z_2; z_1, z_2\}_{z_1} = 0, \\ \{z_1, z_2; z_1, z_2\}_{z_2} = 0, \\ [z_1, z_2; z_1, z_2]_{z_1} = 0, \\ [z_1, z_2; z_1, z_2]_{z_2} = 0. \end{cases}$$

The above Proposition gives the desired result.

□

Proposition 2.5 (Main Proposition 1). *The change of independent variables obeys the laws:*

$$\{u_1, u_2; x, y\}_x$$
$$= \frac{(\frac{\partial w_1}{\partial x})^3}{\frac{\partial(w_1,w_2)}{\partial(x,y)}}\{u_1, u_2; w_1, w_2\}_{w_1} + \frac{(\frac{\partial w_1}{\partial x})^2 \frac{\partial w_2}{\partial x}}{\frac{\partial(w_1,w_2)}{\partial(x,y)}}[u_1, u_2; w_1, w_2]_{w_1}$$
$$+ \frac{(\frac{\partial w_2}{\partial x})^3}{\frac{\partial(w_2,w_1)}{\partial(x,y)}}\{u_1, u_2; w_1, w_2\}_{w_2} + \frac{\frac{\partial w_1}{\partial x}(\frac{\partial w_2}{\partial x})^2}{\frac{\partial(w_2,w_1)}{\partial(x,y)}}[u_1, u_2; w_1, w_2]_{w_2} + \{w_1, w_2; x, y\}_x. \tag{2.13}$$

$$\begin{aligned}
&\{u_1, u_2; x, y\}_y \\
&= \frac{\left(\frac{\partial w_1}{\partial y}\right)^3}{\frac{\partial(w_2,w_1)}{\partial(x,y)}}\{u_1, u_2; w_1, w_2\}_{w_1} + \frac{\left(\frac{\partial w_1}{\partial y}\right)^2 \frac{\partial w_2}{\partial y}}{\frac{\partial(w_2,w_1)}{\partial(x,y)}}[u_1, u_2; w_1, w_2]_{w_1} \\
&\quad + \frac{\left(\frac{\partial w_2}{\partial y}\right)^3}{\frac{\partial(w_1,w_2)}{\partial(x,y)}}\{u_1, u_2; w_1, w_2\}_{w_2} + \frac{\frac{\partial w_1}{\partial y}\left(\frac{\partial w_2}{\partial y}\right)^2}{\frac{\partial(w_1,w_2)}{\partial(x,y)}}[u_1, u_2; w_1, w_2]_{w_2} + \{w_1, w_2; x, y\}_y.
\end{aligned} \tag{2.14}$$

$$\begin{aligned}
&[u_1, u_2; x, y]_x \\
&= \frac{3\left(\frac{\partial w_1}{\partial x}\right)^2 \frac{\partial w_1}{\partial y}}{\frac{\partial(w_1,w_2)}{\partial(x,y)}}\{u_1, u_2; w_1, w_2\}_{w_1} \\
&\quad + \frac{\left(\frac{\partial w_1}{\partial x}\right)^2 \frac{\partial w_2}{\partial y} + 2\frac{\partial w_1}{\partial x}\frac{\partial w_2}{\partial x}\frac{\partial w_1}{\partial y}}{\frac{\partial(w_1,w_2)}{\partial(x,y)}}[u_1, u_2; w_1, w_2]_{w_1} \\
&\quad + \frac{3\left(\frac{\partial w_2}{\partial x}\right)^2 \frac{\partial w_2}{\partial y}}{\frac{\partial(w_2,w_1)}{\partial(x,y)}}\{u_1, u_2; w_1, w_2\}_{w_2} \\
&\quad + \frac{\left(\frac{\partial w_2}{\partial x}\right)^2 \frac{\partial w_1}{\partial y} + 2\frac{\partial w_1}{\partial x}\frac{\partial w_2}{\partial x}\frac{\partial w_2}{\partial y}}{\frac{\partial(w_2,w_1)}{\partial(x,y)}}[u_1, u_2; w_1, w_2]_{w_2} \\
&\quad + [w_1, w_2; x, y]_x.
\end{aligned} \tag{2.15}$$

$$\begin{aligned}
&[u_1, u_2; x, y]_y \\
&= \frac{3\frac{\partial w_1}{\partial x}\left(\frac{\partial w_1}{\partial y}\right)^2}{\frac{\partial(w_2,w_1)}{\partial(x,y)}}\{u_1, u_2; w_1, w_2\}_{w_1} \\
&\quad + \frac{\frac{\partial w_2}{\partial x}\left(\frac{\partial w_1}{\partial y}\right)^2 + 2\frac{\partial w_1}{\partial x}\frac{\partial w_1}{\partial y}\frac{\partial w_2}{\partial y}}{\frac{\partial(w_2,w_1)}{\partial(x,y)}}[u_1, u_2; w_1, w_2]_{w_1} \\
&\quad + \frac{3\frac{\partial w_2}{\partial x}\left(\frac{\partial w_2}{\partial y}\right)^2}{\frac{\partial(w_1,w_2)}{\partial(x,y)}}\{u_1, u_2; w_1, w_2\}_{w_2} \\
&\quad + \frac{\frac{\partial w_1}{\partial x}\left(\frac{\partial w_2}{\partial y}\right)^2 + 2\frac{\partial w_2}{\partial x}\frac{\partial w_1}{\partial y}\frac{\partial w_2}{\partial y}}{\frac{\partial(w_1,w_2)}{\partial(x,y)}}[u_1, u_2; w_1, w_2]_{w_2} \\
&\quad + [w_1, w_2; x, y]_y.
\end{aligned} \tag{2.16}$$

Proof. We have

$$\frac{\partial u_1}{\partial x} = \frac{\partial u_1}{\partial w_1}\frac{\partial w_1}{\partial x} + \frac{\partial u_1}{\partial w_2}\frac{\partial w_2}{\partial x}, \quad \frac{\partial u_2}{\partial x} = \frac{\partial u_2}{\partial w_1}\frac{\partial w_1}{\partial x} + \frac{\partial u_2}{\partial w_2}\frac{\partial w_2}{\partial x}.$$

$$\frac{\partial u_1}{\partial y} = \frac{\partial u_1}{\partial w_1}\frac{\partial w_1}{\partial y} + \frac{\partial u_1}{\partial w_2}\frac{\partial w_2}{\partial y}, \quad \frac{\partial u_2}{\partial y} = \frac{\partial u_2}{\partial w_1}\frac{\partial w_1}{\partial y} + \frac{\partial u_2}{\partial w_2}\frac{\partial w_2}{\partial y}.$$

$$\begin{aligned}\frac{\partial^2 u_1}{\partial x^2} =& \frac{\partial^2 u_1}{\partial w_1^2}\left(\frac{\partial w_1}{\partial x}\right)^2 + 2\frac{\partial^2 u_1}{\partial w_1 \partial w_2}\frac{\partial w_1}{\partial x}\frac{\partial w_2}{\partial x} \\ &+ \frac{\partial^2 u_1}{\partial w_2^2}\left(\frac{\partial w_2}{\partial x}\right)^2 + \frac{\partial u_1}{\partial w_1}\frac{\partial^2 w_1}{\partial x^2} + \frac{\partial u_1}{\partial w_2}\frac{\partial^2 w_2}{\partial x^2}.\end{aligned}$$

$$\begin{aligned}\frac{\partial^2 u_2}{\partial x^2} =& \frac{\partial^2 u_2}{\partial w_1^2}\left(\frac{\partial w_1}{\partial x}\right)^2 + 2\frac{\partial^2 u_2}{\partial w_1 \partial w_2}\frac{\partial w_1}{\partial x}\frac{\partial w_2}{\partial x} \\ &+ \frac{\partial^2 u_2}{\partial w_2^2}\left(\frac{\partial w_2}{\partial x}\right)^2 + \frac{\partial u_2}{\partial w_1}\frac{\partial^2 w_1}{\partial x^2} + \frac{\partial u_2}{\partial w_2}\frac{\partial^2 w_2}{\partial x^2}.\end{aligned}$$

$$\begin{aligned}\frac{\partial^2 u_1}{\partial y^2} =& \frac{\partial^2 u_1}{\partial w_1^2}\left(\frac{\partial w_1}{\partial y}\right)^2 + 2\frac{\partial^2 u_1}{\partial w_1 \partial w_2}\frac{\partial w_1}{\partial y}\frac{\partial w_2}{\partial y} \\ &+ \frac{\partial^2 u_1}{\partial w_2^2}\left(\frac{\partial w_2}{\partial y}\right)^2 + \frac{\partial u_1}{\partial w_1}\frac{\partial^2 w_1}{\partial y^2} + \frac{\partial u_1}{\partial w_2}\frac{\partial^2 w_2}{\partial y^2}.\end{aligned}$$

$$\begin{aligned}\frac{\partial^2 u_2}{\partial y^2} =& \frac{\partial^2 u_2}{\partial w_1^2}\left(\frac{\partial w_1}{\partial y}\right)^2 + 2\frac{\partial^2 u_2}{\partial w_1 \partial w_2}\frac{\partial w_1}{\partial y}\frac{\partial w_2}{\partial y} \\ &+ \frac{\partial^2 u_2}{\partial w_2^2}\left(\frac{\partial w_2}{\partial y}\right)^2 + \frac{\partial u_2}{\partial w_1}\frac{\partial^2 w_1}{\partial y^2} + \frac{\partial u_2}{\partial w_2}\frac{\partial^2 w_2}{\partial y^2}.\end{aligned}$$

$$\begin{aligned}\frac{\partial^2 u_1}{\partial x \partial y} =& \frac{\partial^2 u_1}{\partial w_1^2}\frac{\partial w_1}{\partial x}\frac{\partial w_1}{\partial y} + \frac{\partial^2 u_1}{\partial w_1 \partial w_2}\left(\frac{\partial w_1}{\partial x}\frac{\partial w_2}{\partial y} + \frac{\partial w_2}{\partial x}\frac{\partial w_1}{\partial y}\right) \\ &+ \frac{\partial^2 u_1}{\partial w_2^2}\frac{\partial w_2}{\partial x}\frac{\partial w_2}{\partial y} + \frac{\partial u_1}{\partial w_1}\frac{\partial^2 w_1}{\partial x \partial y} + \frac{\partial u_1}{\partial w_2}\frac{\partial^2 w_2}{\partial x \partial y}.\end{aligned}$$

$$\begin{aligned}\frac{\partial^2 u_2}{\partial x \partial y} =& \frac{\partial^2 u_2}{\partial w_1^2}\frac{\partial w_1}{\partial x}\frac{\partial w_1}{\partial y} + \frac{\partial^2 u_2}{\partial w_1 \partial w_2}\left(\frac{\partial w_1}{\partial x}\frac{\partial w_2}{\partial y} + \frac{\partial w_2}{\partial x}\frac{\partial w_1}{\partial y}\right) \\ &+ \frac{\partial^2 u_2}{\partial w_2^2}\frac{\partial w_2}{\partial x}\frac{\partial w_2}{\partial y} + \frac{\partial u_2}{\partial w_1}\frac{\partial^2 w_1}{\partial x \partial y} + \frac{\partial u_2}{\partial w_2}\frac{\partial^2 w_2}{\partial x \partial y}.\end{aligned}$$

Hence,

$$\begin{aligned}
&\begin{vmatrix} \frac{\partial u_1}{\partial x} & \frac{\partial u_2}{\partial x} \\ \frac{\partial^2 u_1}{\partial x^2} & \frac{\partial^2 u_2}{\partial x^2} \end{vmatrix} \\
&= \left(\frac{\partial w_1}{\partial x}\right)^3 \begin{vmatrix} \frac{\partial u_1}{\partial w_1} & \frac{\partial u_2}{\partial w_1} \\ \frac{\partial^2 u_1}{\partial w_1^2} & \frac{\partial^2 u_2}{\partial w_1^2} \end{vmatrix} \\
&\quad + \left(\frac{\partial w_1}{\partial x}\right)^2 \frac{\partial w_2}{\partial x} \left[\begin{vmatrix} \frac{\partial u_1}{\partial w_2} & \frac{\partial u_2}{\partial w_2} \\ \frac{\partial^2 u_1}{\partial w_1^2} & \frac{\partial^2 u_2}{\partial w_1^2} \end{vmatrix} + 2 \begin{vmatrix} \frac{\partial u_1}{\partial w_1} & \frac{\partial u_2}{\partial w_1} \\ \frac{\partial^2 u_1}{\partial w_1 \partial w_2} & \frac{\partial^2 u_2}{\partial w_1 \partial w_2} \end{vmatrix} \right] \\
&\quad + \left(\frac{\partial w_2}{\partial x}\right)^3 \begin{vmatrix} \frac{\partial u_1}{\partial w_2} & \frac{\partial u_2}{\partial w_2} \\ \frac{\partial^2 u_1}{\partial w_2^2} & \frac{\partial^2 u_2}{\partial w_2^2} \end{vmatrix} \\
&\quad + \frac{\partial w_1}{\partial x} \left(\frac{\partial w_2}{\partial x}\right)^2 \left[\begin{vmatrix} \frac{\partial u_1}{\partial w_1} & \frac{\partial u_2}{\partial w_1} \\ \frac{\partial^2 u_1}{\partial w_2^2} & \frac{\partial^2 u_2}{\partial w_2^2} \end{vmatrix} + 2 \begin{vmatrix} \frac{\partial u_1}{\partial w_2} & \frac{\partial u_2}{\partial w_2} \\ \frac{\partial^2 u_1}{\partial w_1 \partial w_2} & \frac{\partial^2 u_2}{\partial w_1 \partial w_2} \end{vmatrix} \right] \\
&\quad + \begin{vmatrix} \frac{\partial u_1}{\partial w_1} & \frac{\partial u_2}{\partial w_1} \\ \frac{\partial u_1}{\partial w_2} & \frac{\partial u_2}{\partial w_2} \end{vmatrix} \begin{vmatrix} \frac{\partial w_1}{\partial x} & \frac{\partial w_2}{\partial x} \\ \frac{\partial^2 w_1}{\partial x^2} & \frac{\partial^2 w_2}{\partial x^2} \end{vmatrix}.
\end{aligned}$$

Note that

$$\frac{\partial(u_1,u_2)}{\partial(x,y)} = \frac{\partial(u_1,u_2)}{\partial(w_1,w_2)} \frac{\partial(w_1,w_2)}{\partial(x,y)}.$$

We find that

$$\begin{aligned}
&\{u_1,u_2;x,y\}_x \\
&= \frac{\left(\frac{\partial w_1}{\partial x}\right)^3}{\frac{\partial(w_1,w_2)}{\partial(x,y)}} \{u_1,u_2;w_1,w_2\}_{w_1} + \frac{\left(\frac{\partial w_1}{\partial x}\right)^2 \frac{\partial w_2}{\partial x}}{\frac{\partial(w_1,w_2)}{\partial(x,y)}} [u_1,u_2;w_1,w_2]_{w_1} \\
&\quad + \frac{\left(\frac{\partial w_2}{\partial x}\right)^3}{\frac{\partial(w_2,w_1)}{\partial(x,y)}} \{u_1,u_2;w_1,w_2\}_{w_2} + \frac{\frac{\partial w_1}{\partial x}\left(\frac{\partial w_2}{\partial x}\right)^2}{\frac{\partial(w_2,w_1)}{\partial(x,y)}} [u_1,u_2;w_1,w_2]_{w_2} + \{w_1,w_2;x,y\}_x.
\end{aligned}$$

On the other hand,

$$\begin{aligned}
&\begin{vmatrix} \frac{\partial u_1}{\partial y} & \frac{\partial u_2}{\partial y} \\ \frac{\partial^2 u_1}{\partial x^2} & \frac{\partial^2 u_2}{\partial x^2} \end{vmatrix} \\
&= \left(\frac{\partial w_1}{\partial x}\right)^2 \frac{\partial w_1}{\partial y} \begin{vmatrix} \frac{\partial u_1}{\partial w_1} & \frac{\partial u_2}{\partial w_1} \\ \frac{\partial^2 u_1}{\partial w_1^2} & \frac{\partial^2 u_2}{\partial w_1^2} \end{vmatrix} + \left(\frac{\partial w_1}{\partial x}\right)^2 \frac{\partial w_2}{\partial y} \begin{vmatrix} \frac{\partial u_1}{\partial w_2} & \frac{\partial u_2}{\partial w_2} \\ \frac{\partial^2 u_1}{\partial w_1^2} & \frac{\partial^2 u_2}{\partial w_1^2} \end{vmatrix} \\
&\quad + 2\frac{\partial w_1}{\partial x}\frac{\partial w_2}{\partial x}\frac{\partial w_1}{\partial y} \begin{vmatrix} \frac{\partial u_1}{\partial w_1} & \frac{\partial u_2}{\partial w_1} \\ \frac{\partial^2 u_1}{\partial w_1 \partial w_2} & \frac{\partial^2 u_2}{\partial w_1 \partial w_2} \end{vmatrix} + 2\frac{\partial w_1}{\partial x}\frac{\partial w_2}{\partial x}\frac{\partial w_2}{\partial y} \begin{vmatrix} \frac{\partial u_1}{\partial w_2} & \frac{\partial u_2}{\partial w_2} \\ \frac{\partial^2 u_1}{\partial w_1 \partial w_2} & \frac{\partial^2 u_2}{\partial w_1 \partial w_2} \end{vmatrix} \\
&\quad + \left(\frac{\partial w_2}{\partial x}\right)^2 \frac{\partial w_1}{\partial y} \begin{vmatrix} \frac{\partial u_1}{\partial w_1} & \frac{\partial u_2}{\partial w_1} \\ \frac{\partial^2 u_1}{\partial w_2^2} & \frac{\partial^2 u_2}{\partial w_2^2} \end{vmatrix} + \left(\frac{\partial w_2}{\partial x}\right)^2 \frac{\partial w_2}{\partial y} \begin{vmatrix} \frac{\partial u_1}{\partial w_2} & \frac{\partial u_2}{\partial w_2} \\ \frac{\partial^2 u_1}{\partial w_2^2} & \frac{\partial^2 u_2}{\partial w_2^2} \end{vmatrix} \\
&\quad + \begin{vmatrix} \frac{\partial w_1}{\partial y} & \frac{\partial w_2}{\partial y} \\ \frac{\partial^2 w_1}{\partial x^2} & \frac{\partial^2 w_2}{\partial x^2} \end{vmatrix} \begin{vmatrix} \frac{\partial u_1}{\partial w_1} & \frac{\partial u_2}{\partial w_1} \\ \frac{\partial u_1}{\partial w_2} & \frac{\partial u_2}{\partial w_2} \end{vmatrix}.
\end{aligned}$$

$$\begin{vmatrix} \frac{\partial u_1}{\partial x} & \frac{\partial u_2}{\partial x} \\ \frac{\partial^2 u_1}{\partial x \partial y} & \frac{\partial^2 u_2}{\partial x \partial y} \end{vmatrix}$$

$$= \left(\frac{\partial w_1}{\partial x}\right)^2 \frac{\partial w_1}{\partial y} \begin{vmatrix} \frac{\partial u_1}{\partial w_1} & \frac{\partial u_2}{\partial w_1} \\ \frac{\partial^2 u_1}{\partial w_1^2} & \frac{\partial^2 u_2}{\partial w_1^2} \end{vmatrix} + \frac{\partial w_1}{\partial x}\frac{\partial w_1}{\partial y}\frac{\partial w_2}{\partial x} \begin{vmatrix} \frac{\partial u_1}{\partial w_2} & \frac{\partial u_2}{\partial w_2} \\ \frac{\partial^2 u_1}{\partial w_1^2} & \frac{\partial^2 u_2}{\partial w_1^2} \end{vmatrix}$$

$$+ \frac{\partial w_1}{\partial x}\left(\frac{\partial w_1}{\partial x}\frac{\partial w_2}{\partial y} + \frac{\partial w_2}{\partial x}\frac{\partial w_1}{\partial y}\right) \begin{vmatrix} \frac{\partial u_1}{\partial w_1} & \frac{\partial u_2}{\partial w_1} \\ \frac{\partial^2 u_1}{\partial w_1 \partial w_2} & \frac{\partial^2 u_2}{\partial w_1 \partial w_2} \end{vmatrix}$$

$$+ \frac{\partial w_2}{\partial x}\left(\frac{\partial w_1}{\partial x}\frac{\partial w_2}{\partial y} + \frac{\partial w_2}{\partial x}\frac{\partial w_1}{\partial y}\right) \begin{vmatrix} \frac{\partial u_1}{\partial w_2} & \frac{\partial u_2}{\partial w_2} \\ \frac{\partial^2 u_1}{\partial w_1 \partial w_2} & \frac{\partial^2 u_2}{\partial w_1 \partial w_2} \end{vmatrix}$$

$$+ \frac{\partial w_1}{\partial x}\frac{\partial w_2}{\partial x}\frac{\partial w_2}{\partial y} \begin{vmatrix} \frac{\partial u_1}{\partial w_1} & \frac{\partial u_2}{\partial w_1} \\ \frac{\partial^2 u_1}{\partial w_2^2} & \frac{\partial^2 u_2}{\partial w_2^2} \end{vmatrix} + \left(\frac{\partial w_2}{\partial x}\right)^2 \frac{\partial w_2}{\partial y} \begin{vmatrix} \frac{\partial u_1}{\partial w_2} & \frac{\partial u_2}{\partial w_2} \\ \frac{\partial^2 u_1}{\partial w_2^2} & \frac{\partial^2 u_2}{\partial w_2^2} \end{vmatrix}$$

$$+ \begin{vmatrix} \frac{\partial w_1}{\partial x} & \frac{\partial w_2}{\partial x} \\ \frac{\partial^2 w_1}{\partial x \partial y} & \frac{\partial^2 w_2}{\partial x \partial y} \end{vmatrix} \begin{vmatrix} \frac{\partial u_1}{\partial w_1} & \frac{\partial u_2}{\partial w_1} \\ \frac{\partial u_1}{\partial w_2} & \frac{\partial u_2}{\partial w_2} \end{vmatrix}.$$

We find that

$$\begin{aligned} &[u_1, u_2; x, y]_x \\ &= \frac{3(\frac{\partial w_1}{\partial x})^2 \frac{\partial w_1}{\partial y}}{\frac{\partial(w_1,w_2)}{\partial(x,y)}} \{u_1, u_2; w_1, w_2\}_{w_1} \\ &+ \frac{(\frac{\partial w_1}{\partial x})^2 \frac{\partial w_2}{\partial y} + 2\frac{\partial w_1}{\partial x}\frac{\partial w_2}{\partial x}\frac{\partial w_1}{\partial y}}{\frac{\partial(w_1,w_2)}{\partial(x,y)}} [u_1, u_2; w_1, w_2]_{w_1} \\ &+ \frac{3(\frac{\partial w_2}{\partial x})^2 \frac{\partial w_2}{\partial y}}{\frac{\partial(w_2,w_1)}{\partial(x,y)}} \{u_1, u_2; w_1, w_2\}_{w_2} \\ &+ \frac{(\frac{\partial w_2}{\partial x})^2 \frac{\partial w_1}{\partial y} + 2\frac{\partial w_1}{\partial x}\frac{\partial w_2}{\partial x}\frac{\partial w_2}{\partial y}}{\frac{\partial(w_2,w_1)}{\partial(x,y)}} [u_1, u_2; w_1, w_2]_{w_2} \\ &+ [w_1, w_2; x, y]_x. \end{aligned}$$

Exchange the position of x and y, we get the other identities.

□

Corollary 2.6. *The exchange of the first and the second arguments obeys the laws:*

$$\begin{aligned} &\{w_1, w_2; x, y\}_x \\ &= -\frac{(\frac{\partial w_1}{\partial x})^3}{\frac{\partial(w_1,w_2)}{\partial(x,y)}} \{x, y; w_1, w_2\}_{w_1} - \frac{(\frac{\partial w_1}{\partial x})^2 \frac{\partial w_2}{\partial x}}{\frac{\partial(w_1,w_2)}{\partial(x,y)}} [x, y; w_1, w_2]_{w_1} \\ &+ \frac{(\frac{\partial w_2}{\partial x})^3}{\frac{\partial(w_1,w_2)}{\partial(x,y)}} \{x, y; w_1, w_2\}_{w_2} + \frac{\frac{\partial w_1}{\partial x}(\frac{\partial w_2}{\partial x})^2}{\frac{\partial(w_1,w_2)}{\partial(x,y)}} [x, y; w_1, w_2]_{w_2}, \end{aligned} \tag{2.17}$$

$$\{w_1, w_2; x, y\}_y$$
$$= \frac{\left(\frac{\partial w_1}{\partial y}\right)^3}{\frac{\partial(w_1,w_2)}{\partial(x,y)}} \{x, y; w_1, w_2\}_{w_1} + \frac{\left(\frac{\partial w_1}{\partial y}\right)^2 \frac{\partial w_2}{\partial y}}{\frac{\partial(w_1,w_2)}{\partial(x,y)}} [x, y; w_1, w_2]_{w_1}$$
$$- \frac{\left(\frac{\partial w_2}{\partial y}\right)^3}{\frac{\partial(w_1,w_2)}{\partial(x,y)}} \{x, y; w_1, w_2\}_{w_2} - \frac{\frac{\partial w_1}{\partial y}\left(\frac{\partial w_2}{\partial y}\right)^2}{\frac{\partial(w_1,w_2)}{\partial(x,y)}} [x, y; w_1, w_2]_{w_2}, \tag{2.18}$$

$$[w_1, w_2; x, y]_x$$
$$= - \frac{3\left(\frac{\partial w_1}{\partial x}\right)^2 \frac{\partial w_1}{\partial y}}{\frac{\partial(w_1,w_2)}{\partial(x,y)}} \{x, y; w_1, w_2\}_{w_1} - \frac{\left(\frac{\partial w_1}{\partial x}\right)^2 \frac{\partial w_2}{\partial y} + 2\frac{\partial w_1}{\partial x}\frac{\partial w_2}{\partial x}\frac{\partial w_1}{\partial y}}{\frac{\partial(w_1,w_2)}{\partial(x,y)}} [x, y; w_1, w_2]_{w_1}$$
$$+ \frac{3\left(\frac{\partial w_2}{\partial x}\right)^2 \frac{\partial w_2}{\partial y}}{\frac{\partial(w_1,w_2)}{\partial(x,y)}} \{x, y; w_1, w_2\}_{w_2} + \frac{\left(\frac{\partial w_2}{\partial x}\right)^2 \frac{\partial w_1}{\partial y} + 2\frac{\partial w_1}{\partial x}\frac{\partial w_2}{\partial x}\frac{\partial w_2}{\partial y}}{\frac{\partial(w_1,w_2)}{\partial(x,y)}} [x, y; w_1, w_2]_{w_2}, \tag{2.19}$$

$$[w_1, w_2; x, y]_y$$
$$= \frac{3\frac{\partial w_1}{\partial x}\left(\frac{\partial w_1}{\partial y}\right)^2}{\frac{\partial(w_1,w_2)}{\partial(x,y)}} \{x, y; w_1, w_2\}_{w_1} + \frac{\frac{\partial w_2}{\partial x}\left(\frac{\partial w_1}{\partial y}\right)^2 + 2\frac{\partial w_1}{\partial x}\frac{\partial w_1}{\partial y}\frac{\partial w_2}{\partial y}}{\frac{\partial(w_1,w_2)}{\partial(x,y)}} [x, y; w_1, w_2]_{w_1}$$
$$- \frac{3\frac{\partial w_2}{\partial x}\left(\frac{\partial w_2}{\partial y}\right)^2}{\frac{\partial(w_1,w_2)}{\partial(x,y)}} \{x, y; w_1, w_2\}_{w_2} - \frac{\frac{\partial w_1}{\partial x}\left(\frac{\partial w_2}{\partial y}\right)^2 + 2\frac{\partial w_2}{\partial x}\frac{\partial w_1}{\partial y}\frac{\partial w_2}{\partial y}}{\frac{\partial(w_1,w_2)}{\partial(x,y)}} [x, y; w_1, w_2]_{w_2}. \tag{2.20}$$

Proof. Put $u_1 = x$ and $u_2 = y$ in the Main Proposition 1. Note that

$$\begin{cases} \{x, y; x, y\}_x = 0, \\ \{x, y; x, y\}_y = 0, \\ [x, y; x, y]_x = 0, \\ [x, y; x, y]_y = 0. \end{cases}$$

We get the desired result.

□

We may apply a linear fractional transformation to the second arguments instead of the first. The result is as follows:

Corollary 2.7. *For*

$$(w_1, w_2) = \gamma(x, y) = \left(\frac{a_1x + a_2y + a_3}{c_1x + c_2y + c_3}, \frac{b_1x + b_2y + b_3}{c_1x + c_2y + c_3} \right),$$

where $\gamma = \begin{pmatrix} a_1 & a_2 & a_3 \\ b_1 & b_2 & b_3 \\ c_1 & c_2 & c_3 \end{pmatrix} \in GL(3,\mathbb{C})$, *the following four identities hold:*

$$\begin{aligned}
&\{u_1,u_2;x,y\}_x \\
=&\frac{A_1^3}{\Delta(c_1x+c_2y+c_3)^3}\left\{u_1,u_2;\frac{a_1x+a_2y+a_3}{c_1x+c_2y+c_3},\frac{b_1x+b_2y+b_3}{c_1x+c_2y+c_3}\right\}_{\frac{a_1x+a_2y+a_3}{c_1x+c_2y+c_3}} \\
&+\frac{A_1^2B_1}{\Delta(c_1x+c_2y+c_3)^3}\left[u_1,u_2;\frac{a_1x+a_2y+a_3}{c_1x+c_2y+c_3},\frac{b_1x+b_2y+b_3}{c_1x+c_2y+c_3}\right]_{\frac{a_1x+a_2y+a_3}{c_1x+c_2y+c_3}} \\
&-\frac{B_1^3}{\Delta(c_1x+c_2y+c_3)^3}\left\{u_1,u_2;\frac{a_1x+a_2y+a_3}{c_1x+c_2y+c_3},\frac{b_1x+b_2y+b_3}{c_1x+c_2y+c_3}\right\}_{\frac{b_1x+b_2y+b_3}{c_1x+c_2y+c_3}} \\
&-\frac{A_1B_1^2}{\Delta(c_1x+c_2y+c_3)^3}\left[u_1,u_2;\frac{a_1x+a_2y+a_3}{c_1x+c_2y+c_3},\frac{b_1x+b_2y+b_3}{c_1x+c_2y+c_3}\right]_{\frac{b_1x+b_2y+b_3}{c_1x+c_2y+c_3}},
\end{aligned} \tag{2.21}$$

$$\begin{aligned}
&\{u_1,u_2;x,y\}_y \\
=&-\frac{A_2^3}{\Delta(c_1x+c_2y+c_3)^3}\left\{u_1,u_2;\frac{a_1x+a_2y+a_3}{c_1x+c_2y+c_3},\frac{b_1x+b_2y+b_3}{c_1x+c_2y+c_3}\right\}_{\frac{a_1x+a_2y+a_3}{c_1x+c_2y+c_3}} \\
&-\frac{A_2^2B_2}{\Delta(c_1x+c_2y+c_3)^3}\left[u_1,u_2;\frac{a_1x+a_2y+a_3}{c_1x+c_2y+c_3},\frac{b_1x+b_2y+b_3}{c_1x+c_2y+c_3}\right]_{\frac{a_1x+a_2y+a_3}{c_1x+c_2y+c_3}} \\
&+\frac{B_2^3}{\Delta(c_1x+c_2y+c_3)^3}\left\{u_1,u_2;\frac{a_1x+a_2y+a_3}{c_1x+c_2y+c_3},\frac{b_1x+b_2y+b_3}{c_1x+c_2y+c_3}\right\}_{\frac{b_1x+b_2y+b_3}{c_1x+c_2y+c_3}} \\
&+\frac{A_2B_2^2}{\Delta(c_1x+c_2y+c_3)^3}\left[u_1,u_2;\frac{a_1x+a_2y+a_3}{c_1x+c_2y+c_3},\frac{b_1x+b_2y+b_3}{c_1x+c_2y+c_3}\right]_{\frac{b_1x+b_2y+b_3}{c_1x+c_2y+c_3}},
\end{aligned} \tag{2.22}$$

$$\begin{aligned}
&[u_1,u_2;x,y]_x \\
=&\frac{3A_1^2A_2}{\Delta(c_1x+c_2y+c_3)^3}\left\{u_1,u_2;\frac{a_1x+a_2y+a_3}{c_1x+c_2y+c_3},\frac{b_1x+b_2y+b_3}{c_1x+c_2y+c_3}\right\}_{\frac{a_1x+a_2y+a_3}{c_1x+c_2y+c_3}} \\
&+\frac{A_1^2B_2+2A_1B_1A_2}{\Delta(c_1x+c_2y+c_3)^3}\left[u_1,u_2;\frac{a_1x+a_2y+a_3}{c_1x+c_2y+c_3},\frac{b_1x+b_2y+b_3}{c_1x+c_2y+c_3}\right]_{\frac{a_1x+a_2y+a_3}{c_1x+c_2y+c_3}} \\
&-\frac{3B_1^2B_2}{\Delta(c_1x+c_2y+c_3)^3}\left\{u_1,u_2;\frac{a_1x+a_2y+a_3}{c_1x+c_2y+c_3},\frac{b_1x+b_2y+b_3}{c_1x+c_2y+c_3}\right\}_{\frac{b_1x+b_2y+b_3}{c_1x+c_2y+c_3}} \\
&-\frac{B_1^2A_2+2A_1B_1B_2}{\Delta(c_1x+c_2y+c_3)^3}\left[u_1,u_2;\frac{a_1x+a_2y+a_3}{c_1x+c_2y+c_3},\frac{b_1x+b_2y+b_3}{c_1x+c_2y+c_3}\right]_{\frac{b_1x+b_2y+b_3}{c_1x+c_2y+c_3}},
\end{aligned} \tag{2.23}$$

$$
\begin{aligned}
&[u_1, u_2; x, y]_y \\
&= -\frac{3A_1A_2^2}{\Delta(c_1x+c_2y+c_3)^3}\left\{u_1, u_2; \frac{a_1x+a_2y+a_3}{c_1x+c_2y+c_3}, \frac{b_1x+b_2y+b_3}{c_1x+c_2y+c_3}\right\}_{\frac{a_1x+a_2y+a_3}{c_1x+c_2y+c_3}} \\
&\quad - \frac{B_1A_2^2+2A_1A_2B_2}{\Delta(c_1x+c_2y+c_3)^3}\left[u_1, u_2; \frac{a_1x+a_2y+a_3}{c_1x+c_2y+c_3}, \frac{b_1x+b_2y+b_3}{c_1x+c_2y+c_3}\right]_{\frac{a_1x+a_2y+a_3}{c_1x+c_2y+c_3}} \\
&\quad + \frac{3B_1B_2^2}{\Delta(c_1x+c_2y+c_3)^3}\left\{u_1, u_2; \frac{a_1x+a_2y+a_3}{c_1x+c_2y+c_3}, \frac{b_1x+b_2y+b_3}{c_1x+c_2y+c_3}\right\}_{\frac{b_1x+b_2y+b_3}{c_1x+c_2y+c_3}} \\
&\quad + \frac{A_1B_2^2+2B_1A_2B_2}{\Delta(c_1x+c_2y+c_3)^3}\left[u_1, u_2; \frac{a_1x+a_2y+a_3}{c_1x+c_2y+c_3}, \frac{b_1x+b_2y+b_3}{c_1x+c_2y+c_3}\right]_{\frac{b_1x+b_2y+b_3}{c_1x+c_2y+c_3}}.
\end{aligned}
\tag{2.24}
$$

Here, $\Delta = \det(\gamma)$ *and*

$$
\begin{cases}
A_1 = (a_1c_2 - a_2c_1)y + (a_1c_3 - a_3c_1), \\
A_2 = (a_2c_1 - a_1c_2)x + (a_2c_3 - a_3c_2), \\
B_1 = (b_1c_2 - b_2c_1)y + (b_1c_3 - b_3c_1), \\
B_2 = (b_2c_1 - b_1c_2)x + (b_2c_3 - b_3c_2).
\end{cases}
$$

Proof. We have

$$
\begin{cases}
\dfrac{\partial w_1}{\partial x} = (c_1x+c_2y+c_3)^{-2}A_1, \\
\dfrac{\partial w_2}{\partial x} = (c_1x+c_2y+c_3)^{-2}B_1, \\
\dfrac{\partial w_1}{\partial y} = (c_1x+c_2y+c_3)^{-2}A_2, \\
\dfrac{\partial w_2}{\partial y} = (c_1x+c_2y+c_3)^{-2}B_2.
\end{cases}
$$

Note that

$$
\begin{cases}
\{w_1, w_2; x, y\}_x = 0, \\
\{w_1, w_2; x, y\}_y = 0, \\
[w_1, w_2; x, y]_x = 0, \\
[w_1, w_2; x, y]_y = 0.
\end{cases}
$$

A straightforward calculation gives the result.

□

The Main Proposition 1 implies that

$$\begin{pmatrix} \{u_1, u_2; x, y\}_x \\ \{u_1, u_2; x, y\}_y \\ [u_1, u_2; x, y]_x \\ [u_1, u_2; x, y]_y \end{pmatrix} = \frac{M(w_1, w_2; x, y)}{\frac{\partial(w_1,w_2)}{\partial(x,y)}} \begin{pmatrix} \{u_1, u_2; w_1, w_2\}_{w_1} \\ \{u_1, u_2; w_1, w_2\}_{w_2} \\ [u_1, u_2; w_1, w_2]_{w_1} \\ [u_1, u_2; w_1, w_2]_{w_2} \end{pmatrix} + \begin{pmatrix} \{w_1, w_2; x, y\}_x \\ \{w_1, w_2; x, y\}_y \\ [w_1, w_2; x, y]_x \\ [w_1, w_2; x, y]_y \end{pmatrix}, \tag{2.25}$$

where

$$M(w_1, w_2; x, y) :=$$
$$\begin{pmatrix} (\frac{\partial w_1}{\partial x})^3 & -(\frac{\partial w_2}{\partial x})^3 & (\frac{\partial w_1}{\partial x})^2 \frac{\partial w_2}{\partial x} & -\frac{\partial w_1}{\partial x}(\frac{\partial w_2}{\partial x})^2 \\ -(\frac{\partial w_1}{\partial y})^3 & (\frac{\partial w_2}{\partial y})^3 & -(\frac{\partial w_1}{\partial y})^2 \frac{\partial w_2}{\partial y} & \frac{\partial w_1}{\partial y}(\frac{\partial w_2}{\partial y})^2 \\ 3(\frac{\partial w_1}{\partial x})^2 \frac{\partial w_1}{\partial y} & -3(\frac{\partial w_2}{\partial x})^2 \frac{\partial w_2}{\partial y} & \begin{matrix} (\frac{\partial w_1}{\partial x})^2 \frac{\partial w_2}{\partial y} \\ +2\frac{\partial w_1}{\partial x}\frac{\partial w_2}{\partial x}\frac{\partial w_1}{\partial y} \end{matrix} & \begin{matrix} -(\frac{\partial w_2}{\partial x})^2 \frac{\partial w_1}{\partial y} \\ -2\frac{\partial w_1}{\partial x}\frac{\partial w_2}{\partial x}\frac{\partial w_2}{\partial y} \end{matrix} \\ -3\frac{\partial w_1}{\partial x}(\frac{\partial w_1}{\partial y})^2 & 3\frac{\partial w_2}{\partial x}(\frac{\partial w_2}{\partial y})^2 & \begin{matrix} -\frac{\partial w_2}{\partial x}(\frac{\partial w_1}{\partial y})^2 \\ -2\frac{\partial w_1}{\partial x}\frac{\partial w_1}{\partial y}\frac{\partial w_2}{\partial y} \end{matrix} & \begin{matrix} \frac{\partial w_1}{\partial x}(\frac{\partial w_2}{\partial y})^2 \\ +2\frac{\partial w_2}{\partial x}\frac{\partial w_1}{\partial y}\frac{\partial w_2}{\partial y} \end{matrix} \end{pmatrix}. \tag{2.26}$$

Proposition 2.8.

$$M(w_1, w_2; x, y) \cdot M(u_1, u_2; w_1, w_2) = M(u_1, u_2; x, y). \tag{2.27}$$

Proof. Put

$$(a_{ij}) = M(w_1, w_2; x, y) \cdot M(u_1, u_2; w_1, w_2).$$

We find that

$$\begin{aligned} a_{1,1} &= \left(\frac{\partial w_1}{\partial x}\right)^3 \left(\frac{\partial u_1}{\partial w_1}\right)^3 + \left(\frac{\partial w_2}{\partial x}\right)^3 \left(\frac{\partial u_1}{\partial w_2}\right)^3 \\ &\quad + \left(\frac{\partial w_1}{\partial x}\right)^2 \frac{\partial w_2}{\partial x} \cdot 3 \left(\frac{\partial u_1}{\partial w_1}\right)^2 \frac{\partial u_1}{\partial w_2} + \frac{\partial w_1}{\partial x} \left(\frac{\partial w_2}{\partial x}\right)^2 \cdot 3 \frac{\partial u_1}{\partial w_1} \left(\frac{\partial u_1}{\partial w_2}\right)^2 \\ &= \left(\frac{\partial u_1}{\partial w_1}\frac{\partial w_1}{\partial x} + \frac{\partial u_1}{\partial w_2}\frac{\partial w_2}{\partial x}\right)^3 \\ &= \left(\frac{\partial u_1}{\partial x}\right)^3. \end{aligned}$$

$$\begin{aligned}
a_{1,3} =& \left(\frac{\partial w_1}{\partial x}\right)^3 \left(\frac{\partial u_1}{\partial w_1}\right)^2 \frac{\partial u_2}{\partial w_1} + \left(\frac{\partial w_2}{\partial x}\right)^3 \left(\frac{\partial u_1}{\partial w_2}\right)^2 \frac{\partial u_2}{\partial w_2} \\
&+ \left(\frac{\partial w_1}{\partial x}\right)^2 \frac{\partial w_2}{\partial x} \left[\left(\frac{\partial u_1}{\partial w_1}\right)^2 \frac{\partial u_2}{\partial w_2} + 2\frac{\partial u_1}{\partial w_1}\frac{\partial u_2}{\partial w_1}\frac{\partial u_1}{\partial w_2}\right] \\
&+ \frac{\partial w_1}{\partial x} \left(\frac{\partial w_2}{\partial x}\right)^2 \left[\frac{\partial u_2}{\partial w_1}\left(\frac{\partial u_1}{\partial w_2}\right)^2 + 2\frac{\partial u_1}{\partial w_1}\frac{\partial u_1}{\partial w_2}\frac{\partial u_2}{\partial w_2}\right] \\
=& \left(\frac{\partial u_1}{\partial w_1}\frac{\partial w_1}{\partial x} + \frac{\partial u_1}{\partial w_2}\frac{\partial w_2}{\partial x}\right)^2 \left(\frac{\partial u_2}{\partial w_1}\frac{\partial w_1}{\partial x} + \frac{\partial u_2}{\partial w_2}\frac{\partial w_2}{\partial x}\right) \\
=& \left(\frac{\partial u_1}{\partial x}\right)^2 \frac{\partial u_2}{\partial x}.
\end{aligned}$$

$$\begin{aligned}
a_{3,1} =& 3\left(\frac{\partial w_1}{\partial x}\right)^2 \frac{\partial w_1}{\partial y}\left(\frac{\partial u_1}{\partial w_1}\right)^3 + 3\left(\frac{\partial w_2}{\partial x}\right)^2 \frac{\partial w_2}{\partial y}\left(\frac{\partial u_1}{\partial w_2}\right)^3 \\
&+ \left[\left(\frac{\partial w_1}{\partial x}\right)^2 \frac{\partial w_2}{\partial y} + 2\frac{\partial w_1}{\partial x}\frac{\partial w_2}{\partial x}\frac{\partial w_1}{\partial y}\right] \cdot 3\left(\frac{\partial u_1}{\partial w_1}\right)^2 \frac{\partial u_1}{\partial w_2} \\
&+ \left[\left(\frac{\partial w_2}{\partial x}\right)^2 \frac{\partial w_1}{\partial y} + 2\frac{\partial w_1}{\partial x}\frac{\partial w_2}{\partial x}\frac{\partial w_2}{\partial y}\right] \cdot 3\frac{\partial u_1}{\partial w_1}\left(\frac{\partial u_1}{\partial w_2}\right)^2 \\
=& 3\left(\frac{\partial u_1}{\partial w_1}\frac{\partial w_1}{\partial x} + \frac{\partial u_1}{\partial w_2}\frac{\partial w_2}{\partial x}\right)^2 \left(\frac{\partial u_1}{\partial w_1}\frac{\partial w_1}{\partial y} + \frac{\partial u_1}{\partial w_2}\frac{\partial w_2}{\partial y}\right) \\
=& 3\left(\frac{\partial u_1}{\partial x}\right)^2 \frac{\partial u_1}{\partial y}.
\end{aligned}$$

$$\begin{aligned}
a_{3,3} =& 3\left(\frac{\partial w_1}{\partial x}\right)^2 \frac{\partial w_1}{\partial y}\left(\frac{\partial u_1}{\partial w_1}\right)^2 \frac{\partial u_2}{\partial w_1} + 3\left(\frac{\partial w_2}{\partial x}\right)^2 \frac{\partial w_2}{\partial y}\left(\frac{\partial u_1}{\partial w_2}\right)^2 \frac{\partial u_2}{\partial w_2} \\
&+ \left[\left(\frac{\partial w_1}{\partial x}\right)^2 \frac{\partial w_2}{\partial y} + 2\frac{\partial w_1}{\partial x}\frac{\partial w_2}{\partial x}\frac{\partial w_1}{\partial y}\right]\left[\left(\frac{\partial u_1}{\partial w_1}\right)^2 \frac{\partial u_2}{\partial w_2} + 2\frac{\partial u_1}{\partial w_1}\frac{\partial u_2}{\partial w_1}\frac{\partial u_1}{\partial w_2}\right] \\
&+ \left[\left(\frac{\partial w_2}{\partial x}\right)^2 \frac{\partial w_1}{\partial y} + 2\frac{\partial w_1}{\partial x}\frac{\partial w_2}{\partial x}\frac{\partial w_2}{\partial y}\right]\left[\frac{\partial u_2}{\partial w_1}\left(\frac{\partial u_1}{\partial w_2}\right)^2 + 2\frac{\partial u_1}{\partial w_1}\frac{\partial u_1}{\partial w_2}\frac{\partial u_2}{\partial w_2}\right] \\
=& \left(\frac{\partial u_1}{\partial w_1}\frac{\partial w_1}{\partial x} + \frac{\partial u_1}{\partial w_2}\frac{\partial w_2}{\partial x}\right)^2 \left(\frac{\partial u_2}{\partial w_1}\frac{\partial w_1}{\partial y} + \frac{\partial u_2}{\partial w_2}\frac{\partial w_2}{\partial y}\right) \\
&+ 2\left(\frac{\partial u_1}{\partial w_1}\frac{\partial w_1}{\partial x} + \frac{\partial u_1}{\partial w_2}\frac{\partial w_2}{\partial x}\right)\left(\frac{\partial u_2}{\partial w_1}\frac{\partial w_1}{\partial x} + \frac{\partial u_2}{\partial w_2}\frac{\partial w_2}{\partial x}\right) \\
&\times \left(\frac{\partial u_1}{\partial w_1}\frac{\partial w_1}{\partial y} + \frac{\partial u_1}{\partial w_2}\frac{\partial w_2}{\partial y}\right) \\
=& \left(\frac{\partial u_1}{\partial x}\right)^2 \frac{\partial u_2}{\partial y} + 2\frac{\partial u_1}{\partial x}\frac{\partial u_2}{\partial x}\frac{\partial u_1}{\partial y}.
\end{aligned}$$

By the symmetry property, we can get the other terms in the matrix (a_{ij}).

□

In particular,

$$M(w_1, w_2; x, y) \cdot M(x, y; w_1, w_2) = M(x, y; x, y) = I.$$

Hence,

$$M(w_1, w_2; x, y)^{-1} = M(x, y; w_1, w_2). \tag{2.28}$$

Set

$$U_{(w_1,w_2;x,y)} := \begin{pmatrix} M(w_1, w_2; x, y) & c\frac{\partial(w_1,w_2)}{\partial(x,y)} \begin{pmatrix} \{w_1, w_2; x, y\}_x \\ \{w_1, w_2; x, y\}_y \\ [w_1, w_2; x, y]_x \\ [w_1, w_2; x, y]_y \end{pmatrix} \\ 0 & \frac{\partial(w_1,w_2)}{\partial(x,y)} \end{pmatrix}. \tag{2.29}$$

Then the above proposition implies that

$$U_{(w_1,w_2;x,y)} \cdot U_{(u_1,u_2;w_1,w_2)} = U_{(u_1,u_2;x,y)}. \tag{2.30}$$

The equation

$$U_{W,Z} U_{U,W} = U_{U,Z}$$

with $W = (w_1, w_2)$, $Z = (z_1, z_2)$ and $U = (u_1, u_2)$ is a cocycle condition, which means that the U's can be interpreted as transition functions of a vector bundle. More precisely, when we choose an open cover $\Sigma = \bigcup_i \Sigma_i$, with a local coordinate Z_i on Σ_i, then the above equation permits us to interpret the U_{Z_i,Z_j} as transition functions on the intersection $\Sigma_i \cap \Sigma_j$. Let $\mathcal{V}_0$ be this vector bundle. It is a five dimensional sub-bundle of the infinite dimensional vector bundle $\mathbb{V}$.

We have

$$U_{(w_1,w_2;z_1,z_2)} \begin{pmatrix} \Lambda \\ f \end{pmatrix} = \begin{pmatrix} M(w_1, w_2; z_1, z_2)\Lambda + c\frac{\partial(w_1,w_2)}{\partial(z_1,z_2)} f \begin{pmatrix} \{w_1, w_2; z_1, z_2\}_{z_1} \\ \{w_1, w_2; z_1, z_2\}_{z_2} \\ [w_1, w_2; z_1, z_2]_{z_1} \\ [w_1, w_2; z_1, z_2]_{z_2} \end{pmatrix} \\ \frac{\partial(w_1,w_2)}{\partial(z_1,z_2)} f \end{pmatrix},$$

where $\Lambda = \begin{pmatrix} \lambda_1 \\ \lambda_2 \\ \lambda_3 \\ \lambda_4 \end{pmatrix}$. In particular, when $c = 0$, one has

$$U_{(w_1,w_2;z_1,z_2)} \begin{pmatrix} \Lambda \\ f \end{pmatrix} = \begin{pmatrix} M(w_1, w_2; z_1, z_2)\Lambda \\ \frac{\partial(w_1,w_2)}{\partial(z_1,z_2)} f \end{pmatrix}.$$

At $c = 0$, the pair (Λ, f) is a section of the bundle $N \oplus T$, where N is the rank four vector bundle and T is the line bundle. At $c \neq 0$, the pair (Λ, f) is a section of a deformation $\mathcal{V}_0$ of $N \oplus T$.

The transition matrices $U_{(w_1,w_2;z_1,z_2)}$ are triangular, this means that there is an exact sequence

$$0 \to N \to \mathcal{V}_0 \to T \to 0.$$

There is a natural map α from sections of N to sections of $\mathcal{V}_0$, namely $\Lambda \mapsto (\Lambda, 0)$; and a natural map β from sections of $\mathcal{V}_0$ to sections of T, namely $(\Lambda, f) \mapsto f$. Moreover, $\beta \circ \alpha = 0$ and the image of α is the kernel of β.

Let us consider the compact complex surfaces with $c_1^2 = 3c_2$, where c_i is the i-th Chern class, i.e., the ball quotient surfaces. A uniformization of a ball quotient surface Σ gives a covering by open sets Σ_i with local parameters $Z_i = (z_{1,i}, z_{2,i})$ such that on $\Sigma_i \cap \Sigma_j$ the transformations from Z_i to Z_j are $U(2,1)$ transformations

$$(z_{1,j}, z_{2,j}) = \gamma(z_{1,i}, z_{2,i}) = \left(\frac{a_1 z_{1,i} + a_2 z_{2,i} + a_3}{c_1 z_{1,i} + c_2 z_{2,i} + c_3}, \frac{b_1 z_{1,i} + b_2 z_{2,i} + b_3}{c_1 z_{1,i} + c_2 z_{2,i} + c_3} \right)$$

with $\gamma = \begin{pmatrix} a_1 & a_2 & a_3 \\ b_1 & b_2 & b_3 \\ c_1 & c_2 & c_3 \end{pmatrix} \in U(2,1)$. For such transformations, our four derivatives

$$\{z_{1,i}, z_{2,i}; z_{1,j}, z_{2,j}\}_{z_{1,j}}, \quad \{z_{1,i}, z_{2,i}; z_{1,j}, z_{2,j}\}_{z_{2,j}}$$

and

$$[z_{1,i}, z_{2,i}; z_{1,j}, z_{2,j}]_{z_{1,j}}, \quad [z_{1,i}, z_{2,i}; z_{1,j}, z_{2,j}]_{z_{2,j}}$$

vanish, so the transition matrices become diagonal, and the extension $\mathcal{V}_0$ is split as $N \oplus T$.

Recall that

$$\{z_1, z_2; w_1, w_2\}_{w_1} := \frac{\begin{vmatrix} \frac{\partial z_1}{\partial w_1} & \frac{\partial z_2}{\partial w_1} \\ \frac{\partial^2 z_1}{\partial w_1^2} & \frac{\partial^2 z_2}{\partial w_1^2} \end{vmatrix}}{\begin{vmatrix} \frac{\partial z_1}{\partial w_1} & \frac{\partial z_2}{\partial w_1} \\ \frac{\partial z_1}{\partial w_2} & \frac{\partial z_2}{\partial w_2} \end{vmatrix}}, \quad \{z_1, z_2; w_1, w_2\}_{w_2} := \frac{\begin{vmatrix} \frac{\partial z_1}{\partial w_2} & \frac{\partial z_2}{\partial w_2} \\ \frac{\partial^2 z_1}{\partial w_2^2} & \frac{\partial^2 z_2}{\partial w_2^2} \end{vmatrix}}{\begin{vmatrix} \frac{\partial z_1}{\partial w_2} & \frac{\partial z_2}{\partial w_2} \\ \frac{\partial z_1}{\partial w_1} & \frac{\partial z_2}{\partial w_1} \end{vmatrix}},$$

$$[z_1, z_2; w_1, w_2]_{w_1} := \frac{\begin{vmatrix} \frac{\partial z_1}{\partial w_2} & \frac{\partial z_2}{\partial w_2} \\ \frac{\partial^2 z_1}{\partial w_1^2} & \frac{\partial^2 z_2}{\partial w_1^2} \end{vmatrix} + 2 \begin{vmatrix} \frac{\partial z_1}{\partial w_1} & \frac{\partial z_2}{\partial w_1} \\ \frac{\partial^2 z_1}{\partial w_1 \partial w_2} & \frac{\partial^2 z_2}{\partial w_1 \partial w_2} \end{vmatrix}}{\begin{vmatrix} \frac{\partial z_1}{\partial w_1} & \frac{\partial z_2}{\partial w_1} \\ \frac{\partial z_1}{\partial w_2} & \frac{\partial z_2}{\partial w_2} \end{vmatrix}},$$

$$[z_1, z_2; w_1, w_2]_{w_2} := \frac{\begin{vmatrix} \frac{\partial z_1}{\partial w_1} & \frac{\partial z_2}{\partial w_1} \\ \frac{\partial^2 z_1}{\partial w_2^2} & \frac{\partial^2 z_2}{\partial w_2^2} \end{vmatrix} + 2 \begin{vmatrix} \frac{\partial z_1}{\partial w_2} & \frac{\partial z_2}{\partial w_2} \\ \frac{\partial^2 z_1}{\partial w_1 \partial w_2} & \frac{\partial^2 z_2}{\partial w_1 \partial w_2} \end{vmatrix}}{\begin{vmatrix} \frac{\partial z_1}{\partial w_2} & \frac{\partial z_2}{\partial w_2} \\ \frac{\partial z_1}{\partial w_1} & \frac{\partial z_2}{\partial w_1} \end{vmatrix}}.$$

The next Main Proposition 2 gives the relation between the deformation of the Jacobian and our four derivatives:

Proposition 2.9 (Main Proposition 2). *The following identity holds:*

$$\begin{aligned}
&\frac{\partial(f_1, f_2)}{\partial(z_1, z_2)}\\
&= \frac{\partial(\widehat{f}_1, \widehat{f}_2)}{\partial(w_1, w_2)} - \widehat{f}_1 \frac{\partial \widehat{f}_1}{\partial w_2}\{z_1, z_2; w_1, w_2\}_{w_1} - \widehat{f}_2 \frac{\partial \widehat{f}_2}{\partial w_1}\{z_1, z_2; w_1, w_2\}_{w_2}\\
&\quad - \widehat{f}_1 \frac{\partial \widehat{f}_2}{\partial w_2} \frac{\begin{vmatrix} \frac{\partial z_1}{\partial w_2} & \frac{\partial z_2}{\partial w_2} \\ \frac{\partial^2 z_1}{\partial w_1^2} & \frac{\partial^2 z_2}{\partial w_1^2} \end{vmatrix}}{\begin{vmatrix} \frac{\partial z_1}{\partial w_1} & \frac{\partial z_2}{\partial w_1} \\ \frac{\partial z_1}{\partial w_2} & \frac{\partial z_2}{\partial w_2} \end{vmatrix}} - \left(\widehat{f}_2 \frac{\partial \widehat{f}_1}{\partial w_2} - \widehat{f}_1 \frac{\partial \widehat{f}_1}{\partial w_1}\right) \frac{\begin{vmatrix} \frac{\partial z_1}{\partial w_1} & \frac{\partial z_2}{\partial w_1} \\ \frac{\partial^2 z_1}{\partial w_1 \partial w_2} & \frac{\partial^2 z_2}{\partial w_1 \partial w_2} \end{vmatrix}}{\begin{vmatrix} \frac{\partial z_1}{\partial w_1} & \frac{\partial z_2}{\partial w_1} \\ \frac{\partial z_1}{\partial w_2} & \frac{\partial z_2}{\partial w_2} \end{vmatrix}}\\
&\quad - \widehat{f}_2 \frac{\partial \widehat{f}_1}{\partial w_1} \frac{\begin{vmatrix} \frac{\partial z_1}{\partial w_1} & \frac{\partial z_2}{\partial w_1} \\ \frac{\partial^2 z_1}{\partial w_2^2} & \frac{\partial^2 z_2}{\partial w_2^2} \end{vmatrix}}{\begin{vmatrix} \frac{\partial z_1}{\partial w_2} & \frac{\partial z_2}{\partial w_2} \\ \frac{\partial z_1}{\partial w_1} & \frac{\partial z_2}{\partial w_1} \end{vmatrix}} - \left(\widehat{f}_1 \frac{\partial \widehat{f}_2}{\partial w_1} - \widehat{f}_2 \frac{\partial \widehat{f}_2}{\partial w_2}\right) \frac{\begin{vmatrix} \frac{\partial z_1}{\partial w_2} & \frac{\partial z_2}{\partial w_2} \\ \frac{\partial^2 z_1}{\partial w_1 \partial w_2} & \frac{\partial^2 z_2}{\partial w_1 \partial w_2} \end{vmatrix}}{\begin{vmatrix} \frac{\partial z_1}{\partial w_2} & \frac{\partial z_2}{\partial w_2} \\ \frac{\partial z_1}{\partial w_1} & \frac{\partial z_2}{\partial w_1} \end{vmatrix}}\\
&\quad + \widehat{f}_1^{\,2} \frac{\begin{vmatrix} \frac{\partial^2 z_1}{\partial w_1^2} & \frac{\partial^2 z_2}{\partial w_1^2} \\ \frac{\partial^2 z_1}{\partial w_1 \partial w_2} & \frac{\partial^2 z_2}{\partial w_1 \partial w_2} \end{vmatrix}}{\begin{vmatrix} \frac{\partial z_1}{\partial w_1} & \frac{\partial z_2}{\partial w_1} \\ \frac{\partial z_1}{\partial w_2} & \frac{\partial z_2}{\partial w_2} \end{vmatrix}} + \widehat{f}_2^{\,2} \frac{\begin{vmatrix} \frac{\partial^2 z_1}{\partial w_2^2} & \frac{\partial^2 z_2}{\partial w_2^2} \\ \frac{\partial^2 z_1}{\partial w_1 \partial w_2} & \frac{\partial^2 z_2}{\partial w_1 \partial w_2} \end{vmatrix}}{\begin{vmatrix} \frac{\partial z_1}{\partial w_2} & \frac{\partial z_2}{\partial w_2} \\ \frac{\partial z_1}{\partial w_1} & \frac{\partial z_2}{\partial w_1} \end{vmatrix}} + \widehat{f}_1 \widehat{f}_2 \frac{\begin{vmatrix} \frac{\partial^2 z_1}{\partial w_1^2} & \frac{\partial^2 z_2}{\partial w_1^2} \\ \frac{\partial^2 z_1}{\partial w_2^2} & \frac{\partial^2 z_2}{\partial w_2^2} \end{vmatrix}}{\begin{vmatrix} \frac{\partial z_1}{\partial w_1} & \frac{\partial z_2}{\partial w_1} \\ \frac{\partial z_1}{\partial w_2} & \frac{\partial z_2}{\partial w_2} \end{vmatrix}},
\end{aligned} \tag{2.31}$$

where

$$(f_1, f_2) = (\widehat{f}_1, \widehat{f}_2) \begin{pmatrix} \frac{\partial z_1}{\partial w_1} & \frac{\partial z_2}{\partial w_1} \\ \frac{\partial z_1}{\partial w_2} & \frac{\partial z_2}{\partial w_2} \end{pmatrix}.$$

Proof. Note that

$$\frac{\partial(f_1, f_2)}{\partial(z_1, z_2)} = \frac{\partial(f_1, f_2)}{\partial(w_1, w_2)} \frac{\partial(w_1, w_2)}{\partial(z_1, z_2)}.$$

Here,

$$f_1 = \widehat{f}_1 \frac{\partial z_1}{\partial w_1} + \widehat{f}_2 \frac{\partial z_1}{\partial w_2}, \quad f_2 = \widehat{f}_1 \frac{\partial z_2}{\partial w_1} + \widehat{f}_2 \frac{\partial z_2}{\partial w_2}.$$

We find that

$$
\begin{aligned}
&\frac{\partial(f_1,f_2)}{\partial(w_1,w_2)}\\
&=\frac{\partial(\widehat{f}_1,\widehat{f}_2)}{\partial(w_1,w_2)}\frac{\partial(z_1,z_2)}{\partial(w_1,w_2)}-\widehat{f}_1\frac{\partial\widehat{f}_1}{\partial w_2}\begin{vmatrix}\frac{\partial z_1}{\partial w_1} & \frac{\partial z_2}{\partial w_1}\\ \frac{\partial^2 z_1}{\partial w_1^2} & \frac{\partial^2 z_2}{\partial w_1^2}\end{vmatrix}+\widehat{f}_2\frac{\partial\widehat{f}_2}{\partial w_1}\begin{vmatrix}\frac{\partial z_1}{\partial w_2} & \frac{\partial z_2}{\partial w_2}\\ \frac{\partial^2 z_1}{\partial w_2^2} & \frac{\partial^2 z_2}{\partial w_2^2}\end{vmatrix}\\
&\quad-\widehat{f}_1\frac{\partial\widehat{f}_2}{\partial w_2}\begin{vmatrix}\frac{\partial z_1}{\partial w_2} & \frac{\partial z_2}{\partial w_2}\\ \frac{\partial^2 z_1}{\partial w_1^2} & \frac{\partial^2 z_2}{\partial w_1^2}\end{vmatrix}-\left(\widehat{f}_2\frac{\partial\widehat{f}_1}{\partial w_2}-\widehat{f}_1\frac{\partial\widehat{f}_1}{\partial w_1}\right)\begin{vmatrix}\frac{\partial z_1}{\partial w_1} & \frac{\partial z_2}{\partial w_1}\\ \frac{\partial^2 z_1}{\partial w_1\partial w_2} & \frac{\partial^2 z_2}{\partial w_1\partial w_2}\end{vmatrix}\\
&\quad+\widehat{f}_2\frac{\partial\widehat{f}_1}{\partial w_1}\begin{vmatrix}\frac{\partial z_1}{\partial w_1} & \frac{\partial z_2}{\partial w_1}\\ \frac{\partial^2 z_1}{\partial w_2^2} & \frac{\partial^2 z_2}{\partial w_2^2}\end{vmatrix}+\left(\widehat{f}_1\frac{\partial\widehat{f}_2}{\partial w_1}-\widehat{f}_2\frac{\partial\widehat{f}_2}{\partial w_2}\right)\begin{vmatrix}\frac{\partial z_1}{\partial w_2} & \frac{\partial z_2}{\partial w_2}\\ \frac{\partial^2 z_1}{\partial w_1\partial w_2} & \frac{\partial^2 z_2}{\partial w_1\partial w_2}\end{vmatrix}\\
&\quad+\widehat{f}_1^{\,2}\begin{vmatrix}\frac{\partial^2 z_1}{\partial w_1^2} & \frac{\partial^2 z_2}{\partial w_1^2}\\ \frac{\partial^2 z_1}{\partial w_1\partial w_2} & \frac{\partial^2 z_2}{\partial w_1\partial w_2}\end{vmatrix}-\widehat{f}_2^{\,2}\begin{vmatrix}\frac{\partial^2 z_1}{\partial w_2^2} & \frac{\partial^2 z_2}{\partial w_2^2}\\ \frac{\partial^2 z_1}{\partial w_1\partial w_2} & \frac{\partial^2 z_2}{\partial w_1\partial w_2}\end{vmatrix}+\widehat{f}_1\widehat{f}_2\begin{vmatrix}\frac{\partial^2 z_1}{\partial w_1^2} & \frac{\partial^2 z_2}{\partial w_1^2}\\ \frac{\partial^2 z_1}{\partial w_2^2} & \frac{\partial^2 z_2}{\partial w_2^2}\end{vmatrix}.
\end{aligned}
$$

Thus, we can obtain the desired result. □

Now, put

$$\phi(f_1,f_2):=\iint\frac{\partial(f_1,f_2)}{\partial(t_1,t_2)}dt_1\wedge dt_2.$$

We define an operation

$$
\begin{aligned}
&\left[f_1\frac{\partial}{\partial t_1}\otimes\frac{\partial}{\partial t_2},f_2\frac{\partial}{\partial t_1}\otimes\frac{\partial}{\partial t_2}\right]^*_{Z_0}\\
&:=c\cdot\phi(f_1,f_2)-f_1\frac{\partial f_1}{\partial t_2}\frac{\partial}{\partial t_1}-f_1\frac{\partial f_2}{\partial t_2}\frac{\partial}{\partial t_1}-\left(f_2\frac{\partial f_1}{\partial t_2}-f_1\frac{\partial f_1}{\partial t_1}\right)\frac{\partial}{\partial t_1}\\
&\quad-f_2\frac{\partial f_2}{\partial t_1}\frac{\partial}{\partial t_2}-f_2\frac{\partial f_1}{\partial t_1}\frac{\partial}{\partial t_2}-\left(f_1\frac{\partial f_2}{\partial t_1}-f_2\frac{\partial f_2}{\partial t_2}\right)\frac{\partial}{\partial t_2}\\
&\quad+f_1^2\frac{\partial^2}{\partial t_1^2}+f_2^2\frac{\partial^2}{\partial t_2^2}+f_1f_2\frac{\partial^2}{\partial t_1\partial t_2}.
\end{aligned}
$$

Note that

$$f_1\frac{\partial f_1}{\partial t_2}+f_1\frac{\partial f_2}{\partial t_2}+\left(f_2\frac{\partial f_1}{\partial t_2}-f_1\frac{\partial f_1}{\partial t_1}\right)=(f_1+f_2)\frac{\partial f_1}{\partial t_2}+f_1\left(\frac{\partial f_2}{\partial t_2}-\frac{\partial f_1}{\partial t_1}\right)$$

and

$$f_2\frac{\partial f_2}{\partial t_1}+f_2\frac{\partial f_1}{\partial t_1}+\left(f_1\frac{\partial f_2}{\partial t_1}-f_2\frac{\partial f_2}{\partial t_2}\right)=(f_1+f_2)\frac{\partial f_2}{\partial t_1}+f_2\left(\frac{\partial f_1}{\partial t_1}-\frac{\partial f_2}{\partial t_2}\right).$$

Modulo the terms of second order partial derivatives, we define the following operation:

$$\begin{aligned}&\left[f_1\frac{\partial}{\partial t_1}\otimes\frac{\partial}{\partial t_2}, f_2\frac{\partial}{\partial t_1}\otimes\frac{\partial}{\partial t_2}\right]_{Z_0}\\ :=&\left[(f_1+f_2)\frac{\partial f_1}{\partial t_2}+f_1\left(\frac{\partial f_2}{\partial t_2}-\frac{\partial f_1}{\partial t_1}\right)\right]\frac{\partial}{\partial t_1}\\ &+\left[(f_1+f_2)\frac{\partial f_2}{\partial t_1}+f_2\left(\frac{\partial f_1}{\partial t_1}-\frac{\partial f_2}{\partial t_2}\right)\right]\frac{\partial}{\partial t_2}+c\cdot\phi(f_1,f_2).\end{aligned} \tag{2.32}$$

3. Nonlinear partial differential equations and modular functions associated to $U(2,1)$

Let $U(2,1)=\{g\in GL(3,\mathbb{C}): g^*Jg=J\}$ where $J=\begin{pmatrix} & & 1\\ & -1 & \\ 1 & & \end{pmatrix}$. The corresponding complex domain $\mathfrak{S}_2=\{(z_1,z_2)\in\mathbb{C}^2: z_1+\overline{z_1}-z_2\overline{z_2}>0\}$ is the simplest example of a bounded symmetric domain which is not a tube domain. Put $\rho(Z):=z_1+\overline{z_1}-z_2\overline{z_2}$ for $Z=(z_1,z_2)\in\mathfrak{S}_2$.

The imaginary quadratic field $K=\mathbb{Q}(\sqrt{-3})$ is called the field of Eisenstein numbers. Its ring of integers

$$\mathcal{O}_K=\{a/2+b\sqrt{-3}/2: a,b\in\mathbb{Z}, a\equiv b(\mathrm{mod} 2)\}=\mathbb{Z}+\mathbb{Z}[\omega]$$

with $\omega=\frac{-1+\sqrt{-3}}{2}$ is called the ring of Eisenstein integers.

Let P be the parabolic subgroup of $U(2,1)$ which consists of upper triangular matrices. This group P has the Langlands decomposition $P=NAM$, where

$$N=\left\{n=[\alpha,\beta]:=\begin{pmatrix}1 & \alpha & \beta\\ & 1 & \overline{\alpha}\\ & & 1\end{pmatrix}: \alpha,\beta\in\mathbb{C}, \beta+\overline{\beta}=\alpha\overline{\alpha}\right\},$$

$$A=\left\{a=\begin{pmatrix}\rho^{-1} & & \\ & 1 & \\ & & \rho\end{pmatrix}: \rho\in\mathbb{R}, \rho>0\right\},$$

$$M=\left\{m=\begin{pmatrix}\beta & & \\ & \beta^{-2} & \\ & & \beta\end{pmatrix}: \beta\in\mathbb{C}, |\beta|^2=1\right\}.$$

Therefore,

$$N(\mathcal{O}_K)=\{n=[\alpha,\beta]: \alpha,\beta\in\mathcal{O}_K, \beta+\overline{\beta}=\alpha\overline{\alpha}\}. \tag{3.1}$$

Put $\Gamma=U(2,1;\mathcal{O}_K)$. Then $\Gamma\cap N=N(\mathcal{O}_K)$.

Let

$$T_1=\begin{pmatrix}1 & 1 & -\omega\\ & 1 & 1\\ & & 1\end{pmatrix},\quad T_2=\begin{pmatrix}1 & \omega & -\omega\\ & 1 & \overline{\omega}\\ & & 1\end{pmatrix},\quad S=-\overline{\omega}J=\begin{pmatrix} & & -\overline{\omega}\\ & \overline{\omega} & \\ -\overline{\omega} & & \end{pmatrix}.$$

Theorem 3.1. *The subgroup $N(\mathcal{O}_K)$ is generated by T_1 and T_2.*

Proof. Suppose that $\beta + \overline{\beta} = \alpha\overline{\alpha}$ with $\alpha, \beta \in \mathcal{O}_K$. Set $\alpha = m + n\omega$, $\beta = b_1 + b_2\omega$, where $m, n, b_1, b_2 \in \mathbb{Z}$. Then $2b_1 - b_2 = m^2 - mn + n^2$. The general solution is given by

$$b_2 = m + n + mn + 2l, \quad b_1 = \frac{1}{2}(m^2 + n^2 + m + n) + l, \quad (l \in \mathbb{Z}).$$

Hence,

$$\beta = \frac{1}{2}(m^2 - mn + n^2) + \left[\frac{1}{2}(m + n + mn) + l\right](\omega - \overline{\omega}).$$

Set

$$[\alpha, \beta] := \begin{pmatrix} 1 & \alpha & \beta \\ & 1 & \overline{\alpha} \\ & & 1 \end{pmatrix}, \quad \beta + \overline{\beta} = \alpha\overline{\alpha}, \quad \alpha, \beta \in \mathcal{O}_K.$$

Then

$$[\alpha_1, \beta_1][\alpha_2, \beta_2] = [\alpha_1 + \alpha_2, \beta_1 + \beta_2 + \alpha_1\overline{\alpha_2}].$$

$$[\alpha, \beta]^{-1} = [-\alpha, \overline{\beta}], \quad [\alpha_1, \beta_1][\alpha_2, \beta_2][\alpha_1, \beta_1]^{-1}[\alpha_2, \beta_2]^{-1} = [0, \alpha_1\overline{\alpha_2} - \overline{\alpha_1}\alpha_2].$$

We have

$$[1, -\omega][\omega, -\omega][1, -\omega]^{-1}[\omega, -\omega]^{-1} = [0, \overline{\omega} - \omega].$$

$${}^{m} = \left[m, \frac{1}{2}m(m-1) - m\omega\right], \quad (m \in \mathbb{Z}),$$

$$[\omega, -\omega]^n = \left[n\omega, \frac{1}{2}n(n-1) - n\omega\right], \quad (n \in \mathbb{Z}),$$

$$[0, \overline{\omega} - \omega]^l = [0, l(\overline{\omega} - \omega)], \quad (l \in \mathbb{Z}).$$

This gives that

$$[1, -\omega]^m[\omega, -\omega]^n[0, \overline{\omega} - \omega]^l$$
$$= \left[m + n\omega, \frac{1}{2}(m^2 - mn + n^2) + \left[\frac{1}{2}(m + n + mn) + l\right](\overline{\omega} - \omega)\right].$$

Thus,

$$[1, -\omega]^m[\omega, -\omega]^n[0, \overline{\omega} - \omega]^{-l-m-n-mn}$$
$$= \left[m + n\omega, \frac{1}{2}(m^2 - mn + n^2) + \left[\frac{1}{2}(m + n + mn) + l\right](\omega - \overline{\omega})\right].$$

Note that $T_1 = [1, -\omega]$ and $T_2 = [\omega, -\omega]$. This completes the proof of Theorem 3.1.

□

In fact, T_1 and T_2 are two non-commutative translations, their commutator $[T_1, T_2] := T_1T_2T_1^{-1}T_2^{-1} = [0, \overline{\omega} - \omega]$ is a commutative translation. We have

$$T_1(z_1, z_2) = (z_1 + z_2 - \omega, z_2 + 1), \quad T_2(z_1, z_2) = (z_1 + \omega z_2 - \omega, z_2 + \overline{\omega}),$$

$$S(z_1, z_2) = \left(\frac{1}{z_1}, -\frac{z_2}{z_1}\right),$$

where $(z_1, z_2) \in \mathfrak{S}_2$.

When $T_1(z_1, z_2) = S(z_1, z_2)$, then

$$z_1 + z_2 - \omega = \frac{1}{z_1}, \quad z_2 + 1 = -\frac{z_2}{z_1}. \tag{3.2}$$

The solutions are given by

$$(z_1, z_2) = (\omega, \overline{\omega}), \quad \left(\omega i, \frac{-\omega i}{\omega i + 1}\right), \quad \left(-\omega i, \frac{\omega i}{-\omega i + 1}\right).$$

Here, $\rho(\omega, \overline{\omega}) = -2 < 0$, $\rho\left(\omega i, \frac{-\omega i}{\omega i+1}\right) = -2 - 2\sqrt{3} < 0$ and $\rho\left(-\omega i, \frac{\omega i}{1-\omega i}\right) = 2(\sqrt{3} - 1) > 0$. Thus, $\left(-\omega i, \frac{\omega i}{1-\omega i}\right) \in \mathfrak{S}_2$.

Let

$$(-\omega, 0) \mapsto 1, \quad (-\overline{\omega}, 0) \mapsto z_2, \quad \infty \mapsto \infty, \quad \left(-\omega i, \frac{\omega i}{1 - \omega i}\right) \mapsto 0. \tag{3.3}$$

The point 0 is of order 3, 1 is of order 1, z_2 is of order 1 and ∞ is a logarithmic branch point.

By

$$S\left(-\omega i, \frac{\omega i}{1 - \omega i}\right) = T_1\left(-\omega i, \frac{\omega i}{1 - \omega i}\right),$$

we have

$$\left(-\omega i, \frac{\omega i}{1 - \omega i}\right) = S^2\left(-\omega i, \frac{\omega i}{1 - \omega i}\right) = ST_1\left(-\omega i, \frac{\omega i}{1 - \omega i}\right).$$

On the other hand, $(ST_1)^4 = \omega I$. This implies that ST_1 is of order four. In fact, S is of order two and

$$S(-\omega, 0) = (-\overline{\omega}, 0), \quad S(-\overline{\omega}, 0) = (-\omega, 0).$$

Note that

$$\frac{\partial(v_i^{\frac{1}{4}})}{\partial w_j} = \frac{1}{4} v_i^{-\frac{3}{4}} \frac{\partial v_i}{\partial w_j}, \quad \frac{\partial(v_i^{\frac{1}{2}})}{\partial w_j} = \frac{1}{2} v_i^{-\frac{1}{2}} \frac{\partial v_i}{\partial w_j}, \quad \frac{\partial(\log v_i)}{\partial w_j} = \frac{1}{v_i} \frac{\partial v_i}{\partial w_j}, \quad i, j = 1, 2.$$

This leads us to consider

$$v_1^{-\frac{3}{4}}(1 - v_1)^{-\frac{1}{2}}(v_2 - v_1)^{-\frac{1}{2}} \begin{vmatrix} \frac{\partial v_1}{\partial w_1} & \frac{\partial v_2}{\partial w_1} \\ \frac{\partial v_1}{\partial w_2} & \frac{\partial v_2}{\partial w_2} \end{vmatrix}. \tag{3.4}$$

When $T_2(z_1, z_2) = S(z_1, z_2)$, then

$$z_1 + \omega z_2 - \omega = \frac{1}{z_1}, \quad z_2 + \overline{\omega} = -\frac{z_2}{z_1}. \tag{3.5}$$

The solutions are given by

$$(z_1, z_2) = (\omega, \omega), \quad \left(\omega i, \frac{-i}{\omega i + 1}\right), \quad \left(-\omega i, \frac{i}{-\omega i + 1}\right).$$

Here, $\rho(\omega, \omega) = -2 < 0$, $\rho\left(\omega i, \frac{-i}{\omega i+1}\right) = -2 - 2\sqrt{3} < 0$ and $\rho\left(-\omega i, \frac{i}{1-\omega i}\right) = 2(\sqrt{3} - 1) > 0$. Thus, $\left(-\omega i, \frac{i}{1-\omega i}\right) \in \mathfrak{S}_2$.

Let

$$(-\omega, 0) \mapsto 1, \quad (-\overline{\omega}, 0) \mapsto z_1, \quad \infty \mapsto \infty, \quad \left(-\omega i, \frac{i}{1 - \omega i}\right) \mapsto 0. \tag{3.6}$$

The point 0 is of order 3, 1 is of order 1, z_1 is of order 1 and ∞ is a logarithmic branch point.

By

$$S\left(-\omega i, \frac{i}{1 - \omega i}\right) = T_2\left(-\omega i, \frac{i}{1 - \omega i}\right),$$

we have

$$\left(-\omega i, \frac{i}{1 - \omega i}\right) = S^2\left(-\omega i, \frac{i}{1 - \omega i}\right) = ST_2\left(-\omega i, \frac{i}{1 - \omega i}\right).$$

On the other hand, $(ST_2)^4 = \omega I$. This implies that ST_2 is of order four. By the same argument as above, this leads us to consider

$$v_2^{-\frac{3}{4}}(1 - v_2)^{-\frac{1}{2}}(v_1 - v_2)^{-\frac{1}{2}} \begin{vmatrix} \frac{\partial v_2}{\partial w_1} & \frac{\partial v_1}{\partial w_1} \\ \frac{\partial v_2}{\partial w_2} & \frac{\partial v_1}{\partial w_2} \end{vmatrix}. \tag{3.7}$$

In fact, for

$$u_1 = \frac{a_1 w_1 + a_2 w_2 + a_3}{c_1 w_1 + c_2 w_2 + c_3}, \quad u_2 = \frac{b_1 w_1 + b_2 w_2 + b_3}{c_1 w_1 + c_2 w_2 + c_3}$$

with $\Delta = \begin{vmatrix} a_1 & a_2 & a_3 \\ b_1 & b_2 & b_3 \\ c_1 & c_2 & c_3 \end{vmatrix}$, we have

$$\begin{vmatrix} \frac{\partial v_1}{\partial w_1} & \frac{\partial v_2}{\partial w_1} \\ \frac{\partial v_1}{\partial w_2} & \frac{\partial v_2}{\partial w_2} \end{vmatrix} = \Delta(c_1 w_1 + c_2 w_2 + c_3)^{-3} \begin{vmatrix} \frac{\partial v_1}{\partial u_1} & \frac{\partial v_2}{\partial u_1} \\ \frac{\partial v_1}{\partial u_2} & \frac{\partial v_2}{\partial u_2} \end{vmatrix}. \tag{3.8}$$

Theorem 3.2 (Main Theorem 1). *For the over-determined system of second order, nonlinear (quasilinear) complex partial differential equations:*

$$\begin{cases} \{w_1, w_2; v_1, v_2\}_{v_1} = F_1(v_1, v_2), \\ \{w_1, w_2; v_1, v_2\}_{v_2} = F_2(v_1, v_2), \\ [w_1, w_2; v_1, v_2]_{v_1} = P_1(v_1, v_2), \\ [w_1, w_2; v_1, v_2]_{v_2} = P_2(v_1, v_2), \end{cases} \tag{3.9}$$

set the Jacobian

$$\Delta := \frac{\partial(v_1, v_2)}{\partial(w_1, w_2)} = \begin{vmatrix} \frac{\partial v_1}{\partial w_1} & \frac{\partial v_2}{\partial w_1} \\ \frac{\partial v_1}{\partial w_2} & \frac{\partial v_2}{\partial w_2} \end{vmatrix}. \tag{3.10}$$

Then $z = \sqrt[3]{\Delta}$ *satisfies the following linear differential equations:*

$$\begin{cases} \frac{\partial^2 z}{\partial v_1^2} + \frac{1}{3} P_1 \frac{\partial z}{\partial v_1} - F_1 \frac{\partial z}{\partial v_2} + \left(\frac{\partial F_1}{\partial v_2} - \frac{1}{3} \frac{\partial P_1}{\partial v_1} - \frac{2}{9} P_1^2 - \frac{2}{3} F_1 P_2 \right) z = 0, \\ \frac{\partial^2 z}{\partial v_1 \partial v_2} - \frac{1}{3} P_2 \frac{\partial z}{\partial v_1} - \frac{1}{3} P_1 \frac{\partial z}{\partial v_2} + \left(\frac{1}{3} \frac{\partial P_2}{\partial v_1} + \frac{1}{3} \frac{\partial P_1}{\partial v_2} + \frac{1}{9} P_1 P_2 - F_1 F_2 \right) z = 0, \\ \frac{\partial^2 z}{\partial v_2^2} + \frac{1}{3} P_2 \frac{\partial z}{\partial v_2} - F_2 \frac{\partial z}{\partial v_1} + \left(\frac{\partial F_2}{\partial v_1} - \frac{1}{3} \frac{\partial P_2}{\partial v_2} - \frac{2}{9} P_2^2 - \frac{2}{3} F_2 P_1 \right) z = 0. \end{cases} \tag{3.11}$$

Proof. Note that

$$\begin{cases} 1 = \frac{\partial v_1}{\partial v_1} = \frac{\partial v_1}{\partial w_1} \frac{\partial w_1}{\partial v_1} + \frac{\partial v_1}{\partial w_2} \frac{\partial w_2}{\partial v_1}, \\ 0 = \frac{\partial v_1}{\partial v_2} = \frac{\partial v_1}{\partial w_1} \frac{\partial w_1}{\partial v_2} + \frac{\partial v_1}{\partial w_2} \frac{\partial w_2}{\partial v_2}, \\ 0 = \frac{\partial v_2}{\partial v_1} = \frac{\partial v_2}{\partial w_1} \frac{\partial w_1}{\partial v_1} + \frac{\partial v_2}{\partial w_2} \frac{\partial w_2}{\partial v_1}, \\ 1 = \frac{\partial v_2}{\partial v_2} = \frac{\partial v_2}{\partial w_1} \frac{\partial w_1}{\partial v_2} + \frac{\partial v_2}{\partial w_2} \frac{\partial w_2}{\partial v_2}. \end{cases}$$

Thus,

$$\frac{\partial v_1}{\partial w_1} = \frac{\frac{\partial w_2}{\partial v_2}}{\begin{vmatrix} \frac{\partial w_1}{\partial v_1} & \frac{\partial w_2}{\partial v_1} \\ \frac{\partial w_1}{\partial v_2} & \frac{\partial w_2}{\partial v_2} \end{vmatrix}}, \quad \frac{\partial v_1}{\partial w_2} = \frac{-\frac{\partial w_1}{\partial v_2}}{\begin{vmatrix} \frac{\partial w_1}{\partial v_1} & \frac{\partial w_2}{\partial v_1} \\ \frac{\partial w_1}{\partial v_2} & \frac{\partial w_2}{\partial v_2} \end{vmatrix}},$$

$$\frac{\partial v_2}{\partial w_1} = \frac{-\frac{\partial w_2}{\partial v_1}}{\begin{vmatrix} \frac{\partial w_1}{\partial v_1} & \frac{\partial w_2}{\partial v_1} \\ \frac{\partial w_1}{\partial v_2} & \frac{\partial w_2}{\partial v_2} \end{vmatrix}}, \quad \frac{\partial v_2}{\partial w_2} = \frac{\frac{\partial w_1}{\partial v_1}}{\begin{vmatrix} \frac{\partial w_1}{\partial v_1} & \frac{\partial w_2}{\partial v_1} \\ \frac{\partial w_1}{\partial v_2} & \frac{\partial w_2}{\partial v_2} \end{vmatrix}}.$$

Hence,

$$\begin{vmatrix} \frac{\partial v_1}{\partial w_1} & \frac{\partial v_2}{\partial w_1} \\ \frac{\partial v_1}{\partial w_2} & \frac{\partial v_2}{\partial w_2} \end{vmatrix} = \frac{1}{\begin{vmatrix} \frac{\partial w_1}{\partial v_1} & \frac{\partial w_2}{\partial v_1} \\ \frac{\partial w_1}{\partial v_2} & \frac{\partial w_2}{\partial v_2} \end{vmatrix}}.$$

We have

$$\frac{\partial}{\partial v_1}\left(\frac{\partial v_2}{\partial w_2}\right)=\frac{\partial}{\partial v_1}\left(\frac{\frac{\partial w_1}{\partial v_1}}{\begin{vmatrix}\frac{\partial w_1}{\partial v_1} & \frac{\partial w_2}{\partial v_1}\\ \frac{\partial w_1}{\partial v_2} & \frac{\partial w_2}{\partial v_2}\end{vmatrix}}\right)$$

$$=\begin{vmatrix}\frac{\partial w_1}{\partial v_1} & \frac{\partial w_2}{\partial v_1}\\ \frac{\partial w_1}{\partial v_2} & \frac{\partial w_2}{\partial v_2}\end{vmatrix}^{-2}\left[-\left(\frac{\partial w_1}{\partial v_1}\right)^3\frac{\partial}{\partial v_2}\left(\frac{\frac{\partial w_2}{\partial v_1}}{\frac{\partial w_1}{\partial v_1}}\right)+\frac{\partial w_1}{\partial v_2}\begin{vmatrix}\frac{\partial w_1}{\partial v_1} & \frac{\partial w_2}{\partial v_1}\\ \frac{\partial^2 w_1}{\partial v_1^2} & \frac{\partial^2 w_2}{\partial v_1^2}\end{vmatrix}\right]$$

$$=\left(\frac{\partial v_2}{\partial w_2}\right)^3\begin{vmatrix}\frac{\partial v_1}{\partial w_1} & \frac{\partial v_2}{\partial w_1}\\ \frac{\partial v_1}{\partial w_2} & \frac{\partial v_2}{\partial w_2}\end{vmatrix}^{-1}\frac{\partial}{\partial v_2}\left(\frac{\frac{\partial v_2}{\partial w_1}}{\frac{\partial v_2}{\partial w_2}}\right)-\frac{\partial v_1}{\partial w_2}F_1(v_1,v_2).$$

Similarly,

$$\frac{\partial}{\partial v_2}\left(\frac{\partial v_1}{\partial w_1}\right)=\left(\frac{\partial v_1}{\partial w_1}\right)^3\begin{vmatrix}\frac{\partial v_1}{\partial w_1} & \frac{\partial v_2}{\partial w_1}\\ \frac{\partial v_1}{\partial w_2} & \frac{\partial v_2}{\partial w_2}\end{vmatrix}^{-1}\frac{\partial}{\partial v_1}\left(\frac{\frac{\partial v_1}{\partial w_2}}{\frac{\partial v_1}{\partial w_1}}\right)-\frac{\partial v_2}{\partial w_1}F_2(v_1,v_2).$$

$$\frac{\partial}{\partial v_1}\left(\frac{\partial v_2}{\partial w_1}\right)=-\frac{\partial v_1}{\partial w_1}F_1(v_1,v_2)+\begin{vmatrix}\frac{\partial v_1}{\partial w_1} & \frac{\partial v_2}{\partial w_1}\\ \frac{\partial v_1}{\partial w_2} & \frac{\partial v_2}{\partial w_2}\end{vmatrix}^{-1}\frac{\partial v_2}{\partial w_1}\left(\frac{\partial v_2}{\partial w_2}\right)^2\frac{\partial}{\partial v_2}\left(\frac{\frac{\partial v_2}{\partial w_1}}{\frac{\partial v_2}{\partial w_2}}\right).$$

$$\frac{\partial}{\partial v_2}\left(\frac{\partial v_1}{\partial w_2}\right)=-\frac{\partial v_2}{\partial w_2}F_2(v_1,v_2)+\begin{vmatrix}\frac{\partial v_1}{\partial w_1} & \frac{\partial v_2}{\partial w_1}\\ \frac{\partial v_1}{\partial w_2} & \frac{\partial v_2}{\partial w_2}\end{vmatrix}^{-1}\frac{\partial v_1}{\partial w_2}\left(\frac{\partial v_1}{\partial w_1}\right)^2\frac{\partial}{\partial v_1}\left(\frac{\frac{\partial v_1}{\partial w_2}}{\frac{\partial v_1}{\partial w_1}}\right).$$

Put

$$\frac{\partial v_1}{\partial w_1}=f_{1,1}(v_1,v_2),\quad \frac{\partial v_1}{\partial w_2}=f_{1,2}(v_1,v_2),$$

$$\frac{\partial v_2}{\partial w_1}=f_{2,1}(v_1,v_2),\quad \frac{\partial v_2}{\partial w_2}=f_{2,2}(v_1,v_2).$$

Set $\Delta=f_{1,1}f_{2,2}-f_{1,2}f_{2,1}$, then

$$\begin{cases} f_{1,2}F_1=\dfrac{f_{2,2}^2}{\Delta}\dfrac{\partial f_{2,1}}{\partial v_2}-\dfrac{f_{2,1}f_{2,2}}{\Delta}\dfrac{\partial f_{2,2}}{\partial v_2}-\dfrac{\partial f_{2,2}}{\partial v_1},\\[2ex] f_{2,1}F_2=\dfrac{f_{1,1}^2}{\Delta}\dfrac{\partial f_{1,2}}{\partial v_1}-\dfrac{f_{1,1}f_{1,2}}{\Delta}\dfrac{\partial f_{1,1}}{\partial v_1}-\dfrac{\partial f_{1,1}}{\partial v_2},\\[2ex] f_{1,1}F_1=\dfrac{f_{2,1}f_{2,2}}{\Delta}\dfrac{\partial f_{2,1}}{\partial v_2}-\dfrac{f_{2,1}^2}{\Delta}\dfrac{\partial f_{2,2}}{\partial v_2}-\dfrac{\partial f_{2,1}}{\partial v_1},\\[2ex] f_{2,2}F_2=\dfrac{f_{1,1}f_{1,2}}{\Delta}\dfrac{\partial f_{1,2}}{\partial v_1}-\dfrac{f_{1,2}^2}{\Delta}\dfrac{\partial f_{1,1}}{\partial v_1}-\dfrac{\partial f_{1,2}}{\partial v_2}, \end{cases}$$

which are equivalent to

$$\begin{cases} \Delta F_1 = f_{2,1}\dfrac{\partial f_{2,2}}{\partial v_1} - f_{2,2}\dfrac{\partial f_{2,1}}{\partial v_1}, \\ \Delta F_2 = f_{1,2}\dfrac{\partial f_{1,1}}{\partial v_2} - f_{1,1}\dfrac{\partial f_{1,2}}{\partial v_2}, \\ f_{1,1}\dfrac{\partial f_{2,2}}{\partial v_1} - f_{1,2}\dfrac{\partial f_{2,1}}{\partial v_1} = f_{2,2}\dfrac{\partial f_{2,1}}{\partial v_2} - f_{2,1}\dfrac{\partial f_{2,2}}{\partial v_2}, \\ f_{1,1}\dfrac{\partial f_{1,2}}{\partial v_1} - f_{1,2}\dfrac{\partial f_{1,1}}{\partial v_1} = f_{2,2}\dfrac{\partial f_{1,1}}{\partial v_2} - f_{2,1}\dfrac{\partial f_{1,2}}{\partial v_2}. \end{cases}$$

We have

$$\{w_1, w_2; v_1, v_2\}_{v_1} = \frac{\partial v_2}{\partial w_2}\frac{\partial}{\partial v_1}\left(\frac{\partial w_2}{\partial v_1}\right) + \frac{\partial v_2}{\partial w_1}\frac{\partial}{\partial v_1}\left(\frac{\partial w_1}{\partial v_1}\right)$$

$$= f_{2,2}\frac{\partial}{\partial v_1}\left(-\frac{f_{2,1}}{\Delta}\right) + f_{2,1}\frac{\partial}{\partial v_1}\left(\frac{f_{2,2}}{\Delta}\right) = \frac{1}{\Delta}\left(f_{2,1}\frac{\partial f_{2,2}}{\partial v_1} - f_{2,2}\frac{\partial f_{2,1}}{\partial v_1}\right).$$

Similarly,

$$\{w_1, w_2; v_1, v_2\}_{v_2} = \frac{1}{\Delta}\left(f_{1,2}\frac{\partial f_{1,1}}{\partial v_2} - f_{1,1}\frac{\partial f_{1,2}}{\partial v_2}\right).$$

$$[w_1, w_2; v_1, v_2]_{v_1}$$
$$= \frac{1}{\Delta}\left[\left(f_{1,2}\frac{\partial f_{2,1}}{\partial v_1} - f_{1,1}\frac{\partial f_{2,2}}{\partial v_1}\right) + 2\left(f_{2,2}\frac{\partial f_{1,1}}{\partial v_1} - f_{2,1}\frac{\partial f_{1,2}}{\partial v_1}\right) - \frac{\partial \Delta}{\partial v_1}\right].$$

$$[w_1, w_2; v_1, v_2]_{v_2}$$
$$= \frac{1}{\Delta}\left[\left(f_{2,1}\frac{\partial f_{1,2}}{\partial v_2} - f_{2,2}\frac{\partial f_{1,1}}{\partial v_2}\right) + 2\left(f_{1,1}\frac{\partial f_{2,2}}{\partial v_2} - f_{1,2}\frac{\partial f_{2,1}}{\partial v_2}\right) - \frac{\partial \Delta}{\partial v_2}\right].$$

By

$$\frac{\partial \Delta}{\partial v_1} = \frac{\partial}{\partial v_1}(f_{1,1}f_{2,2} - f_{1,2}f_{2,1}) = \frac{\partial f_{1,1}}{\partial v_1}f_{2,2} + f_{1,1}\frac{\partial f_{2,2}}{\partial v_1} - \frac{\partial f_{1,2}}{\partial v_1}f_{2,1} - f_{1,2}\frac{\partial f_{2,1}}{\partial v_1},$$

$$\frac{\partial \Delta}{\partial v_2} = \frac{\partial}{\partial v_2}(f_{1,1}f_{2,2} - f_{1,2}f_{2,1}) = \frac{\partial f_{1,1}}{\partial v_2}f_{2,2} + f_{1,1}\frac{\partial f_{2,2}}{\partial v_2} - \frac{\partial f_{1,2}}{\partial v_2}f_{2,1} - f_{1,2}\frac{\partial f_{2,1}}{\partial v_2},$$

we get

$$[w_1, w_2; v_1, v_2]_{v_1} = \frac{1}{\Delta}\left[2\left(f_{1,2}\frac{\partial f_{2,1}}{\partial v_1} - f_{1,1}\frac{\partial f_{2,2}}{\partial v_1}\right) + \left(f_{2,2}\frac{\partial f_{1,1}}{\partial v_1} - f_{2,1}\frac{\partial f_{1,2}}{\partial v_1}\right)\right],$$

$$[w_1, w_2; v_1, v_2]_{v_2} = \frac{1}{\Delta}\left[2\left(f_{2,1}\frac{\partial f_{1,2}}{\partial v_2} - f_{2,2}\frac{\partial f_{1,1}}{\partial v_2}\right) + \left(f_{1,1}\frac{\partial f_{2,2}}{\partial v_2} - f_{1,2}\frac{\partial f_{2,1}}{\partial v_2}\right)\right].$$

The equations

$$\begin{cases} \{w_1, w_2; v_1, v_2\}_{v_1} = F_1(v_1, v_2), \\ \{w_1, w_2; v_1, v_2\}_{v_2} = F_2(v_1, v_2), \\ [w_1, w_2; v_1, v_2]_{v_1} = P_1(v_1, v_2), \\ [w_1, w_2; v_1, v_2]_{v_2} = P_2(v_1, v_2) \end{cases}$$

give that

$$\begin{cases} \Delta F_1 = f_{2,1}\dfrac{\partial f_{2,2}}{\partial v_1} - f_{2,2}\dfrac{\partial f_{2,1}}{\partial v_1}, \\ \Delta F_2 = f_{1,2}\dfrac{\partial f_{1,1}}{\partial v_2} - f_{1,1}\dfrac{\partial f_{1,2}}{\partial v_2}, \\ \Delta P_1 = 2\left(f_{1,2}\dfrac{\partial f_{2,1}}{\partial v_1} - f_{1,1}\dfrac{\partial f_{2,2}}{\partial v_1}\right) + \left(f_{2,2}\dfrac{\partial f_{1,1}}{\partial v_1} - f_{2,1}\dfrac{\partial f_{1,2}}{\partial v_1}\right), \\ \Delta P_2 = 2\left(f_{2,1}\dfrac{\partial f_{1,2}}{\partial v_2} - f_{2,2}\dfrac{\partial f_{1,1}}{\partial v_2}\right) + \left(f_{1,1}\dfrac{\partial f_{2,2}}{\partial v_2} - f_{1,2}\dfrac{\partial f_{2,1}}{\partial v_2}\right). \end{cases}$$

By

$$\begin{cases} \dfrac{\partial \Delta}{\partial v_1} = \left(f_{2,2}\dfrac{\partial f_{1,1}}{\partial v_1} - f_{2,1}\dfrac{\partial f_{1,2}}{\partial v_1}\right) - \left(f_{1,2}\dfrac{\partial f_{2,1}}{\partial v_1} - f_{1,1}\dfrac{\partial f_{2,2}}{\partial v_1}\right), \\ \Delta P_1 = \left(f_{2,2}\dfrac{\partial f_{1,1}}{\partial v_1} - f_{2,1}\dfrac{\partial f_{1,2}}{\partial v_1}\right) + 2\left(f_{1,2}\dfrac{\partial f_{2,1}}{\partial v_1} - f_{1,1}\dfrac{\partial f_{2,2}}{\partial v_1}\right), \end{cases}$$

we have

$$\begin{cases} f_{2,2}\dfrac{\partial f_{1,1}}{\partial v_1} - f_{2,1}\dfrac{\partial f_{1,2}}{\partial v_1} = \dfrac{2}{3}\dfrac{\partial \Delta}{\partial v_1} + \dfrac{1}{3}\Delta P_1, \\ f_{1,2}\dfrac{\partial f_{2,1}}{\partial v_1} - f_{1,1}\dfrac{\partial f_{2,2}}{\partial v_1} = \dfrac{1}{3}\Delta P_1 - \dfrac{1}{3}\dfrac{\partial \Delta}{\partial v_1}. \end{cases}$$

By

$$\begin{cases} \dfrac{\partial \Delta}{\partial v_2} = \left(f_{1,1}\dfrac{\partial f_{2,2}}{\partial v_2} - f_{1,2}\dfrac{\partial f_{2,1}}{\partial v_2}\right) - \left(f_{2,1}\dfrac{\partial f_{1,2}}{\partial v_2} - f_{2,2}\dfrac{\partial f_{1,1}}{\partial v_2}\right), \\ \Delta P_2 = \left(f_{1,1}\dfrac{\partial f_{2,2}}{\partial v_2} - f_{1,2}\dfrac{\partial f_{2,1}}{\partial v_2}\right) + 2\left(f_{2,1}\dfrac{\partial f_{1,2}}{\partial v_2} - f_{2,2}\dfrac{\partial f_{1,1}}{\partial v_2}\right), \end{cases}$$

we have

$$\begin{cases} f_{1,1}\dfrac{\partial f_{2,2}}{\partial v_2} - f_{1,2}\dfrac{\partial f_{2,1}}{\partial v_2} = \dfrac{2}{3}\dfrac{\partial \Delta}{\partial v_2} + \dfrac{1}{3}\Delta P_2, \\ f_{2,1}\dfrac{\partial f_{1,2}}{\partial v_2} - f_{2,2}\dfrac{\partial f_{1,1}}{\partial v_2} = \dfrac{1}{3}\Delta P_2 - \dfrac{1}{3}\dfrac{\partial \Delta}{\partial v_2}. \end{cases}$$

The above two equations

$$\begin{cases} f_{1,1}\dfrac{\partial f_{2,2}}{\partial v_1} - f_{1,2}\dfrac{\partial f_{2,1}}{\partial v_1} = f_{2,2}\dfrac{\partial f_{2,1}}{\partial v_2} - f_{2,1}\dfrac{\partial f_{2,2}}{\partial v_2}, \\ f_{1,1}\dfrac{\partial f_{1,2}}{\partial v_1} - f_{1,2}\dfrac{\partial f_{1,1}}{\partial v_1} = f_{2,2}\dfrac{\partial f_{1,1}}{\partial v_2} - f_{2,1}\dfrac{\partial f_{1,2}}{\partial v_2} \end{cases}$$

imply that

$$\begin{cases} f_{2,1}\dfrac{\partial f_{2,2}}{\partial v_2} - f_{2,2}\dfrac{\partial f_{2,1}}{\partial v_2} = \dfrac{1}{3}\Delta P_1 - \dfrac{1}{3}\dfrac{\partial \Delta}{\partial v_1}, \\ f_{1,2}\dfrac{\partial f_{1,1}}{\partial v_1} - f_{1,1}\dfrac{\partial f_{1,2}}{\partial v_1} = \dfrac{1}{3}\Delta P_2 - \dfrac{1}{3}\dfrac{\partial \Delta}{\partial v_2}. \end{cases}$$

By

$$\begin{cases} f_{1,1}\dfrac{\partial f_{2,2}}{\partial v_2} - f_{1,2}\dfrac{\partial f_{2,1}}{\partial v_2} = \dfrac{2}{3}\dfrac{\partial \Delta}{\partial v_2} + \dfrac{1}{3}\Delta P_2, \\ f_{2,1}\dfrac{\partial f_{2,2}}{\partial v_2} - f_{2,2}\dfrac{\partial f_{2,1}}{\partial v_2} = -\dfrac{1}{3}\dfrac{\partial \Delta}{\partial v_1} + \dfrac{1}{3}\Delta P_1, \end{cases}$$

we have

$$\begin{cases} \dfrac{\partial f_{2,2}}{\partial v_2} = \dfrac{1}{3}\dfrac{f_{1,2}}{\Delta}\dfrac{\partial \Delta}{\partial v_1} + \dfrac{2}{3}\dfrac{f_{2,2}}{\Delta}\dfrac{\partial \Delta}{\partial v_2} - \dfrac{1}{3}f_{1,2}P_1 + \dfrac{1}{3}f_{2,2}P_2, \\ \dfrac{\partial f_{2,1}}{\partial v_2} = \dfrac{1}{3}\dfrac{f_{1,1}}{\Delta}\dfrac{\partial \Delta}{\partial v_1} + \dfrac{2}{3}\dfrac{f_{2,1}}{\Delta}\dfrac{\partial \Delta}{\partial v_2} - \dfrac{1}{3}f_{1,1}P_1 + \dfrac{1}{3}f_{2,1}P_2. \end{cases}$$

By

$$\begin{cases} f_{2,1}\dfrac{\partial f_{2,2}}{\partial v_1} - f_{2,2}\dfrac{\partial f_{2,1}}{\partial v_1} = \Delta F_1, \\ f_{1,1}\dfrac{\partial f_{2,2}}{\partial v_1} - f_{1,2}\dfrac{\partial f_{2,1}}{\partial v_1} = \dfrac{1}{3}\dfrac{\partial \Delta}{\partial v_1} - \dfrac{1}{3}\Delta P_1, \end{cases}$$

we have

$$\begin{cases} \dfrac{\partial f_{2,2}}{\partial v_1} = \dfrac{1}{3}\dfrac{f_{2,2}}{\Delta}\dfrac{\partial \Delta}{\partial v_1} - f_{1,2}F_1 - \dfrac{1}{3}f_{2,2}P_1, \\ \dfrac{\partial f_{2,1}}{\partial v_1} = \dfrac{1}{3}\dfrac{f_{2,1}}{\Delta}\dfrac{\partial \Delta}{\partial v_1} - f_{1,1}F_1 - \dfrac{1}{3}f_{2,1}P_1. \end{cases}$$

Differentiating the equation

$$\frac{1}{3}\left(\frac{\partial \Delta}{\partial v_1} - \Delta P_1\right) = f_{2,2}\frac{\partial f_{2,1}}{\partial v_2} - f_{2,1}\frac{\partial f_{2,2}}{\partial v_2}$$

with respect to v_1, we get

$$\begin{aligned} &\frac{1}{3}\left(\frac{\partial^2 \Delta}{\partial v_1^2} - \frac{\partial \Delta}{\partial v_1}P_1 - \Delta\frac{\partial P_1}{\partial v_1}\right) \\ =&\frac{\partial f_{2,2}}{\partial v_1}\frac{\partial f_{2,1}}{\partial v_2} - \frac{\partial f_{2,1}}{\partial v_1}\frac{\partial f_{2,2}}{\partial v_2} + f_{2,2}\frac{\partial^2 f_{2,1}}{\partial v_1 \partial v_2} - f_{2,1}\frac{\partial^2 f_{2,2}}{\partial v_1 \partial v_2}. \end{aligned}$$

Differentiating the equation

$$\Delta F_1 = f_{2,1}\frac{\partial f_{2,2}}{\partial v_1} - f_{2,2}\frac{\partial f_{2,1}}{\partial v_1}$$

with respect to v_2, we get

$$\frac{\partial \Delta}{\partial v_2}F_1 + \Delta\frac{\partial F_1}{\partial v_2} = \frac{\partial f_{2,1}}{\partial v_2}\frac{\partial f_{2,2}}{\partial v_1} - \frac{\partial f_{2,2}}{\partial v_2}\frac{\partial f_{2,1}}{\partial v_1} + f_{2,1}\frac{\partial^2 f_{2,2}}{\partial v_1 \partial v_2} - f_{2,2}\frac{\partial^2 f_{2,1}}{\partial v_1 \partial v_2}.$$

Combine the above two equalities, we find that

$$\frac{1}{3}\left(\frac{\partial^2 \Delta}{\partial v_1^2} - \frac{\partial \Delta}{\partial v_1}P_1 - \Delta\frac{\partial P_1}{\partial v_1}\right) + \frac{\partial \Delta}{\partial v_2}F_1 + \Delta\frac{\partial F_1}{\partial v_2} = 2\left(\frac{\partial f_{2,2}}{\partial v_1}\frac{\partial f_{2,1}}{\partial v_2} - \frac{\partial f_{2,1}}{\partial v_1}\frac{\partial f_{2,2}}{\partial v_2}\right).$$

A straightforward calculation gives that

$$\frac{\partial f_{2,2}}{\partial v_1}\frac{\partial f_{2,1}}{\partial v_2}-\frac{\partial f_{2,1}}{\partial v_1}\frac{\partial f_{2,2}}{\partial v_2}=\frac{1}{9}\frac{1}{\Delta}\left(\frac{\partial \Delta}{\partial v_1}\right)^2-\frac{2}{9}P_1\frac{\partial \Delta}{\partial v_1}+\frac{2}{3}F_1\frac{\partial \Delta}{\partial v_2}+\frac{1}{9}\Delta P_1^2+\frac{1}{3}\Delta F_1P_2.$$

This gives the following equality:

$$\begin{aligned}&\frac{1}{3}\frac{\partial^2\Delta}{\partial v_1^2}-\frac{2}{9}\frac{1}{\Delta}\left(\frac{\partial \Delta}{\partial v_1}\right)^2+\frac{1}{9}P_1\frac{\partial \Delta}{\partial v_1}-\frac{1}{3}F_1\frac{\partial \Delta}{\partial v_2}\\&+\left(\frac{\partial F_1}{\partial v_2}-\frac{1}{3}\frac{\partial P_1}{\partial v_1}-\frac{2}{9}P_1^2-\frac{2}{3}F_1P_2\right)\Delta=0,\end{aligned}$$

which is equivalent to

$$\frac{\partial^2(\sqrt[3]{\Delta})}{\partial v_1^2}+\frac{1}{3}P_1\frac{\partial(\sqrt[3]{\Delta})}{\partial v_1}-F_1\frac{\partial(\sqrt[3]{\Delta})}{\partial v_2}+\left(\frac{\partial F_1}{\partial v_2}-\frac{1}{3}\frac{\partial P_1}{\partial v_1}-\frac{2}{9}P_1^2-\frac{2}{3}F_1P_2\right)\sqrt[3]{\Delta}=0.$$

By the symmetry property, we get the other equation:

$$\frac{\partial^2(\sqrt[3]{\Delta})}{\partial v_2^2}+\frac{1}{3}P_2\frac{\partial(\sqrt[3]{\Delta})}{\partial v_2}-F_2\frac{\partial(\sqrt[3]{\Delta})}{\partial v_1}+\left(\frac{\partial F_2}{\partial v_1}-\frac{1}{3}\frac{\partial P_2}{\partial v_2}-\frac{2}{9}P_2^2-\frac{2}{3}F_2P_1\right)\sqrt[3]{\Delta}=0.$$

Here,

$$\begin{cases}\dfrac{\partial f_{1,1}}{\partial v_1}=\dfrac{1}{3}\dfrac{f_{2,1}}{\Delta}\dfrac{\partial \Delta}{\partial v_2}+\dfrac{2}{3}\dfrac{f_{1,1}}{\Delta}\dfrac{\partial \Delta}{\partial v_1}-\dfrac{1}{3}f_{2,1}P_2+\dfrac{1}{3}f_{1,1}P_1,\\[2mm]\dfrac{\partial f_{1,2}}{\partial v_1}=\dfrac{1}{3}\dfrac{f_{2,2}}{\Delta}\dfrac{\partial \Delta}{\partial v_2}+\dfrac{2}{3}\dfrac{f_{1,2}}{\Delta}\dfrac{\partial \Delta}{\partial v_1}-\dfrac{1}{3}f_{2,2}P_2+\dfrac{1}{3}f_{1,2}P_1.\end{cases}$$

$$\begin{cases}\dfrac{\partial f_{1,1}}{\partial v_2}=\dfrac{1}{3}\dfrac{f_{1,1}}{\Delta}\dfrac{\partial \Delta}{\partial v_2}-f_{2,1}F_2-\dfrac{1}{3}f_{1,1}P_2,\\[2mm]\dfrac{\partial f_{1,2}}{\partial v_2}=\dfrac{1}{3}\dfrac{f_{1,2}}{\Delta}\dfrac{\partial \Delta}{\partial v_2}-f_{2,2}F_2-\dfrac{1}{3}f_{1,2}P_2.\end{cases}$$

Differentiating the equation

$$f_{2,2}\frac{\partial f_{2,1}}{\partial v_2}-f_{2,1}\frac{\partial f_{2,2}}{\partial v_2}=\frac{1}{3}\left(\frac{\partial \Delta}{\partial v_1}-\Delta P_1\right)$$

with respect to v_2, we get

$$f_{2,2}\frac{\partial^2 f_{2,1}}{\partial v_2^2}-f_{2,1}\frac{\partial^2 f_{2,2}}{\partial v_2^2}=\frac{1}{3}\left(\frac{\partial^2\Delta}{\partial v_1\partial v_2}-\frac{\partial \Delta}{\partial v_2}P_1-\Delta\frac{\partial P_1}{\partial v_2}\right).$$

Differentiating the equation

$$f_{1,1}\frac{\partial f_{2,2}}{\partial v_2}-f_{1,2}\frac{\partial f_{2,1}}{\partial v_2}=\frac{2}{3}\frac{\partial \Delta}{\partial v_2}+\frac{1}{3}\Delta P_2$$

with respect to v_1, we get

$$\begin{aligned}&\frac{\partial f_{1,1}}{\partial v_1}\frac{\partial f_{2,2}}{\partial v_2}-\frac{\partial f_{1,2}}{\partial v_1}\frac{\partial f_{2,1}}{\partial v_2}+f_{1,1}\frac{\partial^2 f_{2,2}}{\partial v_1\partial v_2}-f_{1,2}\frac{\partial^2 f_{2,1}}{\partial v_1\partial v_2}\\=&\frac{2}{3}\frac{\partial^2\Delta}{\partial v_1\partial v_2}+\frac{1}{3}\frac{\partial\Delta}{\partial v_1}P_2+\frac{1}{3}\Delta\frac{\partial P_2}{\partial v_1}.\end{aligned}$$

Differentiating the equation

$$f_{1,1}\frac{\partial f_{2,2}}{\partial v_1}-f_{1,2}\frac{\partial f_{2,1}}{\partial v_1}=f_{2,2}\frac{\partial f_{2,1}}{\partial v_2}-f_{2,1}\frac{\partial f_{2,2}}{\partial v_2}$$

with respect to v_2, we get

$$f_{2,2}\frac{\partial^2 f_{2,1}}{\partial v_2^2}-f_{2,1}\frac{\partial^2 f_{2,2}}{\partial v_2^2}=\frac{\partial f_{1,1}}{\partial v_2}\frac{\partial f_{2,2}}{\partial v_1}-\frac{\partial f_{1,2}}{\partial v_2}\frac{\partial f_{2,1}}{\partial v_1}+f_{1,1}\frac{\partial^2 f_{2,2}}{\partial v_1\partial v_2}-f_{1,2}\frac{\partial^2 f_{2,1}}{\partial v_1\partial v_2}.$$

Combining the above three equalities, we find that

$$\begin{aligned}&\frac{1}{3}\frac{\partial^2\Delta}{\partial v_1\partial v_2}+\frac{1}{3}\frac{\partial\Delta}{\partial v_1}P_2+\frac{1}{3}\Delta\frac{\partial P_2}{\partial v_1}+\frac{1}{3}\frac{\partial\Delta}{\partial v_2}P_1+\frac{1}{3}\Delta\frac{\partial P_1}{\partial v_2}\\=&\frac{\partial f_{1,2}}{\partial v_2}\frac{\partial f_{2,1}}{\partial v_1}-\frac{\partial f_{1,1}}{\partial v_2}\frac{\partial f_{2,2}}{\partial v_1}+\frac{\partial f_{1,1}}{\partial v_1}\frac{\partial f_{2,2}}{\partial v_2}-\frac{\partial f_{1,2}}{\partial v_1}\frac{\partial f_{2,1}}{\partial v_2}.\end{aligned}$$

A straightforward calculation gives that the right hand side of the above equation is equal to

$$\frac{2}{9}\frac{1}{\Delta}\frac{\partial\Delta}{\partial v_1}\frac{\partial\Delta}{\partial v_2}+\frac{4}{9}P_2\frac{\partial\Delta}{\partial v_1}+\frac{4}{9}P_1\frac{\partial\Delta}{\partial v_2}+\Delta F_1F_2-\frac{1}{9}\Delta P_1P_2.$$

This implies that

$$\begin{aligned}&\frac{1}{3}\frac{\partial^2\Delta}{\partial v_1\partial v_2}-\frac{2}{9}\frac{1}{\Delta}\frac{\partial\Delta}{\partial v_1}\frac{\partial\Delta}{\partial v_2}-\frac{1}{9}P_2\frac{\partial\Delta}{\partial v_1}-\frac{1}{9}P_1\frac{\partial\Delta}{\partial v_2}\\&+\left(\frac{1}{3}\frac{\partial P_2}{\partial v_1}+\frac{1}{3}\frac{\partial P_1}{\partial v_2}+\frac{1}{9}P_1P_2-F_1F_2\right)\Delta=0,\end{aligned}$$

which is equivalent to

$$\begin{aligned}&\frac{\partial^2(\sqrt[3]{\Delta})}{\partial v_1\partial v_2}-\frac{1}{3}P_2\frac{\partial(\sqrt[3]{\Delta})}{\partial v_1}-\frac{1}{3}P_1\frac{\partial(\sqrt[3]{\Delta})}{\partial v_2}\\&+\left(\frac{1}{3}\frac{\partial P_2}{\partial v_1}+\frac{1}{3}\frac{\partial P_1}{\partial v_2}+\frac{1}{9}P_1P_2-F_1F_2\right)\sqrt[3]{\Delta}=0.\end{aligned}$$

□

Now, we will study the equations:

$$\begin{cases}\{w_1,w_2;v_1,v_2\}_{v_1}=F_1(v_1,v_2),\\\{w_1,w_2;v_1,v_2\}_{v_2}=F_2(v_1,v_2),\\{}[w_1,w_2;v_1,v_2]_{v_1}=P_1(v_1,v_2),\\{}[w_1,w_2;v_1,v_2]_{v_2}=P_2(v_1,v_2).\end{cases}$$

Here $F_1(v_1, v_2)$ is finite except at $v_1 = 0,\ 1,\ v_2,\ \infty$, where respectively, the products

$$v_1^{-\frac{3}{4}} \begin{vmatrix} \frac{\partial v_1}{\partial w_1} & \frac{\partial v_2}{\partial w_1} \\ \frac{\partial v_1}{\partial w_2} & \frac{\partial v_2}{\partial w_2} \end{vmatrix}, \quad (1-v_1)^{-\frac{1}{2}} \begin{vmatrix} \frac{\partial v_1}{\partial w_1} & \frac{\partial v_2}{\partial w_1} \\ \frac{\partial v_1}{\partial w_2} & \frac{\partial v_2}{\partial w_2} \end{vmatrix},$$

$$(v_2 - v_1)^{-\frac{1}{2}} \begin{vmatrix} \frac{\partial v_1}{\partial w_1} & \frac{\partial v_2}{\partial w_1} \\ \frac{\partial v_1}{\partial w_2} & \frac{\partial v_2}{\partial w_2} \end{vmatrix}, \quad v_1^{-1} \begin{vmatrix} \frac{\partial v_1}{\partial w_1} & \frac{\partial v_2}{\partial w_1} \\ \frac{\partial v_1}{\partial w_2} & \frac{\partial v_2}{\partial w_2} \end{vmatrix}$$

are finite and non-zero. Similarly, $F_2(v_1, v_2)$ is finite except at $v_2 = 0,\ 1,\ v_1,\ \infty$, where respectively, the products

$$v_2^{-\frac{3}{4}} \begin{vmatrix} \frac{\partial v_2}{\partial w_1} & \frac{\partial v_1}{\partial w_1} \\ \frac{\partial v_2}{\partial w_2} & \frac{\partial v_1}{\partial w_2} \end{vmatrix}, \quad (1-v_2)^{-\frac{1}{2}} \begin{vmatrix} \frac{\partial v_2}{\partial w_1} & \frac{\partial v_1}{\partial w_1} \\ \frac{\partial v_2}{\partial w_2} & \frac{\partial v_1}{\partial w_2} \end{vmatrix},$$

$$(v_1 - v_2)^{-\frac{1}{2}} \begin{vmatrix} \frac{\partial v_2}{\partial w_1} & \frac{\partial v_1}{\partial w_1} \\ \frac{\partial v_2}{\partial w_2} & \frac{\partial v_1}{\partial w_2} \end{vmatrix}, \quad v_2^{-1} \begin{vmatrix} \frac{\partial v_2}{\partial w_1} & \frac{\partial v_1}{\partial w_1} \\ \frac{\partial v_2}{\partial w_2} & \frac{\partial v_1}{\partial w_2} \end{vmatrix}$$

are finite and non-zero.

Set $v_1^{-\frac{1}{4}} \sqrt[3]{\Delta} = y$, which is finite and non-zero at $v_1 = 0$. Then

$$v_1^2 \frac{\partial^2 y}{\partial v_1^2} + \left(\frac{1}{3}P_1 v_1^2 + \frac{1}{2}v_1\right) \frac{\partial y}{\partial v_1} - F_1 v_1^2 \frac{\partial y}{\partial v_2} + \left[\left(\frac{\partial F_1}{\partial v_2} - \frac{1}{3}\frac{\partial P_1}{\partial v_1} - \frac{2}{9}P_1^2 - \frac{2}{3}F_1 P_2\right) v_1^2 + \frac{1}{12}P_1 v_1 - \frac{3}{16}\right] y = 0$$

and

$$v_1 \frac{\partial^2 y}{\partial v_1 \partial v_2} - \frac{1}{3}P_2 v_1 \frac{\partial y}{\partial v_1} + \left(-\frac{1}{3}P_1 v_1 + \frac{1}{4}\right) \frac{\partial y}{\partial v_2} + \left[\left(\frac{1}{3}\frac{\partial P_2}{\partial v_1} + \frac{1}{3}\frac{\partial P_1}{\partial v_2} + \frac{1}{9}P_1 P_2 - F_1 F_2\right) v_1 - \frac{1}{12}P_2\right] y = 0$$

at $v_1 = 0$.

Set $(1 - v_1)^{-\frac{1}{6}} \sqrt[3]{\Delta} = y$, which is finite and non-zero at $v_1 = 1$. Then

$$(1-v_1)^2 \frac{\partial^2 y}{\partial v_1^2} + \left[\frac{1}{3}P_1(1-v_1)^2 - \frac{1}{3}(1-v_1)\right] \frac{\partial y}{\partial v_1} - F_1(1-v_1)^2 \frac{\partial y}{\partial v_2} + \left[\left(\frac{\partial F_1}{\partial v_2} - \frac{1}{3}\frac{\partial P_1}{\partial v_1} - \frac{2}{9}P_1^2 - \frac{2}{3}F_1 P_2\right)(1-v_1)^2 - \frac{1}{18}P_1(1-v_1) - \frac{5}{36}\right] y = 0$$

and

$$(1-v_1) \frac{\partial^2 y}{\partial v_1 \partial v_2} - \frac{1}{3}P_2(1-v_1) \frac{\partial y}{\partial v_1} + \left[-\frac{1}{3}P_1(1-v_1) - \frac{1}{6}\right] \frac{\partial y}{\partial v_2} + \left[\left(\frac{1}{3}\frac{\partial P_2}{\partial v_1} + \frac{1}{3}\frac{\partial P_1}{\partial v_2} + \frac{1}{9}P_1 P_2 - F_1 F_2\right)(1-v_1) + \frac{1}{18}P_2\right] y = 0$$

at $v_1 = 1$.

Set $(v_2 - v_1)^{-\frac{1}{6}} \sqrt[3]{\Delta} = y$, which is finite and non-zero at $v_1 = v_2$. Then

$$(v_2 - v_1)^2 \frac{\partial^2 y}{\partial v_1^2} + \left[\frac{1}{3}P_1(v_2 - v_1)^2 - \frac{1}{3}(v_2 - v_1)\right] \frac{\partial y}{\partial v_1} - F_1(v_2 - v_1)^2 \frac{\partial y}{\partial v_2}$$
$$+ \left[\left(\frac{\partial F_1}{\partial v_2} - \frac{1}{3}\frac{\partial P_1}{\partial v_1} - \frac{2}{9}P_1^2 - \frac{2}{3}F_1P_2\right)(v_2 - v_1)^2 - \frac{1}{18}P_1(v_2 - v_1)\right.$$
$$\left. - \frac{1}{6}F_1(v_2 - v_1) - \frac{5}{36}\right] y = 0$$

at $v_1 = v_2$.

Set $v_1^{-\frac{1}{3}} \sqrt[3]{\Delta} = y$, which is finite and non-zero at $v_1 = \infty$. Then

$$v_1^2 \frac{\partial^2 y}{\partial v_1^2} + \left(\frac{1}{3}P_1 v_1^2 + \frac{2}{3}v_1\right) \frac{\partial y}{\partial v_1} - F_1 v_1^2 \frac{\partial y}{\partial v_2}$$
$$+ \left[\left(\frac{\partial F_1}{\partial v_2} - \frac{1}{3}\frac{\partial P_1}{\partial v_1} - \frac{2}{9}P_1^2 - \frac{2}{3}F_1P_2\right) v_1^2 + \frac{1}{9}P_1 v_1 - \frac{2}{9}\right] y = 0$$

and

$$v_1 \frac{\partial^2 y}{\partial v_1 \partial v_2} - \frac{1}{3}P_2 v_1 \frac{\partial y}{\partial v_1} + \left(-\frac{1}{3}P_1 v_1 + \frac{1}{3}\right) \frac{\partial y}{\partial v_2}$$
$$+ \left[\left(\frac{1}{3}\frac{\partial P_2}{\partial v_1} + \frac{1}{3}\frac{\partial P_1}{\partial v_2} + \frac{1}{9}P_1P_2 - F_1F_2\right) v_1 - \frac{1}{9}P_2\right] y = 0$$

at $v_1 = \infty$. Put $u_1 = \frac{1}{v_1}$, then

$$u_1^2 \frac{\partial^2 y}{\partial u_1^2} + \left(\frac{4}{3}u_1 - \frac{1}{3}P_1\right) \frac{\partial y}{\partial u_1} - \frac{F_1}{u_1^2}\frac{\partial y}{\partial v_2}$$
$$+ \left[\left(\frac{\partial F_1}{\partial v_2} + \frac{1}{3}u_1^2 \frac{\partial P_1}{\partial u_1} - \frac{2}{9}P_1^2 - \frac{2}{3}F_1P_2\right) \frac{1}{u_1^2} + \frac{1}{9}\frac{P_1}{u_1} - \frac{2}{9}\right] y = 0$$

and

$$- u_1 \frac{\partial^2 y}{\partial u_1 \partial v_2} + \frac{1}{3}P_2 u_1 \frac{\partial y}{\partial u_1} + \left(-\frac{1}{3}\frac{P_1}{u_1} + \frac{1}{3}\right) \frac{\partial y}{\partial v_2}$$
$$+ \left[\left(-\frac{1}{3}u_1^2 \frac{\partial P_2}{\partial u_1} + \frac{1}{3}\frac{\partial P_1}{\partial v_2} + \frac{1}{9}P_1P_2 - F_1F_2\right) \frac{1}{u_1} - \frac{1}{9}P_2\right] y = 0$$

at $u_1 = 0$. Here, $F_1 = F_1(\frac{1}{u_1}, v_2)$, $F_2 = F_2(\frac{1}{u_1}, v_2)$, $P_1 = P_1(\frac{1}{u_1}, v_2)$ and $P_2 = P_2(\frac{1}{u_1}, v_2)$.

Since

$$\begin{cases} \frac{1}{3}P_1 v_1^2 + \frac{1}{2}v_1 = 0 \quad \text{at} \quad v_1 = 0, \\ \frac{1}{3}P_1(1 - v_1)^2 - \frac{1}{3}(1 - v_1) = 0 \quad \text{at} \quad v_1 = 1, \\ \frac{1}{3}P_1(v_2 - v_1)^2 - \frac{1}{3}(v_2 - v_1) = 0 \quad \text{at} \quad v_1 = v_2, \\ \frac{4}{3}u_1 - \frac{1}{3}P_1 = 0 \quad \text{at} \quad u_1 = 0, \end{cases}$$

we have

$$P_1(v_1) = \frac{p_1(v_1)}{v_1(1-v_1)(v_2-v_1)}$$

and

$$\frac{4}{3}u_1 - \frac{1}{3}\frac{u_1^3 p_1\left(\frac{1}{u_1}\right)}{(u_1-1)(u_1v_2-1)} = 0 \quad \text{at} \quad u_1 = 0.$$

Hence, $p_1\left(\frac{1}{u_1}\right) = \frac{a_1}{u_1^2} + \frac{b_1}{u_1} + c_1$ and

$$P_1(v_1) = \frac{a_1v_1^2 + b_1v_1 + c_1}{v_1(1-v_1)(v_2-v_1)}.$$

Moreover,

$$\begin{cases} -\frac{1}{3}P_1v_1 + \frac{1}{4} = 0 & \text{at} \quad v_1 = 0, \\ -\frac{1}{3}P_1(1-v_1) - \frac{1}{6} = 0 & \text{at} \quad v_1 = 1, \\ -\frac{1}{3}\frac{P_1}{u_1} + \frac{1}{3} = 0 & \text{at} \quad u_1 = 0. \end{cases}$$

This implies that

$$c_1 = \frac{3}{4}v_2, \quad a_1 + b_1 + c_1 = -\frac{1}{2}(v_2 - 1), \quad a_1 = 1.$$

Thus,

$$a_1 = 1, \quad b_1 = -\frac{5}{4}v_2 - \frac{1}{2}, \quad c_1 = \frac{3}{4}v_2.$$

Hence,

$$P_1(v_1, v_2) = \frac{\frac{3}{4}}{v_1} + \frac{\frac{1}{2}}{v_1 - 1} + \frac{-\frac{1}{4}}{v_1 - v_2}.$$

For

$$\begin{cases} F_1v_1^2 = 0 & \text{at} \quad v_1 = 0, \\ F_1(1-v_1)^2 = 0 & \text{at} \quad v_1 = 1, \\ F_1(v_2-v_1)^2 = 0 & \text{at} \quad v_1 = v_2, \\ F_1\frac{1}{u_1^2} = 0 & \text{at} \quad u_1 = 0, \end{cases}$$

we have

$$F_1(v_1) = \frac{f_1(v_1)}{v_1(1-v_1)(v_2-v_1)}$$

and

$$\frac{u_1 f_1\left(\frac{1}{u_1}\right)}{(u_1-1)(u_1v_2-1)} = 0 \quad \text{at} \quad u_1 = 0.$$

Hence,

$$F_1(v_1, v_2) = \frac{A(v_2)}{v_1(v_1-1)(v_1-v_2)}.$$

Similarly, we have

$$P_2(v_2) = \frac{a_2 v_2^2 + b_2 v_2 + c_2}{v_2(v_2-1)(v_2-v_1)}.$$

Here, $a_2 = 1$, $b_2 = -\frac{5}{4}v_1 - \frac{1}{2}$ and $c_2 = \frac{3}{4}v_1$. Hence,

$$P_2(v_1, v_2) = \frac{\frac{3}{4}}{v_2} + \frac{\frac{1}{2}}{v_2 - 1} + \frac{-\frac{1}{4}}{v_2 - v_1}.$$

On the other hand,

$$F_2(v_1, v_2) = \frac{B(v_1)}{v_2(v_2-1)(v_2-v_1)}.$$

The equation

$$\left(\frac{\partial F_1}{\partial v_2} - \frac{1}{3}\frac{\partial P_1}{\partial v_1} - \frac{2}{9}P_1^2 - \frac{2}{3}F_1 P_2\right)(v_2 - v_1)^2 - \frac{1}{18}P_1(v_2 - v_1) - \frac{1}{6}F_1(v_2 - v_1) - \frac{5}{36} = 0$$

at $v_1 = v_2$ gives that

$$A(v_2) = \frac{1}{4}v_2(v_2 - 1).$$

Hence,

$$F_1(v_1, v_2) = \frac{\frac{1}{4}v_2(v_2-1)}{v_1(v_1-1)(v_1-v_2)}.$$

Similarly,

$$F_2(v_1, v_2) = \frac{\frac{1}{4}v_1(v_1-1)}{v_2(v_2-1)(v_2-v_1)}.$$

This leads us to study the following equations:

Theorem 3.3 (Main Theorem 2). *For the system of linear partial differential equations:*

$$\begin{cases} \dfrac{\partial^2 z}{\partial v_1^2} + \dfrac{1}{3}P_1\dfrac{\partial z}{\partial v_1} - F_1\dfrac{\partial z}{\partial v_2} + \left(\dfrac{\partial F_1}{\partial v_2} - \dfrac{1}{3}\dfrac{\partial P_1}{\partial v_1} - \dfrac{2}{9}P_1^2 - \dfrac{2}{3}F_1 P_2\right) z = 0, \\ \dfrac{\partial^2 z}{\partial v_1 \partial v_2} - \dfrac{1}{3}P_2\dfrac{\partial z}{\partial v_1} - \dfrac{1}{3}P_1\dfrac{\partial z}{\partial v_2} + \left(\dfrac{1}{3}\dfrac{\partial P_2}{\partial v_1} + \dfrac{1}{3}\dfrac{\partial P_1}{\partial v_2} + \dfrac{1}{9}P_1 P_2 - F_1 F_2\right) z = 0, \\ \dfrac{\partial^2 z}{\partial v_2^2} + \dfrac{1}{3}P_2\dfrac{\partial z}{\partial v_2} - F_2\dfrac{\partial z}{\partial v_1} + \left(\dfrac{\partial F_2}{\partial v_1} - \dfrac{1}{3}\dfrac{\partial P_2}{\partial v_2} - \dfrac{2}{9}P_2^2 - \dfrac{2}{3}F_2 P_1\right) z = 0, \end{cases} \tag{3.12}$$

put

$$P_1 = \frac{\alpha}{v_1} + \frac{\beta}{v_1 - 1} + \frac{\gamma}{v_1 - v_2}, \quad P_2 = \frac{\alpha}{v_2} + \frac{\beta}{v_2 - 1} + \frac{\gamma}{v_2 - v_1}, \tag{3.13}$$

$$F_1 = \frac{-\gamma v_2(v_2-1)}{v_1(v_1-1)(v_1-v_2)}, \quad F_2 = \frac{-\gamma v_1(v_1-1)}{v_2(v_2-1)(v_2-v_1)}. \tag{3.14}$$

Set

$$w = v_1^{-\frac{1}{3}\alpha} v_2^{-\frac{1}{3}\alpha}(v_1-1)^{-\frac{1}{3}\beta}(v_2-1)^{-\frac{1}{3}\beta}(v_1-v_2)^{\frac{2}{3}\gamma} z, \tag{3.15}$$

then the function $w = w(v_1, v_2)$ satisfies the following equations:

$$\begin{cases} \dfrac{\partial^2 w}{\partial v_1^2} + \left(\dfrac{\alpha}{v_1} + \dfrac{\beta}{v_1-1} + \dfrac{-\gamma}{v_1-v_2}\right)\dfrac{\partial w}{\partial v_1} + \dfrac{\gamma v_2(v_2-1)}{v_1(v_1-1)(v_1-v_2)}\dfrac{\partial w}{\partial v_2} \\ + \dfrac{(1-\alpha-\beta)\gamma}{v_1(v_1-1)} w = 0, \\ \dfrac{\partial^2 w}{\partial v_2^2} + \left(\dfrac{\alpha}{v_2} + \dfrac{\beta}{v_2-1} + \dfrac{-\gamma}{v_2-v_1}\right)\dfrac{\partial w}{\partial v_2} + \dfrac{\gamma v_1(v_1-1)}{v_2(v_2-1)(v_2-v_1)}\dfrac{\partial w}{\partial v_1} \\ + \dfrac{(1-\alpha-\beta)\gamma}{v_2(v_2-1)} w = 0, \\ \dfrac{\partial^2 w}{\partial v_1 \partial v_2} = \dfrac{-\gamma}{v_1-v_2}\dfrac{\partial w}{\partial v_1} + \dfrac{\gamma}{v_1-v_2}\dfrac{\partial w}{\partial v_2}. \end{cases} \tag{3.16}$$

A solution is given by

$$\begin{aligned} w = {} & const \cdot F_1(\alpha+\beta-1; -\gamma, -\gamma; \alpha-\gamma; v_1, v_2) \\ & + const \cdot F_1(\alpha+\beta-1; -\gamma, -\gamma; \beta-\gamma; 1-v_1, 1-v_2), \end{aligned} \tag{3.17}$$

where $F_1 = F_1(a; b, b'; c; x, y)$ is the two variable hypergeometric function (see [3]). Consequently,

$$\begin{aligned} z = {} & v_1^{\frac{1}{3}\alpha} v_2^{\frac{1}{3}\alpha}(v_1-1)^{\frac{1}{3}\beta}(v_2-1)^{\frac{1}{3}\beta}(v_1-v_2)^{-\frac{2}{3}\gamma} \\ & \times [const \cdot F_1(\alpha+\beta-1; -\gamma, -\gamma; \alpha-\gamma; v_1, v_2) \\ & + const \cdot F_1(\alpha+\beta-1; -\gamma, -\gamma; \beta-\gamma; 1-v_1, 1-v_2)]. \end{aligned} \tag{3.18}$$

Proof. Note that

$$z = v_1^{\frac{1}{3}\alpha} v_2^{\frac{1}{3}\alpha}(v_1-1)^{\frac{1}{3}\beta}(v_2-1)^{\frac{1}{3}\beta}(v_1-v_2)^{-\frac{2}{3}\gamma} w.$$

We have

$$\frac{\partial z}{\partial v_1} = v_1^{\frac{1}{3}\alpha} v_2^{\frac{1}{3}\alpha}(v_1-1)^{\frac{1}{3}\beta}(v_2-1)^{\frac{1}{3}\beta}(v_1-v_2)^{-\frac{2}{3}\gamma}\left[\frac{\partial w}{\partial v_1} + \left(\frac{1}{3}P_1 - \frac{\gamma}{v_1-v_2}\right) w\right].$$

$$\frac{\partial z}{\partial v_2} = v_1^{\frac{1}{3}\alpha} v_2^{\frac{1}{3}\alpha}(v_1-1)^{\frac{1}{3}\beta}(v_2-1)^{\frac{1}{3}\beta}(v_1-v_2)^{-\frac{2}{3}\gamma}\left[\frac{\partial w}{\partial v_2} + \left(\frac{1}{3}P_2 - \frac{\gamma}{v_2-v_1}\right) w\right].$$

$$\begin{aligned} \frac{\partial^2 z}{\partial v_1^2} = {} & v_1^{\frac{1}{3}\alpha} v_2^{\frac{1}{3}\alpha}(v_1-1)^{\frac{1}{3}\beta}(v_2-1)^{\frac{1}{3}\beta}(v_1-v_2)^{-\frac{2}{3}\gamma} \\ & \times \left[\frac{\partial^2 w}{\partial v_1^2} + 2\left(\frac{1}{3}P_1 - \frac{\gamma}{v_1-v_2}\right)\frac{\partial w}{\partial v_1}\right. \\ & \left. + \left(\left(\frac{1}{3}P_1 - \frac{\gamma}{v_1-v_2}\right)^2 + \frac{1}{3}\frac{\partial P_1}{\partial v_1} + \frac{\gamma}{(v_1-v_2)^2}\right) w\right]. \end{aligned}$$

$$\frac{\partial^2 z}{\partial v_2^2} = v_1^{\frac{1}{3}\alpha} v_2^{\frac{1}{3}\alpha}(v_1-1)^{\frac{1}{3}\beta}(v_2-1)^{\frac{1}{3}\beta}(v_1-v_2)^{-\frac{2}{3}\gamma}$$
$$\times\left[\frac{\partial^2 w}{\partial v_2^2} + 2\left(\frac{1}{3}P_2 - \frac{\gamma}{v_2-v_1}\right)\frac{\partial w}{\partial v_2}\right.$$
$$\left. + \left(\left(\frac{1}{3}P_2 - \frac{\gamma}{v_2-v_1}\right)^2 + \frac{1}{3}\frac{\partial P_2}{\partial v_2} + \frac{\gamma}{(v_2-v_1)^2}\right) w\right].$$

$$\frac{\partial^2 z}{\partial v_1 \partial v_2}$$
$$= v_1^{\frac{1}{3}\alpha} v_2^{\frac{1}{3}\alpha}(v_1-1)^{\frac{1}{3}\beta}(v_2-1)^{\frac{1}{3}\beta}(v_1-v_2)^{-\frac{2}{3}\gamma}$$
$$\times\left[\frac{\partial^2 w}{\partial v_1 \partial v_2} + \left(\frac{1}{3}P_2 - \frac{\gamma}{v_2-v_1}\right)\frac{\partial w}{\partial v_1} + \left(\frac{1}{3}P_1 - \frac{\gamma}{v_1-v_2}\right)\frac{\partial w}{\partial v_2}\right.$$
$$\left. + \left(\left(\frac{1}{3}P_1 - \frac{\gamma}{v_1-v_2}\right)\left(\frac{1}{3}P_2 - \frac{\gamma}{v_2-v_1}\right) + \frac{1}{3}\frac{\partial P_1}{\partial v_2} - \frac{\gamma}{(v_1-v_2)^2}\right) w\right].$$

The equation

$$\frac{\partial^2 z}{\partial v_1^2} + \frac{1}{3}P_1\frac{\partial z}{\partial v_1} - F_1\frac{\partial z}{\partial v_2} + \left(\frac{\partial F_1}{\partial v_2} - \frac{1}{3}\frac{\partial P_1}{\partial v_1} - \frac{2}{9}P_1^2 - \frac{2}{3}F_1P_2\right) z = 0$$

gives that

$$\frac{\partial^2 w}{\partial v_1^2} + \left(P_1 - \frac{2\gamma}{v_1-v_2}\right)\frac{\partial w}{\partial v_1} - F_1\frac{\partial w}{\partial v_2}$$
$$+ \left[-\frac{\gamma}{v_1-v_2}P_1 + \frac{\gamma^2+\gamma}{(v_1-v_2)^2} - F_1P_2 + \frac{\gamma}{v_2-v_1}F_1 + \frac{\partial F_1}{\partial v_2}\right] w = 0.$$

Here,

$$\frac{\partial F_1}{\partial v_2} = \frac{\gamma}{v_1(v_1-1)} - \frac{\gamma}{(v_1-v_2)^2}.$$

Thus,

$$-\frac{\gamma}{v_1-v_2}P_1 + \frac{\gamma^2+\gamma}{(v_1-v_2)^2} - F_1P_2 + \frac{\gamma}{v_2-v_1}F_1 + \frac{\partial F_1}{\partial v_2}$$
$$= \frac{\gamma}{v_1-v_2}\left(-P_1 + \frac{\gamma}{v_1-v_2}\right) + F_1\left(-P_2 + \frac{\gamma}{v_2-v_1}\right) + \frac{\gamma}{(v_1-v_2)^2} + \frac{\partial F_1}{\partial v_2}$$
$$= \frac{\gamma}{v_1-v_2}\left(-\frac{\alpha}{v_1} - \frac{\beta}{v_1-1}\right) + \frac{\gamma v_2(v_2-1)}{v_1(v_1-1)(v_1-v_2)}\left(\frac{\alpha}{v_2} + \frac{\beta}{v_2-1}\right) + \frac{\gamma}{v_1(v_1-1)}$$
$$= \frac{(1-\alpha-\beta)\gamma}{v_1(v_1-1)}.$$

Hence, we get the following equation:

$$\frac{\partial^2 w}{\partial v_1^2} + \left(\frac{\alpha}{v_1} + \frac{\beta}{v_1-1} + \frac{-\gamma}{v_1-v_2}\right)\frac{\partial w}{\partial v_1}$$
$$+ \frac{\gamma v_2(v_2-1)}{v_1(v_1-1)(v_1-v_2)}\frac{\partial w}{\partial v_2} + \frac{(1-\alpha-\beta)\gamma}{v_1(v_1-1)} w = 0.$$

Exchange the position of v_1 and v_2, we get the other equation.

The equation

$$\frac{\partial^2 z}{\partial v_1 \partial v_2} - \frac{1}{3}P_2\frac{\partial z}{\partial v_1} - \frac{1}{3}P_1\frac{\partial z}{\partial v_2} + \left(\frac{1}{3}\frac{\partial P_2}{\partial v_1} + \frac{1}{3}\frac{\partial P_1}{\partial v_2} + \frac{1}{9}P_1P_2 - F_1F_2\right) z = 0$$

gives that

$$\frac{\partial^2 w}{\partial v_1 \partial v_2} + \frac{\gamma}{v_1 - v_2}\frac{\partial w}{\partial v_1} + \frac{\gamma}{v_2 - v_1}\frac{\partial w}{\partial v_2}$$
$$+ \left[-\frac{\gamma^2+\gamma}{(v_1-v_2)^2} + \frac{2}{3}\frac{\partial P_1}{\partial v_2} + \frac{1}{3}\frac{\partial P_2}{\partial v_1} - F_1F_2\right] w = 0.$$

Note that

$$\frac{\partial P_1}{\partial v_2} = \frac{\partial P_2}{\partial v_1} = \frac{\gamma}{(v_1-v_2)^2}, \quad F_1F_2 = \frac{-\gamma^2}{(v_1-v_2)^2},$$

we find that

$$-\frac{\gamma^2+\gamma}{(v_1-v_2)^2} + \frac{2}{3}\frac{\partial P_1}{\partial v_2} + \frac{1}{3}\frac{\partial P_2}{\partial v_1} - F_1F_2 = 0.$$

Thus,

$$\frac{\partial^2 w}{\partial v_1 \partial v_2} + \frac{\gamma}{v_1 - v_2}\frac{\partial w}{\partial v_1} + \frac{\gamma}{v_2 - v_1}\frac{\partial w}{\partial v_2} = 0.$$

It is known that the equations

$$F_1: \begin{cases} x(1-x)\dfrac{\partial^2 z}{\partial x^2} + y(1-x)\dfrac{\partial^2 z}{\partial x \partial y} + [c-(a+b+1)x]\dfrac{\partial z}{\partial x} \\ \quad - by\dfrac{\partial z}{\partial y} - abz = 0, \\ y(1-y)\dfrac{\partial^2 z}{\partial y^2} + x(1-y)\dfrac{\partial^2 z}{\partial x \partial y} + [c-(a+b'+1)y]\dfrac{\partial z}{\partial y} \\ \quad - b'x\dfrac{\partial z}{\partial x} - ab'z = 0 \end{cases}$$

are equivalent to the following three equations:

$$\begin{cases} \dfrac{\partial^2 z}{\partial x^2} + \left(\dfrac{c-b'}{x} + \dfrac{a+b-c+1}{x-1} + \dfrac{b'}{x-y}\right)\dfrac{\partial z}{\partial x} \\ \quad - \dfrac{by(y-1)}{x(x-1)(x-y)}\dfrac{\partial z}{\partial y} + \dfrac{abz}{x(x-1)} = 0, \\ \dfrac{\partial^2 z}{\partial y^2} + \left(\dfrac{c-b}{y} + \dfrac{a+b'-c+1}{y-1} + \dfrac{b}{y-x}\right)\dfrac{\partial z}{\partial y} \\ \quad - \dfrac{b'x(x-1)}{y(y-1)(y-x)}\dfrac{\partial z}{\partial x} + \dfrac{ab'z}{y(y-1)} = 0, \\ \dfrac{\partial^2 z}{\partial x \partial y} = \dfrac{b'}{x-y}\dfrac{\partial z}{\partial x} - \dfrac{b}{x-y}\dfrac{\partial z}{\partial y}, \end{cases}$$

where $F_1 = F_1(a; b, b'; c; x, y)$ is the two variable hypergeometric function (see [3]). Here,

$$c - b' = c - b = \alpha, \quad a + b - c + 1 = a + b' - c + 1 = \beta,$$
$$b = b' = -\gamma, \quad ab = ab' = (1 - \alpha - \beta)\gamma.$$

Thus,

$$a = \alpha + \beta - 1, \quad b = -\gamma, \quad b' = -\gamma, \quad c = \alpha - \gamma.$$

A solution is given by

$$\begin{aligned} w = & \text{const} \cdot F_1(\alpha + \beta - 1; -\gamma, -\gamma; \alpha - \gamma; v_1, v_2) \\ & + \text{const} \cdot F_1(\alpha + \beta - 1; -\gamma, -\gamma; \beta - \gamma; 1 - v_1, 1 - v_2). \end{aligned}$$

Consequently,

$$\begin{aligned} z = & v_1^{\frac{1}{3}\alpha} v_2^{\frac{1}{3}\alpha} (v_1 - 1)^{\frac{1}{3}\beta} (v_2 - 1)^{\frac{1}{3}\beta} (v_1 - v_2)^{-\frac{2}{3}\gamma} \\ & \times [\text{const} \cdot F_1(\alpha + \beta - 1; -\gamma, -\gamma; \alpha - \gamma; v_1, v_2) \\ & + \text{const} \cdot F_1(\alpha + \beta - 1; -\gamma, -\gamma; \beta - \gamma; 1 - v_1, 1 - v_2)]. \end{aligned}$$

□

In our case, $\alpha = \frac{3}{4}$, $\beta = \frac{1}{2}$, $\gamma = -\frac{1}{4}$.

$$w = v_1^{-\frac{1}{4}} v_2^{-\frac{1}{4}} (v_1 - 1)^{-\frac{1}{6}} (v_2 - 1)^{-\frac{1}{6}} (v_1 - v_2)^{-\frac{1}{6}} \sqrt[3]{\Delta} \tag{3.19}$$

satisfies the following equations:

$$\begin{cases} \dfrac{\partial^2 w}{\partial v_1^2} + \left(\dfrac{\frac{3}{4}}{v_1} + \dfrac{\frac{1}{2}}{v_1 - 1} + \dfrac{\frac{1}{4}}{v_1 - v_2} \right) \dfrac{\partial w}{\partial v_1} \\ \quad + \dfrac{-\frac{1}{4} v_2 (v_2 - 1)}{v_1 (v_1 - 1)(v_1 - v_2)} \dfrac{\partial w}{\partial v_2} + \dfrac{\frac{1}{16}}{v_1 (v_1 - 1)} w = 0, \\ \dfrac{\partial^2 w}{\partial v_2^2} + \left(\dfrac{\frac{3}{4}}{v_2} + \dfrac{\frac{1}{2}}{v_2 - 1} + \dfrac{\frac{1}{4}}{v_2 - v_1} \right) \dfrac{\partial w}{\partial v_2} \\ \quad + \dfrac{-\frac{1}{4} v_1 (v_1 - 1)}{v_2 (v_2 - 1)(v_2 - v_1)} \dfrac{\partial w}{\partial v_1} + \dfrac{\frac{1}{16}}{v_2 (v_2 - 1)} w = 0, \\ \dfrac{\partial^2 w}{\partial v_1 \partial v_2} = \dfrac{\frac{1}{4}}{v_1 - v_2} \dfrac{\partial w}{\partial v_1} + \dfrac{-\frac{1}{4}}{v_1 - v_2} \dfrac{\partial w}{\partial v_2}. \end{cases} \tag{3.20}$$

A solution is given by

$$w = \text{const} \cdot F_1 \left(\frac{1}{4}; \frac{1}{4}, \frac{1}{4}; 1; v_1, v_2 \right) + \text{const} \cdot F_1 \left(\frac{1}{4}; \frac{1}{4}, \frac{1}{4}; \frac{3}{4}; 1 - v_1, 1 - v_2 \right). \tag{3.21}$$

Consequently,

$$\begin{aligned} \sqrt[3]{\Delta} = & v_1^{\frac{1}{4}} v_2^{\frac{1}{4}} (v_1 - 1)^{\frac{1}{6}} (v_2 - 1)^{\frac{1}{6}} (v_1 - v_2)^{\frac{1}{6}} \\ & \times \left[\text{const} \cdot F_1 \left(\frac{1}{4}; \frac{1}{4}, \frac{1}{4}; 1; v_1, v_2 \right) + \text{const} \cdot F_1 \left(\frac{1}{4}; \frac{1}{4}, \frac{1}{4}; \frac{3}{4}; 1 - v_1, 1 - v_2 \right) \right]. \end{aligned} \tag{3.22}$$

In fact, when (w_1, w_2) is a solution of the system of nonlinear partial differential equations:

$$\begin{cases} \{w_1, w_2; v_1, v_2\}_{v_1} = F_1(v_1, v_2), \\ \{w_1, w_2; v_1, v_2\}_{v_2} = F_2(v_1, v_2), \\ [w_1, w_2; v_1, v_2]_{v_1} = P_1(v_1, v_2), \\ [w_1, w_2; v_1, v_2]_{v_2} = P_2(v_1, v_2), \end{cases}$$

the general solution has the form

$$\left(\frac{a_1w_1 + a_2w_2 + a_3}{c_1w_1 + c_2w_2 + c_3}, \frac{b_1w_1 + b_2w_2 + b_3}{c_1w_1 + c_2w_2 + c_3}\right),$$

where $\begin{pmatrix} a_1 & a_2 & a_3 \\ b_1 & b_2 & b_3 \\ c_1 & c_2 & c_3 \end{pmatrix} \in GL(3, \mathbb{C})$.

4. η-functions associated to $U(2,1)$ and Picard curves

The η-function associated to the unitary group $U(2,1)$ is given as follows:

$$\eta(w_1, w_2) := v_1^{-\frac{1}{12}} v_2^{-\frac{1}{12}} (1-v_1)^{-\frac{1}{18}} (1-v_2)^{-\frac{1}{18}} (v_1 - v_2)^{-\frac{1}{18}} \left[\frac{\partial(v_1, v_2)}{\partial(w_1, w_2)}\right]^{\frac{1}{9}}. \quad (4.1)$$

It satisfies the following transformation:

$$\eta\left(\frac{a_1w_1 + a_2w_2 + a_3}{c_1w_1 + c_2w_2 + c_3}, \frac{b_1w_1 + b_2w_2 + b_3}{c_1w_1 + c_2w_2 + c_3}\right) = \Delta^{-\frac{1}{9}} (c_1w_1 + c_2w_2 + c_3)^{\frac{1}{3}} \eta(w_1, w_2), \quad (4.2)$$

where $\begin{pmatrix} a_1 & a_2 & a_3 \\ b_1 & b_2 & b_3 \\ c_1 & c_2 & c_3 \end{pmatrix} \in GL(3, \mathbb{C})$ and $\Delta = \begin{vmatrix} a_1 & a_2 & a_3 \\ b_1 & b_2 & b_3 \\ c_1 & c_2 & c_3 \end{vmatrix}$.

Let

$$T_1 = \begin{pmatrix} 1 & 1 & -\omega \\ & 1 & 1 \\ & & 1 \end{pmatrix}, \quad T_2 = \begin{pmatrix} 1 & \omega & -\omega \\ & 1 & \overline{\omega} \\ & & 1 \end{pmatrix}, \quad S = -\overline{\omega} J = \begin{pmatrix} & & -\overline{\omega} \\ & \overline{\omega} & \\ -\overline{\omega} & & \end{pmatrix},$$

$$U_1 = \begin{pmatrix} 1 & & \\ & -\omega & \\ & & 1 \end{pmatrix}, \quad U_2 = \begin{pmatrix} -1 & & \\ & -\omega & \\ & & -1 \end{pmatrix}.$$

The actions of the above five generators on $(w_1, w_2) \in \mathfrak{S}_2$ are given as follows:

$$\begin{cases} T_1(w_1, w_2) = (w_1 + w_2 - \omega, w_2 + 1), \\ T_2(w_1, w_2) = (w_1 + \omega w_2 - \omega, w_2 + \overline{\omega}), \\ S(w_1, w_2) = \left(\frac{1}{w_1}, -\frac{w_2}{w_1}\right), \\ U_1(w_1, w_2) = (w_1, -\omega w_2), \\ U_2(w_1, w_2) = (w_1, \omega w_2). \end{cases} \quad (4.3)$$

We have

$$\begin{aligned}
\eta(T_1(w_1,w_2)) &= 1^{-\frac{1}{9}}1^{\frac{1}{3}}\eta(w_1,w_2) = 1^{\frac{2}{9}}\eta(w_1,w_2) = e^{\frac{4}{9}\pi i}\eta(w_1,w_2).\\
\eta(T_2(w_1,w_2)) &= 1^{-\frac{1}{9}}1^{\frac{1}{3}}\eta(w_1,w_2) = 1^{\frac{2}{9}}\eta(w_1,w_2) = e^{\frac{4}{9}\pi i}\eta(w_1,w_2).\\
\eta(U_1(w_1,w_2)) &= (-\omega)^{-\frac{1}{9}}1^{\frac{1}{3}}\eta(w_1,w_2) = (e^{\frac{5}{3}\pi i})^{-\frac{1}{9}}e^{\frac{2}{3}\pi i}\eta(w_1,w_2)\\
&= e^{\frac{13}{27}\pi i}\eta(w_1,w_2) = 1^{\frac{13}{54}}\eta(w_1,w_2).\\
\eta(U_2(w_1,w_2)) &= (-\omega)^{-\frac{1}{9}}(-1)^{\frac{1}{3}}\eta(w_1,w_2) = (e^{\frac{5}{3}\pi i})^{-\frac{1}{9}}e^{\frac{1}{3}\pi i}\eta(w_1,w_2)\\
&= e^{\frac{4}{27}\pi i}\eta(w_1,w_2) = 1^{\frac{2}{27}}\eta(w_1,w_2).\\
\eta(S(w_1,w_2)) &= (-1)^{-\frac{1}{9}}(-\overline{\omega}w_1)^{\frac{1}{3}}\eta(w_1,w_2) = (e^{\pi i})^{-\frac{1}{9}}(e^{\frac{1}{3}\pi i}w_1)^{\frac{1}{3}}\eta(w_1,w_2)\\
&= w_1^{\frac{1}{3}}\eta(w_1,w_2).
\end{aligned}$$

Note that

$$[T_1,T_2] := T_1T_2T_1^{-1}T_2^{-1} = \begin{pmatrix} 1 & 0 & \overline{\omega}-\omega \\ & 1 & 0 \\ & & 1 \end{pmatrix}.$$

In fact,

$$\eta(T_1^{-1}(w_1,w_2)) = 1^{-\frac{2}{9}}\eta(w_1,w_2), \quad \eta(T_2^{-1}(w_1,w_2)) = 1^{-\frac{2}{9}}\eta(w_1,w_2).$$

Hence,

$$\eta([T_1,T_2](w_1,w_2)) = \eta(w_1,w_2).$$

In fact, Holzapfel (see [17]) proved the following theorem:

Theorem (see [17]). *The monodromy group of the hypergeometric differential equation system considered by Picard is the principal congruence subgroup $U(2,1;\mathcal{O}_K)(1-\omega)$ of the full Eisenstein lattice of the ball $U(2,1;\mathcal{O}_K)$ with respect to the principal ideal $\mathcal{O}_K\cdot(1-\omega)$, $\omega=\exp(2\pi i/3)$, $K=\mathbb{Q}(\sqrt{-3})$.*

According to Picard (see [33] or [17]), the five generators of the principal congruence subgroup $U(2,1;\mathcal{O}_K)(1-\omega)$ are given as follows:

$$g_1 = \begin{pmatrix} 1 & \overline{\omega}-\omega & 1-\overline{\omega} \\ 0 & \overline{\omega} & 1-\omega \\ 0 & 0 & 1 \end{pmatrix}, \quad g_2 = \begin{pmatrix} 1 & \overline{\omega}-1 & 1-\overline{\omega} \\ 0 & \overline{\omega} & 1-\overline{\omega} \\ 0 & 0 & 1 \end{pmatrix}, \quad g_3 = \begin{pmatrix} 1 & 0 & 0 \\ 0 & \omega & 0 \\ 0 & 0 & 1 \end{pmatrix}, \tag{4.4}$$

$$g_4 = \begin{pmatrix} -\omega & 0 & \overline{\omega}-1 \\ 0 & -1 & 0 \\ \overline{\omega}-1 & 0 & 2\omega \end{pmatrix}, \quad g_5 = \begin{pmatrix} 1 & 0 & 0 \\ \overline{\omega}-\omega & \overline{\omega} & 0 \\ 1-\overline{\omega} & 1-\omega & 1 \end{pmatrix}. \tag{4.5}$$

In fact,

$$\begin{cases} g_1 = U_1^{-4}T_1^{-1}T_2^{-2}, \\ g_2 = U_1^{-4}T_1^{-2}T_2^{-1}, \\ g_3 = U_1^4, \\ g_4 = S^3[T_1,T_2]S^3(S^4U_2)^{-1}[T_1,T_2], \\ g_5 = S^3U_1^{-4}T_1^{-1}T_2S^3. \end{cases} \tag{4.6}$$

For $g = \begin{pmatrix} & & \alpha \\ & \beta & \\ \gamma & & \end{pmatrix} \in U(2,1;\mathcal{O}_K)$, we have $\overline{\alpha}\gamma = \beta\overline{\beta} = \alpha\overline{\gamma} = 1$. Hence,

$$g = \begin{pmatrix} & & \alpha \\ & \beta & \\ \overline{\alpha}^{-1} & & \end{pmatrix}, \quad \alpha, \beta \in \{\pm 1, \pm\omega, \pm\overline{\omega}\}.$$

Note that

$$U_1^3 = \begin{pmatrix} & & 1 \\ & -1 & \\ 1 & & \end{pmatrix}, \quad (U_1U_2)^3 = \begin{pmatrix} & & -1 \\ & 1 & \\ -1 & & \end{pmatrix}.$$

We only need to consider the following nine cases:

$$I = U_1^6 = U_2^6 = S^6, \quad \begin{pmatrix} & & \omega \\ & 1 & \\ \omega & & \end{pmatrix} = S^2U_1^2, \quad \begin{pmatrix} & & \overline{\omega} \\ & 1 & \\ \overline{\omega} & & \end{pmatrix} = (S^2U_1^2)^2.$$

$$\begin{pmatrix} & & 1 \\ & \omega & \\ 1 & & \end{pmatrix} = U_1^4 = U_2^4, \quad \begin{pmatrix} & & \omega \\ & \omega & \\ \omega & & \end{pmatrix} = S^2, \quad \begin{pmatrix} & & \overline{\omega} \\ & \omega & \\ \overline{\omega} & & \end{pmatrix} = S^4U_1^2.$$

$$\begin{pmatrix} & & 1 \\ & \overline{\omega} & \\ 1 & & \end{pmatrix} = U_1^2 = U_2^2, \quad \begin{pmatrix} & & \omega \\ & \overline{\omega} & \\ \omega & & \end{pmatrix} = (S^4U_1^2)^2, \quad \begin{pmatrix} & & \overline{\omega} \\ & \overline{\omega} & \\ \overline{\omega} & & \end{pmatrix} = S^4.$$

Put

$$\eta_1(w_1, w_2) := \eta\left(\begin{pmatrix} 3 & & \\ & 1-\omega & \\ & & 1 \end{pmatrix}(w_1, w_2)\right) = \eta(3w_1, (1-\omega)w_2), \qquad (4.7)$$

$$\eta_2(w_1, w_2) := \eta\left(\begin{pmatrix} 1 & & \\ & 1-\omega & \\ & & 3 \end{pmatrix}(w_1, w_2)\right) = \eta\left(\frac{w_1}{3}, \frac{w_2}{1-\overline{\omega}}\right). \qquad (4.8)$$

Proposition 4.1.

$$\begin{cases} \eta_1(T_1(w_1, w_2)) = 1^{\frac{2}{3}}\eta_1(w_1, w_2), \\ \eta_2(T_1^3(w_1, w_2)) = 1^{\frac{2}{3}}\eta_2(w_1, w_2), \\ \eta_1(T_2(w_1, w_2)) = \eta_1(w_1, w_2), \\ \eta_2(T_2^3(w_1, w_2)) = \eta_2(w_1, w_2), \\ \eta_1(S(w_1, w_2)) = \sqrt[3]{\dfrac{w_1}{3}}\eta_2(w_1, w_2). \end{cases} \qquad (4.9)$$

Proof. We have

$$
\begin{aligned}
&\eta_1(T_1(w_1,w_2))\\
=&\eta\left(\begin{pmatrix}3 & & \\ & 1-\omega & \\ & & 1\end{pmatrix} T_1 \begin{pmatrix}\frac{1}{3} & & \\ & \frac{1}{1-\omega} & \\ & & 1\end{pmatrix}(3w_1,(1-\omega)w_2)\right)\\
=&\eta\left(\begin{pmatrix}1 & 1-\overline{\omega} & -3\omega \\ 0 & 1 & 1-\omega \\ 0 & 0 & 1\end{pmatrix}(3w_1,(1-\omega)w_2)\right)\\
=&\eta(T_1^2T_2[T_1,T_2]^{-1}(3w_1,(1-\omega)w_2))\\
=&1^{\frac{2}{3}}\eta(3w_1,(1-\omega)w_2)\\
=&1^{\frac{2}{3}}\eta_1(w_1,w_2).
\end{aligned}
$$

$$
\begin{aligned}
&\eta_2(T_1^3(w_1,w_2))\\
=&\eta\left(\begin{pmatrix}1 & & \\ & 1-\omega & \\ & & 3\end{pmatrix} T_1^3 \begin{pmatrix}1 & & \\ & \frac{1}{1-\omega} & \\ & & \frac{1}{3}\end{pmatrix}\left(\frac{w_1}{3},\frac{w_2}{1-\overline{\omega}}\right)\right)\\
=&\eta\left(\begin{pmatrix}1 & 1-\overline{\omega} & 1-\omega \\ 0 & 1 & 1-\omega \\ 0 & 0 & 1\end{pmatrix}\left(\frac{w_1}{3},\frac{w_2}{1-\overline{\omega}}\right)\right)\\
=&\eta\left(T_1^2T_2[T_1,T_2]^{-2}\left(\frac{w_1}{3},\frac{w_2}{1-\overline{\omega}}\right)\right)\\
=&1^{\frac{2}{3}}\eta\left(\frac{w_1}{3},\frac{w_2}{1-\overline{\omega}}\right)\\
=&1^{\frac{2}{3}}\eta_2(w_1,w_2).
\end{aligned}
$$

$$
\begin{aligned}
&\eta_1(T_2(w_1,w_2))\\
=&\eta\left(\begin{pmatrix}3 & & \\ & 1-\omega & \\ & & 1\end{pmatrix} T_2 \begin{pmatrix}\frac{1}{3} & & \\ & \frac{1}{1-\omega} & \\ & & 1\end{pmatrix}(3w_1,(1-\omega)w_2)\right)\\
=&\eta\left(\begin{pmatrix}1 & \omega-1 & -3\omega \\ 0 & 1 & \overline{\omega}-1 \\ 0 & 0 & 1\end{pmatrix}(3w_1,(1-\omega)w_2)\right)\\
=&\eta(T_1^{-1}T_2[T_1,T_2]^2(3w_1,(1-\omega)w_2))\\
=&\eta(3w_1,(1-\omega)w_2)\\
=&\eta_1(w_1,w_2).
\end{aligned}
$$

$$\begin{aligned}
&\eta_2(T_2^3(w_1,w_2))\\
=&\eta\left(\begin{pmatrix}1&&\\&1-\omega&\\&&3\end{pmatrix}T_2^3\begin{pmatrix}1&&\\&\frac{1}{1-\omega}&\\&&\frac{1}{3}\end{pmatrix}\left(\frac{w_1}{3},\frac{w_2}{1-\overline{\omega}}\right)\right)\\
=&\eta\left(\begin{pmatrix}1&\omega-1&1-\omega\\0&1&\overline{\omega}-1\\0&0&1\end{pmatrix}\left(\frac{w_1}{3},\frac{w_2}{1-\overline{\omega}}\right)\right)\\
=&\eta\left(T_1^{-1}T_2[T_1,T_2]\left(\frac{w_1}{3},\frac{w_2}{1-\overline{\omega}}\right)\right)\\
=&\eta\left(\frac{w_1}{3},\frac{w_2}{1-\overline{\omega}}\right)\\
=&\eta_2(w_1,w_2).
\end{aligned}$$

$$\begin{aligned}
\eta_1(S(w_1,w_2)) &= \eta\left(\frac{3}{w_1},(1-\omega)\left(-\frac{w_2}{w_1}\right)\right)\\
=&\eta\left(S\left(\frac{w_1}{3},\frac{w_2}{1-\overline{\omega}}\right)\right)=\sqrt[3]{\frac{w_1}{3}}\eta\left(\frac{w_1}{3},\frac{w_2}{1-\overline{\omega}}\right)\\
=&\sqrt[3]{\frac{w_1}{3}}\eta_2(w_1,w_2).
\end{aligned}$$

□

Put

$$\eta_3(w_1,w_2) := \eta_2([T_1,T_2]^{-1}(w_1,w_2)), \tag{4.10}$$

$$\eta_4(w_1,w_2) := \eta_1(S^3[T_1,T_2]^{-1}S^3(w_1,w_2)), \tag{4.11}$$

$$\eta_5(w_1,w_2) := \eta_2([T_1,T_2]^{-1}S^3[T_1,T_2]^{-1}S^3(w_1,w_2)). \tag{4.12}$$

Proposition 4.2.

$$\eta_3(g_1(w_1,w_2)) = 1^{\frac{7}{27}}\eta_2(w_1,w_2),\quad \eta_1(g_1(w_1,w_2)) = 1^{-\frac{5}{27}}\eta_1(w_1,w_2). \tag{4.13}$$

$$\eta_3(g_2(w_1,w_2)) = 1^{\frac{1}{27}}\eta_2(w_1,w_2),\quad \eta_1(g_2(w_1,w_2)) = 1^{-\frac{23}{27}}\eta_1(w_1,w_2). \tag{4.14}$$

$$\eta_1(g_3(w_1,w_2)) = 1^{\frac{26}{27}}\eta_1(w_1,w_2),\quad \eta_2(g_3(w_1,w_2)) = 1^{\frac{26}{27}}\eta_2(w_1,w_2). \tag{4.15}$$

$$\eta_4(g_4(w_1,w_2)) = 1^{\frac{19}{27}}\eta_1(w_1,w_2),\quad \eta_5(g_4(w_1,w_2)) = 1^{\frac{19}{27}}\eta_2(w_1,w_2). \tag{4.16}$$

$$\begin{aligned}
\eta_4(g_5(w_1,w_2)) &= 1^{\frac{7}{27}}\sqrt[3]{-3\overline{\omega}w_1+(1-\omega)w_2+1}\,\eta_1(w_1,w_2),\\
\eta_2(g_5(w_1,w_2)) &= 1^{-\frac{5}{27}}\sqrt[3]{(1-\overline{\omega})w_1+(1-\omega)w_2+1}\,\eta_2(w_1,w_2).
\end{aligned} \tag{4.17}$$

Proof.

$$
\begin{aligned}
&\eta_3(g_1(w_1,w_2))\\
=&\eta\left(\begin{pmatrix}1&&\\&1-\omega&\\&&3\end{pmatrix}[T_1,T_2]^{-1}g_1\begin{pmatrix}1&&\\&\frac{1}{1-\omega}&\\&&\frac{1}{3}\end{pmatrix}\left(\frac{w_1}{3},\frac{w_2}{1-\overline{\omega}}\right)\right)\\
=&\eta\left(\begin{pmatrix}1&-\omega&-\overline{\omega}\\0&\overline{\omega}&-\omega\\0&0&1\end{pmatrix}\left(\frac{w_1}{3},\frac{w_2}{1-\overline{\omega}}\right)\right)\\
=&\eta\left(U_1^2T_2^{-1}\left(\frac{w_1}{3},\frac{w_2}{1-\overline{\omega}}\right)\right)\\
=&1^{\frac{7}{27}}\eta_2(w_1,w_2).
\end{aligned}
$$

$$
\begin{aligned}
&\eta_1(g_1(w_1,w_2))\\
=&\eta\left(\begin{pmatrix}3&&\\&1-\omega&\\&&1\end{pmatrix}g_1\begin{pmatrix}\frac{1}{3}&&\\&\frac{1}{1-\omega}&\\&&1\end{pmatrix}(3w_1,(1-\omega)w_2)\right)\\
=&\eta\left(\begin{pmatrix}1&-3\omega&3(1-\overline{\omega})\\0&\overline{\omega}&-3\omega\\0&0&1\end{pmatrix}(3w_1,(1-\omega)w_2)\right)\\
=&\eta(U_1^2T_2^{-3}(3w_1,(1-\omega)w_2))\\
=&1^{-\frac{5}{27}}\eta_1(w_1,w_2).
\end{aligned}
$$

$$
\begin{aligned}
&\eta_3(g_2(w_1,w_2))\\
=&\eta\left(\begin{pmatrix}1&&\\&1-\omega&\\&&3\end{pmatrix}[T_1,T_2]^{-1}g_2\begin{pmatrix}1&&\\&\frac{1}{1-\omega}&\\&&\frac{1}{3}\end{pmatrix}\left(\frac{w_1}{3},\frac{w_2}{1-\overline{\omega}}\right)\right)\\
=&\eta\left(\begin{pmatrix}1&\overline{\omega}&-\overline{\omega}\\0&\overline{\omega}&1\\0&0&1\end{pmatrix}\left(\frac{w_1}{3},\frac{w_2}{1-\overline{\omega}}\right)\right)\\
=&\eta\left(U_1^2T_1^{-1}T_2^{-1}\left(\frac{w_1}{3},\frac{w_2}{1-\overline{\omega}}\right)\right)\\
=&1^{\frac{1}{27}}\eta_2(w_1,w_2).
\end{aligned}
$$

$$\begin{aligned}
&\eta_1(g_2(w_1,w_2))\\
=&\eta\left(\begin{pmatrix}3&&\\&1-\omega&\\&&1\end{pmatrix}g_2\begin{pmatrix}\frac{1}{3}&&\\&\frac{1}{1-\omega}&\\&&1\end{pmatrix}(3w_1,(1-\omega)w_2)\right)\\
=&\eta\left(\begin{pmatrix}1&3\overline{\omega}&3(1-\overline{\omega})\\0&\overline{\omega}&3\\0&0&1\end{pmatrix}(3w_1,(1-\omega)w_2)\right)\\
=&\eta(U_1^2T_1^{-3}T_2^{-3}[T_1,T_2]^{-3}(3w_1,(1-\omega)w_2))\\
=&1^{-\frac{23}{27}}\eta_1(w_1,w_2).
\end{aligned}$$

$$\begin{aligned}
&\eta_1(g_3(w_1,w_2))\\
=&\eta\left(\begin{pmatrix}3&&\\&1-\omega&\\&&1\end{pmatrix}g_3\begin{pmatrix}\frac{1}{3}&&\\&\frac{1}{1-\omega}&\\&&1\end{pmatrix}(3w_1,(1-\omega)w_2)\right)\\
=&\eta(U_1^4(3w_1,(1-\omega)w_2))\\
=&1^{\frac{26}{27}}\eta_1(w_1,w_2).
\end{aligned}$$

$$\begin{aligned}
&\eta_2(g_3(w_1,w_2))\\
=&\eta\left(\begin{pmatrix}1&&\\&1-\omega&\\&&3\end{pmatrix}g_3\begin{pmatrix}1&&\\&\frac{1}{1-\omega}&\\&&\frac{1}{3}\end{pmatrix}\left(\frac{w_1}{3},\frac{w_2}{1-\overline{\omega}}\right)\right)\\
=&\eta\left(U_1^4\left(\frac{w_1}{3},\frac{w_2}{1-\overline{\omega}}\right)\right)\\
=&1^{\frac{26}{27}}\eta_2(w_1,w_2).
\end{aligned}$$

$$\begin{aligned}
&\eta_4(g_4(w_1,w_2))\\
=&\eta\left(\begin{pmatrix}3&&\\&1-\omega&\\&&1\end{pmatrix}S^3[T_1,T_2]^{-1}S^3g_4\begin{pmatrix}\frac{1}{3}&&\\&\frac{1}{1-\omega}&\\&&1\end{pmatrix}(3w_1,(1-\omega)w_2)\right)\\
=&\eta\left(\begin{pmatrix}-\omega&0&3(\overline{\omega}-1)\\0&-1&0\\0&0&-\omega\end{pmatrix}(3w_1,(1-\omega)w_2)\right)\\
=&\eta(U_1^2U_2^3S^2[T_1,T_2]^3(3w_1,(1-\omega)w_2))\\
=&1^{\frac{19}{27}}\eta_1(w_1,w_2).
\end{aligned}$$

$$
\begin{aligned}
&\eta_5(g_4(w_1,w_2))\\
=&\eta\left(\begin{pmatrix}1&&\\&1-\omega&\\&&3\end{pmatrix}[T_1,T_2]^{-1}S^3[T_1,T_2]^{-1}S^3g_4\right.\\
&\left.\times\begin{pmatrix}1&&\\&\frac{1}{1-\omega}&\\&&\frac{1}{3}\end{pmatrix}\left(\frac{w_1}{3},\frac{w_2}{1-\overline{\omega}}\right)\right)\\
=&\eta\left(\begin{pmatrix}-\omega&&\\&-1&\\&&-\omega\end{pmatrix}\left(\frac{w_1}{3},\frac{w_2}{1-\overline{\omega}}\right)\right)\\
=&\eta\left(U_1^2U_2^3S^2\left(\frac{w_1}{3},\frac{w_2}{1-\overline{\omega}}\right)\right)\\
=&1^{\frac{19}{27}}\eta_2(w_1,w_2).
\end{aligned}
$$

$$
\begin{aligned}
&\eta_4(g_5(w_1,w_2))\\
=&\eta\left(\begin{pmatrix}3&&\\&1-\omega&\\&&1\end{pmatrix}S^3[T_1,T_2]^{-1}S^3g_5\begin{pmatrix}\frac{1}{3}&&\\&\frac{1}{1-\omega}&\\&&1\end{pmatrix}(3w_1,(1-\omega)w_2)\right)\\
=&\eta\left(\begin{pmatrix}1&0&0\\\overline{\omega}&\overline{\omega}&0\\-\overline{\omega}&1&1\end{pmatrix}(3w_1,(1-\omega)w_2)\right)\\
=&\eta(U_1^2S^3T_1^{-1}S^3(3w_1,(1-\omega)w_2))\\
=&1^{\frac{7}{27}}\sqrt[3]{-3\overline{\omega}w_1+(1-\omega)w_2+1}\,\eta_1(w_1,w_2).
\end{aligned}
$$

$$
\begin{aligned}
&\eta_2(g_5(w_1,w_2))\\
=&\eta\left(\begin{pmatrix}1&&\\&1-\omega&\\&&3\end{pmatrix}g_5\begin{pmatrix}1&&\\&\frac{1}{1-\omega}&\\&&\frac{1}{3}\end{pmatrix}\left(\frac{w_1}{3},\frac{w_2}{1-\overline{\omega}}\right)\right)\\
=&\eta\left(\begin{pmatrix}1&0&0\\3\overline{\omega}&\overline{\omega}&0\\3(1-\overline{\omega})&3&1\end{pmatrix}\left(\frac{w_1}{3},\frac{w_2}{1-\overline{\omega}}\right)\right)\\
=&\eta\left(U_1^2S^3T_1^{-3}S^3\left(\frac{w_1}{3},\frac{w_2}{1-\overline{\omega}}\right)\right)\\
=&1^{-\frac{5}{27}}\sqrt[3]{(1-\overline{\omega})w_1+(1-\omega)w_2+1}\,\eta_2(w_1,w_2).
\end{aligned}
$$

□

In fact,

$$
\begin{cases}
\eta_3([T_1,T_2](w_1,w_2))=\eta_2(w_1,w_2),\\
\eta_4(S^3[T_1,T_2]S^3(w_1,w_2))=\eta_1(w_1,w_2),\\
\eta_5(S^3[T_1,T_2]S^3[T_1,T_2](w_1,w_2))=\eta_2(w_1,w_2).
\end{cases}
\tag{4.18}
$$

Writing explicitly,

$$\eta_3(w_1,w_2) = \eta\left(\begin{pmatrix}1 & 0 & \omega-\overline{\omega}\\ & 1-\omega & 0\\ & & 3\end{pmatrix}(w_1,w_2)\right) = \eta\left(\frac{w_1+(\omega-\overline{\omega})}{3}, \frac{w_2}{1-\overline{\omega}}\right), \tag{4.19}$$

$$\begin{aligned}\eta_4(w_1,w_2) =&\eta\left(\begin{pmatrix}3 & 0 & 0\\ 0 & 1-\omega & 0\\ \omega-\overline{\omega} & 0 & 1\end{pmatrix}(w_1,w_2)\right)\\ =&\eta\left(\frac{3w_1}{(\omega-\overline{\omega})w_1+1}, \frac{(1-\omega)w_2}{(\omega-\overline{\omega})w_1+1}\right),\end{aligned} \tag{4.20}$$

$$\begin{aligned}\eta_5(w_1,w_2) =&\eta\left(\begin{pmatrix}-2 & 0 & \omega-\overline{\omega}\\ 0 & 1-\omega & 0\\ 3(\omega-\overline{\omega}) & 0 & 3\end{pmatrix}(w_1,w_2)\right)\\ =&\eta\left(\frac{-2w_1+(\omega-\overline{\omega})}{3(\omega-\overline{\omega})w_1+3}, \frac{(1-\omega)w_2}{3(\omega-\overline{\omega})w_1+3}\right).\end{aligned} \tag{4.21}$$

Proposition 4.3.

$$\eta_1([T_1,T_2]^{-1}(w_1,w_2)) = \eta_1(w_1,w_2). \tag{4.22}$$

$$\eta_2([T_1,T_2]^{-3}(w_1,w_2)) = \eta_2(w_1,w_2). \tag{4.23}$$

$$\eta_3([T_1,T_2]^{-3}(w_1,w_2)) = \eta_3(w_1,w_2). \tag{4.24}$$

$$\eta_4(S^3[T_1,T_2]^{-3}S^3(w_1,w_2)) = \sqrt[3]{\frac{4(\omega-\overline{\omega})w_1+1}{(\omega-\overline{\omega})w_1+1}}\,\eta_4(w_1,w_2). \tag{4.25}$$

$$\eta_5(S^3[T_1,T_2]S^3[T_1,T_2]^{-3}S^3[T_1,T_2]^{-1}S^3(w_1,w_2)) = \eta_5(w_1,w_2). \tag{4.26}$$

Proof. We have

$$\begin{aligned}&\eta_1([T_1,T_2]^{-1}(w_1,w_2))\\ =&\eta\left(\begin{pmatrix}3 & & \\ & 1-\omega & \\ & & 1\end{pmatrix}[T_1,T_2]^{-1}(w_1,w_2)\right)\\ =&\eta\left(\begin{pmatrix}3 & 0 & 3(\omega-\overline{\omega})\\ & 1-\omega & 0\\ & & 1\end{pmatrix}(w_1,w_2)\right)\\ =&\eta\left(\begin{pmatrix}1 & 0 & 3(\omega-\overline{\omega})\\ & 1 & 0\\ & & 1\end{pmatrix}\begin{pmatrix}3 & & \\ & 1-\omega & \\ & & 1\end{pmatrix}(w_1,w_2)\right)\\ =&\eta([T_1,T_2]^{-3}(3w_1,(1-\omega)w_2))\\ =&\eta(3w_1,(1-\omega)w_2)\\ =&\eta_1(w_1,w_2).\end{aligned}$$

$$
\begin{aligned}
&\eta_2([T_1,T_2]^{-3}(w_1,w_2))\\
=&\eta\left(\begin{pmatrix}1 & & \\ & 1-\omega & \\ & & 3\end{pmatrix}[T_1,T_2]^{-3}(w_1,w_2)\right)\\
=&\eta\left(\begin{pmatrix}1 & 0 & 3(\omega-\overline{\omega})\\ 0 & 1-\omega & 0\\ 0 & 0 & 3\end{pmatrix}(w_1,w_2)\right)\\
=&\eta\left(\begin{pmatrix}1 & 0 & \omega-\overline{\omega}\\ & 1 & 0\\ & & 1\end{pmatrix}\begin{pmatrix}1 & & \\ & 1-\omega & \\ & & 3\end{pmatrix}(w_1,w_2)\right)\\
=&\eta\left([T_1,T_2]^{-1}\left(\frac{w_1}{3},\frac{w_2}{1-\overline{\omega}}\right)\right)\\
=&\eta\left(\frac{w_1}{3},\frac{w_2}{1-\overline{\omega}}\right)\\
=&\eta_2(w_1,w_2).
\end{aligned}
$$

$$
\begin{aligned}
&\eta_3([T_1,T_2]^{-3}(w_1,w_2))\\
=&\eta_2([T_1,T_2]^{-1}[T_1,T_2]^{-3}(w_1,w_2))\\
=&\eta_2([T_1,T_2]^{-1}(w_1,w_2))\\
=&\eta_3(w_1,w_2).
\end{aligned}
$$

$$
\begin{aligned}
&\eta_4(S^3[T_1,T_2]^{-3}S^3(w_1,w_2))\\
=&\eta\left(\begin{pmatrix}3 & & \\ & 1-\omega & \\ & & 1\end{pmatrix}S^3[T_1,T_2]^{-4}S^3(w_1,w_2)\right)\\
=&\eta\left(\begin{pmatrix}3 & 0 & 0\\ 0 & 1-\omega & 0\\ 4(\omega-\overline{\omega}) & 0 & 1\end{pmatrix}(w_1,w_2)\right)\\
=&\eta\left(\begin{pmatrix}1 & 0 & 0\\ 0 & 1 & 0\\ \omega-\overline{\omega} & 0 & 1\end{pmatrix}\begin{pmatrix}3 & 0 & 0\\ 0 & 1-\omega & 0\\ \omega-\overline{\omega} & 0 & 1\end{pmatrix}(w_1,w_2)\right)\\
=&\eta\left(S^3[T_1,T_2]^{-1}S^3\begin{pmatrix}3 & 0 & 0\\ 0 & 1-\omega & 0\\ \omega-\overline{\omega} & 0 & 1\end{pmatrix}(w_1,w_2)\right)\\
=&\sqrt[3]{\frac{4(\omega-\overline{\omega})w_1+1}{(\omega-\overline{\omega})w_1+1}}\,\eta\left(\begin{pmatrix}3 & 0 & 0\\ 0 & 1-\omega & 0\\ \omega-\overline{\omega} & 0 & 1\end{pmatrix}(w_1,w_2)\right)\\
=&\sqrt[3]{\frac{4(\omega-\overline{\omega})w_1+1}{(\omega-\overline{\omega})w_1+1}}\,\eta_4(w_1,w_2).
\end{aligned}
$$

$$\begin{aligned}
&\eta_5(S^3[T_1,T_2]S^3[T_1,T_2]^{-3}S^3[T_1,T_2]^{-1}S^3(w_1,w_2))\\
=&\eta_2([T_1,T_2]^{-4}S^3[T_1,T_2]^{-1}S^3(w_1,w_2))\\
=&\eta_2([T_1,T_2]^{-1}S^3[T_1,T_2]^{-1}S^3(w_1,w_2))\\
=&\eta_5(w_1,w_2).
\end{aligned}$$

□

Set

$$\begin{aligned}
\varphi_0(w_1,w_2) &:= \frac{\eta_1(w_1,w_2)}{\eta_2(w_1,w_2)}, \quad \varphi_1(w_1,w_2) := \frac{\eta_1(w_1,w_2)}{\eta_3(w_1,w_2)},\\
\varphi_2(w_1,w_2) &:= \frac{\eta_4(w_1,w_2)}{\eta_2(w_1,w_2)}, \quad \varphi_3(w_1,w_2) := \frac{\eta_4(w_1,w_2)}{\eta_5(w_1,w_2)}.
\end{aligned} \tag{4.27}$$

Proposition 4.4.

$$\begin{cases}
\varphi_1(g_1(w_1,w_2)) = 1^{-\frac{4}{9}}\varphi_0(w_1,w_2),\\
\varphi_1(g_2(w_1,w_2)) = 1^{-\frac{8}{9}}\varphi_0(w_1,w_2),\\
\varphi_0(g_3(w_1,w_2)) = \varphi_0(w_1,w_2),\\
\varphi_3(g_4(w_1,w_2)) = \varphi_0(w_1,w_2),\\
\varphi_2(g_5(w_1,w_2)) = 1^{\frac{4}{9}}\sqrt[3]{\dfrac{-3\overline{\omega}w_1+(1-\omega)w_2+1}{(1-\overline{\omega})w_1+(1-\omega)w_2+1}}\varphi_0(w_1,w_2).
\end{cases} \tag{4.28}$$

Proof. This is the corollary of Proposition 4.2.

□

Proposition 4.5.

$$\begin{cases}
\varphi_1([T_1,T_2](w_1,w_2)) = \varphi_0(w_1,w_2),\\
\varphi_2(S^3[T_1,T_2]S^3(w_1,w_2)) = \dfrac{1}{\sqrt[3]{(\overline{\omega}-\omega)w_1+1}}\varphi_0(w_1,w_2),\\
\varphi_3(S^3[T_1,T_2]S^3[T_1,T_2](w_1,w_2)) = \varphi_0(w_1,w_2).
\end{cases} \tag{4.29}$$

Proof. We have

$$\varphi_1([T_1,T_2](w_1,w_2)) = \frac{\eta_1([T_1,T_2](w_1,w_2))}{\eta_3([T_1,T_2](w_1,w_2)} = \frac{\eta_1(w_1,w_2)}{\eta_2(w_1,w_2)} = \varphi_0(w_1,w_2).$$

$$\varphi_2(S^3[T_1,T_2]S^3(w_1,w_2)) = \frac{\eta_4(S^3[T_1,T_2]S^3(w_1,w_2))}{\eta_2(S^3[T_1,T_2]S^3(w_1,w_2))}.$$

Here,

$$\eta_4(S^3[T_1,T_2]S^3(w_1,w_2)) = \eta_1(w_1,w_2).$$

$$\begin{aligned}
&\eta_2(S^3[T_1,T_2]S^3(w_1,w_2))\\
=&\eta\left(\begin{pmatrix}1 & & \\ & 1-\omega & \\ & & 3\end{pmatrix} S^3[T_1,T_2]S^3 \begin{pmatrix}1 & & \\ & \frac{1}{1-\omega} & \\ & & \frac{1}{3}\end{pmatrix}\left(\frac{w_1}{3},\frac{w_2}{1-\overline{\omega}}\right)\right)\\
=&\eta\left(\begin{pmatrix}1 & 0 & 0\\ 0 & 1 & 0\\ 3(\overline{\omega}-\omega) & 0 & 1\end{pmatrix}\left(\frac{w_1}{3},\frac{w_2}{1-\overline{\omega}}\right)\right)\\
=&\eta\left(S^3[T_1,T_2]^3S^3\left(\frac{w_1}{3},\frac{w_2}{1-\overline{\omega}}\right)\right)\\
=&\sqrt[3]{(\overline{\omega}-\omega)w_1+1}\,\eta\left(\frac{w_1}{3},\frac{w_2}{1-\overline{\omega}}\right)\\
=&\sqrt[3]{(\overline{\omega}-\omega)w_1+1}\,\eta_2(w_1,w_2).
\end{aligned}$$

Thus,

$$\varphi_2(S^3[T_1,T_2]S^3(w_1,w_2)) = \frac{1}{\sqrt[3]{(\overline{\omega}-\omega)w_1+1}}\varphi_0(w_1,w_2).$$

Note that

$$\eta_4(S^3[T_1,T_2]S^3[T_1,T_2](w_1,w_2)) = \eta_1([T_1,T_2](w_1,w_2)) = \eta_1(w_1,w_2).$$

$$\eta_5(S^3[T_1,T_2]S^3[T_1,T_2](w_1,w_2)) = \eta_2(w_1,w_2).$$

This gives that

$$\varphi_3(S^3[T_1,T_2]S^3[T_1,T_2](w_1,w_2)) = \frac{\eta_1(w_1,w_2)}{\eta_2(w_1,w_2)} = \varphi_0(w_1,w_2).$$

□

Proposition 4.6.

$$\varphi_0([T_1,T_2]^{-3}(w_1,w_2)) = \varphi_0(w_1,w_2). \tag{4.30}$$

$$\varphi_1([T_1,T_2]^{-3}(w_1,w_2)) = \varphi_1(w_1,w_2). \tag{4.31}$$

$$\varphi_2(S^3[T_1,T_2]^{-3}S^3(w_1,w_2)) = \sqrt[3]{\frac{4(\omega-\overline{\omega})w_1+1}{[(\omega-\overline{\omega})w_1+1][3(\omega-\overline{\omega})w_1+1]}}\varphi_2(w_1,w_2). \tag{4.32}$$

$$\varphi_3(S^3[T_1,T_2]S^3[T_1,T_2]^{-3}S^3[T_1,T_2]^{-1}S^3(w_1,w_2)) = \varphi_3(w_1,w_2). \tag{4.33}$$

Proof. We have

$$\varphi_0([T_1,T_2]^{-3}(w_1,w_2)) = \frac{\eta_1([T_1,T_2]^{-3}(w_1,w_2))}{\eta_2([T_1,T_2]^{-3}(w_1,w_2))} = \frac{\eta_1(w_1,w_2)}{\eta_2(w_1,w_2)} = \varphi_0(w_1,w_2).$$

$$\varphi_1([T_1,T_2]^{-3}(w_1,w_2)) = \frac{\eta_1([T_1,T_2]^{-3}(w_1,w_2))}{\eta_3([T_1,T_2]^{-3}(w_1,w_2))} = \frac{\eta_1(w_1,w_2)}{\eta_3(w_1,w_2)} = \varphi_1(w_1,w_2).$$

$$\varphi_2(S^3[T_1,T_2]^{-3}S^3(w_1,w_2)) = \frac{\eta_4(S^3[T_1,T_2]^{-3}S^3(w_1,w_2))}{\eta_2(S^3[T_1,T_2]^{-3}S^3(w_1,w_2))}.$$

Here,

$$\eta_4(S^3[T_1,T_2]^{-3}S^3(w_1,w_2)) = \sqrt[3]{\frac{4(\omega-\overline{\omega})w_1+1}{(\omega-\overline{\omega})w_1+1}}\eta_4(w_1,w_2).$$

$$\begin{aligned}
&\eta_2(S^3[T_1,T_2]^{-3}S^3(w_1,w_2))\\
=&\eta\left(\begin{pmatrix}1 & & \\ & 1-\omega & \\ & & 3\end{pmatrix} S^3[T_1,T_2]^{-3}S^3 \begin{pmatrix}1 & & \\ & \frac{1}{1-\omega} & \\ & & \frac{1}{3}\end{pmatrix}\left(\frac{w_1}{3},\frac{w_2}{1-\overline{\omega}}\right)\right)\\
=&\eta\left(\begin{pmatrix}1 & 0 & 0\\ 0 & 1 & 0\\ 9(\omega-\overline{\omega}) & 0 & 1\end{pmatrix}\left(\frac{w_1}{3},\frac{w_2}{1-\overline{\omega}}\right)\right)\\
=&\eta\left(S^3[T_1,T_2]^{-9}S^3\left(\frac{w_1}{3},\frac{w_2}{1-\overline{\omega}}\right)\right)\\
=&\sqrt[3]{3(\omega-\overline{\omega})w_1+1}\,\eta\left(\frac{w_1}{3},\frac{w_2}{1-\overline{\omega}}\right)\\
=&\sqrt[3]{3(\omega-\overline{\omega})w_1+1}\,\eta_2(w_1,w_2).
\end{aligned}$$

Thus,

$$\varphi_2(S^3[T_1,T_2]^{-3}S^3(w_1,w_2)) = \sqrt[3]{\frac{4(\omega-\overline{\omega})w_1+1}{[(\omega-\overline{\omega})w_1+1][3(\omega-\overline{\omega})w_1+1]}}\varphi_2(w_1,w_2).$$

Note that

$$\begin{aligned}
&\eta_4(S^3[T_1,T_2]S^3[T_1,T_2]^{-3}S^3[T_1,T_2]^{-1}S^3(w_1,w_2))\\
=&\eta_1([T_1,T_2]^{-3}S^3[T_1,T_2]^{-1}S^3(w_1,w_2))\\
=&\eta_1(S^3[T_1,T_2]^{-1}S^3(w_1,w_2))\\
=&\eta_4(w_1,w_2).
\end{aligned}$$

$$\eta_5(S^3[T_1,T_2]S^3[T_1,T_2]^{-3}S^3[T_1,T_2]^{-1}S^3(w_1,w_2)) = \eta_5(w_1,w_2).$$

This implies that

$$\varphi_3(S^3[T_1,T_2]S^3[T_1,T_2]^{-3}S^3[T_1,T_2]^{-1}S^3(w_1,w_2)) = \frac{\eta_4(w_1,w_2)}{\eta_5(w_1,w_2)} = \varphi_3(w_1,w_2).$$

□

Note that

$$\frac{\varphi_1(w_1,w_2)}{\varphi_2(w_1,w_2)} = \frac{\eta_1(w_1,w_2)\eta_2(w_1,w_2)}{\eta_3(w_1,w_2)\eta_4(w_1,w_2)}.$$

Put

$$\begin{cases} \varphi_0(w_1,w_2) = \dfrac{\eta_1(w_1,w_2)}{\eta_2(w_1,w_2)} =: \kappa_0^{\frac{1}{9}} =: k_0^{\frac{1}{27}}, \\ \varphi_1(w_1,w_2) = \dfrac{\eta_1(w_1,w_2)}{\eta_3(w_1,w_2)} =: \kappa_1^{\frac{1}{9}} =: k_1^{\frac{1}{27}}, \\ \varphi_2(w_1,w_2) = \dfrac{\eta_4(w_1,w_2)}{\eta_2(w_1,w_2)} =: \kappa_2^{\frac{1}{9}} =: k_2^{\frac{1}{27}}, \\ \varphi_3(w_1,w_2) = \dfrac{\eta_4(w_1,w_2)}{\eta_5(w_1,w_2)} =: \kappa_3^{\frac{1}{9}} =: k_3^{\frac{1}{27}}. \end{cases} \tag{4.34}$$

Here, there are four functions φ_0, φ_1, φ_2 and φ_3, $C_4^2 = 6$. In fact, the dimension of the moduli space of algebraic curves of genus three is: $\dim \mathcal{M}_3 = 3\times 3 - 3 = 6$. Namely, we define

$$\begin{cases} K_1 := \displaystyle\int_0^1 \frac{dx}{\sqrt[3]{x^2(1-x)(1-k_2x)(1-k_3x)}}, \\ K_2 := \displaystyle\int_0^1 \frac{dx}{\sqrt[3]{x^2(1-x)(1-k_3x)(1-k_1x)}}, \\ K_3 := \displaystyle\int_0^1 \frac{dx}{\sqrt[3]{x^2(1-x)(1-k_1x)(1-k_2x)}}. \end{cases} \tag{4.35}$$

$$\begin{cases} K_4 := \displaystyle\int_0^1 \frac{dx}{\sqrt[3]{x^2(1-x)(1-k_0x)(1-k_1x)}}, \\ K_5 := \displaystyle\int_0^1 \frac{dx}{\sqrt[3]{x^2(1-x)(1-k_0x)(1-k_2x)}}, \\ K_6 := \displaystyle\int_0^1 \frac{dx}{\sqrt[3]{x^2(1-x)(1-k_0x)(1-k_3x)}}. \end{cases} \tag{4.36}$$

Note that

$$\begin{cases} K_1 = 3\displaystyle\int_0^1 \frac{dt}{\sqrt[3]{(1-t^3)(1-k_2t^3)(1-k_3t^3)}}, \\ K_2 = 3\displaystyle\int_0^1 \frac{dt}{\sqrt[3]{(1-t^3)(1-k_3t^3)(1-k_1t^3)}}, \\ K_3 = 3\displaystyle\int_0^1 \frac{dt}{\sqrt[3]{(1-t^3)(1-k_1t^3)(1-k_2t^3)}}. \end{cases}$$

$$\begin{cases} K_4 = 3\displaystyle\int_0^1 \frac{dt}{\sqrt[3]{(1-t^3)(1-k_0t^3)(1-k_1t^3)}}, \\ K_5 = 3\displaystyle\int_0^1 \frac{dt}{\sqrt[3]{(1-t^3)(1-k_0t^3)(1-k_2t^3)}}, \\ K_6 = 3\displaystyle\int_0^1 \frac{dt}{\sqrt[3]{(1-t^3)(1-k_0t^3)(1-k_3t^3)}}. \end{cases}$$

5. Transform problems and modular equations associated to algebraic surfaces

For the group $GL(3)$, the two integrals are given by

$$\begin{cases} K_3 = \displaystyle\int_0^1 \frac{dx}{\sqrt[3]{x^2(1-x)(1-\kappa_1^3 x)(1-\kappa_2^3 x)}}, \\ K_6 = \displaystyle\int_0^1 \frac{dx}{\sqrt[3]{x^2(1-x)(1-\kappa_0^3 x)(1-\kappa_3^3 x)}}. \end{cases}$$

Theorem 5.1 (Main Theorem 3). *The following identity of differential forms holds:*

$$\begin{aligned} &\frac{dw_1 \wedge dw_2}{\sqrt[3]{w_1^2 w_2^2 (1-w_1)(1-\lambda_1^3 w_1)(1-\lambda_2^3 w_1)}} \\ &= -\frac{5t_1}{t_2^2} \frac{dt_1 \wedge dt_2}{\sqrt[3]{t_1^2 t_2^2 (1-t_1)(1-\kappa_1^3 t_1)(1-\kappa_2^3 t_1)}}, \end{aligned} \tag{5.1}$$

where

$$w_1 = \frac{(\beta+\gamma)t_1 + \alpha}{\beta t_1 + (\alpha+\gamma)}, \quad w_2 = \frac{t_1[(\beta+\gamma)t_1+\alpha]^2[\beta t_1 + (\alpha+\gamma)]}{t_2^5}, \tag{5.2}$$

which is a rational transformation of order five. Here,

$$\begin{cases} \alpha = \dfrac{(v_1-1)(v_2-1)(v_1+v_2-2) - (u_1-1)(u_2-1)(u_1+u_2-2)}{2(u_1-1)(u_2-1)(v_1-1)(v_2-1)}, \\ \beta = \dfrac{(u_1-1)(u_2-1)(2u_1u_2-u_1-u_2) - (v_1-1)(v_2-1)(2v_1v_2-v_1-v_2)}{2(u_1-1)(u_2-1)(v_1-1)(v_2-1)}, \\ \gamma = \dfrac{(v_1-1)(v_2-1)}{(u_1-1)(u_2-1)}, \end{cases} \tag{5.3}$$

with

$$u_1 = \kappa_1^3, \quad u_2 = \kappa_2^3, \quad v_1 = \lambda_1^3, \quad v_2 = \lambda_2^3.$$

Moreover, the moduli (κ_1, κ_2) *and* (λ_1, λ_2) *satisfy the modular equation:*

$$(\kappa_1^3-1)(\kappa_2^3-1)(\kappa_1^3-\kappa_2^3) = (\lambda_1^3-1)(\lambda_2^3-1)(\lambda_1^3-\lambda_2^3), \tag{5.4}$$

which is an algebraic variety of dimension three. The corresponding two algebraic surfaces are:

$$\begin{cases} w_3^3 = w_1^2 w_2^2 (1-w_1)(1-\lambda_1^3 w_1)(1-\lambda_2^3 w_1), \\ t_3^3 = t_1^2 t_2^2 (1-t_1)(1-\kappa_1^3 t_1)(1-\kappa_2^3 t_1), \end{cases} \tag{5.5}$$

which give the homogeneous algebraic equations of degree seven:

$$\begin{cases} w_3^3 w_4^4 = w_1^2 w_2^2 (w_4-w_1)(w_4-\lambda_1^3 w_1)(w_4-\lambda_2^3 w_1), \\ t_3^3 t_4^4 = t_1^2 t_2^2 (t_4-t_1)(t_4-\kappa_1^3 t_1)(t_4-\kappa_2^3 t_1). \end{cases} \tag{5.6}$$

Proof. Let us consider the differential forms:

$$\frac{xdy \wedge dz + ydz \wedge dx + zdx \wedge dy}{\sqrt[3]{f_j(x,y,z)}},$$

where $f_j(x,y,z)$ $(j = 1,2)$ are homogeneous polynomials about x, y and z of degree nine.

Set

$$x = U(\xi,\zeta,\eta), \quad y = V(\xi,\zeta,\eta), \quad z = W(\xi,\zeta,\eta),$$

where U, V and W are homogeneous functions of degree n. By

$$\begin{cases} nU = \xi\dfrac{\partial U}{\partial \xi} + \zeta\dfrac{\partial U}{\partial \zeta} + \eta\dfrac{\partial U}{\partial \eta}, \\ nV = \xi\dfrac{\partial V}{\partial \xi} + \zeta\dfrac{\partial V}{\partial \zeta} + \eta\dfrac{\partial V}{\partial \eta}, \\ nW = \xi\dfrac{\partial W}{\partial \xi} + \zeta\dfrac{\partial W}{\partial \zeta} + \eta\dfrac{\partial W}{\partial \eta} \end{cases}$$

and

$$\begin{cases} dU = d\xi\dfrac{\partial U}{\partial \xi} + d\zeta\dfrac{\partial U}{\partial \zeta} + d\eta\dfrac{\partial U}{\partial \eta}, \\ dV = d\xi\dfrac{\partial V}{\partial \xi} + d\zeta\dfrac{\partial V}{\partial \zeta} + d\eta\dfrac{\partial V}{\partial \eta}, \\ dW = d\xi\dfrac{\partial W}{\partial \xi} + d\zeta\dfrac{\partial W}{\partial \zeta} + d\eta\dfrac{\partial W}{\partial \eta}, \end{cases}$$

we have

$$UdV \wedge dW + VdW \wedge dU + WdU \wedge dV = H(\xi d\zeta \wedge d\eta + \zeta d\eta \wedge d\xi + \eta d\xi \wedge d\zeta),$$

where

$$H = \frac{1}{n}\begin{vmatrix} \frac{\partial U}{\partial \xi} & \frac{\partial V}{\partial \xi} & \frac{\partial W}{\partial \xi} \\ \frac{\partial U}{\partial \zeta} & \frac{\partial V}{\partial \zeta} & \frac{\partial W}{\partial \zeta} \\ \frac{\partial U}{\partial \eta} & \frac{\partial V}{\partial \eta} & \frac{\partial W}{\partial \eta} \end{vmatrix}.$$

Hence,

$$\frac{xdy \wedge dz + ydz \wedge dx + zdx \wedge dy}{\sqrt[3]{f_j(x,y,z)}} = H\frac{\xi d\zeta \wedge d\eta + \zeta d\eta \wedge d\xi + \eta d\xi \wedge d\zeta}{\sqrt[3]{f_j(U,V,W)}}.$$

Here, the functions $f_j(U,V,W)$ $(j = 1,2)$ are of degree $9n$. Let

$$f_j(U,V,W) = T_j^3\varphi_j(\xi,\zeta,\eta),$$

where T_j^3 $(j = 1,2)$ are cubic factors of degree $9n - 9$. Put $M_j = \frac{T_j}{H}$ $(j = 1,2)$. Then

$$\frac{xdy \wedge dz + ydz \wedge dx + zdx \wedge dy}{\sqrt[3]{f_j(x,y,z)}} = \frac{1}{M_j}\frac{\xi d\zeta \wedge d\eta + \zeta d\eta \wedge d\xi + \eta d\xi \wedge d\zeta}{\sqrt[3]{\varphi_j(\xi,\zeta,\eta)}},$$

where $\varphi_j(\xi,\zeta,\eta)$ $(j=1,2)$ are homogeneous functions of degree 9.

Put

$$\begin{cases} f_1(x,y,z) := x^2y^2z^2(z-x)(z-\lambda_1^3x)(z-\lambda_2^3x), \\ f_2(x,y,z) := x^2y^2z^2(z-y)(z-\lambda_0^3y)(z-\lambda_3^3y). \end{cases}$$

Note that

$$xdy\wedge dz + ydz\wedge dx + zdx\wedge dy = z^3 d(\frac{x}{z})\wedge d(\frac{y}{z}).$$

We will study the following two differential forms:

$$\omega_1 := \frac{xdy\wedge dz + ydz\wedge dx + zdx\wedge dy}{\sqrt[3]{f_1(x,y,z)}},$$

$$\omega_2 := \frac{xdy\wedge dz + ydz\wedge dx + zdx\wedge dy}{\sqrt[3]{f_2(x,y,z)}}.$$

In fact,

$$\omega_1 = \frac{d(\frac{x}{z})\wedge d(\frac{y}{z})}{\sqrt[3]{(\frac{x}{z})^2(\frac{y}{z})^2(1-\frac{x}{z})(1-\lambda_1^3\frac{x}{z})(1-\lambda_2^3\frac{x}{z})}},$$

$$\omega_2 = \frac{d(\frac{x}{z})\wedge d(\frac{y}{z})}{\sqrt[3]{(\frac{x}{z})^2(\frac{y}{z})^2(1-\frac{y}{z})(1-\lambda_0^3\frac{y}{z})(1-\lambda_3^3\frac{y}{z})}}.$$

Set $\frac{x}{z}=u^3$ and $\frac{y}{z}=v^3$. Then

$$\omega_1 = \frac{9du\wedge dv}{\sqrt[3]{(1-u^3)(1-\lambda_1^3u^3)(1-\lambda_2^3u^3)}}, \quad \omega_2 = \frac{9du\wedge dv}{\sqrt[3]{(1-v^3)(1-\lambda_0^3v^3)(1-\lambda_3^3v^3)}}.$$

For

$$\begin{cases} f_1(x,y,z) = x^2y^2z^2(z-x)(z-\lambda_1^3x)(z-\lambda_2^3x), \\ f_2(x,y,z) = x^2y^2z^2(z-y)(z-\lambda_0^3y)(z-\lambda_3^3y), \\ \varphi_1(\xi,\zeta,\eta) = \xi^2\zeta^2\eta^2(\eta-\xi)(\eta-\kappa_1^3\xi)(\eta-\kappa_2^3\xi), \\ \varphi_2(\xi,\zeta,\eta) = \xi^2\zeta^2\eta^2(\eta-\zeta)(\eta-\kappa_0^3\zeta)(\eta-\kappa_3^3\zeta), \end{cases}$$

consider the following equations:

$$U^2V^2W^2(W-U)(W-\lambda_1^3U)(W-\lambda_2^3U) = T_1^3\xi^2\zeta^2\eta^2(\eta-\xi)(\eta-\kappa_1^3\xi)(\eta-\kappa_2^3\xi)$$

and

$$U^2V^2W^2(W-V)(W-\lambda_0^3V)(W-\lambda_3^3V) = T_2^3\xi^2\zeta^2\eta^2(\eta-\zeta)(\eta-\kappa_0^3\zeta)(\eta-\kappa_3^3\zeta).$$

The exponents of U, V, W at $f_1(U,V,W)$ are $5:2:5$, the exponents of U, V, W at $f_2(U,V,W)$ are $2:5:5$. On the other hand, T_1^3 and T_2^3 are of degree $9n-9$. This leads us to consider the system of equations:

$$\begin{aligned} &U^2V^2W^2(W-U)(W-\lambda_1^3U)(W-\lambda_2^3U) \\ =&(a\xi+b\eta)^6\zeta^{24}(c\xi+d\eta)^6\xi^2\zeta^2\eta^2(\eta-\xi)(\eta-\kappa_1^3\xi)(\eta-\kappa_2^3\xi) \end{aligned}$$

and

$$U^2V^2W^2(W-V)(W-\lambda_0^3V)(W-\lambda_3^3V)$$
$$=\xi^{24}(a\zeta+b\eta)^6(c\zeta+d\eta)^6\xi^2\zeta^2\eta^2(\eta-\zeta)(\eta-\kappa_0^3\zeta)(\eta-\kappa_3^3\zeta).$$

For the equation

$$U^2V^2W^2(W-U)(W-\lambda_1^3U)(W-\lambda_2^3U)=T_1^3\xi^2\zeta^2\eta^2(\eta-\xi)(\eta-\kappa_1^3\xi)(\eta-\kappa_2^3\xi),$$

put

$$\begin{cases} U=\zeta^6(a\xi+b\eta), \\ V=\xi\zeta\eta(a\xi+b\eta)^2(c\xi+d\eta)^2, \\ W=\zeta^6(c\xi+d\eta). \end{cases}$$

Here,

$$\left(\frac{U}{W},\frac{V}{W}\right)=\left(\frac{a\xi+b\eta}{c\xi+d\eta},\frac{\xi\eta(a\xi+b\eta)^2(c\xi+d\eta)}{\zeta^5}\right).$$

This is a transform of order 5.

Let

$$\xi\mapsto A\xi+B\eta,\quad \zeta\mapsto\zeta,\quad \eta\mapsto C\xi+D\eta,$$

where

$$\begin{cases} A=-\kappa_1^6\kappa_2^3-\kappa_1^3-\kappa_2^6+3\kappa_1^3\kappa_2^3, \\ B=\kappa_1^6+\kappa_2^6-\kappa_1^3\kappa_2^3-\kappa_1^3-\kappa_2^3+1, \\ C=-\kappa_1^6\kappa_2^6+\kappa_1^6\kappa_2^3+\kappa_1^3\kappa_2^6-\kappa_1^6-\kappa_2^6+\kappa_1^3\kappa_2^3, \\ D=\kappa_1^3\kappa_2^6+\kappa_1^6+\kappa_2^3-3\kappa_1^3\kappa_2^3. \end{cases}$$

Then

$$\begin{cases} \eta-\xi\mapsto(C-A)\xi+(D-B)\eta=(\kappa_1^3-1)(\kappa_2^3-1)^2(\eta-\kappa_1^3\xi), \\ \eta-\kappa_1^3\xi\mapsto(C-\kappa_1^3A)\xi+(D-\kappa_1^3B)\eta=(\kappa_2^3-\kappa_1^3)(\kappa_1^3-1)^2(\eta-\kappa_2^3\xi), \\ \eta-\kappa_2^3\xi\mapsto(C-\kappa_2^3A)\xi+(D-\kappa_2^3B)\eta=(1-\kappa_2^3)(\kappa_1^3-\kappa_2^3)^2(\eta-\xi). \end{cases}$$

Note that

$$A+B-C-D=(\kappa_1^3-1)^2(\kappa_2^3-1)^2.$$

We are interested in the three quotients:

$$\frac{W-U}{\eta-\xi},\quad \frac{W-\lambda_1^3U}{\eta-\kappa_1^3\xi},\quad \frac{W-\lambda_2^3U}{\eta-\kappa_2^3\xi}.$$

In fact, when $d-b=a-c$, i.e., set

$$b=\alpha,\quad d=\alpha+\gamma,\quad c=\beta,\quad a=\beta+\gamma,$$

then

$$\frac{W-U}{\eta-\xi}=\gamma\zeta^6.$$

Under the above action,

$$U \mapsto \zeta^6[a(A\xi + B\eta) + b(C\xi + D\eta)],$$
$$W \mapsto \zeta^6[c(A\xi + B\eta) + d(C\xi + D\eta)].$$

Put

$$\begin{cases} A = (\lambda_2^3 - 1)^2(c - \lambda_1^3 a)(d - \lambda_1^3 b) - (\lambda_2^3 - \lambda_1^3)(\lambda_1^3 - 1)(c - \lambda_2^3 a)(d - b), \\ B = (\lambda_2^3 - 1)^2(d - \lambda_1^3 b)^2 - (\lambda_2^3 - \lambda_1^3)(\lambda_1^3 - 1)(d - b)(d - \lambda_2^3 b), \\ C = (\lambda_2^3 - \lambda_1^3)(\lambda_1^3 - 1)(c - a)(c - \lambda_2^3 a) - (\lambda_2^3 - 1)^2(c - \lambda_1^3 a)^2, \\ D = (\lambda_2^3 - \lambda_1^3)(\lambda_1^3 - 1)(c - a)(d - \lambda_2^3 b) - (\lambda_2^3 - 1)^2(c - \lambda_1^3 a)(d - \lambda_1^3 b). \end{cases}$$

Then

$$\begin{cases} W - U \mapsto (\lambda_1^3 - 1)(\lambda_2^3 - 1)^2(ad - bc)(W - \lambda_1^3 U), \\ W - \lambda_1^3 U \mapsto (\lambda_2^3 - \lambda_1^3)(\lambda_1^3 - 1)^2(ad - bc)(W - \lambda_2^3 U), \\ W - \lambda_2^3 U \mapsto (1 - \lambda_2^3)(\lambda_1^3 - \lambda_2^3)^2(ad - bc)(W - U). \end{cases}$$

When $ad - bc = 1$, we have

$$\begin{cases} \dfrac{W - U}{\eta - \xi} \mapsto \dfrac{(\lambda_1^3 - 1)(\lambda_2^3 - 1)^2}{(\kappa_1^3 - 1)(\kappa_2^3 - 1)^2} \dfrac{W - \lambda_1^3 U}{\eta - \kappa_1^3 \xi}, \\ \dfrac{W - \lambda_1^3 U}{\eta - \kappa_1^3 \xi} \mapsto \dfrac{(\lambda_2^3 - \lambda_1^3)(\lambda_1^3 - 1)^2}{(\kappa_2^3 - \kappa_1^3)(\kappa_1^3 - 1)^2} \dfrac{W - \lambda_2^3 U}{\eta - \kappa_2^3 \xi}, \\ \dfrac{W - \lambda_2^3 U}{\eta - \kappa_2^3 \xi} \mapsto \dfrac{(1 - \lambda_2^3)(\lambda_1^3 - \lambda_2^3)^2}{(1 - \kappa_2^3)(\kappa_1^3 - \kappa_2^3)^2} \dfrac{W - U}{\eta - \xi}. \end{cases}$$

Note that

$$A + B - C - D = (\alpha + \beta + \gamma)^2(\lambda_1^3 - 1)^2(\lambda_2^3 - 1)^2.$$

Two different expressions about $A + B - C - D$ imply that

$$(\alpha + \beta + \gamma)^2(\lambda_1^3 - 1)^2(\lambda_2^3 - 1)^2 = (\kappa_1^3 - 1)^2(\kappa_2^3 - 1)^2.$$

Set

$$u_1 = \kappa_1^3, \quad u_2 = \kappa_2^3, \quad v_1 = \lambda_1^3, \quad v_2 = \lambda_2^3.$$

Then

$$\begin{cases} A = -u_1^2 u_2 - u_1 - u_2^2 + 3u_1 u_2, \\ B = u_1^2 + u_2^2 - u_1 u_2 - u_1 - u_2 + 1, \\ C = -u_1^2 u_2^2 + u_1^2 u_2 + u_1 u_2^2 - u_1^2 - u_2^2 + u_1 u_2, \\ D = u_1 u_2^2 + u_1^2 + u_2 - 3u_1 u_2, \end{cases}$$

and

$$\begin{cases} A = (v_1-1)^2(v_2-1)^2\alpha\beta + v_1(v_1-1)(v_2-1)^2\alpha\gamma \\ \quad -(v_1-1)^2(v_2-1)\beta\gamma - (v_1^2v_2+v_1+v_2^2-3v_1v_2)\gamma^2, \\ B = (v_1-1)^2(v_2-1)^2\alpha^2 - [(v_1-1)^2(v_2-1)+(v_1-1)(v_2-1)^2]\alpha\gamma \\ \quad +(v_1^2+v_2^2-v_1v_2-v_1-v_2+1)\gamma^2, \\ C = -[v_1(v_1-1)(v_2-1)^2+v_2(v_2-1)(v_1-1)^2]\beta\gamma \\ \quad -(v_1-1)^2(v_2-1)^2\beta^2 - (v_1^2v_2^2-v_1^2v_2-v_1v_2^2+v_1^2+v_2^2-v_1v_2)\gamma^2, \\ D = -(v_1-1)^2(v_2-1)^2\alpha\beta + (v_1-1)(v_2-1)^2\beta\gamma \\ \quad -v_2(v_2-1)(v_1-1)^2\alpha\gamma + (v_1v_2^2+v_1^2+v_2-3v_1v_2)\gamma^2. \end{cases}$$

Note that

$$ad - bc = (\beta+\gamma)(\alpha+\gamma) - \alpha\beta = (\alpha+\beta+\gamma)\gamma.$$

The condition $ad - bc = 1$ implies that

$$(\alpha+\beta+\gamma)\gamma = 1.$$

On the other hand,

$$\alpha+\beta+\gamma = \pm\frac{(\kappa_1^3-1)(\kappa_2^3-1)}{(\lambda_1^3-1)(\lambda_2^3-1)}.$$

Without loss of generality, we can assume that

$$\alpha+\beta+\gamma = \frac{(u_1-1)(u_2-1)}{(v_1-1)(v_2-1)}.$$

Hence,

$$\gamma = \frac{(v_1-1)(v_2-1)}{(u_1-1)(u_2-1)}.$$

Substituting the expression of γ to the equations about B and C, we get the following two quadratic equations about α and β respectively:

$$\begin{aligned} &(u_1-1)^2(u_2-1)^2(v_1-1)^2(v_2-1)^2\alpha^2 \\ &-(u_1-1)(u_2-1)(v_1-1)(v_2-1)[(v_1-1)^2(v_2-1)+(v_1-1)(v_2-1)^2]\alpha \\ &+[(v_1-1)^2(v_2-1)^2(v_1^2+v_2^2-v_1v_2-v_1-v_2+1) \\ &-(u_1-1)^2(u_2-1)^2(u_1^2+u_2^2-u_1u_2-u_1-u_2+1)] = 0 \end{aligned}$$

and

$$\begin{aligned} &(u_1-1)^2(u_2-1)^2(v_1-1)^2(v_2-1)^2\beta^2 \\ &+(u_1-1)(u_2-1)(v_1-1)(v_2-1) \\ &\quad\times[v_1(v_1-1)(v_2-1)^2+v_2(v_2-1)(v_1-1)^2]\beta \\ &+[(v_1-1)^2(v_2-1)^2(v_1^2v_2^2-v_1^2v_2-v_1v_2^2+v_1^2+v_2^2-v_1v_2) \\ &-(u_1-1)^2(u_2-1)^2(u_1^2u_2^2-u_1^2u_2-u_1u_2^2+u_1^2+u_2^2-u_1u_2)] = 0. \end{aligned}$$

Thus,

$$\begin{cases} \alpha = \dfrac{(v_1-1)(v_2-1)(v_1+v_2-2) \pm \sqrt{\Delta_1}}{2(u_1-1)(u_2-1)(v_1-1)(v_2-1)}, \\ \beta = \dfrac{-(v_1-1)(v_2-1)(2v_1v_2-v_1-v_2) \pm \sqrt{\Delta_2}}{2(u_1-1)(u_2-1)(v_1-1)(v_2-1)}, \end{cases}$$

where

$$\begin{cases} \Delta_1 = 4(u_1-1)^2(u_2-1)^2(u_1^2+u_2^2-u_1u_2-u_1-u_2+1) \\ \qquad -3(v_1-1)^2(v_2-1)^2(v_1-v_2)^2, \\ \Delta_2 = 4(u_1-1)^2(u_2-1)^2(u_1^2u_2^2-u_1^2u_2-u_1u_2^2+u_1^2+u_2^2-u_1u_2) \\ \qquad -3(v_1-1)^2(v_2-1)^2(v_1-v_2)^2. \end{cases}$$

Substituting the expressions of α, β and γ to the equations about A and D, we get the next two equations about Δ_1 and Δ_2:

$$\begin{aligned} &\pm\sqrt{\Delta_1}\cdot\pm\sqrt{\Delta_2} - (v_1-1)(v_2-1)(v_1-v_2)(\pm\sqrt{\Delta_1}\pm\sqrt{\Delta_2}) \\ &+4(u_1-1)^2(u_2-1)^2(u_1^2u_2+u_1+u_2^2-3u_1u_2) \\ &-3(v_1-1)^2(v_2-1)^2(v_1-v_2)^2 = 0, \end{aligned}$$

$$\begin{aligned} &\pm\sqrt{\Delta_1}\cdot\pm\sqrt{\Delta_2} + (v_1-1)(v_2-1)(v_1-v_2)(\pm\sqrt{\Delta_1}\pm\sqrt{\Delta_2}) \\ &+4(u_1-1)^2(u_2-1)^2(u_1u_2^2+u_1^2+u_2-3u_1u_2) \\ &-3(v_1-1)^2(v_2-1)^2(v_1-v_2)^2 = 0. \end{aligned}$$

Set $X = \pm\sqrt{\Delta_1}$ and $Y = \pm\sqrt{\Delta_2}$. We have

$$\begin{cases} X+Y = \dfrac{2(u_1-1)^3(u_2-1)^3(u_1-u_2)}{(v_1-1)(v_2-1)(v_1-v_2)}, \\ XY = -2(u_1-1)^2(u_2-1)^2(u_1^2u_2+u_1u_2^2+u_1^2+u_2^2-6u_1u_2+u_1+u_2) \\ \qquad +3(v_1-1)^2(v_2-1)^2(v_1-v_2)^2. \end{cases}$$

Note that

$$(X+Y)^2 - 2XY = X^2+Y^2 = \Delta_1+\Delta_2.$$

This gives a equation which connects the variables u_1, u_2 and v_1, v_2:

$$(u_1-1)^2(u_2-1)^2(u_1-u_2)^2 = (v_1-1)^2(v_2-1)^2(v_1-v_2)^2.$$

Similarly, the relation

$$(XY)^2 = X^2Y^2 = \Delta_1\Delta_2$$

gives the other equation about u_1, u_2 and v_1, v_2:

$$\begin{aligned} &4(u_1^2+u_2^2-u_1u_2-u_1-u_2+1)(u_1^2u_2^2-u_1^2u_2-u_1u_2^2+u_1^2+u_2^2-u_1u_2) \\ &-(u_1^2u_2+u_1u_2^2+u_1^2+u_2^2-6u_1u_2+u_1+u_2)^2 \\ =&3(v_1-1)^2(v_2-1)^2(v_1-v_2)^2. \end{aligned}$$

The identity

$$\begin{aligned}&4(u_1^2+u_2^2-u_1u_2-u_1-u_2+1)(u_1^2u_2^2-u_1^2u_2-u_1u_2^2+u_1^2+u_2^2-u_1u_2)\\&-(u_1^2u_2+u_1u_2^2+u_1^2+u_2^2-6u_1u_2+u_1+u_2)^2\\=&3(u_1-1)^2(u_2-1)^2(u_1-u_2)^2\end{aligned}$$

implies that the above two equations about u_1, u_2 and v_1, v_2 are the same one:

$$(u_1-1)^2(u_2-1)^2(u_1-u_2)^2=(v_1-1)^2(v_2-1)^2(v_1-v_2)^2.$$

In fact, the relation $\frac{W-U}{\eta-\xi}=\gamma\zeta^6$ gives that

$$\begin{cases}\dfrac{W-\lambda_1^3U}{\eta-\kappa_1^3\xi}=\gamma\dfrac{(\kappa_1^3-1)(\kappa_2^3-1)^2}{(\lambda_1^3-1)(\lambda_2^3-1)^2}\zeta^6,\\\dfrac{W-\lambda_2^3U}{\eta-\kappa_2^3\xi}=\gamma\dfrac{(\kappa_1^3-1)^3(\kappa_2^3-1)^2(\kappa_1^3-\kappa_2^3)}{(\lambda_1^3-1)^3(\lambda_2^3-1)^2(\lambda_1^3-\lambda_2^3)}\zeta^6,\\\dfrac{W-U}{\eta-\xi}=\gamma\dfrac{(\kappa_1^3-1)^3(\kappa_2^3-1)^3(\kappa_1^3-\kappa_2^3)^3}{(\lambda_1^3-1)^3(\lambda_2^3-1)^3(\lambda_1^3-\lambda_2^3)^3}\zeta^6.\end{cases}$$

This implies that

$$(\kappa_1^3-1)^3(\kappa_2^3-1)^3(\kappa_1^3-\kappa_2^3)^3=(\lambda_1^3-1)^3(\lambda_2^3-1)^3(\lambda_1^3-\lambda_2^3)^3.$$

Therefore, we get the modular equation:

$$(\kappa_1^3-1)(\kappa_2^3-1)(\kappa_1^3-\kappa_2^3)=(\lambda_1^3-1)(\lambda_2^3-1)(\lambda_1^3-\lambda_2^3).$$

The modular equation

$$(u_1-1)(u_2-1)(u_1-u_2)=(v_1-1)(v_2-1)(v_1-v_2)$$

implies that

$$\Delta_1=(u_1-1)^2(u_2-1)^2(u_1+u_2-2)^2,\quad \Delta_2=(u_1-1)^2(u_2-1)^2(2u_1u_2-u_1-u_2)^2.$$

Thus,

$$\alpha=\frac{(v_1-1)(v_2-1)(v_1+v_2-2)\pm(u_1-1)(u_2-1)(u_1+u_2-2)}{2(u_1-1)(u_2-1)(v_1-1)(v_2-1)},$$

$$\beta=\frac{-(v_1-1)(v_2-1)(2v_1v_2-v_1-v_2)\pm(u_1-1)(u_2-1)(2u_1u_2-u_1-u_2)}{2(u_1-1)(u_2-1)(v_1-1)(v_2-1)}.$$

We have

$$\begin{aligned}T_1^3=&\frac{U^2V^2W^2}{\xi^2\zeta^2\eta^2}\cdot\frac{W-U}{\eta-\xi}\cdot\frac{W-\lambda_1^3U}{\eta-\kappa_1^3\xi}\cdot\frac{W-\lambda_2^3U}{\eta-\kappa_2^3\xi}\\=&\zeta^{42}(a\xi+b\eta)^6(c\xi+d\eta)^6\gamma^3\frac{(\kappa_1^3-1)^3(\kappa_2^3-1)^3}{(\lambda_1^3-1)^3(\lambda_2^3-1)^3}.\end{aligned}$$

Hence,

$$T_1 = \gamma \frac{(u_1-1)(u_2-1)}{(v_1-1)(v_2-1)} \zeta^{14}(a\xi + b\eta)^2(c\xi + d\eta)^2 = \zeta^{14}(a\xi + b\eta)^2(c\xi + d\eta)^2.$$

On the other hand,

$$H = \frac{1}{7} \begin{vmatrix} \frac{\partial U}{\partial \xi} & \frac{\partial V}{\partial \xi} & \frac{\partial W}{\partial \xi} \\ \frac{\partial U}{\partial \zeta} & \frac{\partial V}{\partial \zeta} & \frac{\partial W}{\partial \zeta} \\ \frac{\partial U}{\partial \eta} & \frac{\partial V}{\partial \eta} & \frac{\partial W}{\partial \eta} \end{vmatrix} = -5\xi\eta\zeta^{12}(a\xi + b\eta)^2(c\xi + d\eta)^2.$$

Thus,

$$M_1 = \frac{T_1}{H} = -\frac{1}{5}\frac{\zeta^2}{\xi\eta}.$$

Set

$$w_1 = \frac{x}{z}, \quad w_2 = \frac{y}{z}, \quad t_1 = \frac{\xi}{\eta}, \quad t_2 = \frac{\zeta}{\eta}.$$

Then $M = -\frac{1}{5}\frac{t_2^2}{t_1}$ and

$$w_1 = \frac{U}{W} = \frac{(\beta+\gamma)t_1 + \alpha}{\beta t_1 + (\alpha+\gamma)}, \quad w_2 = \frac{V}{W} = \frac{t_1[(\beta+\gamma)t_1 + \alpha]^2[\beta t_1 + (\alpha+\gamma)]}{t_2^5}.$$

Moreover,

$$\frac{xdy \wedge dz + ydz \wedge dx + zdx \wedge dy}{\sqrt[3]{x^2y^2z^2(z-x)(z-\lambda_1^3 x)(z-\lambda_2^3 x)}} = \frac{dw_1 \wedge dw_2}{\sqrt[3]{w_1^2 w_2^2(1-w_1)(1-v_1w_1)(1-v_2w_1)}}.$$

$$\frac{\xi d\zeta \wedge d\eta + \zeta d\eta \wedge d\xi + \eta d\xi \wedge d\zeta}{\sqrt[3]{\xi^2\zeta^2\eta^2(\eta-\xi)(\eta-\kappa_1^3\xi)(\eta-\kappa_2^3\xi)}} = \frac{dt_1 \wedge dt_2}{\sqrt[3]{t_1^2t_2^2(1-t_1)(1-u_1t_1)(1-u_2t_1)}}.$$

Therefore,

$$\frac{dw_1 \wedge dw_2}{\sqrt[3]{w_1^2 w_2^2(1-w_1)(1-v_1w_1)(1-v_2w_1)}} = -\frac{5t_1}{t_2^2}\frac{dt_1 \wedge dt_2}{\sqrt[3]{t_1^2t_2^2(1-t_1)(1-u_1t_1)(1-u_2t_1)}}.$$

Note that the condition $(\alpha+\beta+\gamma)\gamma = 1$ gives that

$$\begin{cases} \alpha = \dfrac{(v_1-1)(v_2-1)(v_1+v_2-2) - (u_1-1)(u_2-1)(u_1+u_2-2)}{2(u_1-1)(u_2-1)(v_1-1)(v_2-1)}, \\ \beta = \dfrac{(u_1-1)(u_2-1)(2u_1u_2-u_1-u_2) - (v_1-1)(v_2-1)(2v_1v_2-v_1-v_2)}{2(u_1-1)(u_2-1)(v_1-1)(v_2-1)}, \\ \gamma = \dfrac{(v_1-1)(v_2-1)}{(u_1-1)(u_2-1)}. \end{cases}$$

For

$$U^2V^2W^2(W-V)(W-\lambda_0^3V)(W-\lambda_3^3V) = T_2^3\xi^2\zeta^2\eta^2(\eta-\zeta)(\eta-\kappa_0^3\zeta)(\eta-\kappa_3^3\zeta),$$

put

$$\begin{cases} U = \xi\zeta\eta(a\zeta + b\eta)^2(c\zeta + d\eta)^2, \\ V = \xi^6(a\zeta + b\eta), \\ W = \xi^6(c\zeta + d\eta). \end{cases}$$

Here,

$$\left(\frac{U}{W}, \frac{V}{W}\right) = \left(\frac{\zeta\eta(a\zeta + b\eta)^2(c\zeta + d\eta)}{\xi^5}, \frac{a\zeta + b\eta}{c\zeta + d\eta}\right).$$

This is a transform of order 5. The other argument is similar as above. This completes the proof of Main Theorem 3.

□

Corollary 5.2. *The following identity holds:*

$$\frac{dy_1 \wedge dy_2}{\sqrt[3]{(1-y_1^3)(1-\lambda_1^3 y_1^3)(1-\lambda_2^3 y_1^3)}} = -\frac{5x_1^3}{x_2^6} \frac{dx_1 \wedge dx_2}{\sqrt[3]{(1-x_1^3)(1-\kappa_1^3 x_1^3)(1-\kappa_2^3 x_1^3)}}, \tag{5.7}$$

where

$$y_1^3 = \frac{(\beta+\gamma)x_1^3 + \alpha}{\beta x_1^3 + (\alpha+\gamma)}, \quad y_2^3 = \frac{x_1^3[(\beta+\gamma)x_1^3 + \alpha]^2[\beta x_1^3 + (\alpha+\gamma)]}{x_2^{15}}, \tag{5.8}$$

and

$$\begin{cases} \alpha = \dfrac{(v_1-1)(v_2-1)(v_1+v_2-2) - (u_1-1)(u_2-1)(u_1+u_2-2)}{2(u_1-1)(u_2-1)(v_1-1)(v_2-1)}, \\ \beta = \dfrac{(u_1-1)(u_2-1)(2u_1u_2-u_1-u_2) - (v_1-1)(v_2-1)(2v_1v_2-v_1-v_2)}{2(u_1-1)(u_2-1)(v_1-1)(v_2-1)}, \\ \gamma = \dfrac{(v_1-1)(v_2-1)}{(u_1-1)(u_2-1)}, \end{cases}$$

with

$$u_1 = \kappa_1^3, \quad u_2 = \kappa_2^3, \quad v_1 = \lambda_1^3, \quad v_2 = \lambda_2^3.$$

Moreover, the moduli (κ_1, κ_2) and (λ_1, λ_2) satisfy the modular equation:

$$(\kappa_1^3 - 1)(\kappa_2^3 - 1)(\kappa_1^3 - \kappa_2^3) = (\lambda_1^3 - 1)(\lambda_2^3 - 1)(\lambda_1^3 - \lambda_2^3).$$

Proof. Put $w_1 = y_1^3$, $w_2 = y_2^3$, $t_1 = x_1^3$ and $t_2 = x_2^3$ in the Main Theorem 3.

□

6. Triangular s-functions associated to $U(2,1)$ and Picard curves

We will study the system of partial differential equations:

$$\begin{cases} \{w_1, w_2; v_1, v_2\}_{v_1} = F_1(\gamma; v_1, v_2), \\ \{w_1, w_2; v_1, v_2\}_{v_2} = F_2(\gamma; v_1, v_2), \\ [w_1, w_2; v_1, v_2]_{v_1} = P_1(\alpha, \beta, \gamma; v_1, v_2), \\ [w_1, w_2; v_1, v_2]_{v_2} = P_2(\alpha, \beta, \gamma; v_1, v_2), \end{cases} \tag{6.1}$$

where

$$\begin{cases} F_1(\gamma; v_1, v_2) = \dfrac{-\gamma v_2(v_2-1)}{v_1(v_1-1)(v_1-v_2)}, \\ F_2(\gamma; v_1, v_2) = \dfrac{-\gamma v_1(v_1-1)}{v_2(v_2-1)(v_2-v_1)}, \\ P_1(\alpha, \beta, \gamma; v_1, v_2) = \dfrac{\alpha}{v_1} + \dfrac{\beta}{v_1-1} + \dfrac{\gamma}{v_1-v_2}, \\ P_2(\alpha, \beta, \gamma; v_1, v_2) = \dfrac{\alpha}{v_2} + \dfrac{\beta}{v_2-1} + \dfrac{\gamma}{v_2-v_1}. \end{cases} \tag{6.2}$$

The solutions are denoted by

$$w_1 = s_1(\alpha, \beta, \gamma; v_1, v_2), \quad w_2 = s_2(\alpha, \beta, \gamma; v_1, v_2). \tag{6.3}$$

Consequently,

$$\begin{cases} \{s_1(\alpha, \beta, \gamma; v_1, v_2), s_2(\alpha, \beta, \gamma; v_1, v_2); v_1, v_2\}_{v_1} = F_1(\gamma; v_1, v_2), \\ \{s_1(\alpha, \beta, \gamma; v_1, v_2), s_2(\alpha, \beta, \gamma; v_1, v_2); v_1, v_2\}_{v_2} = F_2(\gamma; v_1, v_2), \\ [s_1(\alpha, \beta, \gamma; v_1, v_2), s_2(\alpha, \beta, \gamma; v_1, v_2); v_1, v_2]_{v_1} = P_1(\alpha, \beta, \gamma; v_1, v_2), \\ [s_1(\alpha, \beta, \gamma; v_1, v_2), s_2(\alpha, \beta, \gamma; v_1, v_2); v_1, v_2]_{v_2} = P_2(\alpha, \beta, \gamma; v_1, v_2), \end{cases} \tag{6.4}$$

By the above Main Proposition 1, for the next two systems of equations,

$$\begin{cases} \{u_1, u_2; w_1, w_2\}_{w_1} = F_1(\gamma; w_1, w_2), \\ \{u_1, u_2; w_1, w_2\}_{w_2} = F_2(\gamma; w_1, w_2), \\ [u_1, u_2; w_1, w_2]_{w_1} = P_1(\alpha, \beta, \gamma; w_1, w_2), \\ [u_1, u_2; w_1, w_2]_{w_2} = P_2(\alpha, \beta, \gamma; w_1, w_2), \end{cases}$$

and

$$\begin{cases} \{u_1, u_2; x, y\}_x = F_1(\gamma'; x, y), \\ \{u_1, u_2; x, y\}_y = F_2(\gamma'; x, y), \\ [u_1, u_2; x, y]_x = P_1(\alpha', \beta', \gamma'; x, y), \\ [u_1, u_2; x, y]_y = P_2(\alpha', \beta', \gamma'; x, y), \end{cases}$$

we have

$$\begin{pmatrix} F_1(\gamma'; x, y) \\ F_2(\gamma'; x, y) \\ P_1(\alpha', \beta', \gamma'; x, y) \\ P_2(\alpha', \beta', \gamma'; x, y) \end{pmatrix} = \frac{M(w_1, w_2; x, y)}{\frac{\partial(w_1, w_2)}{\partial(x, y)}} \begin{pmatrix} F_1(\gamma; w_1, w_2) \\ F_2(\gamma; w_1, w_2) \\ P_1(\alpha, \beta, \gamma; w_1, w_2) \\ P_2(\alpha, \beta, \gamma; w_1, w_2) \end{pmatrix} + \begin{pmatrix} \{w_1, w_2; x, y\}_x \\ \{w_1, w_2; x, y\}_y \\ [w_1, w_2; x, y]_x \\ [w_1, w_2; x, y]_y \end{pmatrix}.$$

For the integral

$$K_3 = \int_0^1 \frac{dx}{\sqrt[3]{x^2(1-x)(1-k_1x)(1-k_2x)}}$$
$$= \Gamma\left(\frac{1}{3}\right)\Gamma\left(\frac{2}{3}\right)F_1\left(\frac{1}{3};\frac{1}{3},\frac{1}{3};1;k_1,k_2\right),$$

corresponding to the Picard curve, we have

$$\alpha = c-b = \frac{2}{3}, \quad \beta = a+b-c+1 = \frac{2}{3}, \quad \gamma = -b = -\frac{1}{3}.$$

Thus,

$$\begin{cases} F_1 = \dfrac{\frac{1}{3}v_2(v_2-1)}{v_1(v_1-1)(v_1-v_2)}, \\[2ex] F_2 = \dfrac{\frac{1}{3}v_1(v_1-1)}{v_2(v_2-1)(v_2-v_1)}, \\[2ex] P_1 = \dfrac{\frac{2}{3}}{v_1} + \dfrac{\frac{2}{3}}{v_1-1} + \dfrac{-\frac{1}{3}}{v_1-v_2}, \\[2ex] P_2 = \dfrac{\frac{2}{3}}{v_2} + \dfrac{\frac{2}{3}}{v_2-1} + \dfrac{-\frac{1}{3}}{v_2-v_1}. \end{cases}$$

For the following two equations:

$$\begin{cases} \{s_1, s_2; x, y\}_x = F_1(\gamma; x, y), \\ \{s_1, s_2; x, y\}_y = F_2(\gamma; x, y), \\ [s_1, s_2; x, y]_x = P_1(\alpha, \beta, \gamma; x, y), \\ [s_1, s_2; x, y]_y = P_2(\alpha, \beta, \gamma; x, y), \end{cases}$$

and

$$\begin{cases} \{w_1, w_2; x, y\}_x = F_1\left(-\dfrac{1}{3}; x, y\right), \\[2ex] \{w_1, w_2; x, y\}_y = F_2\left(-\dfrac{1}{3}; x, y\right), \\[2ex] [w_1, w_2; x, y]_x = P_1\left(\dfrac{2}{3}, \dfrac{2}{3}, -\dfrac{1}{3}; x, y\right), \\[2ex] [w_1, w_2; x, y]_y = P_2\left(\dfrac{2}{3}, \dfrac{2}{3}, -\dfrac{1}{3}; x, y\right), \end{cases}$$

where $s_1 = s_1(\alpha, \beta, \gamma; x, y)$ and $s_2 = s_2(\alpha, \beta, \gamma; x, y)$, we have

$$\frac{M(w_1, w_2; x, y)}{\frac{\partial(w_1,w_2)}{\partial(x,y)}} \begin{pmatrix} \{s_1, s_2; w_1, w_2\}_{w_1} \\ \{s_1, s_2; w_1, w_2\}_{w_2} \\ [s_1, s_2; w_1, w_2]_{w_1} \\ [s_1, s_2; w_1, w_2]_{w_2} \end{pmatrix} = \begin{pmatrix} F_1(\gamma; x, y) - F_1\left(-\frac{1}{3}; x, y\right) \\ F_2(\gamma; x, y) - F_2\left(-\frac{1}{3}; x, y\right) \\ P_1(\alpha, \beta, \gamma; x, y) - P_1\left(\frac{2}{3}, \frac{2}{3}, -\frac{1}{3}; x, y\right) \\ P_2(\alpha, \beta, \gamma; x, y) - P_2\left(\frac{2}{3}, \frac{2}{3}, -\frac{1}{3}; x, y\right) \end{pmatrix}.$$

Thus,

$$\begin{pmatrix} \{s_1, s_2; w_1, w_2\}_{w_1} \\ \{s_1, s_2; w_1, w_2\}_{w_2} \\ [s_1, s_2; w_1, w_2]_{w_1} \\ [s_1, s_2; w_1, w_2]_{w_2} \end{pmatrix} = \frac{M(x, y; w_1, w_2)}{\frac{\partial(x,y)}{\partial(w_1,w_2)}} \begin{pmatrix} F_1(\gamma; x, y) - F_1\left(-\frac{1}{3}; x, y\right) \\ F_2(\gamma; x, y) - F_2\left(-\frac{1}{3}; x, y\right) \\ P_1(\alpha, \beta, \gamma; x, y) - P_1\left(\frac{2}{3}, \frac{2}{3}, -\frac{1}{3}; x, y\right) \\ P_2(\alpha, \beta, \gamma; x, y) - P_2\left(\frac{2}{3}, \frac{2}{3}, -\frac{1}{3}; x, y\right) \end{pmatrix}.$$

Let us consider the integral

$$z(x, y) = \int_0^1 \frac{dt}{\sqrt[3]{t(t-1)(t-x)(t-y)}},$$

which satisfies the following partial differential equations:

$$\begin{cases} x(1-x)r + y(1-x)s + \left(1 - \frac{5}{3}x\right)p - \frac{1}{3}yq - \frac{1}{9}z = 0, \\ y(1-y)t + x(1-y)s + \left(1 - \frac{5}{3}y\right)q - \frac{1}{3}xp - \frac{1}{9}z = 0, \end{cases}$$

where

$$r = \frac{\partial^2 z}{\partial x^2}, \quad s = \frac{\partial^2 z}{\partial x \partial y}, \quad t = \frac{\partial^2 z}{\partial y^2}, \quad p = \frac{\partial z}{\partial x}, \quad q = \frac{\partial z}{\partial y}.$$

This gives that

$$\begin{cases} r + \left(\frac{\frac{2}{3}}{x} + \frac{\frac{2}{3}}{x-1} + \frac{\frac{1}{3}}{x-y}\right)p + \frac{-\frac{1}{3}y(y-1)}{x(x-1)(x-y)}q + \frac{\frac{1}{9}}{x(x-1)}z = 0, \\ t + \left(\frac{\frac{2}{3}}{y} + \frac{\frac{2}{3}}{y-1} + \frac{\frac{1}{3}}{y-x}\right)q + \frac{-\frac{1}{3}x(x-1)}{y(y-1)(y-x)}p + \frac{\frac{1}{9}}{y(y-1)}z = 0, \\ s = \frac{\frac{1}{3}}{x-y}p + \frac{-\frac{1}{3}}{x-y}q. \end{cases}$$

Here, $\alpha = \frac{2}{3}$, $\beta = \frac{2}{3}$ and $\gamma = -\frac{1}{3}$. Consequently,

$$P_1(x,y) = \frac{\frac{2}{3}}{x} + \frac{\frac{2}{3}}{x-1} + \frac{-\frac{1}{3}}{x-y}, \quad P_2(x,y) = \frac{\frac{2}{3}}{y} + \frac{\frac{2}{3}}{y-1} + \frac{-\frac{1}{3}}{y-x}.$$

$$F_1(x,y) = \frac{\frac{1}{3}y(y-1)}{x(x-1)(x-y)}, \quad F_2(x,y) = \frac{\frac{1}{3}x(x-1)}{y(y-1)(y-x)}.$$

Let us consider the following partial differential equations:

$$\begin{cases} \{w_1, w_2; x, y\}_x = \dfrac{\frac{1}{3}y(y-1)}{x(x-1)(x-y)}, \\ \{w_1, w_2; x, y\}_y = \dfrac{\frac{1}{3}x(x-1)}{y(y-1)(y-x)}, \\ [w_1, w_2; x, y]_x = \dfrac{\frac{2}{3}}{x} + \dfrac{\frac{2}{3}}{x-1} + \dfrac{-\frac{1}{3}}{x-y}, \\ [w_1, w_2; x, y]_y = \dfrac{\frac{2}{3}}{y} + \dfrac{\frac{2}{3}}{y-1} + \dfrac{-\frac{1}{3}}{y-x}. \end{cases}$$

According to Main Theorem 1 and Main Theorem 2, the transformation is given by

$$w = x^{-\frac{2}{9}} y^{-\frac{2}{9}} (x-1)^{-\frac{2}{9}} (y-1)^{-\frac{2}{9}} (x-y)^{-\frac{2}{9}} \sqrt[3]{\Delta}, \tag{6.5}$$

where $\Delta = \frac{\partial(x,y)}{\partial(w_1,w_2)}$ and the function w satisfies the following partial differential equations:

$$\begin{cases} \dfrac{\partial^2 w}{\partial x^2} + \left(\dfrac{\frac{2}{3}}{x} + \dfrac{\frac{2}{3}}{x-1} + \dfrac{\frac{1}{3}}{x-y}\right)\dfrac{\partial w}{\partial x} + \dfrac{-\frac{1}{3}y(y-1)}{x(x-1)(x-y)}\dfrac{\partial w}{\partial y} + \dfrac{\frac{1}{9}}{x(x-1)}w = 0, \\ \dfrac{\partial^2 w}{\partial x \partial y} = \dfrac{\frac{1}{3}}{x-y}\dfrac{\partial w}{\partial x} + \dfrac{-\frac{1}{3}}{x-y}\dfrac{\partial w}{\partial y}, \\ \dfrac{\partial^2 w}{\partial y^2} + \left(\dfrac{\frac{2}{3}}{y} + \dfrac{\frac{2}{3}}{y-1} + \dfrac{\frac{1}{3}}{y-x}\right)\dfrac{\partial w}{\partial y} + \dfrac{-\frac{1}{3}x(x-1)}{y(y-1)(y-x)}\dfrac{\partial w}{\partial x} + \dfrac{\frac{1}{9}}{y(y-1)}w = 0. \end{cases}$$

The above argument shows that a solution is given by

$$w = z(x,y) = \int_0^1 \frac{dt}{\sqrt[3]{t(t-1)(t-x)(t-y)}}. \tag{6.6}$$

Set

$$f_1(x,y) = \frac{y(y-1)}{x(x-1)(x-y)}, \quad f_2(x,y) = \frac{x(x-1)}{y(y-1)(y-x)}. \tag{6.7}$$

$$P_1(\alpha,\beta,\gamma;x,y) = \frac{\alpha}{x} + \frac{\beta}{x-1} + \frac{\gamma}{x-y}, \quad P_2(\alpha,\beta,\gamma;x,y) = \frac{\alpha}{y} + \frac{\beta}{y-1} + \frac{\gamma}{y-x}. \tag{6.8}$$

Let

$$T=\begin{pmatrix}-1 & 0 & 1\\ & -1 & 1\\ & & 1\end{pmatrix},\quad S_1=\begin{pmatrix}1 & 0 & 0\\ 0 & 0 & 1\\ 0 & 1 & 0\end{pmatrix},\quad S_2=\begin{pmatrix} & & 1\\ & 1 & \\ 1 & & \end{pmatrix}. \tag{6.9}$$

Then

$$T(x,y)=(1-x,1-y). \tag{6.10}$$

$$\begin{cases} S_1(x,y)=\left(\dfrac{x}{y},\dfrac{1}{y}\right),\\ S_1T(x,y)=\left(\dfrac{1-x}{1-y},\dfrac{1}{1-y}\right),\\ TS_1(x,y)=\left(\dfrac{y-x}{y},\dfrac{y-1}{y}\right),\\ S_1TS_1(x,y)=\left(\dfrac{y-x}{y-1},\dfrac{y}{y-1}\right). \end{cases} \tag{6.11}$$

$$\begin{cases} S_2(x,y)=\left(\dfrac{1}{x},\dfrac{y}{x}\right),\\ S_2T(x,y)=\left(\dfrac{1}{1-x},\dfrac{1-y}{1-x}\right),\\ TS_2(x,y)=\left(\dfrac{x-1}{x},\dfrac{x-y}{x}\right),\\ S_2TS_2(x,y)=\left(\dfrac{x}{x-1},\dfrac{x-y}{x-1}\right). \end{cases} \tag{6.12}$$

We have

$$\begin{cases} f_1(T(x,y))=-f_1(x,y),\\ f_1(S_1(x,y))=-f_1(x,y),\\ f_1(S_1T(x,y))=f_1(x,y),\\ f_1(TS_1(x,y))=f_1(x,y),\\ f_1(S_1TS_1(x,y))=-f_1(x,y), \end{cases} \tag{6.13}$$

$$\begin{cases} f_2(T(x,y))=-f_2(x,y),\\ f_2(S_2(x,y))=-f_2(x,y),\\ f_2(S_2T(x,y))=f_2(x,y),\\ f_2(TS_2(x,y))=f_2(x,y),\\ f_2(S_2TS_2(x,y))=-f_2(x,y), \end{cases} \tag{6.14}$$

and

$$\begin{cases} P_1(\alpha,\beta,\gamma;T(x,y))=-P_1(\beta,\alpha,\gamma;x,y),\\ P_1(\alpha,\beta,\gamma;S_1(x,y))=yP_1(\alpha,\gamma,\beta;x,y),\\ P_1(\alpha,\beta,\gamma;S_1T(x,y))=(y-1)P_1(\gamma,\alpha,\beta;x,y),\\ P_1(\alpha,\beta,\gamma;TS_1(x,y))=-yP_1(\beta,\gamma,\alpha;x,y),\\ P_1(\alpha,\beta,\gamma;S_1TS_1(x,y))=(1-y)P_1(\gamma,\beta,\alpha;x,y), \end{cases} \tag{6.15}$$

$$\begin{cases} P_2(\alpha,\beta,\gamma;T(x,y)) = -P_2(\beta,\alpha,\gamma;x,y), \\ P_2(\alpha,\beta,\gamma;S_2(x,y)) = xP_2(\alpha,\gamma,\beta;x,y), \\ P_2(\alpha,\beta,\gamma;S_2T(x,y)) = (x-1)P_2(\gamma,\alpha,\beta;x,y), \\ P_2(\alpha,\beta,\gamma;TS_2(x,y)) = -xP_2(\beta,\gamma,\alpha;x,y), \\ P_2(\alpha,\beta,\gamma;S_2TS_2(x,y)) = (1-x)P_2(\gamma,\beta,\alpha;x,y). \end{cases} \tag{6.16}$$

Hence, we find that

$$\begin{cases} \{s_1(\beta,\alpha,\gamma;v_1,v_2), s_2(\beta,\alpha,\gamma;v_1,v_2);v_1,v_2\}_{v_1} = F_1(-\gamma;T(v_1,v_2)), \\ \{s_1(\beta,\alpha,\gamma;v_1,v_2), s_2(\beta,\alpha,\gamma;v_1,v_2);v_1,v_2\}_{v_2} = F_2(-\gamma;T(v_1,v_2)), \\ [s_1(\beta,\alpha,\gamma;v_1,v_2), s_2(\beta,\alpha,\gamma;v_1,v_2);v_1,v_2]_{v_1} = -P_1(\alpha,\beta,\gamma;T(v_1,v_2)), \\ [s_1(\beta,\alpha,\gamma;v_1,v_2), s_2(\beta,\alpha,\gamma;v_1,v_2);v_1,v_2]_{v_2} = -P_2(\alpha,\beta,\gamma;T(v_1,v_2)). \end{cases} \tag{6.17}$$

$$\begin{cases} \{s_1(\alpha,\gamma,\beta;v_1,v_2), s_2(\alpha,\gamma,\beta;v_1,v_2);v_1,v_2\}_{v_1} = F_1(-\beta;S_1(v_1,v_2)), \\ \{s_1(\alpha,\gamma,\beta;v_1,v_2), s_2(\alpha,\gamma,\beta;v_1,v_2);v_1,v_2\}_{v_2} = F_2(-\beta;S_2(v_1,v_2)), \\ [s_1(\alpha,\gamma,\beta;v_1,v_2), s_2(\alpha,\gamma,\beta;v_1,v_2);v_1,v_2]_{v_1} = \dfrac{1}{v_2}P_1(\alpha,\beta,\gamma;S_1(v_1,v_2)), \\ [s_1(\alpha,\gamma,\beta;v_1,v_2), s_2(\alpha,\gamma,\beta;v_1,v_2);v_1,v_2]_{v_2} = \dfrac{1}{v_1}P_2(\alpha,\beta,\gamma;S_2(v_1,v_2)). \end{cases} \tag{6.18}$$

$$\begin{cases} \{s_1(\gamma,\alpha,\beta;v_1,v_2), s_2(\gamma,\alpha,\beta;v_1,v_2);v_1,v_2\}_{v_1} = F_1(\beta;S_1T(v_1,v_2)), \\ \{s_1(\gamma,\alpha,\beta;v_1,v_2), s_2(\gamma,\alpha,\beta;v_1,v_2);v_1,v_2\}_{v_2} = F_2(\beta;S_2T(v_1,v_2)), \\ [s_1(\gamma,\alpha,\beta;v_1,v_2), s_2(\gamma,\alpha,\beta;v_1,v_2);v_1,v_2]_{v_1} = \dfrac{P_1(\alpha,\beta,\gamma;S_1T(v_1,v_2))}{v_2-1}, \\ [s_1(\gamma,\alpha,\beta;v_1,v_2), s_2(\gamma,\alpha,\beta;v_1,v_2);v_1,v_2]_{v_2} = \dfrac{P_2(\alpha,\beta,\gamma;S_2T(v_1,v_2))}{v_1-1}. \end{cases} \tag{6.19}$$

$$\begin{cases} \{s_1(\beta,\gamma,\alpha;v_1,v_2), s_2(\beta,\gamma,\alpha;v_1,v_2);v_1,v_2\}_{v_1} = F_1(\alpha;TS_1(v_1,v_2)), \\ \{s_1(\beta,\gamma,\alpha;v_1,v_2), s_2(\beta,\gamma,\alpha;v_1,v_2);v_1,v_2\}_{v_2} = F_2(\alpha;TS_2(v_1,v_2)), \\ [s_1(\beta,\gamma,\alpha;v_1,v_2), s_2(\beta,\gamma,\alpha;v_1,v_2);v_1,v_2]_{v_1} = -\dfrac{1}{v_2}P_1(\alpha,\beta,\gamma;TS_1(v_1,v_2)), \\ [s_1(\beta,\gamma,\alpha;v_1,v_2), s_2(\beta,\gamma,\alpha;v_1,v_2);v_1,v_2]_{v_2} = -\dfrac{1}{v_1}P_2(\alpha,\beta,\gamma;TS_2(v_1,v_2)). \end{cases} \tag{6.20}$$

$$\begin{cases} \{s_1(\gamma,\beta,\alpha;v_1,v_2), s_2(\gamma,\beta,\alpha;v_1,v_2);v_1,v_2\}_{v_1} = F_1(-\alpha;S_1TS_1(v_1,v_2)), \\ \{s_1(\gamma,\beta,\alpha;v_1,v_2), s_2(\gamma,\beta,\alpha;v_1,v_2);v_1,v_2\}_{v_2} = F_2(-\alpha;S_2TS_2(v_1,v_2)), \\ [s_1(\gamma,\beta,\alpha;v_1,v_2), s_2(\gamma,\beta,\alpha;v_1,v_2);v_1,v_2]_{v_1} = \dfrac{P_1(\alpha,\beta,\gamma;S_1TS_1(v_1,v_2))}{1-v_2}, \\ [s_1(\gamma,\beta,\alpha;v_1,v_2), s_2(\gamma,\beta,\alpha;v_1,v_2);v_1,v_2]_{v_2} = \dfrac{P_2(\alpha,\beta,\gamma;S_2TS_2(v_1,v_2))}{1-v_1}. \end{cases} \tag{6.21}$$

Proposition 6.1 (Main Proposition 3). *The triangular s-functions have the following properties of transformations:*

(1)

$$\begin{cases} \{s_1(\alpha,\beta,\gamma;T(v_1,v_2)),s_2(\alpha,\beta,\gamma;T(v_1,v_2));v_1,v_2\}_{v_1} = F_1(\gamma;v_1,v_2), \\ \{s_1(\alpha,\beta,\gamma;T(v_1,v_2)),s_2(\alpha,\beta,\gamma;T(v_1,v_2));v_1,v_2\}_{v_2} = F_2(\gamma;v_1,v_2), \\ [s_1(\alpha,\beta,\gamma;T(v_1,v_2)),s_2(\alpha,\beta,\gamma;T(v_1,v_2));v_1,v_2]_{v_1} = P_1(\beta,\alpha,\gamma;v_1,v_2), \\ [s_1(\alpha,\beta,\gamma;T(v_1,v_2)),s_2(\alpha,\beta,\gamma;T(v_1,v_2));v_1,v_2]_{v_2} = P_2(\beta,\alpha,\gamma;v_1,v_2). \end{cases} \tag{6.22}$$

(2)

$$\begin{cases} \{s_1(\alpha,\beta,\gamma;S_1(v_1,v_2)),s_2(\alpha,\beta,\gamma;S_1(v_1,v_2));v_1,v_2\}_{v_1} = F_1(\gamma;v_1,v_2), \\ \{s_1(\alpha,\beta,\gamma;S_2(v_1,v_2)),s_2(\alpha,\beta,\gamma;S_2(v_1,v_2));v_1,v_2\}_{v_2} = F_2(\gamma;v_1,v_2), \\ [s_1(\alpha,\beta,\gamma;S_1(v_1,v_2)),s_2(\alpha,\beta,\gamma;S_1(v_1,v_2));v_1,v_2]_{v_1} \\ = P_1(\alpha,-2\gamma,\beta+3\gamma;v_1,v_2), \\ [s_1(\alpha,\beta,\gamma;S_2(v_1,v_2)),s_2(\alpha,\beta,\gamma;S_2(v_1,v_2));v_1,v_2]_{v_2} \\ = P_2(\alpha,-2\gamma,\beta+3\gamma;v_1,v_2). \end{cases} \tag{6.23}$$

(3)

$$\begin{cases} \{s_1(\alpha,\beta,\gamma;S_1T(v_1,v_2)),s_2(\alpha,\beta,\gamma;S_1T(v_1,v_2));v_1,v_2\}_{v_1} = F_1(\gamma;v_1,v_2), \\ \{s_1(\alpha,\beta,\gamma;S_2T(v_1,v_2)),s_2(\alpha,\beta,\gamma;S_2T(v_1,v_2));v_1,v_2\}_{v_2} = F_2(\gamma;v_1,v_2), \\ [s_1(\alpha,\beta,\gamma;S_1T(v_1,v_2)),s_2(\alpha,\beta,\gamma;S_1T(v_1,v_2));v_1,v_2]_{v_1} \\ = P_1(-2\gamma,\alpha,\beta+3\gamma;v_1,v_2), \\ [s_1(\alpha,\beta,\gamma;S_2T(v_1,v_2)),s_2(\alpha,\beta,\gamma;S_2T(v_1,v_2));v_1,v_2]_{v_2} \\ = P_2(-2\gamma,\alpha,\beta+3\gamma;v_1,v_2). \end{cases} \tag{6.24}$$

(4)

$$\begin{cases} \{s_1(\alpha,\beta,\gamma;TS_1(v_1,v_2)),s_2(\alpha,\beta,\gamma;TS_1(v_1,v_2));v_1,v_2\}_{v_1} = F_1(\gamma;v_1,v_2), \\ \{s_1(\alpha,\beta,\gamma;TS_2(v_1,v_2)),s_2(\alpha,\beta,\gamma;TS_2(v_1,v_2));v_1,v_2\}_{v_2} = F_2(\gamma;v_1,v_2), \\ [s_1(\alpha,\beta,\gamma;TS_1(v_1,v_2)),s_2(\alpha,\beta,\gamma;TS_1(v_1,v_2));v_1,v_2]_{v_1} \\ = P_1(\beta,-2\gamma,\alpha+3\gamma;v_1,v_2), \\ [s_1(\alpha,\beta,\gamma;TS_2(v_1,v_2)),s_2(\alpha,\beta,\gamma;TS_2(v_1,v_2));v_1,v_2]_{v_2} \\ = P_2(\beta,-2\gamma,\alpha+3\gamma;v_1,v_2). \end{cases} \tag{6.25}$$

(5)

$$\begin{cases} \{s_1(\alpha,\beta,\gamma;S_1TS_1(v_1,v_2)),s_2(\alpha,\beta,\gamma;S_1TS_1(v_1,v_2));v_1,v_2\}_{v_1} \\ = F_1(\gamma;v_1,v_2), \\ \{s_1(\alpha,\beta,\gamma;S_2TS_2(v_1,v_2)),s_2(\alpha,\beta,\gamma;S_2TS_2(v_1,v_2));v_1,v_2\}_{v_2} \\ = F_2(\gamma;v_1,v_2), \\ [s_1(\alpha,\beta,\gamma;S_1TS_1(v_1,v_2)),s_2(\alpha,\beta,\gamma;S_1TS_1(v_1,v_2));v_1,v_2]_{v_1} \\ = P_1(-2\gamma,\beta,\alpha+3\gamma;v_1,v_2), \\ [s_1(\alpha,\beta,\gamma;S_2TS_2(v_1,v_2)),s_2(\alpha,\beta,\gamma;S_2TS_2(v_1,v_2));v_1,v_2]_{v_2} \\ = P_2(-2\gamma,\beta,\alpha+3\gamma;v_1,v_2). \end{cases} \tag{6.26}$$

Proof.
(1) Let $(u_1,u_2)=T(v_1,v_2)=(1-v_1,1-v_2)$.
By Main Proposition 1, we have

$$\begin{aligned} &\{s_1(\alpha,\beta,\gamma;T(v_1,v_2)),s_2(\alpha,\beta,\gamma;T(v_1,v_2));v_1,v_2\}_{v_1} \\ =&-\{s_1(\alpha,\beta,\gamma;u_1,u_2),s_2(\alpha,\beta,\gamma;u_1,u_2);u_1,u_2\}_{u_1} \\ =&-F_1(\gamma;u_1,u_2) \\ =&-F_1(\gamma;T(v_1,v_2)) \\ =&F_1(\gamma;v_1,v_2). \end{aligned}$$

$$\begin{aligned} &[s_1(\alpha,\beta,\gamma;T(v_1,v_2)),s_2(\alpha,\beta,\gamma;T(v_1,v_2));v_1,v_2]_{v_1} \\ =&-[s_1(\alpha,\beta,\gamma;u_1,u_2),s_2(\alpha,\beta,\gamma;u_1,u_2);u_1,u_2]_{u_1} \\ =&-P_1(\alpha,\beta,\gamma;u_1,u_2) \\ =&-P_1(\alpha,\beta,\gamma;T(v_1,v_2)) \\ =&P_1(\beta,\alpha,\gamma;v_1,v_2). \end{aligned}$$

(2) Let $(u_1,u_2)=S_1(v_1,v_2)=\left(\frac{v_1}{v_2},\frac{1}{v_2}\right)$.
By Main Proposition 1, we have

$$\begin{aligned} &\{s_1(\alpha,\beta,\gamma;S_1(v_1,v_2)),s_2(\alpha,\beta,\gamma;S_1(v_1,v_2));v_1,v_2\}_{v_1} \\ =&-\{s_1(\alpha,\beta,\gamma;u_1,u_2),s_2(\alpha,\beta,\gamma;u_1,u_2);u_1,u_2\}_{u_1} \\ =&-F_1(\gamma;u_1,u_2) \\ =&-F_1(\gamma;S_1(v_1,v_2)) \\ =&F_1(\gamma;v_1,v_2). \end{aligned}$$

$$
\begin{aligned}
&[s_1(\alpha,\beta,\gamma;S_1(v_1,v_2)),s_2(\alpha,\beta,\gamma;S_1(v_1,v_2));v_1,v_2]_{v_1}\\
=&3\frac{v_1}{v_2}\{s_1(\alpha,\beta,\gamma;u_1,u_2),s_2(\alpha,\beta,\gamma;u_1,u_2);u_1,u_2\}_{u_1}\\
&+\frac{1}{v_2}[s_1(\alpha,\beta,\gamma;u_1,u_2),s_2(\alpha,\beta,\gamma;u_1,u_2);u_1,u_2]_{u_1}\\
=&3\frac{v_1}{v_2}F_1(\gamma;u_1,u_2)+\frac{1}{v_2}P_1(\alpha,\beta,\gamma;u_1,u_2)\\
=&3\frac{v_1}{v_2}F_1(\gamma;S_1(v_1,v_2))+\frac{1}{v_2}P_1(\alpha,\beta,\gamma;S_1(v_1,v_2))\\
=&-3\frac{v_1}{v_2}F_1(\gamma;v_1,v_2)+P_1(\alpha,\gamma,\beta;v_1,v_2)\\
=&\frac{3\gamma(v_2-1)}{(v_1-1)(v_1-v_2)}+\frac{\alpha}{v_1}+\frac{\gamma}{v_1-1}+\frac{\beta}{v_1-v_2}\\
=&\frac{\alpha}{v_1}+\frac{-2\gamma}{v_1-1}+\frac{\beta+3\gamma}{v_1-v_2}\\
=&P_1(\alpha,-2\gamma,\beta+3\gamma;v_1,v_2).
\end{aligned}
$$

(3) Let $(u_1,u_2)=S_1T(v_1,v_2)=\left(\frac{1-v_1}{1-v_2},\frac{1}{1-v_2}\right)$.
By Main Proposition 1, we have

$$
\begin{aligned}
&\{s_1(\alpha,\beta,\gamma;S_1T(v_1,v_2)),s_2(\alpha,\beta,\gamma;S_1T(v_1,v_2));v_1,v_2\}_{v_1}\\
=&\{s_1(\alpha,\beta,\gamma;u_1,u_2),s_2(\alpha,\beta,\gamma;u_1,u_2);u_1,u_2\}_{u_1}\\
=&F_1(\gamma;u_1,u_2)\\
=&F_1(\gamma;S_1T(v_1,v_2))\\
=&F_1(\gamma;v_1,v_2).
\end{aligned}
$$

$$
\begin{aligned}
&[s_1(\alpha,\beta,\gamma;S_1T(v_1,v_2)),s_2(\alpha,\beta,\gamma;S_1T(v_1,v_2));v_1,v_2]_{v_1}\\
=&-3\frac{v_1-1}{v_2-1}\{s_1(\alpha,\beta,\gamma;u_1,u_2),s_2(\alpha,\beta,\gamma;u_1,u_2);u_1,u_2\}_{u_1}\\
&+\frac{1}{v_2-1}[s_1(\alpha,\beta,\gamma;u_1,u_2),s_2(\alpha,\beta,\gamma;u_1,u_2);u_1,u_2]_{u_1}\\
=&-3\frac{v_1-1}{v_2-1}F_1(\gamma;u_1,u_2)+\frac{1}{v_2-1}P_1(\alpha,\beta,\gamma;u_1,u_2)\\
=&-3\frac{v_1-1}{v_2-1}F_1(\gamma;S_1T(v_1,v_2))+\frac{1}{v_2-1}P_1(\alpha,\beta,\gamma;S_1T(v_1,v_2))\\
=&-3\frac{v_1-1}{v_2-1}F_1(\gamma;v_1,v_2)+P_1(\gamma,\alpha,\beta;v_1,v_2)\\
=&\frac{3\gamma v_2}{v_1(v_1-v_2)}+\frac{\gamma}{v_1}+\frac{\alpha}{v_1-1}+\frac{\beta}{v_1-v_2}\\
=&\frac{-2\gamma}{v_1}+\frac{\alpha}{v_1-1}+\frac{\beta+3\gamma}{v_1-v_2}\\
=&P_1(-2\gamma,\alpha,\beta+3\gamma;v_1,v_2).
\end{aligned}
$$

(4) Let $(u_1,u_2)=TS_1(v_1,v_2)=\left(\frac{v_2-v_1}{v_2},\frac{v_2-1}{v_2}\right)$.

By Main Proposition 1, we have

$$\begin{aligned}
&\{s_1(\alpha,\beta,\gamma;TS_1(v_1,v_2)),s_2(\alpha,\beta,\gamma;TS_1(v_1,v_2));v_1,v_2\}_{v_1}\\
=&\{s_1(\alpha,\beta,\gamma;u_1,u_2),s_2(\alpha,\beta,\gamma;u_1,u_2);u_1,u_2\}_{u_1}\\
=&F_1(\gamma;u_1,u_2)\\
=&F_1(\gamma;TS_1(v_1,v_2))\\
=&F_1(\gamma;v_1,v_2).
\end{aligned}$$

$$\begin{aligned}
&[s_1(\alpha,\beta,\gamma;TS_1(v_1,v_2)),s_2(\alpha,\beta,\gamma;TS_1(v_1,v_2));v_1,v_2]_{v_1}\\
=&-3\frac{v_1}{v_2}\{s_1(\alpha,\beta,\gamma;u_1,u_2),s_2(\alpha,\beta,\gamma;u_1,u_2);u_1,u_2\}_{u_1}\\
&-\frac{1}{v_2}[s_1(\alpha,\beta,\gamma;u_1,u_2),s_2(\alpha,\beta,\gamma;u_1,u_2);u_1,u_2]_{u_1}\\
=&-3\frac{v_1}{v_2}F_1(\gamma;u_1,u_2)-\frac{1}{v_2}P_1(\alpha,\beta,\gamma;u_1,u_2)\\
=&-3\frac{v_1}{v_2}F_1(\gamma;TS_1(v_1,v_2))-\frac{1}{v_2}P_1(\alpha,\beta,\gamma;TS_1(v_1,v_2))\\
=&-3\frac{v_1}{v_2}F_1(\gamma;v_1,v_2)+P_1(\beta,\gamma,\alpha;v_1,v_2)\\
=&\frac{3\gamma(v_2-1)}{(v_1-1)(v_1-v_2)}+\frac{\beta}{v_1}+\frac{\gamma}{v_1-1}+\frac{\alpha}{v_1-v_2}\\
=&\frac{\beta}{v_1}+\frac{-2\gamma}{v_1-1}+\frac{\alpha+3\gamma}{v_1-v_2}\\
=&P_1(\beta,-2\gamma,\alpha+3\gamma;v_1,v_2).
\end{aligned}$$

(5) Let $(u_1,u_2)=S_1TS_1(v_1,v_2)=\left(\frac{v_2-v_1}{v_2-1},\frac{v_2}{v_2-1}\right)$.
By Main Proposition 1, we have

$$\begin{aligned}
&\{s_1(\alpha,\beta,\gamma;S_1TS_1(v_1,v_2)),s_2(\alpha,\beta,\gamma;S_1TS_1(v_1,v_2));v_1,v_2\}_{v_1}\\
=&-\{s_1(\alpha,\beta,\gamma;u_1,u_2),s_2(\alpha,\beta,\gamma;u_1,u_2);u_1,u_2\}_{u_1}\\
=&-F_1(\gamma;u_1,u_2)\\
=&-F_1(\gamma;S_1TS_1(v_1,v_2))\\
=&F_1(\gamma;v_1,v_2).
\end{aligned}$$

$$
\begin{aligned}
&[s_1(\alpha,\beta,\gamma;S_1TS_1(v_1,v_2)),s_2(\alpha,\beta,\gamma;S_1TS_1(v_1,v_2));v_1,v_2]_{v_1}\\
=&3\frac{v_1-1}{v_2-1}\{s_1(\alpha,\beta,\gamma;u_1,u_2),s_2(\alpha,\beta,\gamma;u_1,u_2);u_1,u_2\}_{u_1}\\
&-\frac{1}{v_2-1}[s_1(\alpha,\beta,\gamma;u_1,u_2),s_2(\alpha,\beta,\gamma;u_1,u_2);u_1,u_2]_{u_1}\\
=&3\frac{v_1-1}{v_2-1}F_1(\gamma;u_1,u_2)-\frac{1}{v_2-1}P_1(\alpha,\beta,\gamma;u_1,u_2)\\
=&3\frac{v_1-1}{v_2-1}F_1(\gamma;S_1TS_1(v_1,v_2))-\frac{1}{v_2-1}P_1(\alpha,\beta,\gamma;S_1TS_1(v_1,v_2))\\
=&-3\frac{v_1-1}{v_2-1}F_1(\gamma;v_1,v_2)+P_1(\gamma,\beta,\alpha;v_1,v_2)\\
=&\frac{3\gamma v_2}{v_1(v_1-v_2)}+\frac{\gamma}{v_1}+\frac{\beta}{v_1-1}+\frac{\alpha}{v_1-v_2}\\
=&\frac{-2\gamma}{v_1}+\frac{\beta}{v_1-1}+\frac{\alpha+3\gamma}{v_1-v_2}\\
=&P_1(-2\gamma,\beta,\alpha+3\gamma;v_1,v_2).
\end{aligned}
$$

By the symmetry property, we can prove the other cases.

□

Corollary 6.2. *The triangular s-functions have the following properties of transformations:*

(1)

$$
\begin{cases}
\{s_1(\alpha,\beta,\gamma;T(v_1,v_2)),s_2(\alpha,\beta,\gamma;T(v_1,v_2));v_1,v_2\}_{v_1}\\
=\{s_1(\alpha,\beta,\gamma;v_1,v_2),s_2(\alpha,\beta,\gamma;v_1,v_2);v_1,v_2\}_{v_1},\\
\{s_1(\alpha,\beta,\gamma;T(v_1,v_2)),s_2(\alpha,\beta,\gamma;T(v_1,v_2));v_1,v_2\}_{v_2}\\
=\{s_1(\alpha,\beta,\gamma;v_1,v_2),s_2(\alpha,\beta,\gamma;v_1,v_2);v_1,v_2\}_{v_2},\\
[s_1(\alpha,\beta,\gamma;T(v_1,v_2)),s_2(\alpha,\beta,\gamma;T(v_1,v_2));v_1,v_2]_{v_1}\\
=[s_1(\beta,\alpha,\gamma;v_1,v_2),s_2(\beta,\alpha,\gamma;v_1,v_2);v_1,v_2]_{v_1},\\
[s_1(\alpha,\beta,\gamma;T(v_1,v_2)),s_2(\alpha,\beta,\gamma;T(v_1,v_2));v_1,v_2]_{v_2}\\
=[s_1(\beta,\alpha,\gamma;v_1,v_2),s_2(\beta,\alpha,\gamma;v_1,v_2);v_1,v_2]_{v_2}.
\end{cases}
\tag{6.27}
$$

(2)

$$
\begin{cases}
\{s_1(\alpha,\beta,\gamma;S_1(v_1,v_2)),s_2(\alpha,\beta,\gamma;S_1(v_1,v_2));v_1,v_2\}_{v_1}\\
=\{s_1(\alpha,\beta,\gamma;v_1,v_2),s_2(\alpha,\beta,\gamma;v_1,v_2);v_1,v_2\}_{v_1},\\
\{s_1(\alpha,\beta,\gamma;S_2(v_1,v_2)),s_2(\alpha,\beta,\gamma;S_2(v_1,v_2));v_1,v_2\}_{v_2}\\
=\{s_1(\alpha,\beta,\gamma;v_1,v_2),s_2(\alpha,\beta,\gamma;v_1,v_2);v_1,v_2\}_{v_2},\\
[s_1(\alpha,\beta,\gamma;S_1(v_1,v_2)),s_2(\alpha,\beta,\gamma;S_1(v_1,v_2));v_1,v_2]_{v_1}\\
=[s_1(\alpha,-2\gamma,\beta+3\gamma;v_1,v_2),s_2(\alpha,-2\gamma,\beta+3\gamma;v_1,v_2);v_1,v_2]_{v_1},\\
[s_1(\alpha,\beta,\gamma;S_2(v_1,v_2)),s_2(\alpha,\beta,\gamma;S_2(v_1,v_2));v_1,v_2]_{v_2}\\
=[s_1(\alpha,-2\gamma,\beta+3\gamma;v_1,v_2),s_2(\alpha,-2\gamma,\beta+3\gamma;v_1,v_2);v_1,v_2]_{v_2}.
\end{cases}
\tag{6.28}
$$

(3)

$$\begin{cases} \quad\{s_1(\alpha,\beta,\gamma;S_1T(v_1,v_2)),s_2(\alpha,\beta,\gamma;S_1T(v_1,v_2));v_1,v_2\}_{v_1} \\ =\{s_1(\alpha,\beta,\gamma;v_1,v_2),s_2(\alpha,\beta,\gamma;v_1,v_2);v_1,v_2\}_{v_1}, \\ \quad\{s_1(\alpha,\beta,\gamma;S_2T(v_1,v_2)),s_2(\alpha,\beta,\gamma;S_2T(v_1,v_2));v_1,v_2\}_{v_2} \\ =\{s_1(\alpha,\beta,\gamma;v_1,v_2),s_2(\alpha,\beta,\gamma;v_1,v_2);v_1,v_2\}_{v_2}, \\ \quad[s_1(\alpha,\beta,\gamma;S_1T(v_1,v_2)),s_2(\alpha,\beta,\gamma;S_1T(v_1,v_2));v_1,v_2]_{v_1} \\ =[s_1(-2\gamma,\alpha,\beta+3\gamma;v_1,v_2),s_2(-2\gamma,\alpha,\beta+3\gamma;v_1,v_2);v_1,v_2]_{v_1}, \\ \quad[s_1(\alpha,\beta,\gamma;S_2T(v_1,v_2)),s_2(\alpha,\beta,\gamma;S_2T(v_1,v_2));v_1,v_2]_{v_2} \\ =[s_1(-2\gamma,\alpha,\beta+3\gamma;v_1,v_2),s_2(-2\gamma,\alpha,\beta+3\gamma;v_1,v_2);v_1,v_2]_{v_2}. \end{cases} \tag{6.29}$$

(4)

$$\begin{cases} \quad\{s_1(\alpha,\beta,\gamma;TS_1(v_1,v_2)),s_2(\alpha,\beta,\gamma;TS_1(v_1,v_2));v_1,v_2\}_{v_1} \\ =\{s_1(\alpha,\beta,\gamma;v_1,v_2),s_2(\alpha,\beta,\gamma;v_1,v_2);v_1,v_2\}_{v_1}, \\ \quad\{s_1(\alpha,\beta,\gamma;TS_2(v_1,v_2)),s_2(\alpha,\beta,\gamma;TS_2(v_1,v_2));v_1,v_2\}_{v_2} \\ =\{s_1(\alpha,\beta,\gamma;v_1,v_2),s_2(\alpha,\beta,\gamma;v_1,v_2);v_1,v_2\}_{v_2}, \\ \quad[s_1(\alpha,\beta,\gamma;TS_1(v_1,v_2)),s_2(\alpha,\beta,\gamma;TS_1(v_1,v_2));v_1,v_2]_{v_1} \\ =[s_1(\beta,-2\gamma,\alpha+3\gamma;v_1,v_2),s_2(\beta,-2\gamma,\alpha+3\gamma;v_1,v_2);v_1,v_2]_{v_1}, \\ \quad[s_1(\alpha,\beta,\gamma;TS_2(v_1,v_2)),s_2(\alpha,\beta,\gamma;TS_2(v_1,v_2));v_1,v_2]_{v_2} \\ =[s_1(\beta,-2\gamma,\alpha+3\gamma;v_1,v_2),s_2(\beta,-2\gamma,\alpha+3\gamma;v_1,v_2);v_1,v_2]_{v_2}. \end{cases} \tag{6.30}$$

(5)

$$\begin{cases} \quad\{s_1(\alpha,\beta,\gamma;S_1TS_1(v_1,v_2)),s_2(\alpha,\beta,\gamma;S_1TS_1(v_1,v_2));v_1,v_2\}_{v_1} \\ =\{s_1(\alpha,\beta,\gamma;v_1,v_2),s_2(\alpha,\beta,\gamma;v_1,v_2);v_1,v_2\}_{v_1}, \\ \quad\{s_1(\alpha,\beta,\gamma;S_2TS_2(v_1,v_2)),s_2(\alpha,\beta,\gamma;S_2TS_2(v_1,v_2));v_1,v_2\}_{v_2} \\ =\{s_1(\alpha,\beta,\gamma;v_1,v_2),s_2(\alpha,\beta,\gamma;v_1,v_2);v_1,v_2\}_{v_2}, \\ \quad[s_1(\alpha,\beta,\gamma;S_1TS_1(v_1,v_2)),s_2(\alpha,\beta,\gamma;S_1TS_1(v_1,v_2));v_1,v_2]_{v_1} \\ =[s_1(-2\gamma,\beta,\alpha+3\gamma;v_1,v_2),s_2(-2\gamma,\beta,\alpha+3\gamma;v_1,v_2);v_1,v_2]_{v_1}, \\ \quad[s_1(\alpha,\beta,\gamma;S_2TS_2(v_1,v_2)),s_2(\alpha,\beta,\gamma;S_2TS_2(v_1,v_2));v_1,v_2]_{v_2} \\ =[s_1(-2\gamma,\beta,\alpha+3\gamma;v_1,v_2),s_2(-2\gamma,\beta,\alpha+3\gamma;v_1,v_2);v_1,v_2]_{v_2}. \end{cases} \tag{6.31}$$

For $0,1,\lambda_1,\lambda_2,\infty\in\mathbb{P}^1(\mathbb{C})$, $\lambda_1,\lambda_2\notin\{0,1\}$ and $\lambda_1\neq\lambda_2$, set

$$\begin{cases} J_1=J_1(\lambda_1,\lambda_2):=\dfrac{\lambda_2^2(\lambda_2-1)^2}{\lambda_1^2(\lambda_1-1)^2(\lambda_1-\lambda_2)^2}, \\ J_2=J_2(\lambda_1,\lambda_2):=\dfrac{\lambda_1^2(\lambda_1-1)^2}{\lambda_2^2(\lambda_2-1)^2(\lambda_2-\lambda_1)^2}. \end{cases} \tag{6.32}$$

The corresponding Picard curve is

$$y^3=x(x-1)(x-\lambda_1)(x-\lambda_2). \tag{6.33}$$

The functions J_1 and J_2 satisfy the following relations:

$$\begin{aligned} &J_1(\lambda_1,\lambda_2)=J_1(T(\lambda_1,\lambda_2))=J_1(S_1(\lambda_1,\lambda_2))=J_1(S_1T(\lambda_1,\lambda_2))\\ =&J_1(TS_1(\lambda_1,\lambda_2))=J_1(S_1TS_1(\lambda_1,\lambda_2))=\frac{\lambda_2^2(\lambda_2-1)^2}{\lambda_1^2(\lambda_1-1)^2(\lambda_1-\lambda_2)^2}. \end{aligned} \tag{6.34}$$

$$\begin{aligned} &J_2(\lambda_1,\lambda_2)=J_2(T(\lambda_1,\lambda_2))=J_2(S_2(\lambda_1,\lambda_2))=J_2(S_2T(\lambda_1,\lambda_2))\\ =&J_2(TS_2(\lambda_1,\lambda_2))=J_2(S_2TS_2(\lambda_1,\lambda_2))=\frac{\lambda_1^2(\lambda_1-1)^2}{\lambda_2^2(\lambda_2-1)^2(\lambda_2-\lambda_1)^2}. \end{aligned} \tag{6.35}$$

7. Four derivatives and nonlinear evolution equations

Let

$$(w_1,w_2)=\left(\frac{a_1u_1+a_2u_2+a_3}{c_1u_1+c_2u_2+c_3},\frac{b_1u_1+b_2u_2+b_3}{c_1u_1+c_2u_2+c_3}\right),$$

where $\gamma=\begin{pmatrix} a_1 & a_2 & a_3\\ b_1 & b_2 & b_3\\ c_1 & c_2 & c_3\end{pmatrix}\in GL(3,\mathbb{C})$ and $\Delta=\det(\gamma)$. Note that

$$\begin{vmatrix} \frac{\partial w_1}{\partial t} & \frac{\partial w_2}{\partial t}\\ \frac{\partial w_1}{\partial x} & \frac{\partial w_2}{\partial x}\end{vmatrix}=\Delta(c_1u_1+c_2u_2+c_3)^{-3}\begin{vmatrix} \frac{\partial u_1}{\partial t} & \frac{\partial u_2}{\partial t}\\ \frac{\partial u_1}{\partial x} & \frac{\partial u_2}{\partial x}\end{vmatrix},$$

$$\begin{vmatrix} \frac{\partial w_1}{\partial t} & \frac{\partial w_2}{\partial t}\\ \frac{\partial w_1}{\partial y} & \frac{\partial w_2}{\partial y}\end{vmatrix}=\Delta(c_1u_1+c_2u_2+c_3)^{-3}\begin{vmatrix} \frac{\partial u_1}{\partial t} & \frac{\partial u_2}{\partial t}\\ \frac{\partial u_1}{\partial y} & \frac{\partial u_2}{\partial y}\end{vmatrix},$$

$$\begin{vmatrix} \frac{\partial w_1}{\partial x} & \frac{\partial w_2}{\partial x}\\ \frac{\partial w_1}{\partial y} & \frac{\partial w_2}{\partial y}\end{vmatrix}=\Delta(c_1u_1+c_2u_2+c_3)^{-3}\begin{vmatrix} \frac{\partial u_1}{\partial x} & \frac{\partial u_2}{\partial x}\\ \frac{\partial u_1}{\partial y} & \frac{\partial u_2}{\partial y}\end{vmatrix},$$

we have

$$\frac{\begin{vmatrix} \frac{\partial w_1}{\partial t} & \frac{\partial w_2}{\partial t}\\ \frac{\partial w_1}{\partial x} & \frac{\partial w_2}{\partial x}\end{vmatrix}}{\begin{vmatrix} \frac{\partial w_1}{\partial x} & \frac{\partial w_2}{\partial x}\\ \frac{\partial w_1}{\partial y} & \frac{\partial w_2}{\partial y}\end{vmatrix}}=\frac{\begin{vmatrix} \frac{\partial u_1}{\partial t} & \frac{\partial u_2}{\partial t}\\ \frac{\partial u_1}{\partial x} & \frac{\partial u_2}{\partial x}\end{vmatrix}}{\begin{vmatrix} \frac{\partial u_1}{\partial x} & \frac{\partial u_2}{\partial x}\\ \frac{\partial u_1}{\partial y} & \frac{\partial u_2}{\partial y}\end{vmatrix}},\quad \frac{\begin{vmatrix} \frac{\partial w_1}{\partial t} & \frac{\partial w_2}{\partial t}\\ \frac{\partial w_1}{\partial y} & \frac{\partial w_2}{\partial y}\end{vmatrix}}{\begin{vmatrix} \frac{\partial w_1}{\partial y} & \frac{\partial w_2}{\partial y}\\ \frac{\partial w_1}{\partial x} & \frac{\partial w_2}{\partial x}\end{vmatrix}}=\frac{\begin{vmatrix} \frac{\partial u_1}{\partial t} & \frac{\partial u_2}{\partial t}\\ \frac{\partial u_1}{\partial y} & \frac{\partial u_2}{\partial y}\end{vmatrix}}{\begin{vmatrix} \frac{\partial u_1}{\partial y} & \frac{\partial u_2}{\partial y}\\ \frac{\partial u_1}{\partial x} & \frac{\partial u_2}{\partial x}\end{vmatrix}}.$$

On the other hand,

$$\begin{cases} \{w_1,w_2;x,y\}_x=\{u_1,u_2;x,y\}_x,\\ \{w_1,w_2;x,y\}_y=\{u_1,u_2;x,y\}_y,\\ [w_1,w_2;x,y]_x=[u_1,u_2;x,y]_x,\\ [w_1,w_2;x,y]_x=[u_1,u_2;x,y]_y. \end{cases}$$

This leads us to study the following four evolution equations for $GL(3)$:

$$\frac{\begin{vmatrix} \frac{\partial u_1}{\partial t_1} & \frac{\partial u_2}{\partial t_1}\\ \frac{\partial u_1}{\partial x} & \frac{\partial u_2}{\partial x}\end{vmatrix}}{\begin{vmatrix} \frac{\partial u_1}{\partial y} & \frac{\partial u_2}{\partial y}\\ \frac{\partial u_1}{\partial x} & \frac{\partial u_2}{\partial x}\end{vmatrix}}=\{u_1,u_2;x,y\}_x,\quad \frac{\begin{vmatrix} \frac{\partial u_1}{\partial t_1} & \frac{\partial u_2}{\partial t_1}\\ \frac{\partial u_1}{\partial y} & \frac{\partial u_2}{\partial y}\end{vmatrix}}{\begin{vmatrix} \frac{\partial u_1}{\partial y} & \frac{\partial u_2}{\partial y}\\ \frac{\partial u_1}{\partial x} & \frac{\partial u_2}{\partial x}\end{vmatrix}}=[u_1,u_2;x,y]_x,$$

$$\frac{\begin{vmatrix} \frac{\partial u_1}{\partial t_2} & \frac{\partial u_2}{\partial t_2} \\ \frac{\partial u_1}{\partial x} & \frac{\partial u_2}{\partial x} \end{vmatrix}}{\begin{vmatrix} \frac{\partial u_1}{\partial x} & \frac{\partial u_2}{\partial x} \\ \frac{\partial u_1}{\partial y} & \frac{\partial u_2}{\partial y} \end{vmatrix}} = [u_1, u_2; x, y]_y, \quad \frac{\begin{vmatrix} \frac{\partial u_1}{\partial t_2} & \frac{\partial u_2}{\partial t_2} \\ \frac{\partial u_1}{\partial y} & \frac{\partial u_2}{\partial y} \end{vmatrix}}{\begin{vmatrix} \frac{\partial u_1}{\partial x} & \frac{\partial u_2}{\partial x} \\ \frac{\partial u_1}{\partial y} & \frac{\partial u_2}{\partial y} \end{vmatrix}} = \{u_1, u_2; x, y\}_y,$$

where the functions $u_1 = u_1(x, y; t_1, t_2)$ and $u_2 = u_2(x, y; t_1, t_2)$. The most important property of the above equations is that they are invariant with respect to the group $GL(3, \mathbb{C})$ of linear fractional transformations:

$$(u_1, u_2) \mapsto \left(\frac{a_1 u_1 + a_2 u_2 + a_3}{c_1 u_1 + c_2 u_2 + c_3}, \frac{b_1 u_1 + b_2 u_2 + b_3}{c_1 u_1 + c_2 u_2 + c_3} \right)$$

with $\gamma = \begin{pmatrix} a_1 & a_2 & a_3 \\ b_1 & b_2 & b_3 \\ c_1 & c_2 & c_3 \end{pmatrix} \in GL(3, \mathbb{C})$.

Theorem 7.1 (Main Theorem 4). *For the system of partial differential equations:*

$$\frac{\begin{vmatrix} \frac{\partial u_1}{\partial t_1} & \frac{\partial u_2}{\partial t_1} \\ \frac{\partial u_1}{\partial x} & \frac{\partial u_2}{\partial x} \end{vmatrix}}{\begin{vmatrix} \frac{\partial u_1}{\partial y} & \frac{\partial u_2}{\partial y} \\ \frac{\partial u_1}{\partial x} & \frac{\partial u_2}{\partial x} \end{vmatrix}} = \{u_1, u_2; x, y\}_x, \quad \frac{\begin{vmatrix} \frac{\partial u_1}{\partial t_1} & \frac{\partial u_2}{\partial t_1} \\ \frac{\partial u_1}{\partial y} & \frac{\partial u_2}{\partial y} \end{vmatrix}}{\begin{vmatrix} \frac{\partial u_1}{\partial y} & \frac{\partial u_2}{\partial y} \\ \frac{\partial u_1}{\partial x} & \frac{\partial u_2}{\partial x} \end{vmatrix}} = [u_1, u_2; x, y]_x, \tag{7.1}$$

$$\frac{\begin{vmatrix} \frac{\partial u_1}{\partial t_2} & \frac{\partial u_2}{\partial t_2} \\ \frac{\partial u_1}{\partial x} & \frac{\partial u_2}{\partial x} \end{vmatrix}}{\begin{vmatrix} \frac{\partial u_1}{\partial x} & \frac{\partial u_2}{\partial x} \\ \frac{\partial u_1}{\partial y} & \frac{\partial u_2}{\partial y} \end{vmatrix}} = [u_1, u_2; x, y]_y, \quad \frac{\begin{vmatrix} \frac{\partial u_1}{\partial t_2} & \frac{\partial u_2}{\partial t_2} \\ \frac{\partial u_1}{\partial y} & \frac{\partial u_2}{\partial y} \end{vmatrix}}{\begin{vmatrix} \frac{\partial u_1}{\partial x} & \frac{\partial u_2}{\partial x} \\ \frac{\partial u_1}{\partial y} & \frac{\partial u_2}{\partial y} \end{vmatrix}} = \{u_1, u_2; x, y\}_y, \tag{7.2}$$

set

$$\begin{cases} v_1 := \{u_1, u_2; x, y\}_x, \\ w_1 := [u_1, u_2; x, y]_x, \\ v_2 := \{u_1, u_2; x, y\}_y, \\ w_2 := [u_1, u_2; x, y]_y. \end{cases} \tag{7.3}$$

Then v_1, v_2, w_1 and w_2 satisfy the following two equations:

$$\begin{cases} \frac{\partial w_1}{\partial t_2} + \frac{\partial v_2}{\partial t_1} - v_1 \frac{\partial v_2}{\partial y} - v_2 \frac{\partial w_1}{\partial x} + w_1 \frac{\partial v_2}{\partial x} + w_2 \frac{\partial w_1}{\partial y} = 0, \\ \frac{\partial w_2}{\partial t_1} + \frac{\partial v_1}{\partial t_2} - v_1 \frac{\partial w_2}{\partial y} - v_2 \frac{\partial v_1}{\partial x} + w_1 \frac{\partial w_2}{\partial x} + w_2 \frac{\partial v_1}{\partial y} = 0. \end{cases} \tag{7.4}$$

Proof. The above four equations imply that

$$\frac{\partial u_1}{\partial t_1} = \frac{\partial^2 u_1}{\partial x^2} - 2 \frac{\begin{vmatrix} \frac{\partial u_1}{\partial x} & \frac{\partial u_2}{\partial x} \\ \frac{\partial^2 u_1}{\partial x \partial y} & \frac{\partial^2 u_2}{\partial x \partial y} \end{vmatrix}}{\begin{vmatrix} \frac{\partial u_1}{\partial x} & \frac{\partial u_2}{\partial x} \\ \frac{\partial u_1}{\partial y} & \frac{\partial u_2}{\partial y} \end{vmatrix}} \frac{\partial u_1}{\partial x}, \quad \frac{\partial u_2}{\partial t_1} = \frac{\partial^2 u_2}{\partial x^2} - 2 \frac{\begin{vmatrix} \frac{\partial u_1}{\partial x} & \frac{\partial u_2}{\partial x} \\ \frac{\partial^2 u_1}{\partial x \partial y} & \frac{\partial^2 u_2}{\partial x \partial y} \end{vmatrix}}{\begin{vmatrix} \frac{\partial u_1}{\partial x} & \frac{\partial u_2}{\partial x} \\ \frac{\partial u_1}{\partial y} & \frac{\partial u_2}{\partial y} \end{vmatrix}} \frac{\partial u_2}{\partial x},$$

$$\frac{\partial u_1}{\partial t_2}=\frac{\partial^2 u_1}{\partial y^2}-2\frac{\begin{vmatrix}\frac{\partial u_1}{\partial y} & \frac{\partial u_2}{\partial y}\\ \frac{\partial^2 u_1}{\partial x\partial y} & \frac{\partial^2 u_2}{\partial x\partial y}\end{vmatrix}}{\begin{vmatrix}\frac{\partial u_1}{\partial y} & \frac{\partial u_2}{\partial y}\\ \frac{\partial u_1}{\partial x} & \frac{\partial u_2}{\partial x}\end{vmatrix}}\frac{\partial u_1}{\partial y},\quad \frac{\partial u_2}{\partial t_2}=\frac{\partial^2 u_2}{\partial y^2}-2\frac{\begin{vmatrix}\frac{\partial u_1}{\partial y} & \frac{\partial u_2}{\partial y}\\ \frac{\partial^2 u_1}{\partial x\partial y} & \frac{\partial^2 u_2}{\partial x\partial y}\end{vmatrix}}{\begin{vmatrix}\frac{\partial u_1}{\partial y} & \frac{\partial u_2}{\partial y}\\ \frac{\partial u_1}{\partial x} & \frac{\partial u_2}{\partial x}\end{vmatrix}}\frac{\partial u_2}{\partial y}.$$

The equations

$$v_1=\{u_1,u_2;x,y\}_x,\quad w_1=[u_1,u_2;x,y]_x$$

give that

$$\frac{\partial^2 u_1}{\partial x^2}=\begin{vmatrix}v_1 & w_1\\ \frac{\partial u_1}{\partial x} & \frac{\partial u_1}{\partial y}\end{vmatrix}+2\frac{\begin{vmatrix}\frac{\partial u_1}{\partial x} & \frac{\partial u_2}{\partial x}\\ \frac{\partial^2 u_1}{\partial x\partial y} & \frac{\partial^2 u_2}{\partial x\partial y}\end{vmatrix}}{\begin{vmatrix}\frac{\partial u_1}{\partial x} & \frac{\partial u_2}{\partial x}\\ \frac{\partial u_1}{\partial y} & \frac{\partial u_2}{\partial y}\end{vmatrix}}\frac{\partial u_1}{\partial x},$$

$$\frac{\partial^2 u_2}{\partial x^2}=\begin{vmatrix}v_1 & w_1\\ \frac{\partial u_2}{\partial x} & \frac{\partial u_2}{\partial y}\end{vmatrix}+2\frac{\begin{vmatrix}\frac{\partial u_1}{\partial x} & \frac{\partial u_2}{\partial x}\\ \frac{\partial^2 u_1}{\partial x\partial y} & \frac{\partial^2 u_2}{\partial x\partial y}\end{vmatrix}}{\begin{vmatrix}\frac{\partial u_1}{\partial x} & \frac{\partial u_2}{\partial x}\\ \frac{\partial u_1}{\partial y} & \frac{\partial u_2}{\partial y}\end{vmatrix}}\frac{\partial u_2}{\partial x}.$$

The equations

$$v_2=\{u_1,u_2;x,y\}_y,\quad w_2=[u_1,u_2;x,y]_y$$

give that

$$\frac{\partial^2 u_1}{\partial y^2}=\begin{vmatrix}v_2 & w_2\\ \frac{\partial u_1}{\partial y} & \frac{\partial u_1}{\partial x}\end{vmatrix}+2\frac{\begin{vmatrix}\frac{\partial u_1}{\partial y} & \frac{\partial u_2}{\partial y}\\ \frac{\partial^2 u_1}{\partial x\partial y} & \frac{\partial^2 u_2}{\partial x\partial y}\end{vmatrix}}{\begin{vmatrix}\frac{\partial u_1}{\partial y} & \frac{\partial u_2}{\partial y}\\ \frac{\partial u_1}{\partial x} & \frac{\partial u_2}{\partial x}\end{vmatrix}}\frac{\partial u_1}{\partial y},$$

$$\frac{\partial^2 u_2}{\partial y^2}=\begin{vmatrix}v_2 & w_2\\ \frac{\partial u_2}{\partial y} & \frac{\partial u_2}{\partial x}\end{vmatrix}+2\frac{\begin{vmatrix}\frac{\partial u_1}{\partial y} & \frac{\partial u_2}{\partial y}\\ \frac{\partial^2 u_1}{\partial x\partial y} & \frac{\partial^2 u_2}{\partial x\partial y}\end{vmatrix}}{\begin{vmatrix}\frac{\partial u_1}{\partial y} & \frac{\partial u_2}{\partial y}\\ \frac{\partial u_1}{\partial x} & \frac{\partial u_2}{\partial x}\end{vmatrix}}\frac{\partial u_2}{\partial y}.$$

This implies that

$$\frac{\partial u_1}{\partial t_1}=\begin{vmatrix}v_1 & w_1\\ \frac{\partial u_1}{\partial x} & \frac{\partial u_1}{\partial y}\end{vmatrix},\quad \frac{\partial u_2}{\partial t_1}=\begin{vmatrix}v_1 & w_1\\ \frac{\partial u_2}{\partial x} & \frac{\partial u_2}{\partial y}\end{vmatrix}.$$

$$\frac{\partial u_1}{\partial t_2}=\begin{vmatrix}v_2 & w_2\\ \frac{\partial u_1}{\partial y} & \frac{\partial u_1}{\partial x}\end{vmatrix},\quad \frac{\partial u_2}{\partial t_2}=\begin{vmatrix}v_2 & w_2\\ \frac{\partial u_2}{\partial y} & \frac{\partial u_2}{\partial x}\end{vmatrix}.$$

Note that

$$\frac{\partial^2 u_1}{\partial t_1 \partial t_2} = \frac{\partial}{\partial t_2}\left(\frac{\partial u_1}{\partial t_1}\right) = \frac{\partial}{\partial t_2}\left(v_1 \frac{\partial u_1}{\partial y} - w_1 \frac{\partial u_1}{\partial x}\right)$$
$$= \frac{\partial v_1}{\partial t_2}\frac{\partial u_1}{\partial y} - \frac{\partial w_1}{\partial t_2}\frac{\partial u_1}{\partial x} + v_1 \frac{\partial}{\partial y}\left(\frac{\partial u_1}{\partial t_2}\right) - w_1 \frac{\partial}{\partial x}\left(\frac{\partial u_1}{\partial t_2}\right).$$

Here,

$$\frac{\partial}{\partial y}\left(\frac{\partial u_1}{\partial t_2}\right) = \frac{\partial}{\partial y}\left(v_2 \frac{\partial u_1}{\partial x} - w_2 \frac{\partial u_1}{\partial y}\right)$$
$$= \frac{\partial v_2}{\partial y}\frac{\partial u_1}{\partial x} - \frac{\partial w_2}{\partial y}\frac{\partial u_1}{\partial y} + v_2 \frac{\partial^2 u_1}{\partial x \partial y} - w_2 \frac{\partial^2 u_1}{\partial y^2},$$
$$\frac{\partial}{\partial x}\left(\frac{\partial u_1}{\partial t_2}\right) = \frac{\partial}{\partial x}\left(v_2 \frac{\partial u_1}{\partial x} - w_2 \frac{\partial u_1}{\partial y}\right)$$
$$= \frac{\partial v_2}{\partial x}\frac{\partial u_1}{\partial x} - \frac{\partial w_2}{\partial x}\frac{\partial u_1}{\partial y} + v_2 \frac{\partial^2 u_1}{\partial x^2} - w_2 \frac{\partial^2 u_1}{\partial x \partial y}.$$

Hence,

$$\frac{\partial^2 u_1}{\partial t_1 \partial t_2} = \left(-\frac{\partial w_1}{\partial t_2} + v_1 \frac{\partial v_2}{\partial y} - w_1 \frac{\partial v_2}{\partial x}\right)\frac{\partial u_1}{\partial x} + \left(\frac{\partial v_1}{\partial t_2} - v_1 \frac{\partial w_2}{\partial y} + w_1 \frac{\partial w_2}{\partial x}\right)\frac{\partial u_1}{\partial y}$$
$$+ (v_1 v_2 + w_1 w_2)\frac{\partial^2 u_1}{\partial x \partial y} - v_2 w_1 \frac{\partial^2 u_1}{\partial x^2} - v_1 w_2 \frac{\partial^2 u_1}{\partial y^2}.$$

On the other hand,

$$\frac{\partial^2 u_1}{\partial t_1 \partial t_2} = \frac{\partial}{\partial t_1}\left(\frac{\partial u_1}{\partial t_2}\right) = \frac{\partial}{\partial t_1}\left(v_2 \frac{\partial u_1}{\partial x} - w_2 \frac{\partial u_1}{\partial y}\right)$$
$$= \frac{\partial v_2}{\partial t_1}\frac{\partial u_1}{\partial x} - \frac{\partial w_2}{\partial t_1}\frac{\partial u_1}{\partial y} + v_2 \frac{\partial}{\partial x}\left(\frac{\partial u_1}{\partial t_1}\right) - w_2 \frac{\partial}{\partial y}\left(\frac{\partial u_1}{\partial t_1}\right).$$

Here,

$$\frac{\partial}{\partial x}\left(\frac{\partial u_1}{\partial t_1}\right) = \frac{\partial}{\partial x}\left(v_1 \frac{\partial u_1}{\partial y} - w_1 \frac{\partial u_1}{\partial x}\right)$$
$$= \frac{\partial v_1}{\partial x}\frac{\partial u_1}{\partial y} - \frac{\partial w_1}{\partial x}\frac{\partial u_1}{\partial x} + v_1 \frac{\partial^2 u_1}{\partial x \partial y} - w_1 \frac{\partial^2 u_1}{\partial x^2},$$
$$\frac{\partial}{\partial y}\left(\frac{\partial u_1}{\partial t_1}\right) = \frac{\partial}{\partial y}\left(v_1 \frac{\partial u_1}{\partial y} - w_1 \frac{\partial u_1}{\partial x}\right)$$
$$= \frac{\partial v_1}{\partial y}\frac{\partial u_1}{\partial y} - \frac{\partial w_1}{\partial y}\frac{\partial u_1}{\partial x} + v_1 \frac{\partial^2 u_1}{\partial y^2} - w_1 \frac{\partial^2 u_1}{\partial x \partial y}.$$

Hence,

$$\frac{\partial^2 u_1}{\partial t_1 \partial t_2} = \left(\frac{\partial v_2}{\partial t_1} - v_2 \frac{\partial w_1}{\partial x} + w_2 \frac{\partial w_1}{\partial y}\right)\frac{\partial u_1}{\partial x} + \left(-\frac{\partial w_2}{\partial t_1} + v_2 \frac{\partial v_1}{\partial x} - w_2 \frac{\partial v_1}{\partial y}\right)\frac{\partial u_1}{\partial y}$$
$$+ (v_1 v_2 + w_1 w_2)\frac{\partial^2 u_1}{\partial x \partial y} - v_2 w_1 \frac{\partial^2 u_1}{\partial x^2} - v_1 w_2 \frac{\partial^2 u_1}{\partial y^2}.$$

Comparing with the different expressions of $\frac{\partial^2 u_1}{\partial t_1 \partial t_2}$, we get the following equation:

$$\left(\frac{\partial w_1}{\partial t_2}+\frac{\partial v_2}{\partial t_1}-v_1\frac{\partial v_2}{\partial y}-v_2\frac{\partial w_1}{\partial x}+w_1\frac{\partial v_2}{\partial x}+w_2\frac{\partial w_1}{\partial y}\right)\frac{\partial u_1}{\partial x}$$
$$=\left(\frac{\partial w_2}{\partial t_1}+\frac{\partial v_1}{\partial t_2}-v_1\frac{\partial w_2}{\partial y}-v_2\frac{\partial v_1}{\partial x}+w_1\frac{\partial w_2}{\partial x}+w_2\frac{\partial v_1}{\partial y}\right)\frac{\partial u_1}{\partial y}.$$

Similarly, comparing with the different expressions of $\frac{\partial^2 u_2}{\partial t_1 \partial t_2}$, we get the other equation:

$$\left(\frac{\partial w_1}{\partial t_2}+\frac{\partial v_2}{\partial t_1}-v_1\frac{\partial v_2}{\partial y}-v_2\frac{\partial w_1}{\partial x}+w_1\frac{\partial v_2}{\partial x}+w_2\frac{\partial w_1}{\partial y}\right)\frac{\partial u_2}{\partial x}$$
$$=\left(\frac{\partial w_2}{\partial t_1}+\frac{\partial v_1}{\partial t_2}-v_1\frac{\partial w_2}{\partial y}-v_2\frac{\partial v_1}{\partial x}+w_1\frac{\partial w_2}{\partial x}+w_2\frac{\partial v_1}{\partial y}\right)\frac{\partial u_2}{\partial y}.$$

Note that $\frac{\partial(u_1,u_2)}{\partial(x,y)} \neq 0$. The above two equations imply that

$$\begin{cases} \dfrac{\partial w_1}{\partial t_2}+\dfrac{\partial v_2}{\partial t_1}-v_1\dfrac{\partial v_2}{\partial y}-v_2\dfrac{\partial w_1}{\partial x}+w_1\dfrac{\partial v_2}{\partial x}+w_2\dfrac{\partial w_1}{\partial y}=0, \\ \dfrac{\partial w_2}{\partial t_1}+\dfrac{\partial v_1}{\partial t_2}-v_1\dfrac{\partial w_2}{\partial y}-v_2\dfrac{\partial v_1}{\partial x}+w_1\dfrac{\partial w_2}{\partial x}+w_2\dfrac{\partial v_1}{\partial y}=0. \end{cases}$$

□

In fact, The above two equations are invariant under the action of

$$\begin{cases} v_1(x,y;t_1,t_2) \mapsto v_1(x-a_1t_1, y-b_1t_2; t_1,t_2), \\ v_2(x,y;t_1,t_2) \mapsto v_2(x-a_2t_1, y-b_2t_2; t_1,t_2), \\ w_1(x,y;t_1,t_2) \mapsto w_1(x-a_1t_1, y-b_1t_2; t_1,t_2)+a_2, \\ w_2(x,y;t_1,t_2) \mapsto w_2(x-a_2t_1, y-b_2t_2; t_1,t_2)+b_1. \end{cases} \tag{7.5}$$

References

[1] K. Aomoto and M. Kita, *Theory of Hypergeometric Functions*, with an appendix by Toshitake Kohno, Translated from the Japanese by Kenji Iohara, Springer Monographs in Mathematics, Springer-Verlag, Tokyo, 2011.

[2] K. Aomoto, Hypergeometric functions-their past, present, and ... (in Japanese), Sūgaku **45** (1993), 208-220.

[3] P. Appell and J. Kampé de Fériet, *Fonctions Hypergéométriques et Hypersphériques, Polynomes d'Hermite*, Gauthier-Villars, 1926.

[4] A. A. Beilinson, Yu. I. Manin and V. V. Schechtman, Sheaves of the Virasoro and Neveu-Schwarz algebras, In: *K-Theory, Arithmetic and Geometry*, Yu. I. Manin Ed., Lecture Notes in Math. **1289**, 52-66, Springer-Verlag, 1987.

[5] P. B. Cohen and J. Wolfart, Algebraic Appell-Lauricella functions, Analysis **12** (1992), 359-376.

[6] P. B. Cohen and J. Wolfart, Fonctions hypergéométriques en plusieurs variables et espaces des modules de variétés abéliennes, Ann. Sci. Éc. Norm. Sup. **26** (1993), 665-690.

[7] R. Dedekind, Schreiben an Herrn Borchardt über die Theorie der elliptischen Modul-Functionen, J. Reine Angew. Math. **83** (1877), 265-292.

[8] P. Deligne and G. D. Mostow, Monodromy of hypergeometric functions and non-lattice integral monodromy, Publ. Math. I.H.E.S. **63** (1986), 5-89.

[9] P. Deligne and G. D. Mostow, *Commensurabilities among Lattices in* $PU(1,n)$, Annals of Math. Studies **132**, Princeton University Press, 1993.

[10] I. Dorfman, *Dirac Structures and Integrability of Nonlinear Evolution Equations*, John Wiley & Sons, 1993.

[11] S. Gelbart, J. Rogawski and D. Soudry, Endoscopy, theta-liftings, and period integrals for the unitary group in three variables, Ann. of Math. **145** (1997), 419-476.

[12] G. Giraud, Sur certaines fonctions automorphes de deux variables, Ann. Ec. Norm. (3) **38** (1921), 43-164.

[13] J. J. Gray, *Linear Differential Equations and Group Theory from Riemann to Poincaré*, Second Edition, Birkhäuser, 2000.

[14] K. Heegner, Reduzierbare Abelsche Integrale und transformierbare automorphe Funktionen, Math. Ann. **131** (1956), 87-140.

[15] C. Hermite, *Œuvres de Charles Hermite*, Vol. II, E. Picard Ed., Gauthier-Villars, 1908.

[16] E. Hille, *Ordinary Differential Equations in the Complex Domain*, John Wiley & Sons, Inc., 1976.

[17] R. P. Holzapfel, *Geometry and Arithmetic Around Euler Partial Differential Equations*, Dt. Verl. d. Wiss., Berlin/Reidel, Dordrecht 1986.

[18] R. P. Holzapfel, *The Ball and Some Hilbert Problems*, Lectures in Mathematics, ETH Zürich, Birkhäuser, 1995.

[19] C. G. J. Jacobi, *Fundamenta nova theoriae functionum ellipticarum*, Cambridge University Press, 2012.

[20] F. Klein, Über die Transformation der elliptischen Functionen und die Auflösung der Gleichungen fünften Grades, Math. Ann. **14** (1879), 111-172.

[21] F. Klein, *Lectures on the Icosahedron and the Solution of Equations of the Fifth Degree*, Translated by G. G. Morrice, second and revised edition, Dover Publications, Inc., 1956.

[22] F. Klein, *Vorlesungen über die Entwicklung der Mathematik im 19. Jahrhundert*, Berlin, 1926.

[23] F. Klein, *Vorlesungen über die hypergeometrische Funktion*, Grundlehren der mathematischen Wissenschaften, Bd. **39**, reprint, Springer-Verlag, 1981.

[24] I. M. Krichever and S. P. Novikov, Holomorphic bundles on algebraic curves and nonlinear equations, Uspekhi Mat. Nauk **35** (1980), 47-68.

[25] R. P. Langlands, Some contemporary problems with origins in the Jugendtraum, *Mathematical developments arising from Hilbert problems* (Proc. Sympos. Pure Math., Vol. XXVIII, Northern Illinois Univ., De Kalb, Ill., 1974), pp. 401-418, Amer. Math. Soc., Providence, R. I., 1976.

[26] R. P. Langlands and D. Ramakrishnan, *The Zeta Functions of Picard Modular Surfaces*, Les Publications CRM, Université de Montréal, 1992.

[27] F. Lemmermeyer, *Reciprocity Laws from Euler to Eisenstein*, Springer Monographs in Mathematics, Springer-Verlag, 2000.

[28] G. D. Mostow, Generalized Picard lattices arising from half-integral conditions, Publ. Math. I.H.E.S. **63** (1986), 91-106.

[29] A. I. Ovseevič, Abelian extensions of fields of CM type, Funkcional. Anal. i Priložen. **8** (1974), no.1, 16-24.

[30] E. Papperitz, Untersuchungen über die algebraische Transformation der hypergeometrischen Functionen, Math. Ann. **27** (1886), 315-356.

[31] E. Papperitz, Ueber die Darstellung der hypergeometrischen Transcendenten durch eindeutige Functionen, Math. Ann. **34** (1889), 247-296.

[32] E. Picard, Sur des fonctions de deux variables indépendantes analogues aux fonctions modulaires, Acta Math. **2** (1883), 114-135.

[33] E. Picard, Sur les formes quadratiques ternaires indéfinies à indéterminées conjuguées et sur les fonctions hyperfuchsiennes correspondantes, Acta Math. **5** (1884), 121-184.

[34] E. Picard, Sur les fonctions hyperfuchsiennes provenant des séries hypergéométriques de deux variables, Ann. Ecole Norm. Sup. 3e série **62** (1885), 357-384.

[35] E. Picard, Mémoire sur la théorie des fonctions algébriques de deux variables, J. Math. Pures Appl. 4e série **5** (1889), 135-319.

[36] H. Poincaré, *Papers on Fuchsian Functions*, Translated by J. Stillwell, Springer-Verlag, 1985.

[37] J. Polchinski, *String Theory*, Vol. I, II, Cambridge University Press, 1998.

[38] J. D. Rogawski, *Automorphic Representations of Unitary Groups in Three Variables*, Annals of Math. Studies **123**, Princeton University Press, 1990.

[39] H. A. Schwarz, Über diejenigen Fälle, in welchen die Gaussische hypergeometrische Reihe eine algebraische Function ihres vierten Elementes darstellt, J. Reine Angew. Math. **75** (1872), 292-335.

[40] H. Shiga, One attempt to the $K3$ modular function I, Ann. Scuola Norm. Sup. Pisa Cl. Sci. (4) **6** (1979), 609-635.

[41] H. Shiga, One attempt to the $K3$ modular function II, Ann. Scuola Norm. Sup. Pisa Cl. Sci. (4) **8** (1981), 157-182.

[42] H. Shiga, On $K3$ modular functions (Japanese), Sūgaku **38** (1986), 116-132.

[43] H. Shiga, On the representation of the Picard modular function by θ constants, I, II, Publ. Res. Inst. Math. Sci. **24** (1988), 311-360.

[44] K. Matsumoto and H. Shiga, A variant of Jacobi type formula for Picard curves, J. Math. Soc. Japan **62** (2010), 305-319.

[45] G. Shimura, On purely transcendental fields of automorphic functions of several variables, Osaka J. Math. **1** (1964), 1-14.

[46] G. Shimura, Arithmetic of unitary groups, Ann. of Math. **79** (1964), 369-409.

[47] G. Shimura, *Introduction to The Arithmetic Theory of Automorphic Functions*, Publ. Math. Soc. Japan No. **11**, Iwanami Shoten and Princeton University Press, 1971.

[48] G. Shimura, The arithmetic of automorphic forms with respect to a unitary group, Ann. of Math. **107** (1978), 569-605.

[49] G. Shimura, *Abelian Varieties with Complex Multiplication and Modular Functions*, Princeton Mathematical Series **46**, Princeton University Press, 1998.

[50] V. V. Sokolov and A. B. Shabat, Classification of integrable evolution equations, In: Soviet Scientific Reviews, Section C: Mathematical Physics Reviews, Vol. **4**, 221-280, Harwood, New York, 1984.

[51] T. Terada, Fonctions hypergéométriques F_1 et fonctions automorphes, I, J. Math. Soc. Japan **35** (1983), 451-475.

[52] T. Terada, Fonctions hypergéométriques F_1 et fonctions automorphes, II, Groupes discontinus arithmétiquement définis, J. Math. Soc. Japan **37** (1985), 173-185.

[53] S. G. Vlăduţ, *Kronecker's Jugendtraum and Modular Functions*, Translated from the Russian by M. Tsfasman, Studies in the Development of Modern Mathematics, Vol. **2**, Gordon and Breach Science Publishers, New York, 1991.

[54] H. Weber, *Lehrbuch der Algebra*, Vol. III, Chelsea Publishing Company, New York, 1961.

[55] J. Weiss, The Painlevé property for partial differential equations. II: Bäcklund transformation, Lax pairs, and the Schwarzian derivative, J. Math. Phys. **24** (1983), 1405-1413.

[56] J. Weiss, Bäcklund transformation and the Painlevé property, J. Math. Phys. **27** (1986), 1293-1305.

[57] G. Wilson, On the quasi-Hamiltonian formalism of the KdV equation, Phys. Lett. A **132** (1988), 445-450.

[58] E. Witten, Quantum field theory, Grassmannians, and algebraic curves, Comm. Math. Phys. **113** (1988), 529-600.

[59] M. Yoshida, *Fuchsian Differential Equations, With Special Emphasis on the Gauss-Schwarz Theory*, Aspects of Mathematics, E11, Friedr. Vieweg & Sohn, Braunschweig, 1987.

[60] M. Yoshida, The Schwarz program (Japanese), Sūgaku **40** (1988), 36-46.

[61] K. Iwasaki, H. Kimura, S. Shimomura and M. Yoshida, *From Gauss to Painlevé, a Modern Theory of Special Functions*, Aspects of Mathematics, E16, Friedr. Vieweg & Sohn, Braunschweig, 1991.

[62] M. Yoshida, *Hypergeometric Functions, My Love, Modular Interpretations of Configuration Spaces*, Aspects of Mathematics, E32, Friedr. Vieweg & Sohn, Braunschweig, 1997.

[63] T. Sasaki and M. Yoshida, Schwarzian derivatives and uniformization, in *The Kowalevski property (Leeds, 2000)*, 271-286, CRM Proc. Lecture Notes, **32**, Amer. Math. Soc., Providence, RI, 2002.

[64] K. Matsumoto, T. Sasaki and M. Yoshida, Recent progress of Gauss-Schwarz theory and related geometric structures, Mem. Fac. Sci. Kyushu Univ. Ser. A **47** (1993), 283-381.

Part II. Hessian polyhedra, invariant theory and Appell hypergeometric functions

Contents

1. Introduction

Geometry is one of the oldest and most basic branches of mathematics, as is algebra. Nowhere is the interplay between the two more pronounced than in group theory, and that interplay with group theory acting as a mediator between geometry and algebra. It was Felix Klein in his "Erlangen Programme" who put group theory at the basis of geometry (see [32]). Klein advocated considering geometry as invariance properties under a group of automorphisms.

Polyhedra, particularly the Platonic solids, are the most classical objects in geometry, which have been studied by geometers for thousands of years. Furthermore, finding physical applications of polyhedra is a similarly ancient pursuit. For example, Plato was so captivated by the perfect forms of the five regular solids. Kepler also attributed cosmic significance to the Platonic solids (see [61]).

In his famous lecture: Mathematical problems (see [26]), which was delivered before the International Congress of Mathematicians at Paris in 1900, David Hilbert said: "It often happens that the same special problem finds application in the most unlike branches of mathematical knowledge. For example, how convincingly has Felix Klein, in his work on the icosahedron, pictured the significance which attaches to the problem of the regular polyhedra in elementary geometry, in group theory, in the theory of equations and in that of linear differential equations."

Klein's lectures on the icosahedron, published in 1884, soon became one of the most famous books on mathematics for a half-century; no other book so forcefully put in the limelight the fundamental unity of mathematics (see J. Dieudonné's review on P. Slodowy's paper [58] or see [2]). Klein showed that four apparently disjoint theories: the symmetries of the icosahedron (geometry), the resolution of equations of the fifth degree (algebra), the differential equation of hypergeometric functions (analysis) and the modular equations of elliptic modular functions (arithmetic) were in fact dominated by the structure of a single object, the simple group A_5 of 60 elements. It is the group of rotations leaving the icosahedron invariant, the monodromy group of a hypergeometric equation, and the commutator subgroup of the Galois group of the general fifth degree equation. Klein also pointed out the connection with invariant theory and algebraic geometry (see [33], [42] and [43], especially [34], [35], [36], [37], [38], [39], [40] and [41] for the details).

In a letter to J. D. Gray in response to his questions regarding modern treatments of Klein's fascinating but difficult book, J.-P. Serre said (see [56]) that there is so much more in Klein's book, and in Fricke's! Invariants, hypergeometric functions, and everywhere, a wealth of beautiful formulae!

Now, let us recall some basic facts in Klein's lecture.

There is a long history for solving algebraic equations by means of transcendental functions (see [42] or [23], [29]). Put

$$K=\int_0^{\frac{\pi}{2}}\frac{d\theta}{\sqrt{1-k^2\sin^2\theta}},\quad K'=\int_0^{\frac{\pi}{2}}\frac{d\theta}{\sqrt{1-k'^2\sin^2\theta}},\quad k^2+k'^2=1.$$

Let λ be a modulus related to k by a fifth order transformation. In 1829, Jacobi (see [31], p. 28) showed that $\sqrt[4]{k}=u$ and $\sqrt[4]{\lambda}=v$ are related by the modular equation (Jacobian modular equation):

$$u^6-v^6+5u^2v^2(u^2-v^2)+4uv(1-u^4v^4)=0.$$

In 1858, Hermite (see [25], pp.5-12) found the resolvent of fifth degree which derives from Jacobi's equation of sixth degree. He did this as follows: set

$$y=(v_\infty-v_0)(v_1-v_4)(v_2-v_3),$$

where v_i are the roots of Jacobi's equation. The corresponding resolvent is

$$y^5-2^4\cdot 5^3u^4(1-u^8)^2y-2^6\cdot\sqrt{5^5}u^3(1-u^8)^2(1+u^8)=0.$$

This takes the form of a Bring equation

$$t^5-t-A=0$$

by setting

$$y=2\cdot\sqrt[4]{5^3}\cdot u\sqrt{1-u^8}\cdot t,\quad A=\frac{2}{\sqrt[4]{5^5}}\cdot\frac{1+u^8}{u^2\sqrt{1-u^8}}.$$

Hence, by a Tschirnhaus transformation any quintic equation can be brought into the Bring form, which can be solved by the elliptic functions for the transformations of fifth order, this gives a solution of the quintic equation.

The story of the general equation of fifth degree is ultimately culminating in Klein's book [42]. Here geometry, analysis and algebra were brought to bear, showing not only how one can solve an equation of fifth degree, but also the relationships between the different methods which had been used previously. There is a universal Galois resolvent for general quintic equations: the icosahedral equation. Klein built the entire theory of quintic equations on these foundations. He showed that in studying this resolvent one can apply the geometry of the icosahedron, making the algebra very geometric. The solution suggested by Klein had an algebraic and a transcendental part. For the algebraic part, the classical approach is the reduction of equation with solvable Galois group to a pure equation:

$$x^n = X.$$

The icosahedral approach is the reduction of equation of fifth degree to solution of the icosahedral equation:

$$\frac{H^3(x)}{1728 f^5(x)} = X.$$

For the transcendental part, the classical approach is the solution of the pure equation by means of logarithms:

$$x = e^{\frac{1}{n}\log X}.$$

The icosahedral approach is the solution of the icosahedral equation by means of elliptic functions:

$$x = q^{\frac{2}{5}} \frac{\theta_1\left(\frac{2\pi i K'}{K}, q^5\right)}{\theta_1\left(\frac{\pi i K'}{K}, q^5\right)}.$$

In 1888, Klein ([40]) sketched in a letter to Jordan how one could use a similar process to solve the equation of 27th degree determining the 27 lines on a smooth cubic surface. This leads to the development of the invariant theory of several finite collineation groups (unitary reflection groups in projective space). Moreover, this laid the grounds for a theory of arithmetic groups. Following suggestions of Klein and Jordan, Maschke ([51], [52]) and Burkhardt ([5], [6]) solved the equation of 27th degree for the lines of the cubic surface by using hyperelliptic functions. Here, the algebraic part is the reduction to the Maschke form problem:

$$\frac{F_{24}(z)}{F_{12}^2(z)} = \alpha, \qquad \frac{F_{30}(z)}{F_{12}(z)F_{18}(z)} = \beta, \qquad \frac{F_{18}^2(z)}{F_{12}^3(z)} = \gamma.$$

The transcendental part is the solution of the Maschke equation in terms of the hyperelliptic functions $Z_{\alpha,\beta}$ of

$$\mathbb{P}^3 = \left\{ Z_{\alpha,\beta} = \frac{1}{2}(X_{\alpha,\beta} - X_{-\alpha,-\beta}) \right\}.$$

In fact, thinking of each of the 27 lines as defining a point on the Grassmannian $\mathbf{G}(2,4) \subset \mathbb{P}^5$, one can project to 27 points onto a line, that is, giving the roots of a degree 27 equation. In 1890, Burkhardt showed that the equation could be solved involving the calculation of the periods of an abelian surface. The method, due to Klein and Jordan, carried out by Maschke and Burkhardt, uses the representation of $G_{25,920}$ in $\mathbb{P}^3$ and its invariants.

Later on Coble reconsidered the problem, essentially by considering the Galois group $W(E_6)$ not as a reflection group, but as a Cremona group, i.e., as a finite group of birational automorphisms of the plane $\mathbb{P}^2$ (see [7], [8], [9], [10] and [11]). The approach of Coble uses the representation in $\mathbb{P}^4$ and reduces the problem to the form problem for this group.

In the general case, Umemura (see [59]) gave the resolution of algebraic equations by a Siegel modular function which is explicitly expressed by theta constants. For row vectors $n_1, m_2 \in \mathbb{R}^g$, $z \in \mathbb{C}^g$ and a symmetric $g \times g$ matrix τ with positive definite imaginary part, we define the theta function (see [55])

$$\theta \begin{bmatrix} m_1 \\ m_2 \end{bmatrix} (z, \tau) = \sum_{\xi \in \mathbb{Z}^g} e \left[\frac{1}{2}(\xi + m_1)\tau^t(\xi + m_1) + (\xi + m_1)^t(z + m_2) \right],$$

where $e(x) = \exp(2\pi i x)$. In particular, The theta constant $\theta \begin{bmatrix} m_1 \\ m_2 \end{bmatrix} (0, \tau)$ is denoted by $\theta \begin{bmatrix} m_1 \\ m_2 \end{bmatrix} (\tau)$.

Theorem (see [59]). *Let*

$$a_0 X^n + a_1 X^{n-1} + \cdots + a_n = 0, \quad a_0 \neq 0, a_i \in \mathbb{C}(0 \leq i \leq n)$$

be an algebraic equation irreducible over a certain subfield of $\mathbb{C}$, *then a root of it is given by*

$$\begin{aligned}
&(\theta \begin{bmatrix} \frac{1}{2} & 0 & \cdots & 0 \\ 0 & 0 & \cdots & 0 \end{bmatrix} (\Omega)^4 \theta \begin{bmatrix} \frac{1}{2} & \frac{1}{2} & 0 & \cdots & 0 \\ 0 & 0 & 0 & \cdots & 0 \end{bmatrix} (\Omega)^4 \\
&+ \theta \begin{bmatrix} 0 & \cdots & 0 \\ 0 & \cdots & 0 \end{bmatrix} (\Omega)^4 \theta \begin{bmatrix} 0 & \frac{1}{2} & 0 & \cdots & 0 \\ 0 & 0 & 0 & \cdots & 0 \end{bmatrix} (\Omega)^4 \\
&- \theta \begin{bmatrix} 0 & 0 & \cdots & 0 \\ \frac{1}{2} & 0 & \cdots & 0 \end{bmatrix} (\Omega)^4 \theta \begin{bmatrix} 0 & \frac{1}{2} & 0 & \cdots & 0 \\ \frac{1}{2} & 0 & 0 & \cdots & 0 \end{bmatrix} (\Omega)^4) \\
&\times (2\theta \begin{bmatrix} \frac{1}{2} & 0 & \cdots & 0 \\ 0 & 0 & \cdots & 0 \end{bmatrix} (\Omega)^4 \theta \begin{bmatrix} \frac{1}{2} & \frac{1}{2} & 0 & \cdots & 0 \\ 0 & 0 & 0 & \cdots & 0 \end{bmatrix} (\Omega)^4)^{-1},
\end{aligned}$$

where Ω *is the period matrix of a hyperelliptic curve* $C : Y^2 = F(X)$ *with*

$$F(X) = X(X-1)(a_0 X^n + a_1 X^{n-1} + \cdots + a_n)$$

for n *odd and*

$$F(X) = X(X-1)(X-2)(a_0 X^n + a_1 X^{n-1} + \cdots + a_n)$$

for n *even.*

Let us recall some basic facts about the icosahedral group (see [42] or [22], [60]), which is generated by the following homogeneous substitutions:

$$S : \begin{cases} z_1' = \pm\epsilon^3 z_1, \\ z_2' = \pm\epsilon^2 z_2, \end{cases}$$

$$U : \begin{cases} z_1' = \mp z_2, \\ z_2' = \pm z_1, \end{cases}$$

$$T : \begin{cases} \sqrt{5}\cdot z_1' = \mp(\epsilon-\epsilon^4)z_1 \pm (\epsilon^2-\epsilon^3)z_2, \\ \sqrt{5}\cdot z_2' = \pm(\epsilon^2-\epsilon^3)z_1 \pm (\epsilon-\epsilon^4)z_2, \end{cases}$$

where

$$\epsilon = e^{\frac{2\pi i}{5}}.$$

In fact, U can be generated by S and T (see [22]).

The set of forms for the icosahedron is given by

$$f = z_1 z_2 (z_1^{10} + 11 z_1^5 z_2^5 - z_2^{10});$$

$$H = \frac{1}{121}\begin{vmatrix} \frac{\partial^2 f}{\partial z_1^2} & \frac{\partial^2 f}{\partial z_1 \partial z_2} \\ \frac{\partial^2 f}{\partial z_2 \partial z_1} & \frac{\partial^2 f}{\partial z_2^2} \end{vmatrix} = -(z_1^{20}+z_2^{20}) + 228(z_1^{15}z_2^5 - z_1^5 z_2^{15}) - 494 z_1^{10} z_2^{10};$$

$$T = -\frac{1}{20}\begin{vmatrix} \frac{\partial f}{\partial z_1} & \frac{\partial f}{\partial z_2} \\ \frac{\partial H}{\partial z_1} & \frac{\partial H}{\partial z_2} \end{vmatrix} = (z_1^{30}+z_2^{30}) + 522(z_1^{25}z_2^5 - z_1^5 z_2^{25}) - 10005(z_1^{20}z_2^{10} + z_1^{10}z_2^{20}).$$

In fact, $f = 0$ represents the 12 summits of the icosahedron, $H = 0$ represents the 20 summits of the pentagon-dodecahedron and $T = 0$ represents the 30 mid-edge points. Moreover, f, H and T satisfy the identity:

$$T^2 = -H^3 + 1728 f^5.$$

The icosahedral equation is defined by

$$Z = \frac{H^3}{1728 f^5}.$$

In his paper [45], Klein studied the icosahedral irrationality η defined by the equation:

$$\frac{H^3(\eta)}{1728 f^5(\eta)} = J,$$

where J is Klein's j-invariant in the theory of elliptic modular functions.

Let the octahedron be the form:

$$t = z_1^6 + 2z_1^5 z_2 - 5z_1^4 z_2^2 - 5z_1^2 z_2^4 - 2z_1 z_2^5 + z_2^6.$$

The Hessian form of t is

$$W = -z_1^8 + z_1^7 z_2 - 7z_1^6 z_2^2 - 7z_1^5 z_2^3 + 7z_1^3 z_2^5 - 7z_1^2 z_2^6 - z_1 z_2^7 - z_2^8.$$

Put

$$r = \frac{t^2}{f},$$

this gives the function-theoretic resolvent (analytic resolvent):

$$Z : (Z-1) : 1 = (r-3)^3(r^2 - 11r + 64) : r(r^2 - 10r + 45)^2 : -1728.$$

In fact, set

$$\xi = \frac{125}{z^5 - z^{-5} + 11} = \frac{125 z_1^6 z_2^6}{f(z_1, z_2)},$$

we have (see [22])

$$Z : (Z-1) : 1 = (\xi^2 - 10\xi + 5)^3 : (\xi^2 - 22\xi + 125)(\xi^2 - 4\xi - 1)^2 : -1728\xi,$$

which corresponds to the transformation of fifth order for elliptic functions (see [45]):

$$J : (J-1) : 1 = (\tau^2 - 10\tau + 5)^3 : (\tau^2 - 22\tau + 125)(\tau^2 - 4\tau - 1)^2 : -1728\tau.$$

On the other hand, the form-theoretic resolvent (algebraic resolvent) of t is:

$$t^5 - 10f \cdot t^3 + 45f^2 \cdot t - T = 0.$$

The form-theoretic resolvent (algebraic resolvent) of W is:

$$W^5 + 40f^2 \cdot W^2 - 5fH \cdot W + H^2 = 0.$$

Let

$$u = \frac{12tf^2}{T}, \quad v = \frac{12Wf}{H}.$$

The analytic and algebraic resolvent of u is

$$48u^5(1-Z)^2 - 40u^3(1-Z) + 15u - 4 = 0.$$

The analytic and algebraic resolvent of v is

$$Zv^5 + 40v^2 - 60v + 144 = 0.$$

Furthermore,

$$u = \frac{12}{r^2 - 10r + 45}, \quad v = \frac{12}{r-3}.$$

Put

$$\begin{cases} \phi_\infty = 5z_1^2 z_2^2, \\ \phi_\nu = (\epsilon^\nu z_1^2 + z_1 z_2 - \epsilon^{4\nu} z_2^2)^2, \quad \nu = 0, 1, 2, 3, 4. \end{cases}$$

The form-theoretic resolvent (algebraic resolvent) of ϕ is

$$\phi^6 - 10f \cdot \phi^3 + H \cdot \phi + 5f^2 = 0.$$

Set

$$\zeta = \frac{\phi H}{12 f^2}.$$

Then the analytic and algebraic resolvent of ζ is:

$$\zeta^6 - 10Z \cdot \zeta^3 + 12Z^2 \cdot \zeta + 5Z^2 = 0.$$

Klein set

$$\mathbf{A}_0 = z_1 z_2, \quad \mathbf{A}_1 = z_1^2, \quad \mathbf{A}_2 = -z_2^2$$

and

$$A = \mathbf{A}_0^2 + \mathbf{A}_1 \mathbf{A}_2.$$

The invariants f, H and T can be expressed as the functions of $\mathbf{A}_0$, $\mathbf{A}_1$ and $\mathbf{A}_2$:

$$B = 8\mathbf{A}_0^4\mathbf{A}_1\mathbf{A}_2 - 2\mathbf{A}_0^2\mathbf{A}_1^2\mathbf{A}_2^2 + \mathbf{A}_1^3\mathbf{A}_2^3 - \mathbf{A}_0(\mathbf{A}_1^5 + \mathbf{A}_2^5),$$

$$\begin{aligned} C = {} & 320\mathbf{A}_0^6\mathbf{A}_1^2\mathbf{A}_2^2 - 160\mathbf{A}_0^4\mathbf{A}_1^3\mathbf{A}_2^3 + 20\mathbf{A}_0^2\mathbf{A}_1^4\mathbf{A}_2^4 + 6\mathbf{A}_1^5\mathbf{A}_2^5 \\ & - 4\mathbf{A}_0(\mathbf{A}_1^5 + \mathbf{A}_2^5)(32\mathbf{A}_0^4 - 20\mathbf{A}_0^2\mathbf{A}_1\mathbf{A}_2 + 5\mathbf{A}_1^2\mathbf{A}_2^2) + \mathbf{A}_1^{10} + \mathbf{A}_2^{10}, \end{aligned}$$

$$\begin{aligned} D = {} & (\mathbf{A}_1^5 - \mathbf{A}_2^5)[-1024\mathbf{A}_0^{10} + 3840\mathbf{A}_0^8\mathbf{A}_1\mathbf{A}_2 - 3840\mathbf{A}_0^6\mathbf{A}_1^2\mathbf{A}_2^2 \\ & + 1200\mathbf{A}_0^4\mathbf{A}_1^3\mathbf{A}_2^3 - 100\mathbf{A}_0^2\mathbf{A}_1^4\mathbf{A}_2^4 + \mathbf{A}_1^{10} + \mathbf{A}_2^{10} + 2\mathbf{A}_1^5\mathbf{A}_2^5 \\ & + \mathbf{A}_0(\mathbf{A}_1^5 + \mathbf{A}_2^5)(352\mathbf{A}_0^4 - 160\mathbf{A}_0^2\mathbf{A}_1\mathbf{A}_2 + 10\mathbf{A}_1^2\mathbf{A}_2^2)]. \end{aligned}$$

They satisfy the identity

$$D^2 = -1728B^5 + C^3 + 720ACB^3 - 80A^2C^2B + 64A^3(5B^2 - AC)^2.$$

The actions of the icosahedral group on $\mathbf{A}_0$, $\mathbf{A}_1$ and $\mathbf{A}_2$ are given by

$$S: \quad \mathbf{A}_0' = \mathbf{A}_0, \quad \mathbf{A}_1' = \epsilon\mathbf{A}_1, \quad \mathbf{A}_2' = \epsilon^4\mathbf{A}_2;$$

$$T: \begin{cases} \sqrt{5}\mathbf{A}_0' = \mathbf{A}_0 + \mathbf{A}_1 + \mathbf{A}_2, \\ \sqrt{5}\mathbf{A}_1' = 2\mathbf{A}_0 + (\epsilon^2 + \epsilon^3)\mathbf{A}_1 + (\epsilon + \epsilon^4)\mathbf{A}_2, \\ \sqrt{5}\mathbf{A}_2' = 2\mathbf{A}_0 + (\epsilon + \epsilon^4)\mathbf{A}_1 + (\epsilon^2 + \epsilon^3)\mathbf{A}_2; \end{cases}$$

$$U: \quad \mathbf{A}_0' = -\mathbf{A}_0, \quad \mathbf{A}_1' = -\mathbf{A}_2, \quad \mathbf{A}_2' = -\mathbf{A}_1.$$

This leads to the investigation of the binary quadratic forms:

$$\mathbf{A}_1 z_1^2 + 2\mathbf{A}_0 z_1 z_2 - \mathbf{A}_2 z_2^2$$

and Hilbert modular surfaces (see [27]).

It is well-known that the group A_5 is isomorphic to the finite group I of those elements of $SO(3)$ which carry a given icosahedron centered at the origin of the standard Euclidean space $\mathbb{R}^3$ to itself. The group I operates linearly on $\mathbb{R}^3$ (standard coordinates x_0, x_1, x_2) and thus also on $\mathbb{RP}^2$ and $\mathbb{CP}^2$. We are concerned with the action on $\mathbb{CP}^2$ (see [27]). A curve in $\mathbb{CP}^2$ which is mapped to itself by all elements of I is given by a homogeneous polynomial in x_0, x_1, x_2

which is I-invariant up to constant factors and hence I-invariant, because I is a simple group. The graded ring of all I-invariant polynomials in x_0, x_1, x_2 is generated by homogeneous polynomials A, B, C, D of degrees 2, 6, 10, 15 with $A = x_0^2 + x_1^2 + x_2^2$. The action of I on $\mathbb{CP}^2$ has exactly one minimal orbit where "minimal" means that the number of points in the orbit is minimal. This orbit has six points, they are called poles. These are the points of $\mathbb{RP}^2 \subset \mathbb{CP}^2$ which are represented by the six axes through the vertices of the icosahedron. Klein uses coordinates

$$A_0 = x_0, \quad A_1 = x_1 + ix_2, \quad A_2 = x_1 - ix_2$$

and puts the icosahedron in such a position that the six poles are given by

$$(A_0, A_1, A_2) = \left(\frac{\sqrt{5}}{2}, 0, 0\right), \quad \left(\frac{1}{2}, \varepsilon^\nu, \varepsilon^{-\nu}\right)$$

with $\varepsilon = \exp(2\pi i/5)$ and $0 \le \nu \le 4$.

The invariant curve $A = 0$ does not pass through the poles. There is exactly one invariant curve $B = 0$ of degree 6 which passes through the poles, exactly one invariant curve $C = 0$ of degree 10 which has higher multiplicity than the curve $B = 0$ in the poles and exactly one invariant curve $D = 0$ of degree 15. In fact, $B = 0$ has an ordinary double point (multiplicity 2) in each pole, $C = 0$ has a double cusp (multiplicity 4) in each pole and $D = 0$ is the union of the 15 lines connecting the six poles. Klein gives formulas for the homogeneous polynomials A, B, C, D (determined up to constant factors). They generate the ring of all I-invariant polynomials. According to Klein the ring of I-invariant polynomials is given as follows:

$$\mathbb{C}[A_0, A_1, A_2]^I = \mathbb{C}[A, B, C, D]/(R(A, B, C, D) = 0).$$

The relation $R(A, B, C, D) = 0$ is of degree 30.

The equations for B and C show that the two tangents of $B = 0$ in the pole $(\sqrt{5}/2, 0, 0)$ are given by $A_1 = 0$, $A_2 = 0$. They coincide with the tangents of $C = 0$ in that pole. Therefore the curves $B = 0$ and $C = 0$ have in each pole the intersection multiplicity 10. Thus they intersect only in the poles.

When we restrict the action of I to the conic $A = 0$, we get the well-known action of I on $\mathbb{CP}^1$. The curves $B = 0$, $C = 0$, $D = 0$ intersect $A = 0$ transversally in 12, 20, 30 points respectively.

It is well-known that Klein's lectures ([42]) on the icosahedron and the solution of equations of fifth degree is one of the most important and influential books of 19th-century mathematics. Since the time (1884) of the publication of Klein's lectures, there has been a standing interest in the mathematics of the icosahedron, i.e. the mathematics where the geometry and the symmetry of the icosahedron play an important role (see [44]).

This interest has increased in the last few decades. Let us mention some recent developments: the study of the Klein singularities (Arnold and Brieskorn), the study of the Hilbert modular surfaces (Hirzebruch), the connection between the platonic solids (or the finite subgroups of $SU(2)$) and the complex, simple Lie groups of the type A_r, D_r, E_6, E_7, E_8, discovered by Grothendieck and Brieskorn, the connection between singularities and string duality in the M-theory (Witten, Vafa), the study of the theory of Dessins d'Enfants called by its founder Grothendieck (see [24]).

From the point of view of representation theory, by the regular polyhedra, we can obtain dihedral, tetrahedral, octahedral and icosahedral representations. These lead to Artin's conjecture and Langlands programme (see [47] and [48]). Among them, the icosahedral equations remain intractable in general. The equations with icosahedral Galois groups has a greater historical appeal since they are the simplest equations completely inaccessible to the classical theory of equations with abelian Galois groups.

These results have shown how the icosahedron and the platonic solids have thrown new light in a surprising way on previously separate developments.

Recently, polyhedra, which include the Platonic solids, were found to arise naturally in a number of diverse problems in physics, chemistry, biology and a variety of other disciplines. In his book [61], Weyl gave a general discussion of symmetry in the natural world. Atiyah and Sutcliffe ([3]) gave some examples which appear in physics, chemistry and mathematics, such as the study of electrons on a sphere, cages of carbon atoms, central configurations of gravitating point particles, rare gas microclusters, soliton models of nuclei (the Skyrme model), magnetic monopole scattering and geometrical problems concerning point particles. Moreover, a particularly interesting application of polyhedra in biology is provided by the structure of spherical virus shells, such as HIV which is built around a trivalent polyhedron with icosahedral symmetry.

In the present paper, we will give the complex counterpart of Klein's book [42], i.e., a story about complex regular polyhedra (which were also called regular pseudo-polyhedra defined over $\mathbb{C}$ by Grothendieck in [24]). We will show that the following four apparently disjoint theories: the symmetries of the Hessian polyhedra (geometry), the resolution of some system of algebraic equations (algebra), the system of partial differential equations of Appell hypergeometric functions (analysis) and the modular equation of Picard modular functions (arithmetic) are in fact dominated by the structure of a single object, the Hessian group $\mathfrak{G}'_{216}$. There are two finite unitary groups generated by reflections corresponding to this collineation group. One, of order 648, is the symmetry group of the complex regular polyhedron $3\{3\}3\{3\}3$, and the other, of order 1296, is the symmetry group of the regular complex polyhedron $2\{4\}3\{3\}3$ or its reciprocal $3\{3\}3\{4\}2$. Their beautiful figures (which come from [18], p.121, p.125 and p.126) are as follows:

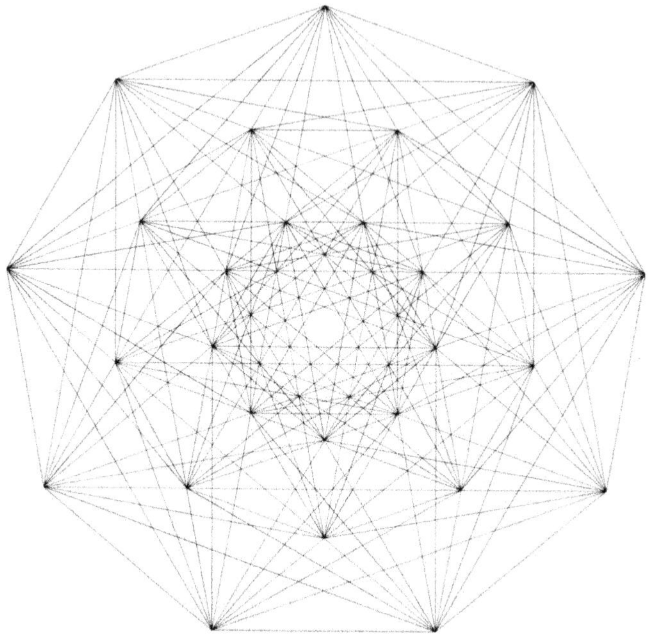

Figure 1. The Hessian polyhedron 3{3}3{3}3

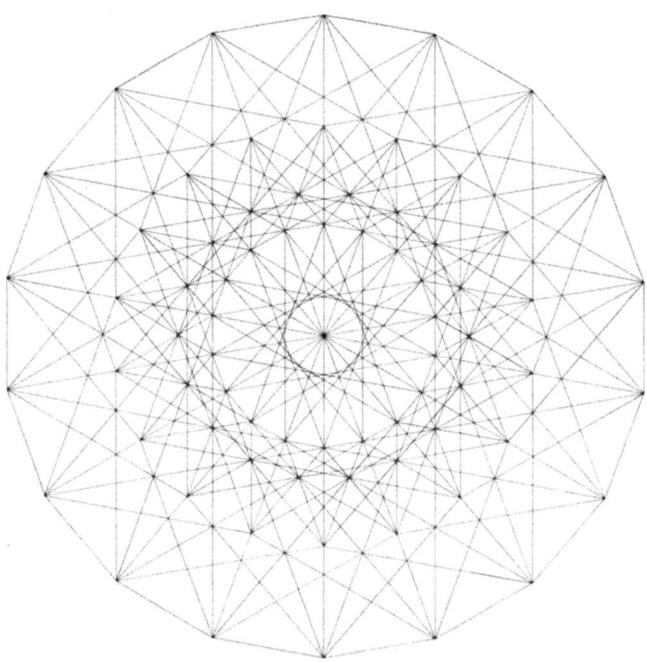

Figure 2. The Hessian polyhedron 2{4}3{3}3

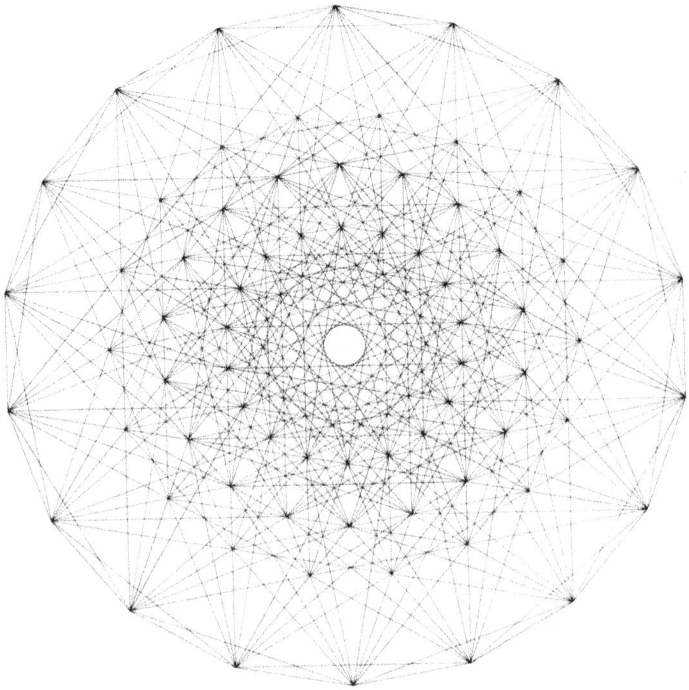

Figure 3. The Hessian polyhedron 3{3}3{4}2

The significance of the Hessian polyhedron comes from the sequence of polytopes (see [18])

$$3\{3\}3, \quad 3\{3\}3\{3\}3, \quad 3\{3\}3\{3\}3\{3\}3$$

(having 8, 27, 240 vertices) which culminates in the four-dimensional honeycomb

$$3\{3\}3\{3\}3\{3\}3\{3\}3.$$

These are very important because of their connection with the celebrated configuration of 27 lines on the general cubic surface in projective 3-space. In fact, the binary tetrahedral group 3[3]3 is a subgroup of index 9 in the Hessian group of order 216, which is a factor group of the symmetry group 3[3]3[3]3 of the Hessian polyhedron 3{3}3{3}3; this group of order 648 is a subgroup of index 40 in the simple group of order 25, 920 which is a factor group of the symmetry group 3[3]3[3]3[3]3 of the Witting polytope 3{3}3{3}3{3}3, whose beautiful figure (which comes from [18], p.ii) is given as follows:

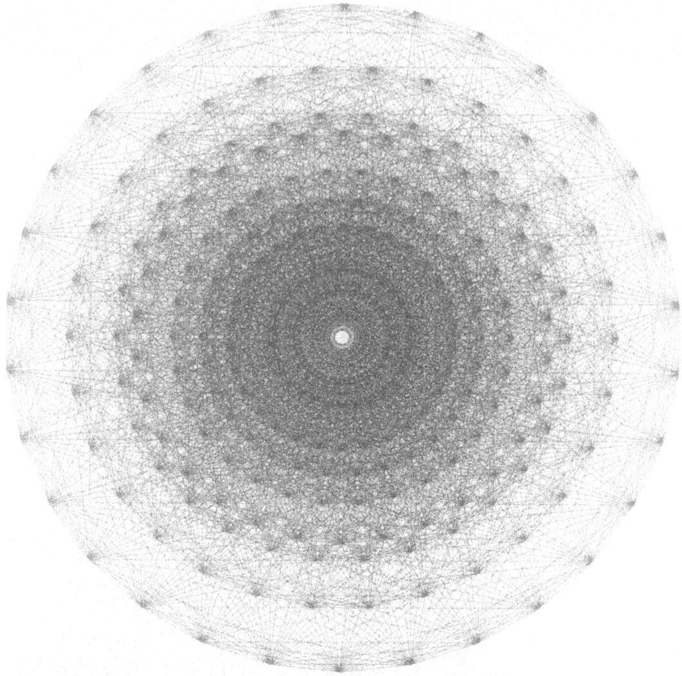

Figure 4. the Witting polytope 3{3}3{3}3{3}3

There are two fundamental reasons for the significance of the triple $W(E_6)$, $G_{25,920}$ and G_{648}:

(1) $W(E_6)$ is a reflection group in $\mathbb{P}^5$, $G_{25,920}$ is a unitary reflection group in $\mathbb{P}^4$ and $\mathbb{P}^3$, G_{648} is a unitary reflection group in $\mathbb{P}^2$.

(2) $W(E_6)$ is the Galois group of the algebraic equation for the 27 lines on the cubic surface, $G_{25,920}$ is the Galois group of the algebraic equation which results after extracting the discriminant of the above, G_{648} is the Galois group of our Hessian polyhedral equations.

In the theory of algebraic curves, if one wants to go beyond the category of hyperelliptic curves, the first non-hyperelliptic curves are plane quartic curves of genus three, such as Klein quartic curve (see [46]), Fermat quartic curve $x^4+y^4=1$ and Picard curves (see [28]). Just as the hyperelliptic modular functions (Siegel modular functions) correspond to the general algebraic equation (one variable), we find that the Picard modular functions correspond to some system of algebraic equations (two variables).

In fact, we have the following dictionaries about algebraic equations, groups, integrals of algebraic functions, genus of algebraic curves and transcendental functions (automorphic functions):

(1)

$GL(1)$,
pure equation, $GL(1)$,
rational integrals, rational functions, $g = 0$,
exponential functions,
rational curves.

(2)

$GL(2), SL(2, \mathbb{R})$,
icosahedral equation, quintic equation, $GL(2)$,
elliptic integrals, elliptic functions, $g = 1$,
elliptic modular functions, $SL(2, \mathbb{Z})$,
Dedekind η-function,
Klein J-invariant,
elliptic curves, $SL(2, \mathbb{Z})(2)$,
$y^2 = x(x-1)(x-\lambda)$,
icosahedron, $PSL(2, \mathbb{F}_4) \cong PSL(2, \mathbb{F}_5) \cong A_5$.

(3)

$GL(4), Sp(4, \mathbb{R})$,
27 lines on the cubic surfaces,
hyperelliptic integrals, hyperelliptic functions, $g = 2$,
Siegel modular functions of genus two, $Sp(4, \mathbb{Z})$,
Igusa's J-invariants J_2, J_4, J_6, J_8 and J_{10} (see [30]),
hyperelliptic curves of genus two, $Sp(4, \mathbb{Z})(2)$,
$y^2 = x(x-1)(x-\lambda_1)(x-\lambda_2)(x-\lambda_3)$,
$PSp(4, \mathbb{F}_3)$, order 25,920,
Witting polytope, 3[3]3[3]3[3]3.

(4)

$Sp(2g, \mathbb{R}), g > 2$
general algebraic equation,
hyperelliptic integrals of genus $g > 2$,
Siegel modular functions of genus $g > 2$,
hyperelliptic curves of genus $g > 2$.

(5)

$GL(3), U(2,1)$,
Hessian polyhedral equations, system of algebraic equations, $GL(3)$,
Picard integrals, $g = 3$,
Picard modular functions, $U(2,1;\mathbb{Z}[\omega])$,
our η-function (see [63]),
our J-invariants J_1 and J_2 (see [63]),
Picard curves, $U(2,1;\mathbb{Z}[\omega])(1-\omega)$,
$y^3 = x(x-1)(x-\lambda_1)(x-\lambda_2)$,
Hessian polyhedra, 3[3]3[3]3, 2[4]3[3]3, 3[3]3[4]2.

Main Results.

The Reciprocity Law about the Appell hypergeometric functions

The Appell hypergeometric partial differential equations

$$\begin{cases} x(1-x)\dfrac{\partial^2 z}{\partial x^2} + y(1-x)\dfrac{\partial^2 z}{\partial x \partial y} + [c-(a+b+1)x]\dfrac{\partial z}{\partial x} - by\dfrac{\partial z}{\partial y} - abz = 0, \\ y(1-y)\dfrac{\partial^2 z}{\partial y^2} + x(1-y)\dfrac{\partial^2 z}{\partial x \partial y} + [c-(a+b'+1)y]\dfrac{\partial z}{\partial y} - b'x\dfrac{\partial z}{\partial x} - ab'z = 0, \end{cases} \tag{1.1}$$

are equivalent to the following three equations:

$$\begin{cases} \dfrac{\partial^2 z}{\partial x^2} + \left(\dfrac{c-b'}{x} + \dfrac{a+b-c+1}{x-1} + \dfrac{b'}{x-y}\right)\dfrac{\partial z}{\partial x} \\ \quad - \dfrac{by(y-1)}{x(x-1)(x-y)}\dfrac{\partial z}{\partial y} + \dfrac{abz}{x(x-1)} = 0, \\ \dfrac{\partial^2 z}{\partial y^2} + \left(\dfrac{c-b}{y} + \dfrac{a+b'-c+1}{y-1} + \dfrac{b}{y-x}\right)\dfrac{\partial z}{\partial y} \\ \quad - \dfrac{b'x(x-1)}{y(y-1)(y-x)}\dfrac{\partial z}{\partial x} + \dfrac{ab'z}{y(y-1)} = 0, \\ \dfrac{\partial^2 z}{\partial x \partial y} = \dfrac{b'}{x-y}\dfrac{\partial z}{\partial x} - \dfrac{b}{x-y}\dfrac{\partial z}{\partial y}. \end{cases} \tag{1.2}$$

There are three linearly independent solutions z_1, z_2 and z_3. Set $u_1 = \frac{z_1}{z_3}$ and $u_2 = \frac{z_2}{z_3}$.

According to [63], we define the following four derivatives associated to the group $GL(3)$:

$$\{u_1, u_2; x, y\}_x := \frac{\begin{vmatrix} \frac{\partial u_1}{\partial x} & \frac{\partial u_2}{\partial x} \\ \frac{\partial^2 u_1}{\partial x^2} & \frac{\partial^2 u_2}{\partial x^2} \end{vmatrix}}{\begin{vmatrix} \frac{\partial u_1}{\partial x} & \frac{\partial u_2}{\partial x} \\ \frac{\partial u_1}{\partial y} & \frac{\partial u_2}{\partial y} \end{vmatrix}}, \qquad \{u_1, u_2; x, y\}_y := \frac{\begin{vmatrix} \frac{\partial u_1}{\partial y} & \frac{\partial u_2}{\partial y} \\ \frac{\partial^2 u_1}{\partial y^2} & \frac{\partial^2 u_2}{\partial y^2} \end{vmatrix}}{\begin{vmatrix} \frac{\partial u_1}{\partial y} & \frac{\partial u_2}{\partial y} \\ \frac{\partial u_1}{\partial x} & \frac{\partial u_2}{\partial x} \end{vmatrix}},$$

$$[u_1, u_2; x, y]_x := \frac{\begin{vmatrix} \frac{\partial u_1}{\partial y} & \frac{\partial u_2}{\partial y} \\ \frac{\partial^2 u_1}{\partial x^2} & \frac{\partial^2 u_2}{\partial x^2} \end{vmatrix} + 2 \begin{vmatrix} \frac{\partial u_1}{\partial x} & \frac{\partial u_2}{\partial x} \\ \frac{\partial^2 u_1}{\partial x \partial y} & \frac{\partial^2 u_2}{\partial x \partial y} \end{vmatrix}}{\begin{vmatrix} \frac{\partial u_1}{\partial x} & \frac{\partial u_2}{\partial x} \\ \frac{\partial u_1}{\partial y} & \frac{\partial u_2}{\partial y} \end{vmatrix}}, \tag{1.3}$$

$$[u_1, u_2; x, y]_y := \frac{\begin{vmatrix} \frac{\partial u_1}{\partial x} & \frac{\partial u_2}{\partial x} \\ \frac{\partial^2 u_1}{\partial y^2} & \frac{\partial^2 u_2}{\partial y^2} \end{vmatrix} + 2 \begin{vmatrix} \frac{\partial u_1}{\partial y} & \frac{\partial u_2}{\partial y} \\ \frac{\partial^2 u_1}{\partial x \partial y} & \frac{\partial^2 u_2}{\partial x \partial y} \end{vmatrix}}{\begin{vmatrix} \frac{\partial u_1}{\partial y} & \frac{\partial u_2}{\partial y} \\ \frac{\partial u_1}{\partial x} & \frac{\partial u_2}{\partial x} \end{vmatrix}},$$

where x and y are complex numbers, u_1 and u_2 are complex-valued functions of x and y.

We find that (see [63])

$$\begin{cases} \{u_1, u_2; x, y\}_x = \dfrac{by(y-1)}{x(x-1)(x-y)}, \\ \{u_1, u_2; x, y\}_y = \dfrac{b'x(x-1)}{y(y-1)(y-x)}, \\ [u_1, u_2; x, y]_x = \dfrac{c-b'}{x} + \dfrac{a+b-c+1}{x-1} + \dfrac{b'-2b}{x-y}, \\ [u_1, u_2; x, y]_y = \dfrac{c-b}{y} + \dfrac{a+b'-c+1}{y-1} + \dfrac{b-2b'}{y-x}. \end{cases} \tag{1.4}$$

Theorem (see [63], Main Theorem 1) (The explicit solutions of nonlinear partial differential system of equations). *For the over-determined system of second order, nonlinear (quasilinear) complex partial differential equations*:

$$\begin{cases} \{w_1, w_2; v_1, v_2\}_{v_1} = F_1(v_1, v_2), \\ \{w_1, w_2; v_1, v_2\}_{v_2} = F_2(v_1, v_2), \\ [w_1, w_2; v_1, v_2]_{v_1} = P_1(v_1, v_2), \\ [w_1, w_2; v_1, v_2]_{v_2} = P_2(v_1, v_2), \end{cases} \tag{1.5}$$

set the Jacobian

$$\Delta := \frac{\partial(v_1, v_2)}{\partial(w_1, w_2)} = \begin{vmatrix} \frac{\partial v_1}{\partial w_1} & \frac{\partial v_2}{\partial w_1} \\ \frac{\partial v_1}{\partial w_2} & \frac{\partial v_2}{\partial w_2} \end{vmatrix}. \tag{1.6}$$

Then $z = \sqrt[3]{\Delta}$ satisfies the following linear differential equations:

$$\begin{cases} \dfrac{\partial^2 z}{\partial v_1^2} + \dfrac{1}{3} P_1 \dfrac{\partial z}{\partial v_1} - F_1 \dfrac{\partial z}{\partial v_2} + \left(\dfrac{\partial F_1}{\partial v_2} - \dfrac{1}{3} \dfrac{\partial P_1}{\partial v_1} - \dfrac{2}{9} P_1^2 - \dfrac{2}{3} F_1 P_2 \right) z = 0, \\ \dfrac{\partial^2 z}{\partial v_1 \partial v_2} - \dfrac{1}{3} P_2 \dfrac{\partial z}{\partial v_1} - \dfrac{1}{3} P_1 \dfrac{\partial z}{\partial v_2} + \left(\dfrac{1}{3} \dfrac{\partial P_2}{\partial v_1} + \dfrac{1}{3} \dfrac{\partial P_1}{\partial v_2} + \dfrac{1}{9} P_1 P_2 - F_1 F_2 \right) z = 0, \\ \dfrac{\partial^2 z}{\partial v_2^2} + \dfrac{1}{3} P_2 \dfrac{\partial z}{\partial v_2} - F_2 \dfrac{\partial z}{\partial v_1} + \left(\dfrac{\partial F_2}{\partial v_1} - \dfrac{1}{3} \dfrac{\partial P_2}{\partial v_2} - \dfrac{2}{9} P_2^2 - \dfrac{2}{3} F_2 P_1 \right) z = 0. \end{cases} \tag{1.7}$$

Moreover, the above system (1.7) of linear partial differential equations are also solvable in the following, which improve Theorem 1.11 of [63].

Theorem 1.1 (Main Theorem 1) (The explicit solutions of linear partial differential system of equations). *For the system of linear partial differential equations:*

$$\begin{cases} \dfrac{\partial^2 z}{\partial v_1^2} + \dfrac{1}{3} P_1 \dfrac{\partial z}{\partial v_1} - F_1 \dfrac{\partial z}{\partial v_2} + \left(\dfrac{\partial F_1}{\partial v_2} - \dfrac{1}{3} \dfrac{\partial P_1}{\partial v_1} - \dfrac{2}{9} P_1^2 - \dfrac{2}{3} F_1 P_2 \right) z = 0, \\ \dfrac{\partial^2 z}{\partial v_1 \partial v_2} - \dfrac{1}{3} P_2 \dfrac{\partial z}{\partial v_1} - \dfrac{1}{3} P_1 \dfrac{\partial z}{\partial v_2} + \left(\dfrac{1}{3} \dfrac{\partial P_2}{\partial v_1} + \dfrac{1}{3} \dfrac{\partial P_1}{\partial v_2} + \dfrac{1}{9} P_1 P_2 - F_1 F_2 \right) z = 0, \\ \dfrac{\partial^2 z}{\partial v_2^2} + \dfrac{1}{3} P_2 \dfrac{\partial z}{\partial v_2} - F_2 \dfrac{\partial z}{\partial v_1} + \left(\dfrac{\partial F_2}{\partial v_1} - \dfrac{1}{3} \dfrac{\partial P_2}{\partial v_2} - \dfrac{2}{9} P_2^2 - \dfrac{2}{3} F_2 P_1 \right) z = 0, \end{cases}$$

put

$$\begin{cases} P_1 = \dfrac{c - b'}{v_1} + \dfrac{a + b - c + 1}{v_1 - 1} + \dfrac{b' - 2b}{v_1 - v_2}, \\ P_2 = \dfrac{c - b}{v_2} + \dfrac{a + b' - c + 1}{v_2 - 1} + \dfrac{b - 2b'}{v_2 - v_1}, \\ F_1 = \dfrac{b v_2 (v_2 - 1)}{v_1 (v_1 - 1)(v_1 - v_2)}, \\ F_2 = \dfrac{b' v_1 (v_1 - 1)}{v_2 (v_2 - 1)(v_2 - v_1)}. \end{cases} \tag{1.8}$$

Set

$$w = v_1^{-\frac{1}{3}(c-b')} v_2^{-\frac{1}{3}(c-b)} (v_1 - 1)^{-\frac{1}{3}(a+b-c+1)} (v_2 - 1)^{-\frac{1}{3}(a+b'-c+1)} (v_1 - v_2)^{-\frac{1}{3}(b+b')} z, \tag{1.9}$$

then the function $w = w(v_1, v_2)$ satisfies the following equations:

$$\begin{cases} \dfrac{\partial^2 w}{\partial v_1^2} + \left(\dfrac{c-b'}{v_1} + \dfrac{a+b-c+1}{v_1-1} + \dfrac{b'}{v_1-v_2}\right)\dfrac{\partial w}{\partial v_1} - \dfrac{bv_2(v_2-1)}{v_1(v_1-1)(v_1-v_2)}\dfrac{\partial w}{\partial v_2} \\ + \dfrac{ab}{v_1(v_1-1)}w = 0, \\ \dfrac{\partial^2 w}{\partial v_2^2} + \left(\dfrac{c-b}{v_2} + \dfrac{a+b'-c+1}{v_2-1} + \dfrac{b}{v_2-v_1}\right)\dfrac{\partial w}{\partial v_2} - \dfrac{b'v_1(v_1-1)}{v_2(v_2-1)(v_2-v_1)}\dfrac{\partial w}{\partial v_1} \\ + \dfrac{ab'}{v_2(v_2-1)}w = 0, \\ \dfrac{\partial^2 w}{\partial v_1 \partial v_2} = \dfrac{b'}{v_1-v_2}\dfrac{\partial w}{\partial v_1} - \dfrac{b}{v_1-v_2}\dfrac{\partial w}{\partial v_2}. \end{cases} \tag{1.10}$$

A solution is given by

$$w = F_1(a; b, b'; c; v_1, v_2), \tag{1.11}$$

where $F_1 = F_1(a; b, b'; c; x, y)$ is the Appell hypergeometric function (see [1]). Consequently,

$$\begin{aligned} z =& v_1^{\frac{1}{3}(c-b')} v_2^{\frac{1}{3}(c-b)} (v_1-1)^{\frac{1}{3}(a+b-c+1)} (v_2-1)^{\frac{1}{3}(a+b'-c+1)} (v_1-v_2)^{\frac{1}{3}(b+b')} \\ & \times F_1(a; b, b'; c; v_1, v_2). \end{aligned} \tag{1.12}$$

In fact, the above reciprocity law is valid not only for the Appell hypergeometric partial differential equations, but also for the system of nonlinear partial differential equations involving our four derivatives (1.4).

A complete system of invariants for the Hessian groups (the corresponding geometric objects are Hessian polyhedra) has degrees 6, 9, 12, 12 and 18 and can be given explicitly by the following forms:

$$\begin{cases} C_6(z_1, z_2, z_3) = z_1^6 + z_2^6 + z_3^6 - 10(z_1^3 z_2^3 + z_2^3 z_3^3 + z_3^3 z_1^3), \\ C_9(z_1, z_2, z_3) = (z_1^3 - z_2^3)(z_2^3 - z_3^3)(z_3^3 - z_1^3), \\ C_{12}(z_1, z_2, z_3) = (z_1^3 + z_2^3 + z_3^3)[(z_1^3 + z_2^3 + z_3^3)^3 + 216 z_1^3 z_2^3 z_3^3], \\ \mathfrak{C}_{12}(z_1, z_2, z_3) = z_1 z_2 z_3 [27 z_1^3 z_2^3 z_3^3 - (z_1^3 + z_2^3 + z_3^3)^3], \\ C_{18}(z_1, z_2, z_3) = (z_1^3 + z_2^3 + z_3^3)^6 - 540 z_1^3 z_2^3 z_3^3 (z_1^3 + z_2^3 + z_3^3)^3 - 5832 z_1^6 z_2^6 z_3^6. \end{cases} \tag{1.13}$$

They satisfy the following relations:

$$\begin{cases} 432 C_9^2 = C_6^3 - 3C_6 C_{12} + 2C_{18}, \\ 1728 \mathfrak{C}_{12}^3 = C_{18}^2 - C_{12}^3. \end{cases} \tag{1.14}$$

For the Hessian group 3[3]3[3]3, of order $6 \times 9 \times 12$, the three forms are C_6, C_9, C_{12}. For the closely related Hessian group 2[4]3[3]3, of order $6 \times 12 \times 18$, they are C_6, C_{12}, C_9^2 (see [18]).

In the affine coordinates $\xi = z_1/z_3$ and $\eta = z_2/z_3$, we have

$$\begin{cases} C_6 = C_6(\xi, \eta) = \xi^6 + \eta^6 + 1 - 10(\xi^3\eta^3 + \xi^3 + \eta^3), \\ C_9 = C_9(\xi, \eta) = (\xi^3 - \eta^3)(\eta^3 - 1)(1 - \xi^3), \\ C_{12} = C_{12}(\xi, \eta) = (\xi^3 + \eta^3 + 1)[(\xi^3 + \eta^3 + 1)^3 + 216\xi^3\eta^3], \\ \mathfrak{C}_{12} = \mathfrak{C}_{12}(\xi, \eta) = \xi\eta[27\xi^3\eta^3 - (\xi^3 + \eta^3 + 1)^3], \\ C_{18} = C_{18}(\xi, \eta) = (\xi^3 + \eta^3 + 1)^6 - 540\xi^3\eta^3(\xi^3 + \eta^3 + 1)^3 - 5832\xi^6\eta^6. \end{cases} \tag{1.15}$$

In [63], we gave the modular equation associated to the Picard curves:
Theorem (see [63], Main Theorem 3) (Higher order transformation of double irrational integrals with modular equation of higher degree). *The following identity of differential forms holds*:

$$\frac{dw_1 \wedge dw_2}{\sqrt[3]{w_1^2 w_2^2 (1 - w_1)(1 - \lambda_1^3 w_1)(1 - \lambda_2^3 w_1)}} = -\frac{5t_1}{t_2^2} \frac{dt_1 \wedge dt_2}{\sqrt[3]{t_1^2 t_2^2 (1 - t_1)(1 - \kappa_1^3 t_1)(1 - \kappa_2^3 t_1)}},$$

where

$$w_1 = \frac{(\beta + \gamma)t_1 + \alpha}{\beta t_1 + (\alpha + \gamma)}, \quad w_2 = \frac{t_1[(\beta + \gamma)t_1 + \alpha]^2[\beta t_1 + (\alpha + \gamma)]}{t_2^5},$$

which is a rational transformation of order five. Here,

$$\begin{cases} \alpha = \dfrac{(v_1 - 1)(v_2 - 1)(v_1 + v_2 - 2) - (u_1 - 1)(u_2 - 1)(u_1 + u_2 - 2)}{2(u_1 - 1)(u_2 - 1)(v_1 - 1)(v_2 - 1)}, \\ \beta = \dfrac{(u_1 - 1)(u_2 - 1)(2u_1u_2 - u_1 - u_2) - (v_1 - 1)(v_2 - 1)(2v_1v_2 - v_1 - v_2)}{2(u_1 - 1)(u_2 - 1)(v_1 - 1)(v_2 - 1)}, \\ \gamma = \dfrac{(v_1 - 1)(v_2 - 1)}{(u_1 - 1)(u_2 - 1)}, \end{cases}$$

with

$$u_1 = \kappa_1^3, \quad u_2 = \kappa_2^3, \quad v_1 = \lambda_1^3, \quad v_2 = \lambda_2^3.$$

Moreover, the moduli (κ_1, κ_2) *and* (λ_1, λ_2) *satisfy the modular equation*:

$$(\kappa_1^3 - 1)(\kappa_2^3 - 1)(\kappa_1^3 - \kappa_2^3) = (\lambda_1^3 - 1)(\lambda_2^3 - 1)(\lambda_1^3 - \lambda_2^3),$$

which is an algebraic variety of dimension three. The corresponding two algebraic surfaces are:

$$\begin{cases} w_3^3 = w_1^2 w_2^2 (1 - w_1)(1 - \lambda_1^3 w_1)(1 - \lambda_2^3 w_1), \\ t_3^3 = t_1^2 t_2^2 (1 - t_1)(1 - \kappa_1^3 t_1)(1 - \kappa_2^3 t_1), \end{cases}$$

which give the homogeneous algebraic equations of degree seven:

$$\begin{cases} w_3^3 w_4^4 = w_1^2 w_2^2 (w_4 - w_1)(w_4 - \lambda_1^3 w_1)(w_4 - \lambda_2^3 w_1), \\ t_3^3 t_4^4 = t_1^2 t_2^2 (t_4 - t_1)(t_4 - \kappa_1^3 t_1)(t_4 - \kappa_2^3 t_1). \end{cases}$$

In fact, the above modular equation can be written as follows:

$$C_9(\kappa_1, \kappa_2, 1) = C_9(\lambda_1, \lambda_2, 1). \tag{1.16}$$

This shows that our modular equation is intimately connected with the Hessian polyhedra. Hence, the above theorem can be considered not only as a higher dimensional analogue of Jacobi's celebrated work [31], §§1–34 on the transformation property of elliptic integrals, but also as a higher dimensional analogue of Klein's celebrated work [44] (or [42]) on the relation between the icosahedron and the Jacobian modular equation of the sixth degree.

Now, we give some basic properties of the invariants associated to the Hessian polyhedra.

Proposition 1.2 (Main Proposition 1). *The following identities hold*:

$$\begin{vmatrix} \frac{\partial C_6}{\partial \xi} & \frac{\partial C_6}{\partial \eta} & 2C_6 \\ \frac{\partial C_9}{\partial \xi} & \frac{\partial C_9}{\partial \eta} & 3C_9 \\ \frac{\partial C_{12}}{\partial \xi} & \frac{\partial C_{12}}{\partial \eta} & 4C_{12} \end{vmatrix} = -864\mathfrak{C}_{12}^2. \tag{1.17}$$

$$\begin{vmatrix} \frac{\partial C_6}{\partial \xi} & \frac{\partial C_6}{\partial \eta} & 2C_6 \\ \frac{\partial C_9}{\partial \xi} & \frac{\partial C_9}{\partial \eta} & 3C_9 \\ \frac{\partial C_{18}}{\partial \xi} & \frac{\partial C_{18}}{\partial \eta} & 6C_{18} \end{vmatrix} = -1296C_6\mathfrak{C}_{12}^2. \tag{1.18}$$

$$\begin{vmatrix} \frac{\partial C_6}{\partial \xi} & \frac{\partial C_6}{\partial \eta} & 2C_6 \\ \frac{\partial C_{12}}{\partial \xi} & \frac{\partial C_{12}}{\partial \eta} & 4C_{12} \\ \frac{\partial C_{18}}{\partial \xi} & \frac{\partial C_{18}}{\partial \eta} & 6C_{18} \end{vmatrix} = 432 \cdot 864C_9\mathfrak{C}_{12}^2. \tag{1.19}$$

$$\begin{vmatrix} \frac{\partial C_9}{\partial \xi} & \frac{\partial C_9}{\partial \eta} & 3C_9 \\ \frac{\partial C_{12}}{\partial \xi} & \frac{\partial C_{12}}{\partial \eta} & 4C_{12} \\ \frac{\partial C_{18}}{\partial \xi} & \frac{\partial C_{18}}{\partial \eta} & 6C_{18} \end{vmatrix} = 1296(C_6^2 - C_{12})\mathfrak{C}_{12}^2. \tag{1.20}$$

$$\begin{vmatrix} \frac{\partial C_6}{\partial \xi} & \frac{\partial C_6}{\partial \eta} & 2C_6 \\ \frac{\partial C_9}{\partial \xi} & \frac{\partial C_9}{\partial \eta} & 3C_9 \\ \frac{\partial \mathfrak{C}_{12}}{\partial \xi} & \frac{\partial \mathfrak{C}_{12}}{\partial \eta} & 4\mathfrak{C}_{12} \end{vmatrix} = \frac{1}{2}(C_{12}^2 - C_6C_{18}). \tag{1.21}$$

$$\begin{vmatrix} \frac{\partial C_6}{\partial \xi} & \frac{\partial C_6}{\partial \eta} & 2C_6 \\ \frac{\partial C_{12}}{\partial \xi} & \frac{\partial C_{12}}{\partial \eta} & 4C_{12} \\ \frac{\partial \mathfrak{C}_{12}}{\partial \xi} & \frac{\partial \mathfrak{C}_{12}}{\partial \eta} & 4\mathfrak{C}_{12} \end{vmatrix} = 144C_9C_{18}. \tag{1.22}$$

$$\begin{vmatrix} \frac{\partial C_9}{\partial \xi} & \frac{\partial C_9}{\partial \eta} & 3C_9 \\ \frac{\partial C_{12}}{\partial \xi} & \frac{\partial C_{12}}{\partial \eta} & 4C_{12} \\ \frac{\partial \mathfrak{C}_{12}}{\partial \xi} & \frac{\partial \mathfrak{C}_{12}}{\partial \eta} & 4\mathfrak{C}_{12} \end{vmatrix} = \frac{1}{2}(C_6^2 - C_{12})C_{18}. \tag{1.23}$$

$$\begin{vmatrix} \frac{\partial C_6}{\partial \xi} & \frac{\partial C_6}{\partial \eta} & 2C_6 \\ \frac{\partial C_{18}}{\partial \xi} & \frac{\partial C_{18}}{\partial \eta} & 6C_{18} \\ \frac{\partial \mathfrak{C}_{12}}{\partial \xi} & \frac{\partial \mathfrak{C}_{12}}{\partial \eta} & 4\mathfrak{C}_{12} \end{vmatrix} = 216C_9C_{12}^2. \tag{1.24}$$

$$\begin{vmatrix} \frac{\partial C_9}{\partial \xi} & \frac{\partial C_9}{\partial \eta} & 3C_9 \\ \frac{\partial C_{18}}{\partial \xi} & \frac{\partial C_{18}}{\partial \eta} & 6C_{18} \\ \frac{\partial \mathfrak{C}_{12}}{\partial \xi} & \frac{\partial \mathfrak{C}_{12}}{\partial \eta} & 4\mathfrak{C}_{12} \end{vmatrix} = \frac{3}{4}(C_6^2 - C_{12})C_{12}^2. \tag{1.25}$$

$$\begin{vmatrix} \frac{\partial C_{12}}{\partial \xi} & \frac{\partial C_{12}}{\partial \eta} & 4C_{12} \\ \frac{\partial C_{18}}{\partial \xi} & \frac{\partial C_{18}}{\partial \eta} & 6C_{18} \\ \frac{\partial \mathfrak{C}_{12}}{\partial \xi} & \frac{\partial \mathfrak{C}_{12}}{\partial \eta} & 4\mathfrak{C}_{12} \end{vmatrix} = 0. \tag{1.26}$$

We define the Hessian polyhedral equations as follows:

$$\begin{cases} 432\dfrac{C_9^2}{C_6^3} = R_1(x, y), \\ 3\dfrac{C_{12}}{C_6^2} = R_2(x, y), \end{cases} \tag{1.27}$$

where R_1 and R_2 are rational functions of x and y. They are the higher dimensional analogue of the icosahedral equation in [44] (or [42]).

Theorem 1.3 (Main Theorem 2) (Explicit construction of algebraic solutions for Appell hypergeometric partial differential equations). *The relations between Hessian polyhedra and Appell hypergeometric partial differential equations are given by the following identities*:

$$\begin{aligned} & z_1^6 + z_2^6 + z_3^6 - 10(z_1^3 z_2^3 + z_2^3 z_3^3 + z_3^3 z_1^3) \\ =& 2^6 \cdot 3^7 \cdot C^2 \cdot R_1 \left[\frac{1}{4}(R_1 + R_2 - 1)^2 - \frac{1}{27}R_2^3\right]^{\frac{4}{3}} \left[\frac{\partial(R_1, R_2)}{\partial(x, y)}\right]^{-2} H(x, y)^2. \end{aligned} \tag{1.28}$$

$$\begin{aligned} & (z_1^3 - z_2^3)(z_2^3 - z_3^3)(z_3^3 - z_1^3) \\ =& 2^7 \cdot 3^9 \cdot C^3 \cdot R_1^2 \left[\frac{1}{4}(R_1 + R_2 - 1)^2 - \frac{1}{27}R_2^3\right]^2 \left[\frac{\partial(R_1, R_2)}{\partial(x, y)}\right]^{-3} H(x, y)^3. \end{aligned} \tag{1.29}$$

$$\begin{aligned} & (z_1^3 + z_2^3 + z_3^3)[(z_1^3 + z_2^3 + z_3^3)^3 + 216 z_1^3 z_2^3 z_3^3] \\ =& 2^{12} \cdot 3^{13} \cdot C^4 \cdot R_1^2 R_2 \left[\frac{1}{4}(R_1 + R_2 - 1)^2 - \frac{1}{27}R_2^3\right]^{\frac{8}{3}} \left[\frac{\partial(R_1, R_2)}{\partial(x, y)}\right]^{-4} H(x, y)^4. \end{aligned} \tag{1.30}$$

$$\begin{aligned} & z_1 z_2 z_3 [27 z_1^3 z_2^3 z_3^3 - (z_1^3 + z_2^3 + z_3^3)^3] \\ =& 2^{10} \cdot 3^{13} \cdot C^4 \cdot R_1^2 \left[\frac{1}{4}(R_1 + R_2 - 1)^2 - \frac{1}{27}R_2^3\right]^3 \left[\frac{\partial(R_1, R_2)}{\partial(x, y)}\right]^{-4} H(x, y)^4. \end{aligned} \tag{1.31}$$

$$\begin{aligned} & (z_1^3 + z_2^3 + z_3^3)^6 - 540 z_1^3 z_2^3 z_3^3 (z_1^3 + z_2^3 + z_3^3)^3 - 5832 z_1^6 z_2^6 z_3^6 \\ =& 2^{17} \cdot 3^{21} \cdot C^6 \cdot R_1^3 (R_1 + R_2 - 1) \left[\frac{1}{4}(R_1 + R_2 - 1)^2 - \frac{1}{27}R_2^3\right]^4 \\ & \times \left[\frac{\partial(R_1, R_2)}{\partial(x, y)}\right]^{-6} H(x, y)^6. \end{aligned} \tag{1.32}$$

Here,

$$H(x,y) := x^{b'-c}y^{b-c}(x-1)^{c-a-b-1}(y-1)^{c-a-b'-1}(x-y)^{-b-b'}. \tag{1.33}$$

The above five identities give the relations between the Hessian polyhedra and Appell hypergeometric partial differential equations. More precisely, given the Hessian polyhedra, we can obtain the rational functions R_1, R_2 and the function $H(x,y)$ with parameters a, b, b', c. Then three linearly independent solutions

$$z_1 = z_1(x,y), \quad z_2 = z_2(x,y), \quad z_3 = z_3(x,y),$$

which are the algebraic solutions of the Appell hypergeometric partial differential equations (1.1) with the same parameters a, b, b', c, satisfy the above five algebraic equations (1.27), (1.28), (1.29), (1.30), (1.31). Moreover, the monodromy groups of these Appell hypergeometric partial differential equations are just the Hessian groups.

The Hessian polyhedral irrationality (ξ, η) is defined by the equations:

$$\begin{cases} 432\dfrac{C_9^2}{C_6^3} = J_1, \\ 3\dfrac{C_{12}}{C_6^2} = J_2, \end{cases} \tag{1.34}$$

where J_1 and J_2 are j-invariants (see [63]):

$$\begin{cases} J_1 = J_1(\lambda_1,\lambda_2) = \dfrac{\lambda_2^2(\lambda_2-1)^2}{\lambda_1^2(\lambda_1-1)^2(\lambda_1-\lambda_2)^2}, \\ J_2 = J_2(\lambda_1,\lambda_2) = \dfrac{\lambda_1^2(\lambda_1-1)^2}{\lambda_2^2(\lambda_2-1)^2(\lambda_2-\lambda_1)^2}. \end{cases} \tag{1.35}$$

The corresponding Picard curve is

$$y^3 = x(x-1)(x-\lambda_1)(x-\lambda_2). \tag{1.36}$$

Here, $\lambda_1 = \lambda_1(\xi,\eta)$ and $\lambda_2 = \lambda_2(\xi,\eta)$ are Picard modular functions.

Denote by

$$(F,G,H) := \frac{\partial(F,G,H)}{\partial(z_1,z_2,z_3)}. \tag{1.37}$$

We have the following differential properties on the invariants associated with the Hessian polyhedra.

Proposition 1.4 (Main Proposition 2) (Duality for the invariants associated with the Hessian polyhedra). *In the projective coordinates, the following duality formulas hold:*

$$\begin{cases} (C_6, C_9, C_{12}) = -2^5 \cdot 3^4 \cdot \mathfrak{C}_{12}^2, \\ (C_6, C_9, C_{18}) = -2^4 \cdot 3^5 \cdot C_6\mathfrak{C}_{12}^2, \\ (C_6, C_{12}, C_{18}) = 2^9 \cdot 3^7 \cdot C_9\mathfrak{C}_{12}^2, \\ (C_9, C_{12}, C_{18}) = 2^4 \cdot 3^5 \cdot (C_6^2 - C_{12})\mathfrak{C}_{12}^2. \end{cases} \tag{1.38}$$

$$\begin{cases}(C_6, C_9, \mathfrak{C}_{12}) = \frac{3}{2}(C_{12}^2 - C_6 C_{18}), \\ (C_6, C_{12}, \mathfrak{C}_{12}) = 432 C_9 C_{18}, \\ (C_9, C_{12}, \mathfrak{C}_{12}) = \frac{3}{2} C_{18}(C_6^2 - C_{12}). \end{cases} \tag{1.39}$$

$$\begin{cases}(C_6, C_{18}, \mathfrak{C}_{12}) = 648 C_9 C_{12}^2, \\ (C_9, C_{18}, \mathfrak{C}_{12}) = \frac{9}{4} C_{12}^2 (C_6^2 - C_{12}), \\ (C_{12}, C_{18}, \mathfrak{C}_{12}) = 0. \end{cases} \tag{1.40}$$

In particular, Proposition 1.4 shows that there exist not only algebraic relations among the invariants C_6, C_9, C_{12}, C_{18} and $\mathfrak{C}_{12}$, but also differential relations among them. Moreover, we will show that there exist algebraic relations even among the Jacobian of (F, G, H), where (F, G, H) are the triples which come from C_6, C_9, C_{12}, C_{18} and $\mathfrak{C}_{12}$.

Put

$$\begin{cases} Y_1 = (C_6, C_{12}, \mathfrak{C}_{12}), \\ Y_2 = (C_9, C_{12}, \mathfrak{C}_{12}), \\ Y_3 = (C_6, C_{12}, C_{18}), \\ Y_4 = (C_9, C_{12}, C_{18}). \end{cases} \tag{1.41}$$

$$\begin{cases} X_1 = (C_6, C_9, C_{12}), \\ X_2 = (C_6, C_9, \mathfrak{C}_{12}), \\ X_3 = (C_6, C_9, C_{18}), \\ X_4 = (C_6, C_{18}, \mathfrak{C}_{12}), \\ X_5 = (C_9, C_{18}, \mathfrak{C}_{12}). \end{cases} \tag{1.42}$$

Theorem 1.5 (Main Theorem 3) (Algebraic relations among the Jacobian). *The functions X_1, X_2, X_3, X_4, X_5 and Y_1, Y_2, Y_3, Y_4 satisfy the system of algebraic equations of the sixth degree*:

$$\begin{cases} 18 X_1^2 Y_4^3 + 24 X_1 X_3^2 Y_4^2 + 8 X_3^4 Y_4 + 27 X_1^5 X_5 = 0, \\ Y_2 Y_3 = Y_1 Y_4, \\ X_4 Y_4 = X_5 Y_3, \\ X_3 Y_1 - X_2 Y_3 = X_1 X_4. \end{cases} \tag{1.43}$$

Let

$$\mathcal{A} = \{(C_6, C_9, C_{12}, \mathfrak{C}_{12}, C_{18}) \in \mathbb{C}^5\} \tag{1.44}$$

be the configuration space of the Hessian invariants. Let

$$\mathcal{B} = \{(X_1, X_2, X_3, X_4, X_5, Y_1, Y_2, Y_3, Y_4) \in \mathbb{C}^9\} \tag{1.45}$$

be the dual configuration space of $\mathcal{A}$. Then

$$\dim \mathcal{A} = 3, \quad \dim \mathcal{B} = 5. \tag{1.46}$$

In the following we will resolve the algebraic system of equations which come from the Hessian polyhedra, i.e., we will decompose the invariants C_6, C_9, C_{12} and $\mathfrak{C}_{12}$ by introducing some new invariants associated with the Hessian polyhedra. This can be considered as a higher dimensional analogue of Klein's celebrated work [44] (or [42]) on the relation between the icosahedron and the equations of the fifth degree.

For the Hessian polyhedral equations

$$Z_1 = 432\mathfrak{C}_9^2 C_6^3, \quad Z_2 = 3\frac{C_{12}}{C_6^2}, \tag{1.47}$$

we find that

$$\begin{aligned}
&Z_1^2 : Z_2^3 : (Z_1 + Z_2 - 1)^2 : [27(Z_1 + Z_2 - 1)^2 - 4Z_2^3] : 1 \\
=&432^2 C_9^4 : 27C_{12}^3 : 4C_{18}^2 : 432^2 \mathfrak{C}_{12}^3 : C_6^6 \\
=&432^2(\xi^3 - \eta^3)^4(\eta^3 - 1)^4(1 - \xi^3)^4 \\
&:27(\xi^3 + \eta^3 + 1)^3[(\xi^3 + \eta^3 + 1)^3 + 216\xi^3\eta^3]^3 \\
&:4[(\xi^3 + \eta^3 + 1)^6 - 540\xi^3\eta^3(\xi^3 + \eta^3 + 1)^3 - 5832\xi^6\eta^6]^2 \\
&:432^2\xi^3\eta^3[27\xi^3\eta^3 - (\xi^3 + \eta^3 + 1)^3]^3 \\
&:[\xi^6 + \eta^6 + 1 - 10(\xi^3\eta^3 + \xi^3 + \eta^3)]^6.
\end{aligned} \tag{1.48}$$

Put

$$\tau_1 = \xi^3 + \eta^3, \quad \tau_2 = \xi\eta. \tag{1.49}$$

Then

$$\begin{cases}
C_6 = \tau_1^2 - 12\tau_2^3 - 10\tau_1 + 1, \\
C_{12} = (\tau_1 + 1)[(\tau_1 + 1)^3 + 216\tau_2^3], \\
\mathfrak{C}_{12} = \tau_2[27\tau_2^3 - (\tau_1 + 1)^3], \\
C_{18} = (\tau_1 + 1)^6 - 540\tau_2^3(\tau_1 + 1)^3 - 5832\tau_2^6, \\
C_9^2 = (\tau_1^2 - 4\tau_2^3)(\tau_2^3 - \tau_1 + 1)^2.
\end{cases} \tag{1.50}$$

Thus,

$$\begin{aligned}
&Z_1^2 : Z_2^3 : (Z_1 + Z_2 - 1)^2 : [27(Z_1 + Z_2 - 1)^2 - 4Z_2^3] : 1 \\
=&432^2(\tau_1^2 - 4\tau_2^3)^2(\tau_2^3 - \tau_1 + 1)^4 \\
&:27(\tau_1 + 1)^3[(\tau_1 + 1)^3 + 216\tau_2^3]^3 \\
&:4[(\tau_1 + 1)^6 - 540\tau_2^3(\tau_1 + 1)^3 - 5832\tau_2^6]^2 \\
&:432^2\tau_2^3[27\tau_2^3 - (\tau_1 + 1)^3]^3 \\
&:(\tau_1^2 - 12\tau_2^3 - 10\tau_1 + 1)^6.
\end{aligned} \tag{1.51}$$

Now, we will introduce some basic algebraic invariants G, H and K associated with the Hessian polyhedra, which play a central role in the resolution of the algebraic system of equations associated with the Hessian polyhedra. Put

$$\varphi = z_1 z_2 z_3, \quad \psi = z_1^3 + z_2^3 + z_3^3, \quad \chi = z_1^3 z_2^3 + z_2^3 z_3^3 + z_3^3 z_1^3. \tag{1.52}$$

Let

$$G(z_1, z_2, z_3) := (z_1 - z_2)(z_2 - z_3)(z_3 - z_1). \tag{1.53}$$

$$H(z_1, z_2, z_3) := \frac{1}{3}[(z_1+z_2+z_3)^3+(z_1+\omega z_2+\omega^2 z_3)^3+(z_1+\omega^2 z_2+\omega z_3)^3]. \tag{1.54}$$

$$K(z_1, z_2, z_3) := (z_1 + z_2 + z_3)(z_1 + \omega z_2 + \omega^2 z_3)(z_1 + \omega^2 z_2 + \omega z_3). \tag{1.55}$$

The Hessians of G, K and H are:

$$\text{Hessian}(G) = 0, \quad \text{Hessian}(K) = -54K, \quad \text{Hessian}(H) = -216K. \tag{1.56}$$

The significance of G, H and K come from the following solvability, which has eight different aspects:

Theorem 1.6 (Main Theorem 4) (The resolution of algebraic system of equations associated with the Hessian polyhedra, I). *The invariants G, H and K satisfy the following algebraic equations, which are the form-theoretic resolvents (algebraic resolvents) of G, H, K:*

$$\begin{cases} 4G^3 + H^2 G - C_6 G - 4C_9 = 0, \\ H(H^3 + 8K^3) - 9C_{12} = 0, \\ K(K^3 - H^3) - 27\mathfrak{C}_{12} = 0. \end{cases} \tag{1.57}$$

Let

$$r_1 = \frac{G^2}{C_6}, \quad r_2 = \frac{H^2}{C_6}, \quad r_3 = \frac{K^2}{C_6}. \tag{1.58}$$

Proposition 1.7 (The resolution of algebraic system of equations associated with the Hessian polyhedra, II). *The following identities hold:*

$$\begin{cases} Z_1 = 27r_1(4r_1 + r_2 - 1)^2, \\ Z_2 = \frac{1}{3}r_2^2 + \frac{8}{3}r_3 \cdot \sqrt{r_2 r_3}, \\ \sqrt[3]{\frac{1}{4}(Z_1 + Z_2 - 1)^2 - \frac{1}{27}Z_2^3} = \frac{4}{9}(r_3^2 - r_2 \cdot \sqrt{r_2 r_3}). \end{cases} \tag{1.59}$$

Theorem 1.8 (Main Theorem 5) (The resolution of algebraic system of equations associated with the Hessian polyhedra, III). *The function-theoretic resolvents (analytic resolvents) of Z_1 and Z_2 are given by:*

$$\begin{aligned} & Z_1^2 : Z_2^3 : (Z_1 + Z_2 - 1)^2 : [27(Z_1 + Z_2 - 1)^2 - 4Z_2^3] : 1 \\ =& 729^2 r_1^2 (4r_1 + r_2 - 1)^4 : 27(r_2^2 + 8r_3 \cdot \sqrt{r_2 r_3})^3 \\ :& [256(r_3^2 - r_2 \cdot \sqrt{r_2 r_3})^3 + 4(r_2^2 + 8r_3 \cdot \sqrt{r_2 r_3})^3] \\ :& 6912(r_3^2 - r_2 \cdot \sqrt{r_2 r_3})^3 : 729. \end{aligned} \tag{1.60}$$

For the Hessian polyhedral equations, the corresponding j-invariants (J_1, J_2) satisfy:

$$\begin{aligned} &J_1^2 : J_2^3 : (J_1 + J_2 - 1)^2 : [27(J_1 + J_2 - 1)^2 - 4J_2^3] : 1 \\ =&729^2 v_1^2 (4v_1 + v_2 - 1)^4 : 27(v_2^2 + 8v_3 \cdot \sqrt{v_2 v_3})^3 \\ :&[256(v_3^2 - v_2 \cdot \sqrt{v_2 v_3})^3 + 4(v_2^2 + 8v_3 \cdot \sqrt{v_2 v_3})^3] \\ :&6912(v_3^2 - v_2 \cdot \sqrt{v_2 v_3})^3 : 729, \end{aligned} \tag{1.61}$$

where

$$v_1 = v_1(\omega_1, \omega_2), \quad v_2 = v_2(\omega_1, \omega_2), \quad v_3 = v_3(\omega_1, \omega_2)$$

are three Picard modular functions with $(\omega_1, \omega_2) \in \mathfrak{S}_2 = \{(z_1, z_2) \in \mathbb{C}^2 : z_1 + \overline{z_1} - z_2 \overline{z_2} > 0\}$.

Put

$$u = \frac{C_6 G}{C_9}, \quad v = \frac{C_9 H}{C_{12}}, \quad w = \frac{C_9 K}{\mathfrak{C}_{12}}. \tag{1.62}$$

We will denote the next three equations as the resolvents of the u, v and w's.

Theorem 1.9 (Main Theorem 6) (The resolution of algebraic system of equations associated with the Hessian polyhedra, IV). *The functions u, v and w satisfy the following algebraic equations, which are the analytic and algebraic resolvents of u, v and w:*

$$\begin{cases} Z_1^2 u^3 + 5184 Z_2^2 u v^2 - 108 Z_1 u - 432 Z_1 = 0, \\ 6912 Z_2^3 v^4 + 8[27(Z_1 + Z_2 - 1)^2 - 4Z_2^3] v w^3 - 9 Z_1^2 = 0, \\ [27(Z_1 + Z_2 - 1)^2 - 4Z_2^3] w^4 - 6912 Z_2^3 v^3 w - 27 Z_1^2 = 0. \end{cases} \tag{1.63}$$

Theorem 1.10 (Main Theorem 7) (The resolution of algebraic system of equations associated with the Hessian polyhedra, V). *The functions u, v and w can be expressed as the algebraic functions of r_1, r_2 and r_3:*

$$\begin{cases} u = \dfrac{4}{4r_1 + r_2 - 1}, \\ v = \dfrac{9}{4} \dfrac{(4r_1 + r_2 - 1)\sqrt{r_1}}{r_2\sqrt{r_2} + 8r_3\sqrt{r_3}}, \\ w = \dfrac{27}{4} \dfrac{(4r_1 + r_2 - 1)\sqrt{r_1}}{r_3\sqrt{r_3} - r_2\sqrt{r_2}}. \end{cases} \tag{1.64}$$

Set

$$M := M_\infty = 81\sqrt{-3}\varphi^3, \quad N := N_\infty = 3\sqrt{-3}\psi^3. \tag{1.65}$$

$$M_0 := M(E(z_1, z_2, z_3)) = K^3, \quad N_0 := N(E(z_1, z_2, z_3)) = H^3. \tag{1.66}$$

They satisfy:

$$N_0(N_0 + 8M_0)^3 = 9^3 C_{12}^3, \quad M_0(M_0 - N_0)^3 = 27^3 \mathfrak{C}_{12}^3. \tag{1.67}$$

Let

$$\begin{cases} M_i := M(ED^i(z_1, z_2, z_3)) = (\psi - 3\omega^{2i}\varphi)^3, \\ N_i := N(ED^i(z_1, z_2, z_3)) = (\psi + 6\omega^{2i}\varphi)^3. \end{cases} \tag{1.68}$$

Then

$$M_\infty M_0 M_1 M_2 = -81\sqrt{-3}\mathfrak{C}_{12}^3, \quad N_\infty N_0 N_1 N_2 = 3\sqrt{-3}C_{12}^3. \tag{1.69}$$

Let

$$L(z_1, z_2, z_3) := (z_1 - z_2)^3 (z_2 - z_3)^3 (z_3 - z_1)^3. \tag{1.70}$$

Set

$$\begin{cases} L_{1,0,0} := L(E(z_1, z_2, z_3)) = (z_2^3 - z_3^3)^3, \\ L_{1,1,0} := L(EA(z_1, z_2, z_3)) = (z_3^3 - z_1^3)^3, \\ L_{1,2,0} := L(EA^2(z_1, z_2, z_3)) = (z_1^3 - z_2^3)^3, \end{cases} \tag{1.71}$$

and

$$\begin{cases} L_{0,0,0} := L(z_1, z_2, z_3) = (z_1 - z_2)^3 (z_2 - z_3)^3 (z_3 - z_1)^3, \\ L_{0,0,1} := L(C(z_1, z_2, z_3)) = (z_1 - \omega z_2)^3 (\omega z_2 - \omega^2 z_3)^3 (\omega^2 z_3 - z_1)^3, \\ L_{0,0,2} := L(C^2(z_1, z_2, z_3)) = (z_1 - \omega^2 z_2)^3 (\omega^2 z_2 - \omega z_3)^3 (\omega z_3 - z_1)^3. \end{cases} \tag{1.72}$$

Then

$$L_{1,0,0} L_{1,1,0} L_{1,2,0} = L_{0,0,0} L_{0,0,1} L_{0,0,2} = C_9^3. \tag{1.73}$$

Put

$$\zeta_1 = \frac{M_0}{C_9}, \quad \zeta_2 = \frac{N_0}{C_9}. \tag{1.74}$$

Theorem 1.11 (Main Theorem 8) (The resolution of algebraic system of equations associated with the Hessian polyhedra, VI). *The functions ζ_1 and ζ_2 satisfy the following equations*:

$$\begin{cases} Z_1^2 \zeta_2 (\zeta_2 + 8\zeta_1)^3 = 2^8 \cdot 3^9 Z_2^3, \\ Z_1^2 \zeta_1 (\zeta_1 - \zeta_2)^3 = 27^3 [27(Z_1 + Z_2 - 1)^2 - 4Z_2^3]. \end{cases} \tag{1.75}$$

Put

$$\xi_1 := \frac{M_0^{\frac{1}{3}}}{C_6^{\frac{1}{2}}}, \quad \xi_2 := \frac{N_0^{\frac{1}{3}}}{C_6^{\frac{1}{2}}}. \tag{1.76}$$

Theorem 1.12 (Main Theorem 9) (The resolution of algebraic system of equations associated with the Hessian polyhedra, VII). *The functions ξ_1 and ξ_2 satisfy the following equations*:

$$\begin{cases} \xi_2^3 (\xi_2^3 + 8\xi_1^3)^3 = 27 Z_2^3, \\ 256 \xi_1^3 (\xi_1^3 - \xi_2^3)^3 = 27[27(Z_1 + Z_2 - 1)^2 - 4Z_2^3]. \end{cases} \tag{1.77}$$

The functions ξ_1, ξ_2 and ζ_1, ζ_2 satisfy the following relations:

$$\begin{cases} \frac{4}{729}\zeta_1^4[\xi_2^3(\xi_2^3+8\xi_1^3)^3+64\xi_1^3(\xi_1^3-\xi_2^3)^3]=[432\xi_1^6+\frac{1}{3}\zeta_1^2\xi_2(\xi_2^3+8\xi_1^3)-\zeta_1^2]^2, \\ \frac{4}{729}\zeta_2^4[\xi_2^3(\xi_2^3+8\xi_1^3)^3+64\xi_1^3(\xi_1^3-\xi_2^3)^3]=[432\xi_2^6+\frac{1}{3}\zeta_2^2\xi_2(\xi_2^3+8\xi_1^3)-\zeta_2^2]^2. \end{cases} \tag{1.78}$$

Theorem 1.13 (Main Theorem 10) (The resolution of algebraic system of equations associated with the Hessian polyhedra, VIII). *The function-theoretic resolvents (analytic resolvents) of Z_1 and Z_2 are also given by:*

$$\begin{aligned} &Z_1^2 : Z_2^3 : (Z_1+Z_2-1)^2 : [27(Z_1+Z_2-1)^2-4Z_2^3] : 1 \\ =&[2(\xi_2^3(\xi_2^3+8\xi_1^3)^3+64\xi_1^3(\xi_1^3-\xi_2^3)^3)^{\frac{1}{2}}-9\xi_2(\xi_2^3+8\xi_1^3)+27]^2 \\ &:27\xi_2^3(\xi_2^3+8\xi_1^3)^3 : 4[\xi_2^3(\xi_2^3+8\xi_1^3)^3+64\xi_1^3(\xi_1^3-\xi_2^3)^3] \\ &:6912\xi_1^3(\xi_1^3-\xi_2^3)^3 : 729. \end{aligned} \tag{1.79}$$

In fact,

$$\begin{cases} \sqrt[3]{M_0}+\sqrt[3]{M_1}+\sqrt[3]{M_2}=3\psi, \\ \sqrt{-3}\sqrt[3]{M_\infty}+\sqrt[3]{M_0}+\omega\sqrt[3]{M_1}+\omega^2\sqrt[3]{M_2}=0, \\ \sqrt[3]{M_0}+\omega^2\sqrt[3]{M_1}+\omega\sqrt[3]{M_2}=0. \end{cases} \tag{1.80}$$

$$\begin{cases} -\sqrt{-3}\sqrt[3]{N_\infty}+\sqrt[3]{N_0}+\sqrt[3]{N_1}+\sqrt[3]{N_2}=0, \\ \sqrt[3]{N_0}+\omega\sqrt[3]{N_1}+\omega^2\sqrt[3]{N_2}=18\varphi, \\ \sqrt[3]{N_0}+\omega^2\sqrt[3]{N_1}+\omega\sqrt[3]{N_2}=0. \end{cases} \tag{1.81}$$

$$\begin{cases} \sqrt[3]{L_{1,0,0}}+\sqrt[3]{L_{1,1,0}}+\sqrt[3]{L_{1,2,0}}=0, \\ \sqrt[3]{L_{0,0,0}}+\sqrt[3]{L_{0,0,1}}+\sqrt[3]{L_{0,0,2}}=0. \end{cases} \tag{1.82}$$

$$\begin{cases} L_{1,0,0}+L_{1,1,0}+L_{1,2,0}=3C_9, \\ L_{0,0,0}+L_{0,0,1}+L_{0,0,2}=3C_9. \end{cases} \tag{1.83}$$

$$\sqrt{-3}(M_0+M_1+M_2)=N_\infty-M_\infty, \quad \sqrt{-3}(N_0+N_1+N_2)=N_\infty+8M_\infty. \tag{1.84}$$

$$\begin{aligned} (L_{1,0,0}+L_{1,1,0}+L_{1,2,0})^3 &= 27L_{1,0,0}L_{1,1,0}L_{1,2,0}, \\ (L_{0,0,0}+L_{0,0,1}+L_{0,0,2})^3 &= 27L_{0,0,0}L_{0,0,1}L_{0,0,2}. \end{aligned} \tag{1.85}$$

The above eight aspects of the resolution of algebraic system of equations associated with the Hessian polyhedra give a higher dimensional analogue of Klein's celebrated work [44] (or [42]) on the resolution of quintic equations by the icosahedron.

Now, we will give some further relations between the invariants G, H, K and C_6, C_9, C_{12}, C_{18}, $\mathfrak{C}_{12}$. Put

$$X(z_1,z_2,z_3)=z_1^3, \quad Y(z_1,z_2,z_3)=z_2^3, \quad Z(z_1,z_2,z_3)=z_3^3. \tag{1.86}$$

$$Q_1(z_1, z_2, z_3) = z_1 z_2^2 + z_2 z_3^2 + z_3 z_1^2, \quad Q_2(z_1, z_2, z_3) = z_1^2 z_2 + z_2^2 z_3 + z_3^2 z_1. \tag{1.87}$$

The functions φ, ψ, χ, X, Y, Z and Q_1, Q_2 satisfy the relations:

$$\begin{cases} \varphi^3 = XYZ, \\ \psi = X + Y + Z, \\ \chi = XY + YZ + ZX, \\ \chi + 3\varphi^2 + \varphi\psi = Q_1 Q_2, \\ \psi\chi + 6\varphi\chi + 6\varphi^2\psi + 9\varphi^3 = Q_1^3 + Q_2^3, \\ (X - Y)(Y - Z)(Z - X) = Q_1^3 - Q_2^3. \end{cases} \tag{1.88}$$

Let

$$\begin{cases} W_2 = (X + Y + Z)^2 - 12(XY + YZ + ZX), \\ W_3 = (X - Y)(Y - Z)(Z - X), \\ \mathfrak{W}_3 = XYZ - \varphi^3, \\ W_4 = (X + Y + Z)[(X + Y + Z)^3 + 216\varphi^3], \\ \mathfrak{W}_4 = \varphi[27\varphi^3 - (X + Y + Z)^3], \\ W_6 = (X + Y + Z)^6 - 540\varphi^3(X + Y + Z)^3 - 5832\varphi^6. \end{cases} \tag{1.89}$$

We will regard (φ, X, Y, Z) as a point in the complex projective space $\mathbb{CP}^3$. In fact,

$$1728\mathfrak{W}_4^3 = W_6^2 - W_4^3 \tag{1.90}$$

and

$$\frac{\partial(\mathfrak{W}_3, W_4, \mathfrak{W}_4, W_6)}{\partial(\varphi, X, Y, Z)} = 0. \tag{1.91}$$

For the ternary cubic form

$$F_1(x_1, x_2, x_3) = Xx_1^3 + Yx_2^3 + Zx_3^3 - 3\varphi x_1 x_2 x_3, \tag{1.92}$$

we have

$$\begin{cases} \sqrt[3]{M_\infty} = -3\sqrt{-3}\varphi, \\ \sqrt[3]{M_0} = -3\varphi + X + Y + Z, \\ \sqrt[3]{M_1} = -3\omega^2\varphi + X + Y + Z, \\ \sqrt[3]{M_2} = -3\omega\varphi + X + Y + Z. \end{cases} \tag{1.93}$$

Correspondingly,

$$\begin{cases} \sqrt[3]{M_\infty}: & (\sqrt{-3}, 0, 0, 0) \in \mathbb{CP}^3, \\ \sqrt[3]{M_0}: & (1, 1, 1, 1) \in \mathbb{CP}^3, \\ \sqrt[3]{M_1}: & (\omega^2, 1, 1, 1) \in \mathbb{CP}^3, \\ \sqrt[3]{M_2}: & (\omega, 1, 1, 1) \in \mathbb{CP}^3. \end{cases} \tag{1.94}$$

For the ternary cubic form

$$F_2(x_1, x_2, x_3) = Xx_1^3 + Yx_2^3 + Zx_3^3 + 6\varphi x_1 x_2 x_3, \tag{1.95}$$

we have

$$\begin{cases} \sqrt[3]{N_\infty} = -\sqrt{-3}(X + Y + Z), \\ \sqrt[3]{N_0} = 6\varphi + X + Y + Z, \\ \sqrt[3]{N_1} = 6\omega^2\varphi + X + Y + Z, \\ \sqrt[3]{N_2} = 6\omega\varphi + X + Y + Z. \end{cases} \tag{1.96}$$

Correspondingly,

$$\begin{cases} \sqrt[3]{N_\infty}: & (0, -\sqrt{-3}, -\sqrt{-3}, -\sqrt{-3}) \in \mathbb{CP}^3, \\ \sqrt[3]{N_0}: & (1, 1, 1, 1) \in \mathbb{CP}^3, \\ \sqrt[3]{N_1}: & (\omega^2, 1, 1, 1) \in \mathbb{CP}^3, \\ \sqrt[3]{N_2}: & (\omega, 1, 1, 1) \in \mathbb{CP}^3. \end{cases} \tag{1.97}$$

When $\mathfrak{W}_3 = 0$, we set

$$\begin{cases} W_2 = (X + Y + Z)^2 - 12(XY + YZ + ZX), \\ W_3 = (X - Y)(Y - Z)(Z - X), \\ W_4 = (X + Y + Z)[(X + Y + Z)^3 + 216XYZ], \\ W_6 = (X + Y + Z)^6 - 540XYZ(X + Y + Z)^3 - 5832X^2Y^2Z^2, \\ W_{12} = XYZ[27XYZ - (X + Y + Z)^3]^3. \end{cases} \tag{1.98}$$

The invariants W_2, W_3, W_4, W_6 and W_{12} satisfy the relations:

$$\begin{cases} 432W_3^2 = W_2^3 - 3W_2W_4 + 2W_6, \\ 1728W_{12} = W_6^2 - W_4^3. \end{cases} \tag{1.99}$$

The algebraic curves in $\mathbb{CP}^2$: $W_2 = 0$, $W_4 = 0$, $W_6 = 0$, $W_{12} = 0$ intersect $W_3 = 0$ in 6, 12, 18, 15 points respectively.

$$\{W_2 = 0\} \cap \{W_3 = 0\} = \{(\alpha_i, 1, 1), (1, \alpha_i, 1), (1, 1, \alpha_i), i = 1, 2\},$$

where α_1 and α_2 are two roots of the quadratic equation:

$$\alpha^2 - 20\alpha - 8 = 0.$$

$$\begin{aligned} &\{W_4 = 0\} \cap \{W_3 = 0\} \\ =&\{(-2, 1, 1), (1, -2, 1), (1, 1, -2), (\beta_i, 1, 1), (1, \beta_i, 1), (1, 1, \beta_i), i = 1, 2, 3\}, \end{aligned}$$

where β_1, β_2 and β_3 are the roots of the cubic equation:

$$\beta^3 + 6\beta^2 + 228\beta + 8 = 0.$$

$$\{W_6 = 0\} \cap \{W_3 = 0\} = \{(\gamma_i, 1, 1), (1, \gamma_i, 1), (1, 1, \gamma_i), 1 \leq i \leq 6\},$$

where γ_i $(1 \leq i \leq 6)$ are the roots of the equation of the sixth degree:

$$[\gamma^3 + 6\gamma^2 - (258 + 162\sqrt{3})\gamma + 8][\gamma^3 + 6\gamma^2 - (258 - 162\sqrt{3})\gamma + 8] = 0.$$

$$\begin{aligned}\{W_{12} = 0\} \cap \{W_3 = 0\} = \{&(1,0,0), (0,1,0), (0,0,1), (1,1,0), (1,0,1), (0,1,1),\\ &(\delta_i, 1, 1), (1, \delta_i, 1), (1, 1, \delta_i), i = 1, 2, 3\},\end{aligned}$$

where δ_1, δ_2 and δ_3 are the roots of the cubic equation:

$$\delta^3 + 6\delta^2 - 15\delta + 8 = 0.$$

In fact,

$$\mathbb{C}[W_2, W_3, W_4, W_6, W_{12}]/I \cong \mathbb{C}[W_2, W_3, W_4], \tag{1.100}$$

where the ideal

$$I = (W_2^3 - 3W_2W_4 - 432W_3^2 + 2W_6, W_6^2 - W_4^3 - 1728W_{12}).$$

We find that

$$\begin{cases} (\sqrt{-3})^3 \varphi(E(z_1, z_2, z_3)) = \psi - 3\varphi, \\ -\sqrt{-3}\psi(E(z_1, z_2, z_3)) = \psi + 6\varphi, \\ (\sqrt{-3})^3 X(E(z_1, z_2, z_3)) = \psi + 6\varphi + 3Q_1 + 3Q_2, \\ (\sqrt{-3})^3 Y(E(z_1, z_2, z_3)) = \psi + 6\varphi + 3\omega^2 Q_1 + 3\omega Q_2, \\ (\sqrt{-3})^3 Z(E(z_1, z_2, z_3)) = \psi + 6\varphi + 3\omega Q_1 + 3\omega^2 Q_2, \\ -\sqrt{-3}Q_1(E(z_1, z_2, z_3)) = X + \omega^2 Y + \omega Z, \\ -\sqrt{-3}Q_2(E(z_1, z_2, z_3)) = X + \omega Y + \omega^2 Z. \end{cases} \tag{1.101}$$

$$H_1 := \text{Hessian}(Q_1) = \begin{vmatrix} \frac{\partial^2 Q_1}{\partial z_1^2} & \frac{\partial^2 Q_1}{\partial z_1 \partial z_2} & \frac{\partial^2 Q_1}{\partial z_1 \partial z_3} \\ \frac{\partial^2 Q_1}{\partial z_1 \partial z_2} & \frac{\partial^2 Q_1}{\partial z_2^2} & \frac{\partial^2 Q_1}{\partial z_2 \partial z_3} \\ \frac{\partial^2 Q_1}{\partial z_1 \partial z_3} & \frac{\partial^2 Q_1}{\partial z_2 \partial z_3} & \frac{\partial^2 Q_1}{\partial z_3^2} \end{vmatrix} = -8(\psi - 3\varphi). \tag{1.102}$$

$$H_2 := \text{Hessian}(Q_2) = \begin{vmatrix} \frac{\partial^2 Q_2}{\partial z_1^2} & \frac{\partial^2 Q_2}{\partial z_1 \partial z_2} & \frac{\partial^2 Q_2}{\partial z_1 \partial z_3} \\ \frac{\partial^2 Q_2}{\partial z_1 \partial z_2} & \frac{\partial^2 Q_2}{\partial z_2^2} & \frac{\partial^2 Q_2}{\partial z_2 \partial z_3} \\ \frac{\partial^2 Q_2}{\partial z_1 \partial z_3} & \frac{\partial^2 Q_2}{\partial z_2 \partial z_3} & \frac{\partial^2 Q_2}{\partial z_3^2} \end{vmatrix} = -8(\psi - 3\varphi). \tag{1.103}$$

$$H_3 := \text{Hessian}(\varphi) = 2\varphi. \tag{1.104}$$

$$H_4 := \text{Hessian}(\psi) = 216\varphi. \tag{1.105}$$

We have

$$J_1 := \begin{vmatrix} \frac{\partial^2 Q_1}{\partial z_1^2} & \frac{\partial^2 Q_1}{\partial z_1 \partial z_2} & \frac{\partial^2 Q_1}{\partial z_1 \partial z_3} & \frac{\partial H_1}{\partial z_1} \\ \frac{\partial^2 Q_1}{\partial z_1 \partial z_2} & \frac{\partial^2 Q_1}{\partial z_2^2} & \frac{\partial^2 Q_1}{\partial z_2 \partial z_3} & \frac{\partial H_1}{\partial z_2} \\ \frac{\partial^2 Q_1}{\partial z_1 \partial z_3} & \frac{\partial^2 Q_1}{\partial z_2 \partial z_3} & \frac{\partial^2 Q_1}{\partial z_3^2} & \frac{\partial H_1}{\partial z_3} \\ \frac{\partial H_1}{\partial z_1} & \frac{\partial H_1}{\partial z_2} & \frac{\partial H_1}{\partial z_3} & 0 \end{vmatrix} = 6912[3Q_1^2 - (\psi + 6\varphi)Q_2]. \tag{1.106}$$

$$J_2 := \begin{vmatrix} \frac{\partial^2 Q_2}{\partial z_1^2} & \frac{\partial^2 Q_2}{\partial z_1 \partial z_2} & \frac{\partial^2 Q_2}{\partial z_1 \partial z_3} & \frac{\partial H_2}{\partial z_1} \\ \frac{\partial^2 Q_2}{\partial z_1 \partial z_2} & \frac{\partial^2 Q_2}{\partial z_2^2} & \frac{\partial^2 Q_2}{\partial z_2 \partial z_3} & \frac{\partial H_2}{\partial z_2} \\ \frac{\partial^2 Q_2}{\partial z_1 \partial z_3} & \frac{\partial^2 Q_2}{\partial z_2 \partial z_3} & \frac{\partial^2 Q_2}{\partial z_3^2} & \frac{\partial H_2}{\partial z_3} \\ \frac{\partial H_2}{\partial z_1} & \frac{\partial H_2}{\partial z_2} & \frac{\partial H_2}{\partial z_3} & 0 \end{vmatrix} = 6912[3Q_2^2 - (\psi + 6\varphi)Q_1]. \quad (1.107)$$

$$J_3 := \begin{vmatrix} \frac{\partial^2 \varphi}{\partial z_1^2} & \frac{\partial^2 \varphi}{\partial z_1 \partial z_2} & \frac{\partial^2 \varphi}{\partial z_1 \partial z_3} & \frac{\partial H_3}{\partial z_1} \\ \frac{\partial^2 \varphi}{\partial z_1 \partial z_2} & \frac{\partial^2 \varphi}{\partial z_2^2} & \frac{\partial^2 \varphi}{\partial z_2 \partial z_3} & \frac{\partial H_3}{\partial z_2} \\ \frac{\partial^2 \varphi}{\partial z_1 \partial z_3} & \frac{\partial^2 \varphi}{\partial z_2 \partial z_3} & \frac{\partial^2 \varphi}{\partial z_3^2} & \frac{\partial H_3}{\partial z_3} \\ \frac{\partial H_3}{\partial z_1} & \frac{\partial H_3}{\partial z_2} & \frac{\partial H_3}{\partial z_3} & 0 \end{vmatrix} = -12\varphi^2. \quad (1.108)$$

$$J_4 := \begin{vmatrix} \frac{\partial^2 \psi}{\partial z_1^2} & \frac{\partial^2 \psi}{\partial z_1 \partial z_2} & \frac{\partial^2 \psi}{\partial z_1 \partial z_3} & \frac{\partial H_4}{\partial z_1} \\ \frac{\partial^2 \psi}{\partial z_1 \partial z_2} & \frac{\partial^2 \psi}{\partial z_2^2} & \frac{\partial^2 \psi}{\partial z_2 \partial z_3} & \frac{\partial H_4}{\partial z_2} \\ \frac{\partial^2 \psi}{\partial z_1 \partial z_3} & \frac{\partial^2 \psi}{\partial z_2 \partial z_3} & \frac{\partial^2 \psi}{\partial z_3^2} & \frac{\partial H_4}{\partial z_3} \\ \frac{\partial H_4}{\partial z_1} & \frac{\partial H_4}{\partial z_2} & \frac{\partial H_4}{\partial z_3} & 0 \end{vmatrix} = -36^4\chi. \quad (1.109)$$

$$K_1 := \begin{vmatrix} \frac{\partial Q_1}{\partial z_1} & \frac{\partial Q_1}{\partial z_2} & \frac{\partial Q_1}{\partial z_3} \\ \frac{\partial H_1}{\partial z_1} & \frac{\partial H_1}{\partial z_2} & \frac{\partial H_1}{\partial z_3} \\ \frac{\partial J_1}{\partial z_1} & \frac{\partial J_1}{\partial z_2} & \frac{\partial J_1}{\partial z_3} \end{vmatrix} = 6912 \cdot 24[(\psi + 6\varphi)^3 - 27Q_2^3]. \quad (1.110)$$

$$K_2 := \begin{vmatrix} \frac{\partial Q_2}{\partial z_1} & \frac{\partial Q_2}{\partial z_2} & \frac{\partial Q_2}{\partial z_3} \\ \frac{\partial H_2}{\partial z_1} & \frac{\partial H_2}{\partial z_2} & \frac{\partial H_2}{\partial z_3} \\ \frac{\partial J_2}{\partial z_1} & \frac{\partial J_2}{\partial z_2} & \frac{\partial J_2}{\partial z_3} \end{vmatrix} = -6912 \cdot 24[(\psi + 6\varphi)^3 - 27Q_1^3]. \quad (1.111)$$

$$K_3 := \begin{vmatrix} \frac{\partial \varphi}{\partial z_1} & \frac{\partial \varphi}{\partial z_2} & \frac{\partial \varphi}{\partial z_3} \\ \frac{\partial H_3}{\partial z_1} & \frac{\partial H_3}{\partial z_2} & \frac{\partial H_3}{\partial z_3} \\ \frac{\partial J_3}{\partial z_1} & \frac{\partial J_3}{\partial z_2} & \frac{\partial J_3}{\partial z_3} \end{vmatrix} = 0. \quad (1.112)$$

$$K_4 := \begin{vmatrix} \frac{\partial \psi}{\partial z_1} & \frac{\partial \psi}{\partial z_2} & \frac{\partial \psi}{\partial z_3} \\ \frac{\partial H_4}{\partial z_1} & \frac{\partial H_4}{\partial z_2} & \frac{\partial H_4}{\partial z_3} \\ \frac{\partial J_4}{\partial z_1} & \frac{\partial J_4}{\partial z_2} & \frac{\partial J_4}{\partial z_3} \end{vmatrix} = -36^5 \cdot 54(Q_1^3 - Q_2^3). \quad (1.113)$$

$$(G, H, K) := \begin{vmatrix} \frac{\partial G}{\partial z_1} & \frac{\partial G}{\partial z_2} & \frac{\partial G}{\partial z_3} \\ \frac{\partial H}{\partial z_1} & \frac{\partial H}{\partial z_2} & \frac{\partial H}{\partial z_3} \\ \frac{\partial K}{\partial z_1} & \frac{\partial K}{\partial z_2} & \frac{\partial K}{\partial z_3} \end{vmatrix} = 27[3(Q_1^2 + Q_2^2) - (\psi + 6\varphi)(Q_1 + Q_2)]. \quad (1.114)$$

Put

$$U_1 = z_1 z_2^2, \quad U_2 = z_2 z_3^2, \quad U_3 = z_3 z_1^2, \quad (1.115)$$

$$V_1 = z_1^2 z_2, \quad V_2 = z_2^2 z_3, \quad V_3 = z_3^2 z_1. \quad (1.116)$$

Then

$$U_1V_1 = XY, \quad U_2V_2 = YZ, \quad U_3V_3 = ZX. \tag{1.117}$$

$$U_1U_2U_3 = V_1V_2V_3 = XYZ = \varphi^3. \tag{1.118}$$

$$Q_1 = U_1 + U_2 + U_3, \quad Q_2 = V_1 + V_2 + V_3, \quad \psi = X + Y + Z. \tag{1.119}$$

The ternary cubic form

$$\begin{aligned} F(x_1, x_2, x_3) =& Xx_1^3 + Yx_2^3 + Zx_3^3 + 6\varphi x_1x_2x_3 \\ &+ 3(U_1x_1x_2^2 + U_2x_2x_3^2 + U_3x_3x_1^2) + 3(V_1x_1^2x_2 + V_2x_2^2x_3 + V_3x_3^2x_1) \\ =& (z_1x_1 + z_2x_2 + z_3x_3)^3. \end{aligned} \tag{1.120}$$

$$\begin{cases} (\sqrt{-3})^3U_1(E(z_1, z_2, z_3)) =X + \omega^2Y + \omega Z + (\omega - 1)U_1 \\ \qquad + (1 - \omega^2)U_2 + (\omega^2 - \omega)U_3 + (\omega - \omega^2)V_1 \\ \qquad + (1 - \omega)V_2 + (\omega^2 - 1)V_3, \\ (\sqrt{-3})^3U_2(E(z_1, z_2, z_3)) =X + \omega^2Y + \omega Z + (1 - \omega^2)U_1 \\ \qquad + (\omega^2 - \omega)U_2 + (\omega - 1)U_3 + (\omega^2 - 1)V_1 \\ \qquad + (\omega - \omega^2)V_2 + (1 - \omega)V_3, \\ (\sqrt{-3})^3U_3(E(z_1, z_2, z_3)) =X + \omega^2Y + \omega Z + (\omega^2 - \omega)U_1 \\ \qquad + (\omega - 1)U_2 + (1 - \omega^2)U_3 + (1 - \omega)V_1 \\ \qquad + (\omega^2 - 1)V_2 + (\omega - \omega^2)V_3. \end{cases} \tag{1.121}$$

$$\begin{cases} (\sqrt{-3})^3V_1(E(z_1, z_2, z_3)) =X + \omega Y + \omega^2Z + (\omega - \omega^2)U_1 \\ \qquad + (\omega^2 - 1)U_2 + (1 - \omega)U_3 + (1 - \omega^2)V_1 \\ \qquad + (\omega - 1)V_2 + (\omega^2 - \omega)V_3, \\ (\sqrt{-3})^3V_2(E(z_1, z_2, z_3)) =X + \omega Y + \omega^2Z + (1 - \omega)U_1 \\ \qquad + (\omega - \omega^2)U_2 + (\omega^2 - 1)U_3 + (\omega - 1)V_1 \\ \qquad + (\omega^2 - \omega)V_2 + (1 - \omega^2)V_3, \\ (\sqrt{-3})^3V_3(E(z_1, z_2, z_3)) =X + \omega Y + \omega^2Z + (\omega^2 - 1)U_1 \\ \qquad + (1 - \omega)U_2 + (\omega - \omega^2)U_3 + (\omega^2 - \omega)V_1 \\ \qquad + (1 - \omega^2)V_2 + (\omega - 1)V_3. \end{cases} \tag{1.122}$$

Let us define three rational functions (invariants) associated to the complex projective plane $\mathbb{CP}^2$:

$$\begin{cases} f_1^{\text{Hessian}}(\xi,\eta) := 432\dfrac{C_9(\xi,\eta)^2}{C_6(\xi,\eta)^3}, \\ f_2^{\text{Hessian}}(\xi,\eta) := 3\dfrac{C_{12}(\xi,\eta)}{C_6(\xi,\eta)^2}, \\ f_3^{\text{Hessian}}(\xi,\eta) := \dfrac{C_{18}(\xi,\eta)^2}{1728\mathfrak{C}_{12}(\xi,\eta)^3}. \end{cases} \tag{1.123}$$

Put

$$\begin{cases} F_1^{\text{Hessian}}(\xi,\eta) := (f_1^{\text{Hessian}}(\xi,\eta), f_3^{\text{Hessian}}(\xi,\eta)) \\ = \left(\dfrac{432C_9(\xi,\eta)^2}{C_6(\xi,\eta)^3}, \dfrac{C_{18}(\xi,\eta)^2}{1728\mathfrak{C}_{12}(\xi,\eta)^3}\right), \\ F_2^{\text{Hessian}}(\xi,\eta) := (f_2^{\text{Hessian}}(\xi,\eta), f_3^{\text{Hessian}}(\xi,\eta)) \\ = \left(\dfrac{3C_{12}(\xi,\eta)}{C_6(\xi,\eta)^2}, \dfrac{C_{18}(\xi,\eta)^2}{1728\mathfrak{C}_{12}(\xi,\eta)^3}\right). \end{cases} \tag{1.124}$$

The function $F_1^{\text{Hessian}}(\xi,\eta)$ is ramified at the points $(0,0)$, $(\infty,0)$, $(0,1)$, $(\infty,1)$, $(0,\infty)$, (∞,∞), with degrees 162, 108, 108, 72, 108, 72, respectively. Their least common multiple is 648. The function $F_2^{\text{Hessian}}(\xi,\eta)$ is ramified at the points $(0,0)$, $(\infty,0)$, $(0,1)$, $(\infty,1)$, $(0,\infty)$, (∞,∞), with degrees 216, 108, ∞, 72, 144, 72, respectively. Their least common multiple is 1296. Thus, the rational invariants F_1^{Hessian} and F_2^{Hessian} correspond to the Hessian groups G_{648} and G_{1296}, respectively.

This paper consists of seven sections. In section two, we give the reciprocity law about the Appell hypergeometric partial differential equations. In section three, we study the algebraic solutions of Appell hypergeometric partial differential equations. In section four, we give the Hessian polyhedral equations and study their properties. In section five, we study the invariant theory for the system of algebraic equations which can be solved by our Hessian polyhedral equations. In section six, we study the ternary cubic forms associated to the Hessian polyhedra. In section seven, we find some rational invariants on $\mathbb{CP}^2$.

Acknowledgements. The author thanks the referees very much for their helpful comments.

2. The reciprocity law about the Appell hypergeometric functions

Let us consider the Appell hypergeometric partial differential equations (see [1] for more details):

$$\begin{cases} x(1-x)\dfrac{\partial^2 z}{\partial x^2} + y(1-x)\dfrac{\partial^2 z}{\partial x \partial y} + [c-(a+b+1)x]\dfrac{\partial z}{\partial x} - by\dfrac{\partial z}{\partial y} - abz = 0, \\ y(1-y)\dfrac{\partial^2 z}{\partial y^2} + x(1-y)\dfrac{\partial^2 z}{\partial x \partial y} + [c-(a+b'+1)y]\dfrac{\partial z}{\partial y} - b'x\dfrac{\partial z}{\partial x} - ab'z = 0, \end{cases}$$

which are equivalent to the following three equations:

$$\begin{cases} \frac{\partial^2 z}{\partial x^2} + \left(\frac{c-b'}{x} + \frac{a+b-c+1}{x-1} + \frac{b'}{x-y}\right)\frac{\partial z}{\partial x} \\ -\frac{by(y-1)}{x(x-1)(x-y)}\frac{\partial z}{\partial y} + \frac{abz}{x(x-1)} = 0, \\ \frac{\partial^2 z}{\partial y^2} + \left(\frac{c-b}{y} + \frac{a+b'-c+1}{y-1} + \frac{b}{y-x}\right)\frac{\partial z}{\partial y} \\ -\frac{b'x(x-1)}{y(y-1)(y-x)}\frac{\partial z}{\partial x} + \frac{ab'z}{y(y-1)} = 0, \\ \frac{\partial^2 z}{\partial x \partial y} = \frac{b'}{x-y}\frac{\partial z}{\partial x} - \frac{b}{x-y}\frac{\partial z}{\partial y}. \end{cases}$$

There are three linearly independent solutions z_1, z_2 and z_3. Set $u_1 = \frac{z_1}{z_3}$ and $u_2 = \frac{z_2}{z_3}$.

We find that (see [63])

$$\begin{cases} \{u_1, u_2; x, y\}_x = \frac{by(y-1)}{x(x-1)(x-y)}, \\ \{u_1, u_2; x, y\}_y = \frac{b'x(x-1)}{y(y-1)(y-x)}, \\ [u_1, u_2; x, y]_x = \frac{c-b'}{x} + \frac{a+b-c+1}{x-1} + \frac{b'-2b}{x-y}, \\ [u_1, u_2; x, y]_y = \frac{c-b}{y} + \frac{a+b'-c+1}{y-1} + \frac{b-2b'}{y-x}. \end{cases}$$

Put

$$\Delta = \begin{vmatrix} \frac{\partial z_1}{\partial x} & \frac{\partial z_1}{\partial y} & z_1 \\ \frac{\partial z_2}{\partial x} & \frac{\partial z_2}{\partial y} & z_2 \\ \frac{\partial z_3}{\partial x} & \frac{\partial z_3}{\partial y} & z_3 \end{vmatrix}. \tag{2.1}$$

We have

$$\begin{aligned} \frac{\partial \Delta}{\partial x} &= \begin{vmatrix} \frac{\partial^2 z_1}{\partial x^2} & \frac{\partial z_1}{\partial y} & z_1 \\ \frac{\partial^2 z_2}{\partial x^2} & \frac{\partial z_2}{\partial y} & z_2 \\ \frac{\partial^2 z_3}{\partial x^2} & \frac{\partial z_3}{\partial y} & z_3 \end{vmatrix} + \begin{vmatrix} \frac{\partial z_1}{\partial x} & \frac{\partial^2 z_1}{\partial x \partial y} & z_1 \\ \frac{\partial z_2}{\partial x} & \frac{\partial^2 z_2}{\partial x \partial y} & z_2 \\ \frac{\partial z_3}{\partial x} & \frac{\partial^2 z_3}{\partial x \partial y} & z_3 \end{vmatrix} \\ &= -\left(\frac{c-b'}{x} + \frac{a+b-c+1}{x-1} + \frac{b'}{x-y}\right)\Delta - \frac{b}{x-y}\Delta \\ &= -\left(\frac{c-b'}{x} + \frac{a+b-c+1}{x-1} + \frac{b+b'}{x-y}\right)\Delta. \end{aligned} \tag{2.2}$$

Similarly,

$$\frac{\partial \Delta}{\partial y} = -\left(\frac{c-b}{y} + \frac{a+b'-c+1}{y-1} + \frac{b+b'}{y-x}\right)\Delta. \tag{2.3}$$

Hence,

$$\Delta = Cx^{b'-c}y^{b-c}(x-1)^{c-a-b-1}(y-1)^{c-a-b'-1}(x-y)^{-b-b'}. \tag{2.4}$$

On the other hand,

$$\Delta = z_3^2 \begin{vmatrix} \frac{\partial(z_1/z_3)}{\partial x} & \frac{\partial(z_1/z_3)}{\partial y} & z_1 \\ \frac{\partial(z_2/z_3)}{\partial x} & \frac{\partial(z_2/z_3)}{\partial y} & z_2 \\ 0 & 0 & z_3 \end{vmatrix} = z_3^3 \frac{\partial(\xi,\eta)}{\partial(x,y)}, \tag{2.5}$$

where $\xi = z_1/z_3$ and $\eta = z_2/z_3$.

Therefore,

$$z_3^3 \frac{\partial(\xi,\eta)}{\partial(x,y)} = Cx^{b'-c}y^{b-c}(x-1)^{c-a-b-1}(y-1)^{c-a-b'-1}(x-y)^{-b-b'}. \tag{2.6}$$

When ξ, η and $x^{b'-c}y^{b-c}(x-1)^{c-a-b-1}(y-1)^{c-a-b'-1}(x-y)^{-b-b'}$ are algebraic functions, (2.6) implies that z_3 is an algebraic function. Hence, z_1, z_2 and z_3 are all algebraic functions.

Theorem 2.1 (Main Theorem 1). *For the system of linear partial differential equations:*

$$\begin{cases} \dfrac{\partial^2 z}{\partial v_1^2} + \dfrac{1}{3}P_1\dfrac{\partial z}{\partial v_1} - F_1\dfrac{\partial z}{\partial v_2} + \left(\dfrac{\partial F_1}{\partial v_2} - \dfrac{1}{3}\dfrac{\partial P_1}{\partial v_1} - \dfrac{2}{9}P_1^2 - \dfrac{2}{3}F_1P_2\right)z = 0, \\ \dfrac{\partial^2 z}{\partial v_1 \partial v_2} - \dfrac{1}{3}P_2\dfrac{\partial z}{\partial v_1} - \dfrac{1}{3}P_1\dfrac{\partial z}{\partial v_2} + \left(\dfrac{1}{3}\dfrac{\partial P_2}{\partial v_1} + \dfrac{1}{3}\dfrac{\partial P_1}{\partial v_2} + \dfrac{1}{9}P_1P_2 - F_1F_2\right)z = 0, \\ \dfrac{\partial^2 z}{\partial v_2^2} + \dfrac{1}{3}P_2\dfrac{\partial z}{\partial v_2} - F_2\dfrac{\partial z}{\partial v_1} + \left(\dfrac{\partial F_2}{\partial v_1} - \dfrac{1}{3}\dfrac{\partial P_2}{\partial v_2} - \dfrac{2}{9}P_2^2 - \dfrac{2}{3}F_2P_1\right)z = 0, \end{cases} \tag{2.7}$$

put

$$\begin{cases} P_1 = \dfrac{c-b'}{v_1} + \dfrac{a+b-c+1}{v_1-1} + \dfrac{b'-2b}{v_1-v_2}, \\ P_2 = \dfrac{c-b}{v_2} + \dfrac{a+b'-c+1}{v_2-1} + \dfrac{b-2b'}{v_2-v_1}, \\ F_1 = \dfrac{bv_2(v_2-1)}{v_1(v_1-1)(v_1-v_2)}, \\ F_2 = \dfrac{b'v_1(v_1-1)}{v_2(v_2-1)(v_2-v_1)}. \end{cases} \tag{2.8}$$

Set

$$w = v_1^{-\frac{1}{3}(c-b')}v_2^{-\frac{1}{3}(c-b)}(v_1-1)^{-\frac{1}{3}(a+b-c+1)}(v_2-1)^{-\frac{1}{3}(a+b'-c+1)}(v_1-v_2)^{-\frac{1}{3}(b+b')}z, \tag{2.9}$$

then the function $w = w(v_1, v_2)$ satisfies the following equations:

$$\begin{cases} \dfrac{\partial^2 w}{\partial v_1^2} + \left(\dfrac{c-b'}{v_1} + \dfrac{a+b-c+1}{v_1-1} + \dfrac{b'}{v_1-v_2}\right)\dfrac{\partial w}{\partial v_1} - \dfrac{bv_2(v_2-1)}{v_1(v_1-1)(v_1-v_2)}\dfrac{\partial w}{\partial v_2} \\ \quad + \dfrac{ab}{v_1(v_1-1)} w = 0, \\ \dfrac{\partial^2 w}{\partial v_2^2} + \left(\dfrac{c-b}{v_2} + \dfrac{a+b'-c+1}{v_2-1} + \dfrac{b}{v_2-v_1}\right)\dfrac{\partial w}{\partial v_2} - \dfrac{b'v_1(v_1-1)}{v_2(v_2-1)(v_2-v_1)}\dfrac{\partial w}{\partial v_1} \\ \quad + \dfrac{ab'}{v_2(v_2-1)} w = 0, \\ \dfrac{\partial^2 w}{\partial v_1 \partial v_2} = \dfrac{b'}{v_1-v_2}\dfrac{\partial w}{\partial v_1} - \dfrac{b}{v_1-v_2}\dfrac{\partial w}{\partial v_2}. \end{cases} \tag{2.10}$$

A solution is given by

$$w = F_1(a; b, b'; c; v_1, v_2), \tag{2.11}$$

where $F_1 = F_1(a; b, b'; c; x, y)$ is the Appell hypergeometric function (see [1]). Consequently,

$$\begin{aligned} z =& v_1^{\frac{1}{3}(c-b')} v_2^{\frac{1}{3}(c-b)} (v_1-1)^{\frac{1}{3}(a+b-c+1)} (v_2-1)^{\frac{1}{3}(a+b'-c+1)} (v_1-v_2)^{\frac{1}{3}(b+b')} \\ & \times F_1(a; b, b'; c; v_1, v_2). \end{aligned} \tag{2.12}$$

Proof. Note that

$$z = v_1^{\frac{1}{3}(c-b')} v_2^{\frac{1}{3}(c-b)} (v_1-1)^{\frac{1}{3}(a+b-c+1)} (v_2-1)^{\frac{1}{3}(a+b'-c+1)} (v_1-v_2)^{\frac{1}{3}(b+b')} w.$$

We have

$$\begin{aligned} \frac{\partial z}{\partial v_1} =& v_1^{\frac{1}{3}(c-b')} v_2^{\frac{1}{3}(c-b)} (v_1-1)^{\frac{1}{3}(a+b-c+1)} (v_2-1)^{\frac{1}{3}(a+b'-c+1)} (v_1-v_2)^{\frac{1}{3}(b+b')} \\ & \times \left[\frac{\partial w}{\partial v_1} + \left(\frac{1}{3}P_1 + \frac{b}{v_1-v_2}\right) w\right]. \end{aligned}$$

$$\begin{aligned} \frac{\partial z}{\partial v_2} =& v_1^{\frac{1}{3}(c-b')} v_2^{\frac{1}{3}(c-b)} (v_1-1)^{\frac{1}{3}(a+b-c+1)} (v_2-1)^{\frac{1}{3}(a+b'-c+1)} (v_1-v_2)^{\frac{1}{3}(b+b')} \\ & \times \left[\frac{\partial w}{\partial v_2} + \left(\frac{1}{3}P_2 + \frac{b'}{v_2-v_1}\right) w\right]. \end{aligned}$$

$$\begin{aligned} \frac{\partial^2 z}{\partial v_1^2} =& v_1^{\frac{1}{3}(c-b')} v_2^{\frac{1}{3}(c-b)} (v_1-1)^{\frac{1}{3}(a+b-c+1)} (v_2-1)^{\frac{1}{3}(a+b'-c+1)} (v_1-v_2)^{\frac{1}{3}(b+b')} \\ & \times \left[\frac{\partial^2 w}{\partial v_1^2} + 2\left(\frac{1}{3}P_1 + \frac{b}{v_1-v_2}\right)\frac{\partial w}{\partial v_1}\right. \\ & \left. + \left(\left(\frac{1}{3}P_1 + \frac{b}{v_1-v_2}\right)^2 + \frac{1}{3}\frac{\partial P_1}{\partial v_1} + \frac{-b}{(v_1-v_2)^2}\right) w\right]. \end{aligned}$$

$$\begin{aligned}\frac{\partial^2 z}{\partial v_2^2} =& v_1^{\frac{1}{3}(c-b')} v_2^{\frac{1}{3}(c-b)} (v_1-1)^{\frac{1}{3}(a+b-c+1)} (v_2-1)^{\frac{1}{3}(a+b'-c+1)} (v_1-v_2)^{\frac{1}{3}(b+b')} \\ &\times \left[\frac{\partial^2 w}{\partial v_2^2} + 2\left(\frac{1}{3}P_2 + \frac{b'}{v_2-v_1}\right)\frac{\partial w}{\partial v_2} \right. \\ &\left. + \left(\left(\frac{1}{3}P_2 + \frac{b'}{v_2-v_1}\right)^2 + \frac{1}{3}\frac{\partial P_2}{\partial v_2} + \frac{-b'}{(v_2-v_1)^2} \right) w \right].\end{aligned}$$

$$\begin{aligned}&\frac{\partial^2 z}{\partial v_1 \partial v_2} \\ =& v_1^{\frac{1}{3}(c-b')} v_2^{\frac{1}{3}(c-b)} (v_1-1)^{\frac{1}{3}(a+b-c+1)} (v_2-1)^{\frac{1}{3}(a+b'-c+1)} (v_1-v_2)^{\frac{1}{3}(b+b')} \\ &\times \left[\frac{\partial^2 w}{\partial v_1 \partial v_2} + \left(\frac{1}{3}P_2 + \frac{b'}{v_2-v_1}\right)\frac{\partial w}{\partial v_1} + \left(\frac{1}{3}P_1 + \frac{b}{v_1-v_2}\right)\frac{\partial w}{\partial v_2} \right. \\ &\left. + \left(\left(\frac{1}{3}P_1 + \frac{b}{v_1-v_2}\right)\left(\frac{1}{3}P_2 + \frac{b'}{v_2-v_1}\right) + \frac{1}{3}\frac{\partial P_1}{\partial v_2} + \frac{b}{(v_1-v_2)^2} \right) w \right].\end{aligned}$$

The equation

$$\frac{\partial^2 z}{\partial v_1^2} + \frac{1}{3}P_1\frac{\partial z}{\partial v_1} - F_1\frac{\partial z}{\partial v_2} + \left(\frac{\partial F_1}{\partial v_2} - \frac{1}{3}\frac{\partial P_1}{\partial v_1} - \frac{2}{9}P_1^2 - \frac{2}{3}F_1P_2\right) z = 0$$

gives that

$$\begin{aligned}&\frac{\partial^2 w}{\partial v_1^2} + \left(P_1 + \frac{2b}{v_1-v_2}\right)\frac{\partial w}{\partial v_1} - F_1\frac{\partial w}{\partial v_2} \\ &+ \left[\frac{b}{v_1-v_2}P_1 + \frac{b^2-b}{(v_1-v_2)^2} - F_1P_2 - \frac{b'}{v_2-v_1}F_1 + \frac{\partial F_1}{\partial v_2} \right] w = 0.\end{aligned}$$

Here,

$$\frac{\partial F_1}{\partial v_2} = \frac{b}{(v_1-v_2)^2} - \frac{b}{v_1(v_1-1)}.$$

Thus,

$$\begin{aligned}&\frac{b}{v_1-v_2}P_1 + \frac{b^2-b}{(v_1-v_2)^2} - F_1P_2 - \frac{b'}{v_2-v_1}F_1 + \frac{\partial F_1}{\partial v_2} \\ =& \frac{b}{v_1-v_2}\left(P_1 + \frac{b}{v_1-v_2}\right) - F_1\left(P_2 + \frac{b'}{v_2-v_1}\right) + \frac{-b}{(v_1-v_2)^2} + \frac{\partial F_1}{\partial v_2} \\ =& \frac{b}{v_1-v_2}\left(\frac{c-b'}{v_1} + \frac{a+b-c+1}{v_1-1} + \frac{b'-b}{v_1-v_2}\right) \\ &- \frac{bv_2(v_2-1)}{v_1(v_1-1)(v_1-v_2)}\left(\frac{c-b}{v_2} + \frac{a+b'-c+1}{v_2-1} + \frac{b-b'}{v_2-v_1}\right) - \frac{b}{v_1(v_1-1)} \\ =& \frac{(a+1)b}{v_1(v_1-1)} - \frac{b}{v_1(v_1-1)} \\ =& \frac{ab}{v_1(v_1-1)}.\end{aligned}$$

Hence, we get the following equation:

$$\frac{\partial^2 w}{\partial v_1^2}+\left(\frac{c-b'}{v_1}+\frac{a+b-c+1}{v_1-1}+\frac{b'}{v_1-v_2}\right)\frac{\partial w}{\partial v_1}-\frac{bv_2(v_2-1)}{v_1(v_1-1)(v_1-v_2)}\frac{\partial w}{\partial v_2}$$
$$+\frac{ab}{v_1(v_1-1)}w=0.$$

Exchange the position of v_1 and v_2, we get the other equation.

The equation

$$\frac{\partial^2 z}{\partial v_1\partial v_2}-\frac{1}{3}P_2\frac{\partial z}{\partial v_1}-\frac{1}{3}P_1\frac{\partial z}{\partial v_2}+\left(\frac{1}{3}\frac{\partial P_2}{\partial v_1}+\frac{1}{3}\frac{\partial P_1}{\partial v_2}+\frac{1}{9}P_1P_2-F_1F_2\right)z=0$$

gives that

$$\frac{\partial^2 w}{\partial v_1\partial v_2}+\frac{b'}{v_2-v_1}\frac{\partial w}{\partial v_1}+\frac{b}{v_1-v_2}\frac{\partial w}{\partial v_2}$$
$$+\left[\frac{b-bb'}{(v_1-v_2)^2}+\frac{2}{3}\frac{\partial P_1}{\partial v_2}+\frac{1}{3}\frac{\partial P_2}{\partial v_1}-F_1F_2\right]w=0.$$

Note that

$$\frac{\partial P_1}{\partial v_2}=\frac{b'-2b}{(v_1-v_2)^2},\quad \frac{\partial P_2}{\partial v_1}=\frac{b-2b'}{(v_2-v_1)^2},\quad F_1F_2=-\frac{bb'}{(v_1-v_2)^2},$$

we find that

$$\frac{b-bb'}{(v_1-v_2)^2}+\frac{2}{3}\frac{\partial P_1}{\partial v_2}+\frac{1}{3}\frac{\partial P_2}{\partial v_1}-F_1F_2=0.$$

Thus,

$$\frac{\partial^2 w}{\partial v_1\partial v_2}=\frac{b'}{v_1-v_2}\frac{\partial w}{\partial v_1}-\frac{b}{v_1-v_2}\frac{\partial w}{\partial v_2}.$$

It is known that the equations of F_1:

$$\begin{cases} x(1-x)\dfrac{\partial^2 z}{\partial x^2}+y(1-x)\dfrac{\partial^2 z}{\partial x\partial y}+[c-(a+b+1)x]\dfrac{\partial z}{\partial x}-by\dfrac{\partial z}{\partial y}-abz=0, \\ y(1-y)\dfrac{\partial^2 z}{\partial y^2}+x(1-y)\dfrac{\partial^2 z}{\partial x\partial y}+[c-(a+b'+1)y]\dfrac{\partial z}{\partial y}-b'x\dfrac{\partial z}{\partial x}-ab'z=0 \end{cases}$$

are equivalent to the following three equations:

$$\begin{cases} \dfrac{\partial^2 z}{\partial x^2}+\left(\dfrac{c-b'}{x}+\dfrac{a+b-c+1}{x-1}+\dfrac{b'}{x-y}\right)\dfrac{\partial z}{\partial x} \\ \quad -\dfrac{by(y-1)}{x(x-1)(x-y)}\dfrac{\partial z}{\partial y}+\dfrac{abz}{x(x-1)}=0, \\ \dfrac{\partial^2 z}{\partial y^2}+\left(\dfrac{c-b}{y}+\dfrac{a+b'-c+1}{y-1}+\dfrac{b}{y-x}\right)\dfrac{\partial z}{\partial y} \\ \quad -\dfrac{b'x(x-1)}{y(y-1)(y-x)}\dfrac{\partial z}{\partial x}+\dfrac{ab'z}{y(y-1)}=0, \\ \dfrac{\partial^2 z}{\partial x\partial y}=\dfrac{b'}{x-y}\dfrac{\partial z}{\partial x}-\dfrac{b}{x-y}\dfrac{\partial z}{\partial y}, \end{cases}$$

where $F_1 = F_1(a; b, b'; c; x, y)$ is the Appell hypergeometric function (see [1]). A solution is given by

$$w = F_1(a; b, b'; c; v_1, v_2).$$

Consequently,

$$\begin{aligned} z =& v_1^{\frac{1}{3}(c-b')} v_2^{\frac{1}{3}(c-b)} (v_1 - 1)^{\frac{1}{3}(a+b-c+1)} (v_2 - 1)^{\frac{1}{3}(a+b'-c+1)} (v_1 - v_2)^{\frac{1}{3}(b+b')} \\ & \times F_1(a; b, b'; c; v_1, v_2). \end{aligned}$$

□

3. The algebraic solutions of Appell hypergeometric partial differential equations

Let us recall some basic facts about Appell-Lauricella functions (see [13], [14] or [20], [21], [54]). With the usual notation $(a, 0) := 1$ and $(a, n) := a(a + 1) \cdots (a + n - 1)$ for $a \in \mathbb{C}$, $n \in \mathbb{N}$, let us define the Appell-Lauricella function F_1 of N complex variables $x_2, \cdots, x_{N+1}$ with parameters $a, b_2, \cdots, b_{N+1}$ and $c \in \mathbb{C}$, $c \neq 0, -1, -2, \cdots$, by the series:

$$F_1(a, b_2, \cdots, b_{N+1}; c; x_2, \cdots, x_{N+1}) = \sum \frac{(a, \sum_j n_j) \prod_j (b_j, n_j)}{(c, \sum_j n_j) \prod_j (1, n_j)} \prod_j x_j^{n_j},$$

where j runs from 2 to $N+1$, each n_j runs from 0 to ∞, and the series converges if all $|x_j| < 1$. Almost everywhere we will use instead of this series its integral representation:

$$\frac{1}{B(1 - \mu_1, 1 - \mu_{N+2})} \int u^{-\mu_0} (u - 1)^{-\mu_1} \prod_{j=2}^{N+1} (u - x_j)^{-\mu_j} du,$$

where the exponential parameters $\mu_0, \ldots, \mu_{N+2}$ are related to a, $b_2, \ldots, b_{N+1}$, c by

$$\mu_j = b_j \quad \text{for} \quad j = 2, \cdots, N+1,$$

$$\mu_0 = c - \sum b_j, \quad \mu_1 = 1 + a - c,$$

$$\sum_{j=0}^{N+2} \mu_j = 2, \quad \text{i.e.} \quad \mu_{N+2} = 1 - a.$$

Theorem (see [13], Theorem 1). *Assume all $\mu_0, \cdots, \mu_{N+2} \in \mathbb{Q} - \mathbb{Z}$. Then there are no algebraic F_1 in more than three variables. The function $F_1(x, y)$ is algebraic if and only if $\mu_0, \cdots, \mu_4$ are up to permutations and a common change of the sign $\equiv \frac{1}{3}, \frac{1}{6}, \frac{1}{6}, \frac{1}{6}, \frac{1}{6}$ mod $\mathbb{Z}$. Then Δ is the symmetry group of the extended Hesse polytope of order* 1296, *projectively isomorphic to a subgroup of order* 216 *of the group $PSL(3, \mathbb{F}_3)$. The function $F_1(x, y, z)$ is algebraic if and only if $\mu_0, \cdots, \mu_5$ are all $\equiv \frac{1}{6}$ mod $\mathbb{Z}$ or all $\equiv -\frac{1}{6}$ mod $\mathbb{Z}$. Then Δ is the symmetry* *group of the Witting polytope of order* 155, 520*, projectively isomorphic to the group* $PSp(4, \mathbb{F}_3)$ *of order* 25, 920.

In the case $N = 2$ this means that $F_1(a, b, b'; c; x, y)$ is algebraic if and only if up to a common change of sign

$$\begin{array}{ccccccccc} a & \equiv & -\frac{1}{6} & -\frac{1}{6} & -\frac{1}{6} & -\frac{1}{6} & -\frac{1}{3} & \bmod \mathbb{Z} \\ b & \equiv & \frac{1}{6} & \frac{1}{6} & \frac{1}{3} & \frac{1}{6} & \frac{1}{6} & \bmod \mathbb{Z} \\ b' & \equiv & \frac{1}{6} & \frac{1}{6} & \frac{1}{6} & \frac{1}{3} & \frac{1}{6} & \bmod \mathbb{Z} \\ c & \equiv & -\frac{1}{3} & \frac{1}{2} & -\frac{1}{3} & -\frac{1}{3} & \frac{1}{2} & \bmod \mathbb{Z} \end{array}$$

For

$$\begin{cases} P_1 = \dfrac{c-b'}{v_1} + \dfrac{a+b-c+1}{v_1-1} + \dfrac{b'-2b}{v_1-v_2}, \\ P_2 = \dfrac{c-b}{v_2} + \dfrac{a+b'-c+1}{v-2-1} + \dfrac{b-2b'}{v_2-v_1}, \\ F_1 = \dfrac{bv_2(v_2-1)}{v_1(v_1-1)(v_1-v_2)}, \\ F_2 = \dfrac{b'v_1(v_1-1)}{v_2(v_2-1)(v_2-v_1)}, \end{cases} \tag{3.1}$$

set

$$\alpha_1 = c - b', \quad \beta_1 = a + b - c + 1, \quad \gamma_1 = b' - 2b,$$

$$\alpha_2 = c - b, \quad \beta_2 = a + b' - c + 1, \quad \gamma_2 = b - 2b'.$$

We find that

$$\alpha_1 - \alpha_2 = b - b', \quad \beta_1 - \beta_2 = b - b', \quad \gamma_1 - \gamma_2 = -3(b - b').$$

This leads us to consider the double-triangular s-functions:

$$s_1 = s_1(\alpha_1, \beta_1, \gamma_1; \alpha_2, \beta_2, \gamma_2; v_1, v_2), \quad s_2 = s_2(\alpha_1, \beta_1, \gamma_1; \alpha_2, \beta_2, \gamma_2; v_1, v_2), \tag{3.2}$$

where

$$\alpha_1 - \alpha_2 = \beta_1 - \beta_2 = -\frac{1}{3}(\gamma_1 - \gamma_2).$$

In particular, when $\alpha_1 = \alpha_2$ or $\beta_1 = \beta_2$ or $\gamma_1 = \gamma_2$, they reduce the triangular s-functions (see [63]):

$$s_1 = s_1(\alpha, \beta, \gamma; v_1, v_2), \quad s_2 = s_2(\alpha, \beta, \gamma; v_1, v_2). \tag{3.3}$$

This gives the following five cases:

(1) $a = -\frac{1}{6}$, $b = \frac{1}{6}$, $b' = \frac{1}{6}$, $c = -\frac{1}{3}$.

$$\begin{cases} P_1(v_1, v_2) = \dfrac{-\frac{1}{2}}{v_1} + \dfrac{\frac{4}{3}}{v_1 - 1} + \dfrac{-\frac{1}{6}}{v_1 - v_2}, \\ P_2(v_1, v_2) = \dfrac{-\frac{1}{2}}{v_2} + \dfrac{\frac{4}{3}}{v_2 - 1} + \dfrac{-\frac{1}{6}}{v_2 - v_1}, \\ F_1(v_1, v_2) = \dfrac{\frac{1}{6} v_2(v_2 - 1)}{v_1(v_1 - 1)(v_1 - v_2)}, \\ F_2(v_1, v_2) = \dfrac{\frac{1}{6} v_1(v_1 - 1)}{v_2(v_2 - 1)(v_2 - v_1)}. \end{cases} \tag{3.4}$$

Correspondingly,

$$z = v_1^{-\frac{1}{6}} v_2^{-\frac{1}{6}} (v_1 - 1)^{\frac{4}{9}} (v_2 - 1)^{\frac{4}{9}} (v_1 - v_2)^{\frac{1}{9}} F_1 \left(-\frac{1}{6}; \frac{1}{6}, \frac{1}{6}; -\frac{1}{3}; v_1, v_2 \right).$$

The triangular s-functions are:

$$s_1 = s_1 \left(-\frac{1}{2}, \frac{4}{3}, -\frac{1}{6}; v_1, v_2 \right), \quad s_2 = s_2 \left(-\frac{1}{2}, \frac{4}{3}, -\frac{1}{6}; v_1, v_2 \right).$$

(2) $a = -\frac{1}{6}$, $b = \frac{1}{6}$, $b' = \frac{1}{6}$, $c = \frac{1}{2}$.

$$\begin{cases} P_1(v_1, v_2) = \dfrac{\frac{1}{3}}{v_1} + \dfrac{\frac{1}{2}}{v_1 - 1} + \dfrac{-\frac{1}{6}}{v_1 - v_2}, \\ P_2(v_1, v_2) = \dfrac{\frac{1}{3}}{v_2} + \dfrac{\frac{1}{2}}{v_2 - 1} + \dfrac{-\frac{1}{6}}{v_2 - v_1}, \\ F_1(v_1, v_2) = \dfrac{\frac{1}{6} v_2(v_2 - 1)}{v_1(v_1 - 1)(v_1 - v_2)}, \\ F_2(v_1, v_2) = \dfrac{\frac{1}{6} v_1(v_1 - 1)}{v_2(v_2 - 1)(v_2 - v_1)}. \end{cases} \tag{3.5}$$

Correspondingly,

$$z = v_1^{\frac{1}{9}} v_2^{\frac{1}{9}} (v_1 - 1)^{\frac{1}{6}} (v_2 - 1)^{\frac{1}{6}} (v_1 - v_2)^{\frac{1}{9}} F_1 \left(-\frac{1}{6}; \frac{1}{6}, \frac{1}{6}; \frac{1}{2}; v_1, v_2 \right).$$

The triangular s-functions are:

$$s_1 = s_1 \left(\frac{1}{3}, \frac{1}{2}, -\frac{1}{6}; v_1, v_2 \right), \quad s_2 = s_2 \left(\frac{1}{3}, \frac{1}{2}, -\frac{1}{6}; v_1, v_2 \right).$$

(3) $a = -\frac{1}{6}$, $b = \frac{1}{3}$, $b' = \frac{1}{6}$, $c = -\frac{1}{3}$.

$$\begin{cases} P_1(v_1, v_2) = \dfrac{-\frac{1}{2}}{v_1} + \dfrac{\frac{3}{2}}{v_1 - 1} + \dfrac{-\frac{1}{2}}{v_1 - v_2}, \\ P_2(v_1, v_2) = \dfrac{-\frac{2}{3}}{v_2} + \dfrac{\frac{4}{3}}{v_2 - 1}, \\ F_1(v_1, v_2) = \dfrac{\frac{1}{3}v_2(v_2 - 1)}{v_1(v_1 - 1)(v_1 - v_2)}, \\ F_2(v_1, v_2) = \dfrac{\frac{1}{6}v_1(v_1 - 1)}{v_2(v_2 - 1)(v_2 - v_1)}. \end{cases} \tag{3.6}$$

Correspondingly,

$$z = v_1^{-\frac{1}{6}} v_2^{-\frac{2}{9}} (v_1 - 1)^{\frac{1}{2}} (v_2 - 1)^{\frac{4}{9}} (v_1 - v_2)^{\frac{1}{6}} F_1 \left(-\frac{1}{6}; \frac{1}{3}, \frac{1}{6}; -\frac{1}{3}; v_1, v_2 \right).$$

The double-triangular s-functions are:

$$s_1 = s_1 \left(-\frac{1}{2}, \frac{3}{2}, -\frac{1}{2}; -\frac{2}{3}, \frac{4}{3}, 0; v_1, v_2 \right),$$

$$s_2 = s_2 \left(-\frac{1}{2}, \frac{3}{2}, -\frac{1}{2}; -\frac{2}{3}, \frac{4}{3}, 0; v_1, v_2 \right).$$

(4) $a = -\frac{1}{6}$, $b = \frac{1}{6}$, $b' = \frac{1}{3}$, $c = -\frac{1}{3}$.

$$\begin{cases} P_1(v_1, v_2) = \dfrac{-\frac{2}{3}}{v_1} + \dfrac{\frac{4}{3}}{v_1 - 1}, \\ P_2(v_1, v_2) = \dfrac{-\frac{1}{2}}{v_2} + \dfrac{\frac{3}{2}}{v_2 - 1} + \dfrac{-\frac{1}{2}}{v_2 - v_1}, \\ F_1(v_1, v_2) = \dfrac{\frac{1}{6}v_2(v_2 - 1)}{v_1(v_1 - 1)(v_1 - v_2)}, \\ F_2(v_1, v_2) = \dfrac{\frac{1}{3}v_1(v_1 - 1)}{v_2(v_2 - 1)(v_2 - v_1)}. \end{cases} \tag{3.7}$$

Correspondingly,

$$z = v_1^{-\frac{2}{9}} v_2^{-\frac{1}{6}} (v_1 - 1)^{\frac{4}{9}} (v_2 - 1)^{\frac{1}{2}} (v_1 - v_2)^{\frac{1}{6}} F_1 \left(-\frac{1}{6}; \frac{1}{6}, \frac{1}{3}; -\frac{1}{3}; v_1, v_2 \right).$$

The double-triangular s-functions are:

$$s_1 = s_1 \left(-\frac{2}{3}, \frac{4}{3}, 0; -\frac{1}{2}, \frac{3}{2}, -\frac{1}{2}; v_1, v_2 \right),$$

$$s_2 = s_2 \left(-\frac{2}{3}, \frac{4}{3}, 0; -\frac{1}{2}, \frac{3}{2}, -\frac{1}{2}; v_1, v_2 \right).$$

(5) $a=-\frac{1}{3}$, $b=\frac{1}{6}$, $b'=\frac{1}{6}$, $c=\frac{1}{2}$.

$$\begin{cases} P_1(v_1,v_2)=\dfrac{\frac{1}{3}}{v_1}+\dfrac{\frac{1}{3}}{v_1-1}+\dfrac{-\frac{1}{6}}{v_1-v_2}, \\ P_2(v_1,v_2)=\dfrac{\frac{1}{3}}{v_2}+\dfrac{\frac{1}{3}}{v_2-1}+\dfrac{-\frac{1}{6}}{v_2-v_1}, \\ F_1(v_1,v_2)=\dfrac{\frac{1}{6}v_2(v_2-1)}{v_1(v_1-1)(v_1-v_2)}, \\ F_2(v_1,v_2)=\dfrac{\frac{1}{6}v_1(v_1-1)}{v_2(v_2-1)(v_2-v_1)}. \end{cases} \tag{3.8}$$

Correspondingly,

$$z=v_1^{\frac{1}{9}}v_2^{\frac{1}{9}}(v_1-1)^{\frac{1}{9}}(v_2-1)^{\frac{1}{9}}(v_1-v_2)^{\frac{1}{9}}F_1\left(-\frac{1}{3};\frac{1}{6},\frac{1}{6};\frac{1}{2};v_1,v_2\right).$$

The triangular s-functions are:

$$s_1=s_1\left(\frac{1}{3},\frac{1}{3},-\frac{1}{6};v_1,v_2\right),\quad s_2=s_2\left(\frac{1}{3},\frac{1}{3},-\frac{1}{6};v_1,v_2\right).$$

We will study the transformation between two Appell hypergeometric partial differential equations:

$$\begin{cases} x(1-x)\dfrac{\partial^2 z}{\partial x^2}+y(1-x)\dfrac{\partial^2 z}{\partial x\partial y}+[c-(a+b+1)x]\dfrac{\partial z}{\partial x}-by\dfrac{\partial z}{\partial y}-abz=0, \\ y(1-y)\dfrac{\partial^2 z}{\partial y^2}+x(1-y)\dfrac{\partial^2 z}{\partial x\partial y}+[c-(a+b'+1)y]\dfrac{\partial z}{\partial y}-b'x\dfrac{\partial z}{\partial x}-ab'z=0, \end{cases}$$

and

$$\begin{cases} u_1(1-u_1)\dfrac{\partial^2 v}{\partial u_1^2}+u_2(1-u_1)\dfrac{\partial^2 v}{\partial u_1\partial u_2}+[\gamma-(\alpha+\beta+1)u_1]\dfrac{\partial v}{\partial u_1}-\beta u_2\dfrac{\partial v}{\partial u_2} \\ -\alpha\beta v=0, \\ u_2(1-u_2)\dfrac{\partial^2 v}{\partial u_2^2}+u_1(1-u_2)\dfrac{\partial^2 v}{\partial u_1\partial u_2}+[\gamma-(\alpha+\beta'+1)u_2]\dfrac{\partial v}{\partial u_2}-\beta' u_1\dfrac{\partial v}{\partial u_1} \\ -\alpha\beta' v=0. \end{cases}$$

Here, $u_1=u_1(x,y)$, $u_2=u_2(x,y)$ and $z=wv$ with $w=w(x,y)$.

We have

$$\frac{\partial z}{\partial x}=\frac{\partial w}{\partial x}v+w\left(\frac{\partial v}{\partial u_1}\frac{\partial u_1}{\partial x}+\frac{\partial v}{\partial u_2}\frac{\partial u_2}{\partial x}\right),$$

$$\frac{\partial z}{\partial y}=\frac{\partial w}{\partial y}v+w\left(\frac{\partial v}{\partial u_1}\frac{\partial u_1}{\partial y}+\frac{\partial v}{\partial u_2}\frac{\partial u_2}{\partial y}\right).$$

For $z_1 = wv_1$, $z_2 = wv_2$ and $z_3 = wv_3$, we find that

$$\begin{vmatrix} \frac{\partial z_1}{\partial x} & \frac{\partial z_1}{\partial y} & z_1 \\ \frac{\partial z_2}{\partial x} & \frac{\partial z_2}{\partial y} & z_2 \\ \frac{\partial z_3}{\partial x} & \frac{\partial z_3}{\partial y} & z_3 \end{vmatrix} = w^3 \frac{\partial(u_1, u_2)}{\partial(x, y)} \begin{vmatrix} \frac{\partial v_1}{\partial u_1} & \frac{\partial v_1}{\partial u_2} & v_1 \\ \frac{\partial v_2}{\partial u_1} & \frac{\partial v_2}{\partial u_2} & v_2 \\ \frac{\partial v_3}{\partial u_1} & \frac{\partial v_3}{\partial u_2} & v_3 \end{vmatrix}. \tag{3.9}$$

On the other hand,

$$\begin{vmatrix} \frac{\partial z_1}{\partial x} & \frac{\partial z_1}{\partial y} & z_1 \\ \frac{\partial z_2}{\partial x} & \frac{\partial z_2}{\partial y} & z_2 \\ \frac{\partial z_3}{\partial x} & \frac{\partial z_3}{\partial y} & z_3 \end{vmatrix} = C_1 x^{b'-c} y^{b-c} (x-1)^{c-a-b-1} (y-1)^{c-a-b'-1} (x-y)^{-b-b'}. \tag{3.10}$$

$$\begin{vmatrix} \frac{\partial v_1}{\partial u_1} & \frac{\partial v_1}{\partial u_2} & v_1 \\ \frac{\partial v_2}{\partial u_1} & \frac{\partial v_2}{\partial u_2} & v_2 \\ \frac{\partial v_3}{\partial u_1} & \frac{\partial v_3}{\partial u_2} & v_3 \end{vmatrix} = C_2 u_1^{\beta'-\gamma} u_2^{\beta-\gamma} (u_1-1)^{\gamma-\alpha-\beta-1} (u_2-1)^{\gamma-\alpha-\beta'-1} (u_1-u_2)^{-\beta-\beta'}. \tag{3.11}$$

Thus,

$$w^3 = C \frac{x^{b'-c} y^{b-c} (x-1)^{c-a-b-1} (y-1)^{c-a-b'-1} (x-y)^{-b-b'}}{u_1^{\beta'-\gamma} u_2^{\beta-\gamma} (u_1-1)^{\gamma-\alpha-\beta-1} (u_2-1)^{\gamma-\alpha-\beta'-1} (u_1-u_2)^{-\beta-\beta'}} \frac{1}{\frac{\partial(u_1,u_2)}{\partial(x,y)}}. \tag{3.12}$$

Here, C_1, C_2 and C are three constants.

Note that

$$\frac{\partial(\xi, \eta)}{\partial(x, y)} = \frac{1}{z_3^3} C x^{b'-c} y^{b-c} (x-1)^{c-a-b-1} (y-1)^{c-a-b'-1} (x-y)^{-b-b'}. \tag{3.13}$$

We will study the following equations:

$$\frac{u_1(\xi, \eta)}{u_3(\xi, \eta)} = R_1(x, y), \quad \frac{u_2(\xi, \eta)}{u_3(\xi, \eta)} = R_2(x, y). \tag{3.14}$$

We find that

$$\begin{aligned} \frac{\partial(R_1, R_2)}{\partial(x, y)} &= \frac{\partial(u_1/u_3, u_2/u_3)}{\partial(\xi, \eta)} \frac{\partial(\xi, \eta)}{\partial(x, y)} \\ &= \frac{\partial(u_1/u_3, u_2/u_3)}{\partial(\xi, \eta)} \frac{1}{z_3^3} C x^{b'-c} y^{b-c} (x-1)^{c-a-b-1} (y-1)^{c-a-b'-1} (x-y)^{-b-b'}. \end{aligned} \tag{3.15}$$

4. The Hessian polyhedral equations

A polytope is a geometrical figure bounded by portions of lines, planes, or hyperplanes; e.g., in two dimensions it is a polygon, in three a polyhedron. We define a polytope (see [17]) as a finite convex region of n-dimensional space enclosed by a finite number of hyperplanes. The part of the polytope that lies in one of the hyperplanes is called a cell. A polytope Π_n ($n > 2$) is said to be regular if its cells are regular and there is a regular vertex figure at every vertex.

In his great monograph in 1852, Schläfli gave the n-dimensional Schläfli symbol (see [17]), which enables us to read off many properties of a regular polytope at a glance; e.g., the elements of $\{p, q, r, \cdots\}$, besides vertices and edges, are plane faces $\{p\}$, solid faces $\{p, q\}$, and so on. The general regular polytope $\{p, q, \cdots, v, w\}$ has cells $\{p, q, \cdots, v\}$ and vertex figures $\{q, \cdots, v, w\}$. In particular, for any $n > 1$,

$$\alpha_n = \{3, 3, \cdots, 3, 3\} = \{3^{n-1}\},$$

$$\beta_n = \{3, 3, \cdots, 3, 4\} = \{3^{n-2}, 4\},$$

$$\gamma_n = \{4, 3, \cdots, 3, 3\} = \{4, 3^{n-2}\}.$$

The Schläfli's criterion for the existence of a regular figure corresponding to a given symbol $\{p, q, \cdots, v, w\}$ is given by (see [17]):

$$\Delta_{p,q,\cdots,v,w} > 0.$$

Here,

$$\Delta_{p,q,\cdots,v,w} = \begin{vmatrix} 1 & c_1 & 0 & \cdots & 0 & 0 & 0 \\ c_1 & 1 & c_2 & \cdots & 0 & 0 & 0 \\ & \cdots & & & & \cdots & \\ 0 & 0 & 0 & \cdots & c_{n-2} & 1 & c_{n-1} \\ 0 & 0 & 0 & \cdots & 0 & c_{n-1} & 1 \end{vmatrix},$$

where

$$c_1 = \cos\frac{\pi}{p}, \quad c_2 = \cos\frac{\pi}{q}, \cdots, c_{n-2} = \cos\frac{\pi}{v}, \quad c_{n-1} = \cos\frac{\pi}{w}.$$

When $n = 3$, we have $\Delta_{p,q} > 0$, or

$$\frac{1}{p} + \frac{1}{q} > \frac{1}{2},$$

which admits the Platonic solids

$$\{3, 3\}, \quad \{3, 4\}, \quad \{4, 3\}, \quad \{3, 5\}, \quad \{5, 3\}.$$

When $n = 4$, we have $\Delta_{p,q,r} > 0$, or

$$\sin\frac{\pi}{p}\sin\frac{\pi}{r} > \cos\frac{\pi}{q},$$

which admits the six polytopes:

$$\{3, 3, 3\}, \quad \{3, 3, 4\}, \quad \{4, 3, 3\}, \quad \{3, 4, 3\}, \quad \{3, 3, 5\}, \quad \{5, 3, 3\}.$$

When $n = 5$, we have $\Delta_{p,q,r,s} > 0$, or

$$\frac{\cos^2\frac{\pi}{q}}{\sin^2\frac{\pi}{p}} + \frac{\cos^2\frac{\pi}{r}}{\sin^2\frac{\pi}{s}} < 1,$$

which admits the three polytopes:

$$\alpha_5 = \{3, 3, 3, 3\}, \quad \beta_5 = \{3, 3, 3, 4\}, \quad \gamma_5 = \{4, 3, 3, 3\}.$$

Since the only regular polytopes in five dimensions are α_5, β_5, γ_5, it follows by induction that in more than five dimensions the only regular polytopes are α_n, β_n, γ_n.

It is well-known that the regular polyhedra are connected with the Gauss hypergeometric differential equation (see [42] and [43]). In contrast with the real regular polyhedra, we will give a story on the complex regular polyhedra, which are intimately connected with the Appell hypergeometric partial differential equations.

According to [18], in the complex affine plane with a unitary metric, let us define a polygon to be a finite figure consisting of points (not all on one line) called vertices, and lines (not all through one point) called edges, satisfying the following two conditions:

(1) Every edge is incident with at least two vertices, and every vertex with at least two edges.

(2) Any two vertices are connected by a chain of successively incident edges and vertices.

The group of all unitary transformations that preserve the incidences is called the symmetry group of the polygon. The figure consisting of a vertex and an incident edge is called a flag. The polygon is said to be regular if its symmetry group is transitive on its flags.

Extending the above ideas to unitary n-space, let us consider a finite non-empty set of subspaces, μ-flats ($0 \leq \mu < n$), with a relation of incidence, defined by saying that a μ-flat and a ν-flat are incident if one of them is a proper subspace of the other. A subset of this set of flats is said to be connected if any two of its flats can be joined by a chain of successively incident flats. If $\lambda < \nu - 1$, any incident λ-flat and ν-flat determine a medial figure consisting of all the μ-flats ($\lambda < \mu < \nu$) that are incident with both.

Such a set of flats is called a polytope if it satisfies the following two conditions:

(1) Every medial figure includes at least two μ-flats for each μ ($-1 \leq \lambda < \mu < \nu \leq n$).

(1) Every medial figure with $\lambda < \nu - 2$ is connected.

The group of all unitary transformations that permute the μ-flats for each μ and preserve the incidences is called the symmetry group of the polytope.

In an n-dimensional polytope Π_n, a μ-flat and all its incident λ-flats ($\lambda < \mu$) form a sub-polytope Π_μ, called a μ-dimensional element of Π_n. In particular there are vertices Π_0, edges Π_1, faces Π_2, $\cdots$, and cells Π_{n-1}. The figure consisting of one element of each dimension number, all mutually incident, is called a flag. The polytope is called regular if its symmetry group is transitive on its flags.

In fact, the symmetry group of a regular complex polytope is generated by a sequence of unitary reflections of periods p_1, p_2, $\cdots$, p_n, such that any two non-consecutive reflections are commutative. This leads to the generalized Schläfli symbol

$$p_1\{q_1\}p_2\{q_2\}\cdots p_{n-1}\{q_{n-1}\}p_n.$$

Let us recall some basic facts about the Hessian polyhedron (see [18]). In the case of the Hessian polyhedron 3{3}3{3}3, the Hermitian form is

$$x^1\overline{x^1}-\sqrt{\frac{1}{3}}(x^1\overline{x^2}+x^2\overline{x^1})+x^2\overline{x^2}-\sqrt{\frac{1}{3}}(x^2\overline{x^3}+x^3\overline{x^2})+x^3\overline{x^3}=u_1\overline{u_1}+u_2\overline{u_2}+u_3\overline{u_3},$$

where

$$u_1=-x^1+\frac{x^2}{\sqrt{3}},\quad u_2=\frac{x^2}{\sqrt{3}},\quad u_3=\frac{x^2}{\sqrt{3}}-x^3.$$

Thus the three mirrors are

$$u_1=0,\quad \frac{u_1+u_2+u_3}{\sqrt{3}}=0,\quad u_3=0.$$

R_1 and R_3 multiply u_1 and u_3 (respectively) by $\omega=\exp(2\pi i/3)$ and R_2 is

$$u'_\lambda=u_\lambda-\frac{i\omega^2}{\sqrt{3}}(u_1+u_2+u_3).$$

In fact, 3{3}3{3}3 has just the 27 vertices:

$$(0,\omega^\mu,-\omega^\nu),\quad (-\omega^\nu,0,\omega^\mu),\quad (\omega^\mu,-\omega^\nu,0),$$

where μ and ν take (independently) the three values 0, 1, 2 or (equally well) 1, 2, 3. Making the latter choice, we may conveniently use the concise symbols:

$$0\mu\nu,\quad \nu 0\mu,\quad \mu\nu 0$$

for these 27 points.

The symbol 3{3}3{3}3 is a palindrome, so this polyhedron is self-reciprocal. Having 27 vertices, it also has 27 faces 3{3}3. Since the vertex figure is another 3{3}3, there are 8 edges at each vertex, and 72 edges altogether.

Two vertices are naturally said to be adjacent if they belong to the same edge; their symbols agree in an even number of places, that is, in just two places or nowhere. The 72 edges occur in 36 pairs of opposites, such that any two vertices belonging to opposite edges are non-adjacent; their symbols agree in just one place.

Although the edges occur in pairs of opposites, the vertices obviously do not (as 27 is an odd number). In fact, 3{3}3{3}3 is not 2-symmetric but 3-symmetric: the 27 vertices lie by threes on nine diameters. The three vertices on any diameter are derived from one another by cyclically permuting the digits 1, 2, 3 wherever they occur.

Similarly, the 72 edges lie by sixes on twelve planes of symmetry

$$u_1 = 0, u_2 = 0, u_3 = 0, \omega^\lambda u_1 + \omega^\mu u_2 + \omega^\nu u_3 = 0 (\lambda, \mu, \nu = 0, 1, 2 \text{ independently}).$$

These planes include the mirrors for the generating reflections R_1, R_2, R_3; in fact, they are the mirrors for all the reflections that occur in 3[3]3[3]3.

Since

$$R_1R_2R_1 = R_2R_1R_2, \quad R_2R_3R_2 = R_3R_2R_3,$$

the generators R_1, R_2, R_3 are mutually conjugate; and every reflection in the group is conjugate to some $R_\nu^{\pm 1}$. Thus the section of 3{3}3{3}3 by any plane of symmetry is a polygon 3{4}2.

The nine diameters and twelve planes of symmetry are related in a remarkable way: every two diameters determine a plane of symmetry which contains a third diameter.

3{3}3{3}3 is called the Hessian polyhedron because the incidences of the diameters and planes of symmetry are the same as the incidences of the points and lines of the Hessian configuration

$$\begin{pmatrix} 9 & 4 \\ 3 & 12 \end{pmatrix}$$

or $(9_4, 12_3)$ (see [53]): nine points lying by threes on twelve lines, with four of the lines through each point. To make this transition from complex affine 3-space to the complex projective plane, we have to interpret the affine coordinates $(0, \omega^\mu, -\omega^\nu)$, $(-\omega^\nu, 0, \omega^\mu)$, $(\omega^\mu, -\omega^\nu, 0)$ for the 27 points $0\mu\nu$, $\nu 0\mu$, $\mu\nu 0$ as homogeneous coordinates:

$$\begin{array}{ccc} (0, 1, -1) & (-1, 0, 1) & (1, -1, 0) \\ (0, 1, -\omega) & (-\omega, 0, 1) & (1, -\omega, 0) \\ (0, 1, -\omega^2) & (-\omega^2, 0, 1) & (1, -\omega^2, 0) \end{array}$$

for the nine points given by

$$x^3 + y^3 + z^3 = xyz = 0.$$

These are the nine points of inflexion of the plane cubic curve

$$x^3 + y^3 + z^3 - 3axyz = 0 \quad (a \neq 0, 1)$$

whose Hessian is

$$x^3 + y^3 + z^3 + (a - 4a^{-2})xyz = 0.$$

In fact, the 27 lines on the cubic surface represent 27 vertices of 3{3}3{3}3 in such a way that two of the lines are intersecting or skew according to the corresponding vertices are non-adjacent or adjacent. This leads to the question why the group of automorphisms of the 27 lines has order 51, 840 whereas the order of 3[3]3[3]3 is only 648. The explanation is that, among the pairs of non-adjacent vertices of 3{3}3{3}3, we have included pairs that belong to a diameter. In fact, there are altogether $9+36$ triads of mutually non-adjacent vertices: 9 lying on diameters, and 36, such as 120, 210, 330, forming equilateral triangles, three inscribed in each of the twelve γ_2^3's. However, on the cubic surface there is no corresponding distinction: the 45 triads of intersecting lines are all alike; their planes are just the 45 tritangent planes. The nine diameters correspond to one of the forty ways in which the 27 lines can be distributed into nine triads, each triad forming a tritangent plane. The representation of the nine diameters by the points of the configuration $(9_4, 12_3)$ corresponds to the fact that there are 12 ways of distributing these 9 tritangent planes into 3 Steiner trihedra so that each plane belongs to 4 of the trihedra.

Thus the group of automorphisms of the Hessian polyhedron 3{3}3{3}3 is a subgroup of index 40 in the group of automorphisms of the 27 lines.

When the three complex coordinates u_ν in unitary 3-space are replaced by six Cartesian coordinates in Euclidean 6-space, the polyhedron 3{3}3{3}3 yields the uniform polytope 2_{21} whose 27 vertices faithfully represent the 27 lines.

A polyhedron closely related to 3{3}3{3}3 is 2{4}3{3}3, for which the Hermitian form is

$$x^1\overline{x^1}-\sqrt{\frac{1}{2}}(x^1\overline{x^2}+x^2\overline{x^1})+x^2\overline{x^2}-\sqrt{\frac{1}{3}}(x^2\overline{x^3}+x^3\overline{x^2})+x^3\overline{x^3}=u_1\overline{u_1}+u_2\overline{u_2}+u_3\overline{u_3},$$

where

$$u_1=-\frac{i\omega^2x^1}{\sqrt{2}}+\frac{x^2}{\sqrt{3}},\quad u_2=\frac{i\omega x^1}{\sqrt{2}}+\frac{x^2}{\sqrt{3}},\quad u_3=\frac{x^2}{\sqrt{3}}-x^3.$$

The three mirrors are now

$$\frac{u_1-\omega u_2}{\sqrt{2}}=0,\quad \frac{u_1+u_2+u_3}{\sqrt{3}}=0,\quad u_3=0.$$

R_2 and R_3 (of period 3) are the same as above, but R_1 (of period 2) is

$$u_1'=\omega u_2,\quad u_2'=\omega^2u_1,\quad u_3'=u_3.$$

In fact, 2{4}3{3}3 has 54 vertices

$$(0,\pm\omega^\mu,\mp\omega^\nu),\quad (\mp\omega^\nu,0,\pm\omega^\mu),\quad (\pm\omega^\mu,\mp\omega^\nu,0),$$

namely the vertices of two reciprocal 3{3}3{3}3's of the same size, each derivable from the other by the central inversion

$$u_1'=-u_1,\quad u_2'=-u_2,\quad u_3'=-u_3.$$

There are 216 edges, each joining a vertex of one 3{3}3{3}3 to one of the 8 nearest vertices of the other. Since the vertex figure 3{3}3 has 8 vertices, the polyhedron has 8 edges at each vertex, 216 edges altogether. There are 72 faces 2{4}3.

The reciprocal polyhedron 3{3}3{4}2 has 72 vertices, 216 edges and 54 faces 3{3}3. Here, the vertices are:

$$(\omega^\lambda, \omega^\mu, \omega^\nu), \quad (-\omega^\lambda, -\omega^\mu, -\omega^\nu), \quad \text{and} \quad (\pm i\omega^\lambda\sqrt{3}, 0, 0), \quad \text{permuted},$$

where $\lambda, \mu, \nu = 0, 1, 2$, independently.

Notice that 2{4}3{3}3 and 3{3}3{4}2 are 6-symmetric. In the case of 3{3}3{4}2, the six vertices $(\pm i\omega^\lambda\sqrt{3}, 0, 0)$ form a γ_1^6 on one of the twelve diameters.

These diameters distribute themselves into four triads of mutually orthogonal diameters, namely, lines joining the origin to the following triads of the points:

$$\begin{array}{ccc} (1,0,0), & (0,1,0), & (0,0,1); \\ (1,1,1), & (\omega,\omega^2,1), & (\omega^2,\omega,1); \\ (\omega,1,1), & (\omega^2,\omega^2,1), & (1,\omega,1); \\ (\omega^2,1,1), & (1,\omega^2,1), & (\omega,\omega,1). \end{array}$$

The twelve planes of symmetry of 3{3}3{3}3 play the same role for 2{4}3{3}3 and 3{3}3{4}2; but these two reciprocal polyhedra have nine additional planes of symmetry:

$$u_\mu - \omega^\lambda u_\nu = 0 \quad (\mu \neq \nu).$$

The section of 3{3}3{4}2 by any plane of symmetry is either 3{4}3 or 2{4}6.

Now, we need some general results about finite complex reflection groups. In 1954, Shephard and Todd ([57]) published a list of all finite irreducible complex reflection groups up to conjugacy. In their classification they separately studied the imprimitive groups and the primitive groups. In the latter case they used the classification of finite collineation groups containing homologies. Furthermore, Shephard and Todd determined the degrees of the reflection groups, using the invariant theory of the corresponding collineation groups in the primitive case. In 1967, Coxeter ([19]) presented a number of graphs connected with complex reflection groups to systematize the results of Shephard and Todd. In 1976, Cohen ([12]) gave another approach to obtain a systematization of the same results.

According to [57], let $\mathfrak{G}$ be a finite unitary group generated by reflections in a unitary space U_n of n dimensions. The matrices which correspond to the operations of $\mathfrak{G}$ can be regarded as collineation matrices in projective space S_{n-1} of $n-1$ dimensions, and the corresponding collineations form a group $\mathfrak{G}'$ which is isomorphic to the quotient group of $\mathfrak{G}$ by the cyclic subgroup $\mathfrak{Z}$ which consists of the elements of $\mathfrak{G}$ represented by scalar matrices. The m-fold reflections of $\mathfrak{G}$ correspond to collineations of finite period m which fixed all points of a prime of S_{n-1}. Such collineations are known as homologies, and thus $\mathfrak{G}'$ is generated by homologies.

Now we consider the primitive finite unitary groups generated by reflections in U_3. The groups $\mathfrak{G}'$ generated by homologies are of orders 60, 168, 216 or 360 (see also [53]).

$\mathfrak{G}'_{60}$ is isomorphic with the alternating group on five symbols, and the corresponding group $\mathfrak{G}$, of order 120, is $[3,5]$, the symmetry group of the icosahedron.

$\mathfrak{G}'_{216}$ is the Hessian group, which leaves invariant the configuration of inflections of a cubic curve. This group is generated by homologies of period 3, belonging to 12 cyclic subgroups, and in addition contains nine homologies of period 2 which generate an imprimitive subgroup. A finite unitary group generated by reflections corresponding to this collineation group must contain 3-fold reflections, and may, in addition, contain 2-fold reflections corresponding to the homologies of period 2. There are in fact two such groups; one, of order 648, contains only 3-fold reflections and is the symmetry group of the complex regular polyhedron 3{3}3{3}3 and the other, of order 1296, containing both 3-fold and 2-fold reflections, is the symmetry group of the regular complex polyhedron 2{4}3{3}3 or its reciprocal 3{3}3{4}2.

When $n = 4$, the groups $\mathfrak{G}'$ generated by homologies are of orders 576, 1920, 7200, 11,520 and 25,920.

$\mathfrak{G}'_{25,920}$ is unique among the quaternary groups in that all its homologies are of period three, and it gives rise to a finite unitary group generated by reflections of order 155,520, the symmetry group of the complex regular polytope 3{3}3{3}3{3}3.

The primitive groups generated by homologies in more than four variables consist of five cases (see [57]), which fall into two sets. The first set comprises groups in S_5, S_6 and S_7 respectively of orders $72 \cdot 6!$, $4 \cdot 9!$ and $96 \cdot 10!$ to which correspond Euclidean groups of orders $72 \cdot 6!$, $8 \cdot 9!$ and $192 \cdot 10!$ which are the symmetry groups of the polytopes 2_{21}, 3_{21} and 4_{21} respectively. The remaining two groups, of orders $36 \cdot 6!$ and $18 \cdot 9!$ in S_4 and S_5 correspond to finite unitary groups generated by reflections (in U_5 and U_6 respectively) of orders $72 \cdot 6!$ and $108 \cdot 9!$, they are the groups $[21; 2]^3$ and $[21; 3]^3$.

Following [57], the symmetry group of the polyhedron 3{3}3{3}3 contains 3-fold reflections in the 12 planes:

$$x_i = 0, \quad x_1 + \omega^j x_2 + \omega^k x_3 = 0 \quad (i, j, k = 1, 2, 3).$$

It is generated by the reflections

$$\mathbf{R}_1 : \begin{pmatrix} 1 & & \\ & 1 & \\ & & \omega^2 \end{pmatrix}, \quad \mathbf{R}_2 : \frac{1}{\sqrt{-3}} \begin{pmatrix} \omega & \omega^2 & \omega^2 \\ \omega^2 & \omega & \omega^2 \\ \omega^2 & \omega^2 & \omega \end{pmatrix}, \quad \mathbf{R}_3 : \begin{pmatrix} 1 & & \\ & \omega^2 & \\ & & 1 \end{pmatrix}$$

in the planes $x_3 = 0$, $x_1 + x_2 + x_3 = 0$, $x_2 = 0$ respectively. The characteristic roots of $\mathbf{R}_1$ and $\mathbf{R}_3$ are $(1, 1, \omega^2)$ and of $\mathbf{R}_2$ are $(1, 1, \omega)$. The matrix of the product $R_1 R_2^{-1} R_3$ is

$$\frac{-1}{\sqrt{-3}} \begin{pmatrix} \omega^2 & 1 & \omega \\ \omega & \omega & \omega \\ 1 & \omega^2 & \omega \end{pmatrix}.$$

Its characteristic equation is

$$\lambda^3 - \omega^2 \lambda^2 + \omega \lambda - 1 = 0,$$

and its characteristic roots are ζ^5, ζ^8, ζ^{11} where $\zeta = \exp(2\pi i/12)$.

The symmetry group of the regular polyhedron 2{4}3{3}3 or its reciprocal contains in addition to the above 3-fold reflections, 2-fold reflections in the nine planes

$$x_i - \omega^k x_j = 0, \quad (i, j, k = 1, 2, 3).$$

It is generated by the 3-fold reflections $\mathbf{R}_1$, $\mathbf{R}_2$ above together with the 2-fold reflection

$$\mathbf{S}: \begin{pmatrix} 1 & 0 & 0 \\ 0 & 0 & 1 \\ 0 & 1 & 0 \end{pmatrix}$$

in the plane $x_2 - x_3 = 0$. The matrix of the product $SR_1R_2^{-1}$ is

$$\frac{-1}{\sqrt{-3}} \begin{pmatrix} \omega^2 & \omega & \omega \\ 1 & 1 & \omega \\ \omega & \omega^2 & \omega \end{pmatrix}.$$

Its characteristic equation is

$$\lambda^3 + \omega = 0,$$

and its characteristic roots are ζ^5, ζ^{11}, ζ^{17} where $\zeta = \exp(2\pi i/18)$.

According to [29], when we study the 27 lines on a general cubic surface, all the geometry is closely tied up with the automorphism group of the lines, the Weyl group of E_6. The group $W(E_6)$ is particularly interesting because it can be expressed in terms of different Chevalley groups (see [15] for more details):

$$G_{25,920} \cong PU(3, 1; \mathbb{F}_4) \cong PSp(4, \mathbb{Z}/3\mathbb{Z}),$$

where $G_{25,920} \subset W(E_6)$ is the simple subgroup of index two consisting of all even elements.

It is well-known that the group of permutations (see [29] or [50]), $\mathrm{Aut}(\mathcal{L})$, of the 27 lines (or of the 45 planes), by which we mean the permutations of the lines preserving the intersection behavior of the lines,

$$|\mathrm{Aut}(\mathcal{L})| = |S_6| \cdot 2 \cdot 36 = 51840.$$

In fact, $\mathrm{Aut}(\mathcal{L})$ is nothing but the Weyl group $W(E_6)$, and the 36 double sixes correspond to the positive roots. The 27 lines correspond to the 27 fundamental weights of E_6, and the many other sets of objects (lines, tritangents, etc.) correspond to natural sets of objects (roots, weights, etc.) of E_6.

Let us recall some fundamental facts about the unitary reflection groups of order 25, 920 (see [29]). We have actions of $G_{25,920}$ on $\mathbb{P}^3$ and on $\mathbb{P}^4$, both of which are in fact generated by unitary reflections. We will be interested in the arrangement in $\mathbb{P}^3$. The arrangement induced in each of the 40 planes of this arrangement is the extended Hessian pencil, the arrangement of 21 lines which are the 12 lines of the Hessian pencil together with the nine lines joining corners of the four triangles. The 12 lines of the Hessian pencil are the 12 two-fold lines lying in the plane, the nine other lines are the four-fold lines lying in the plane. We also remark that the five-fold points of the arrangement split into two different types of singular points in the planes, namely 36 two-fold points and nine four-fold points which are the base points of the Hessian pencil. The 12-fold points of the arrangement lying in one of the planes are five-fold points of the induced arrangement. For each 12-fold point we can also speak of the induced arrangement, by blowing up the point in $\mathbb{P}^3$ and considering the proper transforms of the 12 planes intersecting the point in the exceptional $\mathbb{P}^2$. In our case, we get the Hessian pencil itself, i.e., the 12 lines of the four degenerate cubics.

The invariants forms under the action of $G_{25,920}$ on $\mathbb{P}^3$ were calculated by Maschke (see [52] and [29]). This is done essentially by reducing the problem to that of the invariants of the Hessian group of order 648 acting on $\mathbb{P}^2$. The 40 planes of the arrangement in $\mathbb{P}^3$ are given explicitly in homogeneous coordinates $(z_0 : z_1 : z_2 : z_3)$:

$$
\begin{aligned}
&(4) \quad z_i = 0 \quad (i = 0, 1, 2, 3).\\
&(9) \quad (z_1^3 + z_2^3 + z_3^3)^3 - 27 z_1^3 z_2^3 z_3^3 = 0.\\
&(9) \quad (z_0^3 + z_1^3 + z_3^3)^3 - 27 z_0^3 z_1^3 z_3^3 = 0.\\
&(9) \quad (z_0^3 + z_1^3 + z_2^3)^3 - 27 z_0^3 z_1^3 z_2^3 = 0.\\
&(9) \quad (z_0^3 + z_2^3 + z_3^3)^3 - 27 z_0^3 z_2^3 z_3^3 = 0.
\end{aligned}
$$

In the $\mathbb{P}^2$ given by $z_0 = 0$ with homogeneous coordinates (z_1, z_2, z_3) the action of G_{648} is generated by five collineations:

$$
\begin{cases}
A(z_1, z_2, z_3) = (z_2, z_3, z_1),\\
B(z_1, z_2, z_3) = (z_1, z_3, z_2),\\
C(z_1, z_2, z_3) = (z_1, \omega z_2, \omega^2 z_3),\\
D(z_1, z_2, z_3) = (z_1, \omega z_2, \omega z_3),\\
E(z_1, z_2, z_3) = \left(\dfrac{z_1 + z_2 + z_3}{\sqrt{-3}}, \dfrac{z_1 + \omega z_2 + \omega^2 z_3}{\sqrt{-3}}, \dfrac{z_1 + \omega^2 z_2 + \omega z_3}{\sqrt{-3}}\right).
\end{cases} \tag{4.1}
$$

Writing in the form of matrices:

$$
A = \begin{pmatrix} 0 & 1 & 0 \\ 0 & 0 & 1 \\ 1 & 0 & 0 \end{pmatrix}, \quad B = \begin{pmatrix} 1 & 0 & 0 \\ 0 & 0 & 1 \\ 0 & 1 & 0 \end{pmatrix},
$$

$$
C = \begin{pmatrix} 1 & 0 & 0 \\ 0 & \omega & 0 \\ 0 & 0 & \omega^2 \end{pmatrix}, \quad D = \begin{pmatrix} 1 & 0 & 0 \\ 0 & \omega & 0 \\ 0 & 0 & \omega \end{pmatrix},
$$

$$
E = \frac{1}{\sqrt{-3}} \begin{pmatrix} 1 & 1 & 1 \\ 1 & \omega & \omega^2 \\ 1 & \omega^2 & \omega \end{pmatrix}.
$$

Note that

$$
AA^* = BB^* = CC^* = DD^* = EE^* = I.
$$

This implies that

$$A, B, C, D, E \in U(3).$$

In fact, the Hessian group $G_{648} = \langle A, B, C, D, E \rangle$, where

$$A^3 = B^2 = C^3 = D^3 = E^4 = I.$$

A complete system of invariants for G_{648} has degrees 6, 9, 12, 12 and 18 and can be given explicitly by the following forms:

$$\begin{cases} C_6 = z_1^6 + z_2^6 + z_3^6 - 10(z_1^3 z_2^3 + z_2^3 z_3^3 + z_3^3 z_1^3), \\ C_9 = (z_1^3 - z_2^3)(z_2^3 - z_3^3)(z_3^3 - z_1^3), \\ C_{12} = (z_1^3 + z_2^3 + z_3^3)[(z_1^3 + z_2^3 + z_3^3)^3 + 216 z_1^3 z_2^3 z_3^3], \\ \mathfrak{C}_{12} = z_1 z_2 z_3 [27 z_1^3 z_2^3 z_3^3 - (z_1^3 + z_2^3 + z_3^3)^3], \\ C_{18} = (z_1^3 + z_2^3 + z_3^3)^6 - 540 z_1^3 z_2^3 z_3^3 (z_1^3 + z_2^3 + z_3^3)^3 - 5832 z_1^6 z_2^6 z_3^6. \end{cases} \tag{4.2}$$

C_{18} is the functional determinant of C_{12} and $\mathfrak{C}_{12}$. $\mathfrak{C}_{12}$ is just the product of the 12 lines of the Hessian arrangement. C_{12} is the Hessian of $\mathfrak{C}_{12}$ and vise versa. C_9 is the so-called difference product of z_1^3, z_2^3, z_3^3. The two relations among these invariants are

$$\begin{cases} 432 C_9^2 = C_6^3 - 3 C_6 C_{12} + 2 C_{18}, \\ 1728 \mathfrak{C}_{12}^3 = C_{18}^2 - C_{12}^3. \end{cases} \tag{4.3}$$

Following [57], the Hessian group of order 216 in S_2 is the collineation group that leaves invariant the nine inflections of a pencil of cubic curves in the plane. The invariant forms of this collineation group consist of a sextic C_6, a form C_9 of degree 9 representing the harmonic polars of the inflections, and two forms $\mathfrak{C}_{12}$, C_{12} representing respectively the four degenerate cubics and the four equianharmonic cubics in the pencil. A syzygy connects $\mathfrak{C}_{12}^3$ with C_6, C_9 and C_{12}:

$$6912 \mathfrak{C}_{12}^3 = (432 C_9^2 - C_6^3 + 3 C_6 C_{12})^2 - 4 C_{12}^3. \tag{4.4}$$

For the two corresponding finite unitary groups generated by reflections in U_3 of order 648 and 1296 respectively, the invariant forms may be taken to be C_6, C_9, C_{12} and C_6, C_{12}, C_{18}, respectively. The Jacobian of C_6, C_9, C_{12} is a multiple of the square of $\mathfrak{C}_{12}$; that of C_6, C_{12}, C_{18} is a multiple of the product $C_9 \mathfrak{C}_{12}^2$.

Maschke proved that the action of $G_{25,920}$ in $\mathbb{P}^3$ is generated by the action of a G_{648} acting on one of the 40 planes and a tetrahedral group of order 24 consisting of the permutation group acting on the z_i. This tetrahedral group stabilizes the tetrahedron consisting of the first four forms in the 40 planes, and $G_{25,290}$ is generated by it and the G_{648} acting on $z_0 = 0$ as above. From this one deduces that each invariant form of $G_{25,920}$ can be written as a polynomial in z_0 with coefficients which are polynomial expressions in the invariants of G_{648}.

A complete system of invariants for $G_{25,920}$ has degrees 12, 18, 24, 30 and 40 and can be given by the following forms:

$$\begin{cases} F_{12} = 6z_0^{12} + 6 \cdot 22C_6 z_0^6 + 6 \cdot 220C_9 z_0^3 + C_6^2 + 5C_{12}, \\ F_{18} = -54z_0^{18} + 17 \cdot 54C_6 z_0^{12} + 54 \cdot 1870C_9 z_0^9 \\ \qquad + \frac{1}{2} \cdot 17 \cdot 27(19C_6^2 - 15C_{12})z_0^6 + 54 \cdot 170C_6C_9 z_0^3 + C_6^3 \\ \qquad - 30C_6C_{12} - 25C_{18}, \\ F_{24} = 1728C_6 z_0^{18} - 36 \cdot 1728C_9 z_0^{15} + 15 \cdot 144(7C_{12} + C_6^2)z_0^{12} \\ \qquad - 10 \cdot 1728C_6C_9 z_0^9 + 36(178C_{18} - 135C_6C_{12} + 5C_6^3)z_0^6 \\ \qquad + 432(41C_{12} - C_6^2)C_9 z_0^3 + C_6^4 + 6C_6^2C_{12} - 16C_6C_{18} + 9C_{12}^2, \\ F_{30} = -2 \cdot 6^4 C_6 z_0^{24} + 312 \cdot 6^4 C_9 z_0^{21} + 216(715C_{12} - 127C_6^2)z_0^{18} \\ \qquad + 272 \cdot 6^4 C_6C_9 z_0^{15} + 18(1306C_{18} + 6045C_6C_{12} - 295C_6^3)z_0^{12} \\ \qquad + 216(73C_6^2 - 5473C_{12})C_9 z_0^9 \\ \qquad + \frac{1}{2} \cdot 3(16648C_6C_{18} + 2334C_6^2C_{12} - 20709C_{12}^2 - C_6^4)z_0^6 \\ \qquad - 36(1370C_{18} - 657C_6C_{12} + 7C_6^3)C_9 z_0^3 + C_6^5 - 19C_6^3C_{12} \\ \qquad + 29C_6^2C_{18} - 6C_6C_{12}^2 - 5C_{12}C_{18}. \end{cases}$$

F_{40} is just the product of the 40 planes defining the arrangement in $\mathbb{P}^3$, i.e., it represents the primes of the 40 homologies of period three contained in the group.

The forms F_{12}, F_{18}, F_{24}, F_{30} and F_{40} satisfy the relation:

$$2^{28} \cdot 3^{15} \cdot 5^{15} \cdot F_{40}^3 \\ = \begin{vmatrix} \Phi_{30} & 2\Phi_{24}^2 & F_{18}\Phi_{24} & F_{12}\Phi_{24} \\ 2\Phi_{24} & 27F_{12}\Phi_{30} - 11F_{18}\Phi_{24} & 3F_{12}\Phi_{24} - 4F_{18}^2 & 3\Phi_{30} - 4F_{12}F_{18} \\ F_{18} & 3F_{12}\Phi_{24} - 4F_{18}^2 & 13F_{12}F_{18} - 3\Phi_{30} & 9F_{12}^2 - 2\Phi_{24} \\ F_{12} & 3\Phi_{30} - 4F_{12}F_{18} & 9F_{12}^2 - 2\Phi_{24} & F_{18} \end{vmatrix},$$

where

$$4\Phi_{24} = 25F_{24} - 9F_{12}^2,$$
$$6\Phi_{30} = 25F_{30} - F_{12}F_{18}.$$

There are two simpler relations involving F_{40}. Firstly, F_{40} is the Hessian determinant of F_{12}:

$$\text{Hessian}(F_{12}) = 2^{14} \cdot 3^8 \cdot 5^3 \cdot 11^4 \cdot F_{40}.$$

Secondly, the Jacobian determinant

$$\frac{\partial(F_{12}, F_{18}, F_{24}, F_{30})}{\partial(z_0, z_1, z_2, z_3)} = 2^{26} \cdot 3^{15} \cdot 5^6 \cdot F_{40}^2.$$

The action of $G_{25,920}$ on $\mathbb{P}^3$ in terms of the hyperelliptic functions $Z_{\alpha,\beta}$ was first treated by Witting (see [62]). That is why the arrangement defined by the action of $G_{25,920}$ on $\mathbb{P}^3$ is called a Witting configuration.

In [63], we gave the modular equation associated to the Picard curves:
Theorem (see [63], Main Theorem 3). *The following identity of differential forms holds*:

$$\frac{dw_1 \wedge dw_2}{\sqrt[3]{w_1^2 w_2^2 (1-w_1)(1-\lambda_1^3 w_1)(1-\lambda_2^3 w_1)}} = -\frac{5t_1}{t_2^2} \frac{dt_1 \wedge dt_2}{\sqrt[3]{t_1^2 t_2^2 (1-t_1)(1-\kappa_1^3 t_1)(1-\kappa_2^3 t_1)}},$$

where

$$w_1 = \frac{(\beta+\gamma)t_1 + \alpha}{\beta t_1 + (\alpha+\gamma)}, \quad w_2 = \frac{t_1[(\beta+\gamma)t_1+\alpha]^2[\beta t_1 + (\alpha+\gamma)]}{t_2^5},$$

which is a rational transformation of order five. Here,

$$\begin{cases} \alpha = \dfrac{(v_1-1)(v_2-1)(v_1+v_2-2)-(u_1-1)(u_2-1)(u_1+u_2-2)}{2(u_1-1)(u_2-1)(v_1-1)(v_2-1)}, \\ \beta = \dfrac{(u_1-1)(u_2-1)(2u_1u_2-u_1-u_2)-(v_1-1)(v_2-1)(2v_1v_2-v_1-v_2)}{2(u_1-1)(u_2-1)(v_1-1)(v_2-1)}, \\ \gamma = \dfrac{(v_1-1)(v_2-1)}{(u_1-1)(u_2-1)}, \end{cases}$$

with

$$u_1 = \kappa_1^3, \quad u_2 = \kappa_2^3, \quad v_1 = \lambda_1^3, \quad v_2 = \lambda_2^3.$$

Moreover, the moduli (κ_1, κ_2) *and* (λ_1, λ_2) *satisfy the modular equation:*

$$(\kappa_1^3 - 1)(\kappa_2^3 - 1)(\kappa_1^3 - \kappa_2^3) = (\lambda_1^3 - 1)(\lambda_2^3 - 1)(\lambda_1^3 - \lambda_2^3),$$

which is an algebraic variety of dimension three. The corresponding two algebraic surfaces are:

$$\begin{cases} w_3^3 = w_1^2 w_2^2 (1-w_1)(1-\lambda_1^3 w_1)(1-\lambda_2^3 w_1), \\ t_3^3 = t_1^2 t_2^2 (1-t_1)(1-\kappa_1^3 t_1)(1-\kappa_2^3 t_1), \end{cases}$$

which give the homogeneous algebraic equations of degree seven:

$$\begin{cases} w_3^3 w_4^4 = w_1^2 w_2^2 (w_4-w_1)(w_4-\lambda_1^3 w_1)(w_4-\lambda_2^3 w_1), \\ t_3^3 t_4^4 = t_1^2 t_2^2 (t_4-t_1)(t_4-\kappa_1^3 t_1)(t_4-\kappa_2^3 t_1). \end{cases}$$

In fact, the above modular equation can be written as follows:

$$C_9(\kappa_1, \kappa_2, 1) = C_9(\lambda_1, \lambda_2, 1). \tag{4.5}$$

This shows that our modular equation is intimately connected with the Hessian polyhedra.

Set $X = z_1^3$, $Y = z_2^3$ and $Z = z_3^3$. Put

$$\begin{cases} \sigma_1 = X + Y + Z, \\ \sigma_2 = XY + YZ + ZX, \\ \sigma_3 = XYZ. \end{cases} \tag{4.6}$$

Then

$$\begin{cases} C_6 = \sigma_1^2 - 12\sigma_2, \\ C_{12} = \sigma_1(\sigma_1^3 + 216\sigma_3), \\ C_{18} = \sigma_1^6 - 540\sigma_1^3\sigma_3 - 5832\sigma_3^2, \\ \mathfrak{C}_{12}^3 = \sigma_3(27\sigma_3 - \sigma_1^3)^3. \end{cases} \tag{4.7}$$

Note that

$$C_9 = (X-Y)(Y-Z)(Z-X) = (XY^2 + YZ^2 + ZX^2) - (X^2Y + Y^2Z + Z^2X),$$

we have

$$\begin{aligned} C_9^2 = &(XY^2 + YZ^2 + ZX^2 + X^2Y + Y^2Z + Z^2X)^2 \\ &- 4(XY^2 + YZ^2 + ZX^2)(X^2Y + Y^2Z + Z^2X). \end{aligned}$$

Here,

$$\begin{aligned} &XY^2 + YZ^2 + ZX^2 + X^2Y + Y^2Z + Z^2X \\ =&(X + Y + Z)(XY + YZ + ZX) - 3XYZ \\ =&\sigma_1\sigma_2 - 3\sigma_3. \end{aligned}$$

$$\begin{aligned} &(XY^2 + YZ^2 + ZX^2)(X^2Y + Y^2Z + Z^2X) \\ =&X^3Y^3 + Y^3Z^3 + Z^3X^3 + XYZ(X^3 + Y^3 + Z^3) + 3X^2Y^2Z^2, \end{aligned}$$

where

$$\begin{aligned} &X^3Y^3 + Y^3Z^3 + Z^3X^3 \\ =&(XY + YZ + ZX)[(XY + YZ + ZX)^2 - 3XYZ(X + Y + Z)] + 3(XYZ)^2 \\ =&\sigma_2(\sigma_2^2 - 3\sigma_1\sigma_3) + 3\sigma_3^2, \end{aligned}$$

$$\begin{aligned} &X^3 + Y^3 + Z^3 \\ =&(X + Y + Z)[(X + Y + Z)^2 - 3(XY + YZ + ZX)] + 3XYZ \\ =&\sigma_1(\sigma_1^2 - 3\sigma_2) + 3\sigma_3. \end{aligned}$$

So,

$$(XY^2 + YZ^2 + ZX^2)(X^2Y + Y^2Z + Z^2X) = \sigma_2^3 - 6\sigma_1\sigma_2\sigma_3 + 9\sigma_3^2 + \sigma_1^3\sigma_3.$$

Thus,

$$C_9^2 = \sigma_1^2\sigma_2^2 - 4\sigma_1^3\sigma_3 - 4\sigma_2^3 - 27\sigma_3^2 + 18\sigma_1\sigma_2\sigma_3. \tag{4.8}$$

In the affine coordinates $\xi = z_1/z_3$ and $\eta = z_2/z_3$, we have

$$\begin{cases} C_6 = C_6(\xi,\eta) = \xi^6 + \eta^6 + 1 - 10(\xi^3\eta^3 + \xi^3 + \eta^3), \\ C_9 = C_9(\xi,\eta) = (\xi^3 - \eta^3)(\eta^3 - 1)(1 - \xi^3), \\ C_{12} = C_{12}(\xi,\eta) = (\xi^3 + \eta^3 + 1)[(\xi^3 + \eta^3 + 1)^3 + 216\xi^3\eta^3], \\ \mathfrak{C}_{12} = \mathfrak{C}_{12}(\xi,\eta) = \xi\eta[27\xi^3\eta^3 - (\xi^3 + \eta^3 + 1)^3], \\ C_{18} = C_{18}(\xi,\eta) = (\xi^3 + \eta^3 + 1)^6 - 540\xi^3\eta^3(\xi^3 + \eta^3 + 1)^3 - 5832\xi^6\eta^6. \end{cases} \tag{4.9}$$

The relation

$$432C_9^2 = C_6^3 - 3C_6C_{12} + 2C_{18}$$

gives that

$$432\frac{C_9^2}{C_6^3} = 1 - 3\frac{C_{12}}{C_6^2} + 2\frac{C_{18}}{C_6^3}. \tag{4.10}$$

We define the Hessian polyhedral equations as follows:

$$\begin{cases} 432\dfrac{C_9^2}{C_6^3} = R_1(x,y), \\ 3\dfrac{C_{12}}{C_6^2} = R_2(x,y), \end{cases} \tag{4.11}$$

where R_1 and R_2 are rational functions of x and y.

Theorem 4.1 (Main Theorem 2). *The relations between Hessian polyhedra and Appell hypergeometric partial differential equations are given by the following identities*:

$$\begin{aligned} &z_1^6 + z_2^6 + z_3^6 - 10(z_1^3z_2^3 + z_2^3z_3^3 + z_3^3z_1^3) \\ =&2^6 \cdot 3^7 \cdot C^2 \cdot R_1 \left[\frac{1}{4}(R_1 + R_2 - 1)^2 - \frac{1}{27}R_2^3\right]^{\frac{4}{3}} \left[\frac{\partial(R_1,R_2)}{\partial(x,y)}\right]^{-2} H(x,y)^2. \end{aligned} \tag{4.12}$$

$$\begin{aligned} &(z_1^3 - z_2^3)(z_2^3 - z_3^3)(z_3^3 - z_1^3) \\ =&2^7 \cdot 3^9 \cdot C^3 \cdot R_1^2 \left[\frac{1}{4}(R_1 + R_2 - 1)^2 - \frac{1}{27}R_2^3\right]^2 \left[\frac{\partial(R_1,R_2)}{\partial(x,y)}\right]^{-3} H(x,y)^3. \end{aligned} \tag{4.13}$$

$$\begin{aligned} &(z_1^3 + z_2^3 + z_3^3)[(z_1^3 + z_2^3 + z_3^3)^3 + 216z_1^3z_2^3z_3^3] \\ =&2^{12} \cdot 3^{13} \cdot C^4 \cdot R_1^2R_2 \left[\frac{1}{4}(R_1 + R_2 - 1)^2 - \frac{1}{27}R_2^3\right]^{\frac{8}{3}} \left[\frac{\partial(R_1,R_2)}{\partial(x,y)}\right]^{-4} H(x,y)^4. \end{aligned} \tag{4.14}$$

$$\begin{aligned} &z_1z_2z_3[27z_1^3z_2^3z_3^3 - (z_1^3 + z_2^3 + z_3^3)^3] \\ =&2^{10} \cdot 3^{13} \cdot C^4 \cdot R_1^2 \left[\frac{1}{4}(R_1 + R_2 - 1)^2 - \frac{1}{27}R_2^3\right]^3 \left[\frac{\partial(R_1,R_2)}{\partial(x,y)}\right]^{-4} H(x,y)^4. \end{aligned} \tag{4.15}$$

$$\begin{aligned}&(z_1^3+z_2^3+z_3^3)^6-540z_1^3z_2^3z_3^3(z_1^3+z_2^3+z_3^3)^3-5832z_1^6z_2^6z_3^6\\ =&2^{17}\cdot 3^{21}\cdot C^6\cdot R_1^3(R_1+R_2-1)\left[\frac{1}{4}(R_1+R_2-1)^2-\frac{1}{27}R_2^3\right]^4\\ &\times\left[\frac{\partial(R_1,R_2)}{\partial(x,y)}\right]^{-6}H(x,y)^6.\end{aligned} \tag{4.16}$$

Here,

$$H(x,y):=x^{b'-c}y^{b-c}(x-1)^{c-a-b-1}(y-1)^{c-a-b'-1}(x-y)^{-b-b'}. \tag{4.17}$$

Proof. We have

$$\frac{\partial(R_1,R_2)}{\partial(\xi,\eta)}=1296\frac{\partial\left(\frac{C_9^2}{C_6^3},\frac{C_{12}}{C_6^2}\right)}{\partial(\xi,\eta)}.$$

Here, we find that

$$\frac{\partial\left(\frac{C_9^2}{C_6^3},\frac{C_{12}}{C_6^2}\right)}{\partial(\xi,\eta)}=\frac{C_9}{C_6^6}\left[2C_6\frac{\partial(C_9,C_{12})}{\partial(\xi,\eta)}+3C_9\frac{\partial(C_{12},C_6)}{\partial(\xi,\eta)}+4C_{12}\frac{\partial(C_6,C_9)}{\partial(\xi,\eta)}\right].$$

Now, we need the following proposition:

Proposition 4.2 (Main Proposition 1). *The following identities hold:*

$$\begin{vmatrix}\frac{\partial C_6}{\partial\xi} & \frac{\partial C_6}{\partial\eta} & 2C_6\\ \frac{\partial C_9}{\partial\xi} & \frac{\partial C_9}{\partial\eta} & 3C_9\\ \frac{\partial C_{12}}{\partial\xi} & \frac{\partial C_{12}}{\partial\eta} & 4C_{12}\end{vmatrix}=-864\mathfrak{C}_{12}^2. \tag{4.18}$$

$$\begin{vmatrix}\frac{\partial C_6}{\partial\xi} & \frac{\partial C_6}{\partial\eta} & 2C_6\\ \frac{\partial C_9}{\partial\xi} & \frac{\partial C_9}{\partial\eta} & 3C_9\\ \frac{\partial C_{18}}{\partial\xi} & \frac{\partial C_{18}}{\partial\eta} & 6C_{18}\end{vmatrix}=-1296C_6\mathfrak{C}_{12}^2. \tag{4.19}$$

$$\begin{vmatrix}\frac{\partial C_6}{\partial\xi} & \frac{\partial C_6}{\partial\eta} & 2C_6\\ \frac{\partial C_{12}}{\partial\xi} & \frac{\partial C_{12}}{\partial\eta} & 4C_{12}\\ \frac{\partial C_{18}}{\partial\xi} & \frac{\partial C_{18}}{\partial\eta} & 6C_{18}\end{vmatrix}=432\cdot 864C_9\mathfrak{C}_{12}^2. \tag{4.20}$$

$$\begin{vmatrix}\frac{\partial C_9}{\partial\xi} & \frac{\partial C_9}{\partial\eta} & 3C_9\\ \frac{\partial C_{12}}{\partial\xi} & \frac{\partial C_{12}}{\partial\eta} & 4C_{12}\\ \frac{\partial C_{18}}{\partial\xi} & \frac{\partial C_{18}}{\partial\eta} & 6C_{18}\end{vmatrix}=1296(C_6^2-C_{12})\mathfrak{C}_{12}^2. \tag{4.21}$$

$$\begin{vmatrix}\frac{\partial C_6}{\partial\xi} & \frac{\partial C_6}{\partial\eta} & 2C_6\\ \frac{\partial C_9}{\partial\xi} & \frac{\partial C_9}{\partial\eta} & 3C_9\\ \frac{\partial \mathfrak{C}_{12}}{\partial\xi} & \frac{\partial \mathfrak{C}_{12}}{\partial\eta} & 4\mathfrak{C}_{12}\end{vmatrix}=\frac{1}{2}(C_{12}^2-C_6C_{18}). \tag{4.22}$$

$$\begin{vmatrix}\frac{\partial C_6}{\partial\xi} & \frac{\partial C_6}{\partial\eta} & 2C_6\\ \frac{\partial C_{12}}{\partial\xi} & \frac{\partial C_{12}}{\partial\eta} & 4C_{12}\\ \frac{\partial \mathfrak{C}_{12}}{\partial\xi} & \frac{\partial \mathfrak{C}_{12}}{\partial\eta} & 4\mathfrak{C}_{12}\end{vmatrix}=144C_9C_{18}. \tag{4.23}$$

$$\begin{vmatrix} \frac{\partial C_9}{\partial \xi} & \frac{\partial C_9}{\partial \eta} & 3C_9 \\ \frac{\partial C_{12}}{\partial \xi} & \frac{\partial C_{12}}{\partial \eta} & 4C_{12} \\ \frac{\partial \mathfrak{C}_{12}}{\partial \xi} & \frac{\partial \mathfrak{C}_{12}}{\partial \eta} & 4\mathfrak{C}_{12} \end{vmatrix} = \frac{1}{2}(C_6^2 - C_{12})C_{18}. \tag{4.24}$$

$$\begin{vmatrix} \frac{\partial C_6}{\partial \xi} & \frac{\partial C_6}{\partial \eta} & 2C_6 \\ \frac{\partial C_{18}}{\partial \xi} & \frac{\partial C_{18}}{\partial \eta} & 6C_{18} \\ \frac{\partial \mathfrak{C}_{12}}{\partial \xi} & \frac{\partial \mathfrak{C}_{12}}{\partial \eta} & 4\mathfrak{C}_{12} \end{vmatrix} = 216C_9C_{12}^2. \tag{4.25}$$

$$\begin{vmatrix} \frac{\partial C_9}{\partial \xi} & \frac{\partial C_9}{\partial \eta} & 3C_9 \\ \frac{\partial C_{18}}{\partial \xi} & \frac{\partial C_{18}}{\partial \eta} & 6C_{18} \\ \frac{\partial \mathfrak{C}_{12}}{\partial \xi} & \frac{\partial \mathfrak{C}_{12}}{\partial \eta} & 4\mathfrak{C}_{12} \end{vmatrix} = \frac{3}{4}(C_6^2 - C_{12})C_{12}^2. \tag{4.26}$$

$$\begin{vmatrix} \frac{\partial C_{12}}{\partial \xi} & \frac{\partial C_{12}}{\partial \eta} & 4C_{12} \\ \frac{\partial C_{18}}{\partial \xi} & \frac{\partial C_{18}}{\partial \eta} & 6C_{18} \\ \frac{\partial \mathfrak{C}_{12}}{\partial \xi} & \frac{\partial \mathfrak{C}_{12}}{\partial \eta} & 4\mathfrak{C}_{12} \end{vmatrix} = 0. \tag{4.27}$$

Proof. Put $X = \xi^3$ and $Y = \eta^3$. We have

$$\frac{\partial C_6}{\partial \xi} = 6\xi^2(X - 5Y - 5),$$

$$\frac{\partial C_6}{\partial \eta} = 6\eta^2(Y - 5X - 5),$$

$$\frac{\partial C_9}{\partial \xi} = 3\xi^2(Y - 1)(1 - 2X + Y),$$

$$\frac{\partial C_9}{\partial \eta} = 3\eta^2(1 - X)(1 - 2Y + X),$$

$$\frac{\partial C_{12}}{\partial \xi} = 12\xi^2[(X + Y + 1)^3 + 54XY + 54Y(X + Y + 1)],$$

$$\frac{\partial C_{12}}{\partial \eta} = 12\eta^2[(X + Y + 1)^3 + 54XY + 54X(X + Y + 1)].$$

Hence,

$$\frac{\partial(C_6, C_9)}{\partial(\xi, \eta)} = 18\xi^2\eta^2[-(X + Y)^3 + 15(X^2 + Y^2) - 24XY + 6(X + Y) - 10].$$

$$\begin{aligned} &\frac{\partial(C_9, C_{12})}{\partial(\xi, \eta)} \\ =&36\xi^2\eta^2\{[(X + Y + 1)^3 + 54XY][X^2 + Y^2 - 4XY + 2(X + Y) - 2] \\ &+ 54(X + Y + 1)[-XY(X + Y) + 2(X^2 + Y^2) - (X + Y)]\}. \end{aligned}$$

$$\frac{\partial(C_{12}, C_6)}{\partial(\xi, \eta)} = 432\xi^2\eta^2(Y - X)[(X + Y + 1)^3 + 54XY + 9(X + Y + 1)(X + Y - 5)].$$

Note that

$$\begin{cases} C_6 = X^2 + Y^2 + 1 - 10(XY + X + Y), \\ C_9 = (X - Y)(Y - 1)(1 - X), \\ C_{12} = (X + Y + 1)[(X + Y + 1)^3 + 216XY]. \end{cases}$$

Put

$$\sigma_1 = X + Y + 1, \quad \sigma_2 = XY.$$

Then

$$\begin{cases} C_6 = \sigma_1^2 - 12\sigma_2 - 12\sigma_1 + 12, \\ C_9 = (Y - X)(\sigma_2 - \sigma_1 + 2), \\ C_{12} = \sigma_1(\sigma_1^3 + 216\sigma_2), \end{cases}$$

and

$$\begin{aligned} \frac{\partial(C_9, C_{12})}{\partial(\xi, \eta)} =& 36\xi^2\eta^2(\sigma_1^5 - 6\sigma_1^3\sigma_2 - 324\sigma_2^2 + 105\sigma_1^3 \\ & - 162\sigma_1\sigma_2 - 270\sigma_1^2 - 162\sigma_2 + 162\sigma_1), \\ \frac{\partial(C_{12}, C_6)}{\partial(\xi, \eta)} =& 432\xi^2\eta^2(Y - X)(\sigma_1^3 + 9\sigma_1^2 + 54\sigma_2 - 54\sigma_1), \\ \frac{\partial(C_6, C_9)}{\partial(\xi, \eta)} =& 18\xi^2\eta^2(-\sigma_1^3 + 18\sigma_1^2 - 54\sigma_2 - 27\sigma_1). \end{aligned}$$

Therefore,

$$\begin{aligned} & 2C_6\frac{\partial(C_9, C_{12})}{\partial(\xi, \eta)} + 3C_9\frac{\partial(C_{12}, C_6)}{\partial(\xi, \eta)} + 4C_{12}\frac{\partial(C_6, C_9)}{\partial(\xi, \eta)} \\ =& 72\xi^2\eta^2[(\sigma_1^2 - 12\sigma_2 - 12\sigma_1 + 12) \\ & \times (\sigma_1^5 - 6\sigma_1^3\sigma_2 - 324\sigma_2^2 + 105\sigma_1^3 - 162\sigma_1\sigma_2 - 270\sigma_1^2 - 162\sigma_2 + 162\sigma_1) \\ & + 18(\sigma_1^2 - 4\sigma_2 - 2\sigma_1 + 1)(\sigma_2 - \sigma_1 + 2)(\sigma_1^3 + 9\sigma_1^2 + 54\sigma_2 - 54\sigma_1) \\ & + \sigma_1(\sigma_1^3 + 216\sigma_2)(-\sigma_1^3 + 18\sigma_1^2 - 54\sigma_2 - 27\sigma_1)] \\ =& 72\xi^2\eta^2 \cdot (-12)(27\sigma_2 - \sigma_1^3)^2 \\ =& -864\xi^2\eta^2(27\sigma_2 - \sigma_1^3)^2. \end{aligned}$$

On the other hand, note that

$$\mathfrak{C}_{12} = \xi\eta[27XY - (X + Y + 1)^3] = \xi\eta(27\sigma_2 - \sigma_1^3).$$

Hence, we find that

$$2C_6\frac{\partial(C_9, C_{12})}{\partial(\xi, \eta)} + 3C_9\frac{\partial(C_{12}, C_6)}{\partial(\xi, \eta)} + 4C_{12}\frac{\partial(C_6, C_9)}{\partial(\xi, \eta)} = -864\mathfrak{C}_{12}^2.$$

By the identity $432C_9^2 = C_6^3 - 3C_6C_{12} + 2C_{18}$, we have

$$C_{18} = \frac{1}{2}(C_{12} - C_6^2)C_6 + 216C_9^2 + C_6C_{12}.$$

$$\frac{\partial C_{18}}{\partial \xi}=\frac{3}{2}(C_{12}-C_6^2)\frac{\partial C_6}{\partial \xi}+432C_9\frac{\partial C_9}{\partial \xi}+\frac{3}{2}C_6\frac{\partial C_{12}}{\partial \xi}.$$

$$\frac{\partial C_{18}}{\partial \eta}=\frac{3}{2}(C_{12}-C_6^2)\frac{\partial C_6}{\partial \eta}+432C_9\frac{\partial C_9}{\partial \eta}+\frac{3}{2}C_6\frac{\partial C_{12}}{\partial \eta}.$$

Thus,

$$\begin{vmatrix} \frac{\partial C_6}{\partial \xi} & \frac{\partial C_6}{\partial \eta} & 2C_6 \\ \frac{\partial C_9}{\partial \xi} & \frac{\partial C_9}{\partial \eta} & 3C_9 \\ \frac{\partial C_{18}}{\partial \xi} & \frac{\partial C_{18}}{\partial \eta} & 6C_{18} \end{vmatrix}=\frac{3}{2}C_6\begin{vmatrix} \frac{\partial C_6}{\partial \xi} & \frac{\partial C_6}{\partial \eta} & 2C_6 \\ \frac{\partial C_9}{\partial \xi} & \frac{\partial C_9}{\partial \eta} & 3C_9 \\ \frac{\partial C_{12}}{\partial \xi} & \frac{\partial C_{12}}{\partial \eta} & 4C_{12} \end{vmatrix}=-1296C_6\mathfrak{C}_{12}^2.$$

$$\begin{vmatrix} \frac{\partial C_6}{\partial \xi} & \frac{\partial C_6}{\partial \eta} & 2C_6 \\ \frac{\partial C_{12}}{\partial \xi} & \frac{\partial C_{12}}{\partial \eta} & 4C_{12} \\ \frac{\partial C_{18}}{\partial \xi} & \frac{\partial C_{18}}{\partial \eta} & 6C_{18} \end{vmatrix}=-432C_9\begin{vmatrix} \frac{\partial C_6}{\partial \xi} & \frac{\partial C_6}{\partial \eta} & 2C_6 \\ \frac{\partial C_9}{\partial \xi} & \frac{\partial C_9}{\partial \eta} & 3C_9 \\ \frac{\partial C_{12}}{\partial \xi} & \frac{\partial C_{12}}{\partial \eta} & 4C_{12} \end{vmatrix}=432\cdot 864C_9\mathfrak{C}_{12}^2.$$

$$\begin{vmatrix} \frac{\partial C_9}{\partial \xi} & \frac{\partial C_9}{\partial \eta} & 3C_9 \\ \frac{\partial C_{12}}{\partial \xi} & \frac{\partial C_{12}}{\partial \eta} & 4C_{12} \\ \frac{\partial C_{18}}{\partial \xi} & \frac{\partial C_{18}}{\partial \eta} & 6C_{18} \end{vmatrix}=\frac{3}{2}(C_{12}-C_6^2)\begin{vmatrix} \frac{\partial C_6}{\partial \xi} & \frac{\partial C_6}{\partial \eta} & 2C_6 \\ \frac{\partial C_9}{\partial \xi} & \frac{\partial C_9}{\partial \eta} & 3C_9 \\ \frac{\partial C_{12}}{\partial \xi} & \frac{\partial C_{12}}{\partial \eta} & 4C_{12} \end{vmatrix}=1296(C_6^2-C_{12})\mathfrak{C}_{12}^2.$$

By the identity $1728\mathfrak{C}_{12}^3=C_{18}^2-C_{12}^3$, we have

$$\mathfrak{C}_{12}=\frac{C_{18}^2}{1728\mathfrak{C}_{12}^2}-\frac{C_{12}^3}{1728\mathfrak{C}_{12}^2}.$$

$$\frac{\partial \mathfrak{C}_{12}}{\partial \xi}=\frac{1}{2592}\frac{C_{18}}{\mathfrak{C}_{12}^2}\frac{\partial C_{18}}{\partial \xi}-\frac{1}{1728}\frac{C_{12}^2}{\mathfrak{C}_{12}^2}\frac{\partial C_{12}}{\partial \xi}.$$

$$\frac{\partial \mathfrak{C}_{12}}{\partial \eta}=\frac{1}{2592}\frac{C_{18}}{\mathfrak{C}_{12}^2}\frac{\partial C_{18}}{\partial \eta}-\frac{1}{1728}\frac{C_{12}^2}{\mathfrak{C}_{12}^2}\frac{\partial C_{12}}{\partial \eta}.$$

Thus,

$$\begin{aligned}&\begin{vmatrix} \frac{\partial C_6}{\partial \xi} & \frac{\partial C_6}{\partial \eta} & 2C_6 \\ \frac{\partial C_9}{\partial \xi} & \frac{\partial C_9}{\partial \eta} & 3C_9 \\ \frac{\partial \mathfrak{C}_{12}}{\partial \xi} & \frac{\partial \mathfrak{C}_{12}}{\partial \eta} & 4\mathfrak{C}_{12} \end{vmatrix}\\ =&\frac{1}{2592}\frac{C_{18}}{\mathfrak{C}_{12}^2}\begin{vmatrix} \frac{\partial C_6}{\partial \xi} & \frac{\partial C_6}{\partial \eta} & 2C_6 \\ \frac{\partial C_9}{\partial \xi} & \frac{\partial C_9}{\partial \eta} & 3C_9 \\ \frac{\partial C_{18}}{\partial \xi} & \frac{\partial C_{18}}{\partial \eta} & 6C_{18} \end{vmatrix}-\frac{1}{1728}\frac{C_{12}^2}{\mathfrak{C}_{12}^2}\begin{vmatrix} \frac{\partial C_6}{\partial \xi} & \frac{\partial C_6}{\partial \eta} & 2C_6 \\ \frac{\partial C_9}{\partial \xi} & \frac{\partial C_9}{\partial \eta} & 3C_9 \\ \frac{\partial C_{12}}{\partial \xi} & \frac{\partial C_{12}}{\partial \eta} & 4C_{12} \end{vmatrix}\\ =&\frac{1}{2}(C_{12}^2-C_6C_{18}).\end{aligned}$$

$$\begin{vmatrix} \frac{\partial C_6}{\partial \xi} & \frac{\partial C_6}{\partial \eta} & 2C_6 \\ \frac{\partial C_{12}}{\partial \xi} & \frac{\partial C_{12}}{\partial \eta} & 4C_{12} \\ \frac{\partial \mathfrak{C}_{12}}{\partial \xi} & \frac{\partial \mathfrak{C}_{12}}{\partial \eta} & 4\mathfrak{C}_{12} \end{vmatrix}=\frac{1}{2592}\frac{C_{18}}{\mathfrak{C}_{12}^2}\begin{vmatrix} \frac{\partial C_6}{\partial \xi} & \frac{\partial C_6}{\partial \eta} & 2C_6 \\ \frac{\partial C_{12}}{\partial \xi} & \frac{\partial C_{12}}{\partial \eta} & 4C_{12} \\ \frac{\partial C_{18}}{\partial \xi} & \frac{\partial C_{18}}{\partial \eta} & 6C_{18} \end{vmatrix}=144C_9C_{18}.$$

$$\begin{vmatrix} \frac{\partial C_9}{\partial \xi} & \frac{\partial C_9}{\partial \eta} & 3C_9 \\ \frac{\partial C_{12}}{\partial \xi} & \frac{\partial C_{12}}{\partial \eta} & 4C_{12} \\ \frac{\partial \mathfrak{C}_{12}}{\partial \xi} & \frac{\partial \mathfrak{C}_{12}}{\partial \eta} & 4\mathfrak{C}_{12} \end{vmatrix} = \frac{1}{2592} \frac{C_{18}}{\mathfrak{C}_{12}^2} \begin{vmatrix} \frac{\partial C_9}{\partial \xi} & \frac{\partial C_9}{\partial \eta} & 3C_9 \\ \frac{\partial C_{12}}{\partial \xi} & \frac{\partial C_{12}}{\partial \eta} & 4C_{12} \\ \frac{\partial C_{18}}{\partial \xi} & \frac{\partial C_{18}}{\partial \eta} & 6C_{18} \end{vmatrix} = \frac{1}{2}(C_6^2 - C_{12})C_{18}.$$

$$\begin{vmatrix} \frac{\partial C_6}{\partial \xi} & \frac{\partial C_6}{\partial \eta} & 2C_6 \\ \frac{\partial C_{18}}{\partial \xi} & \frac{\partial C_{18}}{\partial \eta} & 6C_{18} \\ \frac{\partial \mathfrak{C}_{12}}{\partial \xi} & \frac{\partial \mathfrak{C}_{12}}{\partial \eta} & 4\mathfrak{C}_{12} \end{vmatrix} = \frac{1}{1728} \frac{C_{12}^2}{\mathfrak{C}_{12}^2} \begin{vmatrix} \frac{\partial C_6}{\partial \xi} & \frac{\partial C_6}{\partial \eta} & 2C_6 \\ \frac{\partial C_{12}}{\partial \xi} & \frac{\partial C_{12}}{\partial \eta} & 4C_{12} \\ \frac{\partial C_{18}}{\partial \xi} & \frac{\partial C_{18}}{\partial \eta} & 6C_{18} \end{vmatrix} = 216 C_9 C_{12}^2.$$

$$\begin{vmatrix} \frac{\partial C_9}{\partial \xi} & \frac{\partial C_9}{\partial \eta} & 3C_9 \\ \frac{\partial C_{18}}{\partial \xi} & \frac{\partial C_{18}}{\partial \eta} & 6C_{18} \\ \frac{\partial \mathfrak{C}_{12}}{\partial \xi} & \frac{\partial \mathfrak{C}_{12}}{\partial \eta} & 4\mathfrak{C}_{12} \end{vmatrix} = \frac{1}{1728} \frac{C_{12}^2}{\mathfrak{C}_{12}^2} \begin{vmatrix} \frac{\partial C_9}{\partial \xi} & \frac{\partial C_9}{\partial \eta} & 3C_9 \\ \frac{\partial C_{12}}{\partial \xi} & \frac{\partial C_{12}}{\partial \eta} & 4C_{12} \\ \frac{\partial C_{18}}{\partial \xi} & \frac{\partial C_{18}}{\partial \eta} & 6C_{18} \end{vmatrix} = \frac{3}{4}(C_6^2 - C_{12})C_{12}^2.$$

□

Now, we have

$$\frac{\partial(R_1, R_2)}{\partial(\xi, \eta)} = -2^9 \cdot 3^7 \frac{C_9}{C_6^6} \mathfrak{C}_{12}^2.$$

By (3.15), this implies that

$$\frac{\partial(R_1, R_2)}{\partial(x, y)} = -2^9 \cdot 3^7 \frac{C_9}{C_6^6} \mathfrak{C}_{12}^2 \frac{C}{z_3^3} H(x, y).$$

Hence,

$$-2^9 \cdot 3^7 \cdot C \frac{C_9 \mathfrak{C}_{12}^2}{C_6^6 \frac{\partial(R_1, R_2)}{\partial(x, y)}} H(x, y) = z_3^3.$$

By (4.3) and (4.11), we have

$$\begin{cases} \frac{C_{18}}{C_6^3} = \frac{1}{2}(R_1 + R_2 - 1), \\ \frac{\mathfrak{C}_{12}^3}{C_6^6} = \frac{1}{1728}\left[\frac{1}{4}(R_1 + R_2 - 1)^2 - \frac{1}{27}R_2^3\right]. \end{cases}$$

Note that

$$\sqrt{C_6} \frac{C_9 \mathfrak{C}_{12}^2}{C_6^6} = \left(\frac{C_9^2}{C_6^3}\right)^{\frac{1}{2}} \left(\frac{\mathfrak{C}_{12}^3}{C_6^6}\right)^{\frac{2}{3}} = \frac{1}{144}\left(\frac{R_1}{432}\right)^{\frac{1}{2}} \left[\frac{1}{4}(R_1 + R_2 - 1)^2 - \frac{1}{27}R_2^3\right]^{\frac{2}{3}}.$$

We have

$$\begin{aligned} \sqrt{C_6} z_3^3 = & -2^9 \cdot 3^7 \cdot C \cdot \frac{1}{144}\left(\frac{R_1}{432}\right)^{\frac{1}{2}} \left[\frac{1}{4}(R_1 + R_2 - 1)^2 - \frac{1}{27}R_2^3\right]^{\frac{2}{3}} \\ & \times \left[\frac{\partial(R_1, R_2)}{\partial(x, y)}\right]^{-1} H(x, y). \end{aligned}$$

Hence, we find that

$$\begin{aligned} & z_1^6 + z_2^6 + z_3^6 - 10(z_1^3 z_2^3 + z_2^3 z_3^3 + z_3^3 z_1^3) = C_6(\xi, \eta) \cdot z_3^6 \\ = & 2^6 \cdot 3^7 \cdot C^2 \cdot R_1 \left[\frac{1}{4}(R_1 + R_2 - 1)^2 - \frac{1}{27}R_2^3\right]^{\frac{4}{3}} \left[\frac{\partial(R_1, R_2)}{\partial(x, y)}\right]^{-2} H(x, y)^2. \end{aligned}$$

By

$$\begin{cases} C_9 = \dfrac{1}{\sqrt{432}} \sqrt{R_1} C_6^{\frac{3}{2}}, \\ C_{12} = \dfrac{1}{3} R_2 C_6^2, \\ C_{18} = \dfrac{1}{2}(R_1 + R_2 - 1) C_6^3, \\ \mathfrak{C}_{12} = \dfrac{1}{12} \sqrt[3]{\dfrac{1}{4}(R_1 + R_2 - 1)^2 - \dfrac{1}{27} R_2^3} \cdot C_6^2, \end{cases}$$

we have

$$\begin{aligned} &(z_1^3 - z_2^3)(z_2^3 - z_3^3)(z_3^3 - z_1^3) \\ =&2^7 \cdot 3^9 \cdot C^3 \cdot R_1^2 \left[\frac{1}{4}(R_1 + R_2 - 1)^2 - \frac{1}{27} R_2^3\right]^2 \left[\frac{\partial(R_1, R_2)}{\partial(x, y)}\right]^{-3} H(x, y)^3. \end{aligned}$$

$$\begin{aligned} &(z_1^3 + z_2^3 + z_3^3)[(z_1^3 + z_2^3 + z_3^3)^3 + 216 z_1^3 z_2^3 z_3^3] \\ =&2^{12} \cdot 3^{13} \cdot C^4 \cdot R_1^2 R_2 \left[\frac{1}{4}(R_1 + R_2 - 1)^2 - \frac{1}{27} R_2^3\right]^{\frac{8}{3}} \left[\frac{\partial(R_1, R_2)}{\partial(x, y)}\right]^{-4} H(x, y)^4. \end{aligned}$$

$$\begin{aligned} &z_1 z_2 z_3 [27 z_1^3 z_2^3 z_3^3 - (z_1^3 + z_2^3 + z_3^3)^3] \\ =&2^{10} \cdot 3^{13} \cdot C^4 \cdot R_1^2 \left[\frac{1}{4}(R_1 + R_2 - 1)^2 - \frac{1}{27} R_2^3\right]^3 \left[\frac{\partial(R_1, R_2)}{\partial(x, y)}\right]^{-4} H(x, y)^4. \end{aligned}$$

$$\begin{aligned} &(z_1^3 + z_2^3 + z_3^3)^6 - 540 z_1^3 z_2^3 z_3^3 (z_1^3 + z_2^3 + z_3^3)^3 - 5832 z_1^6 z_2^6 z_3^6 \\ =&2^{17} \cdot 3^{21} \cdot C^6 \cdot R_1^3 (R_1 + R_2 - 1) \left[\frac{1}{4}(R_1 + R_2 - 1)^2 - \frac{1}{27} R_2^3\right]^4 \\ &\times \left[\frac{\partial(R_1, R_2)}{\partial(x, y)}\right]^{-6} H(x, y)^6. \end{aligned}$$

□

Let us consider the particular Hessian polyhedral equations:

$$\begin{cases} 432 \dfrac{C_9^2}{C_6^3} = x, \\ 3 \dfrac{C_{12}}{C_6^2} = y. \end{cases} \tag{4.28}$$

In the projective coordinates, they are given by

$$\begin{cases} 432 \dfrac{C_9(z_1, z_2, z_3)^2}{C_6(z_1, z_2, z_3)^3} = x, \\ 3 \dfrac{C_{12}(z_1, z_2, z_3)}{C_6(z_1, z_2, z_3)^2} = y. \end{cases}$$

Writing explicitly, we have

$$\begin{cases} x = 432\dfrac{(\xi^3-\eta^3)^2(\eta^3-1)^2(1-\xi^3)^2}{[\xi^6+\eta^6+1-10(\xi^3\eta^3+\xi^3+\eta^3)]^3}, \\ y = 3\dfrac{(\xi^3+\eta^3+1)[(\xi^3+\eta^3+1)^3+216\xi^3\eta^3]}{[\xi^6+\eta^6+1-10(\xi^3\eta^3+\xi^3+\eta^3)]^2}. \end{cases}$$

The Hessian polyhedral irrationality (ξ, η) is defined by the equations:

$$\begin{cases} 432\dfrac{C_9^2}{C_6^3} = J_1, \\ 3\dfrac{C_{12}}{C_6^2} = J_2, \end{cases} \tag{4.29}$$

where J_1 and J_2 are J-invariants (see [63]):

$$\begin{cases} J_1 = J_1(\lambda_1, \lambda_2) = \dfrac{\lambda_2^2(\lambda_2-1)^2}{\lambda_1^2(\lambda_1-1)^2(\lambda_1-\lambda_2)^2}, \\ J_2 = J_2(\lambda_1, \lambda_2) = \dfrac{\lambda_1^2(\lambda_1-1)^2}{\lambda_2^2(\lambda_2-1)^2(\lambda_2-\lambda_1)^2}. \end{cases} \tag{4.30}$$

The corresponding Picard curve is

$$y^3 = x(x-1)(x-\lambda_1)(x-\lambda_2). \tag{4.31}$$

Here, $\lambda_1 = \lambda_1(\xi, \eta)$ and $\lambda_2 = \lambda_2(\xi, \eta)$ are Picard modular functions.

Denote by

$$(F, G, H) := \frac{\partial(F, G, H)}{\partial(z_1, z_2, z_3)}. \tag{4.32}$$

Proposition 4.3 (Main Proposition 2). *In the projective coordinates, the following duality formulas hold:*

$$\begin{cases} (C_6, C_9, C_{12}) = -2^5 \cdot 3^4 \cdot \mathfrak{C}_{12}^2, \\ (C_6, C_9, C_{18}) = -2^4 \cdot 3^5 \cdot C_6\mathfrak{C}_{12}^2, \\ (C_6, C_{12}, C_{18}) = 2^9 \cdot 3^7 \cdot C_9\mathfrak{C}_{12}^2, \\ (C_9, C_{12}, C_{18}) = 2^4 \cdot 3^5 \cdot (C_6^2 - C_{12})\mathfrak{C}_{12}^2. \end{cases} \tag{4.33}$$

$$\begin{cases} (C_6, C_9, \mathfrak{C}_{12}) = \dfrac{3}{2}(C_{12}^2 - C_6C_{18}), \\ (C_6, C_{12}, \mathfrak{C}_{12}) = 432C_9C_{18}, \\ (C_9, C_{12}, \mathfrak{C}_{12}) = \dfrac{3}{2}C_{18}(C_6^2 - C_{12}). \end{cases} \tag{4.34}$$

$$\begin{cases} (C_6, C_{18}, \mathfrak{C}_{12}) = 648C_9C_{12}^2, \\ (C_9, C_{18}, \mathfrak{C}_{12}) = \dfrac{9}{4}C_{12}^2(C_6^2 - C_{12}), \\ (C_{12}, C_{18}, \mathfrak{C}_{12}) = 0. \end{cases} \tag{4.35}$$

Proof. Put $X = z_1^3$, $Y = z_2^3$ and $Z = z_3^3$. Then

$$\begin{cases} \frac{\partial C_6}{\partial z_1} = 6z_1^2(X - 5Y - 5Z), \\ \frac{\partial C_6}{\partial z_2} = 6z_2^2(Y - 5Z - 5X), \\ \frac{\partial C_6}{\partial z_3} = 6z_3^2(Z - 5X - 5Y). \end{cases}$$

$$\begin{cases} \frac{\partial C_9}{\partial z_1} = 3z_1^2(Y - Z)(Y + Z - 2X), \\ \frac{\partial C_9}{\partial z_2} = 3z_2^2(Z - X)(Z + X - 2Y), \\ \frac{\partial C_9}{\partial z_3} = 3z_3^2(X - Y)(X + Y - 2Z). \end{cases}$$

$$\begin{cases} \frac{\partial C_{12}}{\partial z_1} = 12z_1^2[(X + Y + Z)^3 + 54YZ(2X + Y + Z)], \\ \frac{\partial C_{12}}{\partial z_2} = 12z_2^2[(X + Y + Z)^3 + 54ZX(2Y + Z + X)], \\ \frac{\partial C_{12}}{\partial z_3} = 12z_3^2[(X + Y + Z)^3 + 54XY(2Z + X + Y)]. \end{cases}$$

We have

$$\frac{\partial(C_6, C_9, C_{12})}{\partial(z_1, z_2, z_3)} = 3 \cdot 6 \cdot 12 \cdot z_1^2 z_2^2 z_3^2 \cdot \Delta.$$

Here,

$$\begin{aligned} \Delta = &[(X + Y + Z)^3 + 54YZ(2X + Y + Z)][-(X + Y + Z)^3 \\ &- 9X(X + Y + Z)^2 + 27X^2(Y + Z) + 27X(Y^2 + Z^2)] \\ &+ [(X + Y + Z)^3 + 54ZX(2Y + Z + X)][-(X + Y + Z)^3 \\ &- 9Y(X + Y + Z)^2 + 27Y^2(Z + X) + 27Y(Z^2 + X^2)] \\ &+ [(X + Y + Z)^3 + 54XY(2Z + X + Y)][-(X + Y + Z)^3 \\ &- 9Z(X + Y + Z)^2 + 27Z^2(X + Y) + 27Z(X^2 + Y^2)]. \end{aligned}$$

In fact,

$$\begin{aligned}\Delta = & -3(X+Y+Z)^6 \\ & -54(X+Y+Z)^3[YZ(2X+Y+Z)+ZX(2Y+Z+X) \\ & \qquad +XY(2Z+X+Y)] \\ & -9(X+Y+Z)^5\cdot(X+Y+Z) \\ & -486XYZ(X+Y+Z)^2[(2X+Y+Z)+(2Y+Z+X) \\ & \qquad +(2Z+X+Y)] \\ & +27(X+Y+Z)^3[X^2(Y+Z)+Y^2(Z+X)+Z^2(X+Y)] \\ & +1458XYZ[X(Y+Z)(2X+Y+Z)+Y(Z+X)(2Y+Z+X) \\ & \qquad +Z(X+Y)(2Z+X+Y)] \\ & +27(X+Y+Z)^3[X(Y^2+Z^2)+Y(Z^2+X^2)+Z(X^2+Y^2)] \\ & +1458XYZ[(2X+Y+Z)(Y^2+Z^2)+(2Y+Z+X)(Z^2+X^2) \\ & \qquad +(2Z+X+Y)(X^2+Y^2)].\end{aligned}$$

Hence,

$$\begin{aligned}\Delta = & -3(X+Y+Z)^6 \\ & -54(X+Y+Z)^3[(X+Y+Z)(XY+YZ+ZX)+3XYZ] \\ & -9(X+Y+Z)^6 \\ & -486XYZ(X+Y+Z)^2\cdot 4(X+Y+Z) \\ & +27(X+Y+Z)^3[(X+Y+Z)(XY+YZ+ZX)-3XYZ] \\ & +1458XYZ[3(X+Y+Z)(XY+YZ+ZX)-3XYZ] \\ & +27(X+Y+Z)^3[(X+Y+Z)(XY+YZ+ZX)-3XYZ] \\ & +1458XYZ[2(X+Y+Z)^3-3(X+Y+Z)(XY+YZ+ZX) \\ & \qquad -3XYZ].\end{aligned}$$

Put

$$\sigma_1 = X+Y+Z, \quad \sigma_2 = XY+YZ+ZX, \quad \sigma_3 = XYZ.$$

Then

$$\Delta = -12\sigma_1^6 + 648\sigma_1^3\sigma_3 - 8748\sigma_3^2 = -12(27\sigma_3 - \sigma_1^3)^2.$$

Thus,

$$\frac{\partial(C_6, C_9, C_{12})}{\partial(z_1, z_2, z_3)} = -3\cdot 6\cdot 12^2\cdot [z_1z_2z_3(27\sigma_3 - \sigma_1^3)]^2 = -2^5\cdot 3^4\cdot \mathfrak{C}_{12}^2.$$

Note that

$$C_{18} = \frac{1}{2}(432C_9^2 - C_6^3 + 3C_6C_{12}).$$

This gives that

$$\frac{\partial C_{18}}{\partial z_j}=\frac{3}{2}(C_{12}-C_6^2)\frac{\partial C_6}{\partial z_j}+432C_9\frac{\partial C_9}{\partial z_j}+\frac{3}{2}C_6\frac{\partial C_{12}}{\partial z_j}, \quad (j=1,2,3).$$

Therefore,

$$\frac{\partial(C_6,C_9,C_{18})}{\partial(z_1,z_2,z_3)}=\frac{3}{2}C_6\frac{\partial(C_6,C_9,C_{12})}{\partial(z_1,z_2,z_3)}=-2^4\cdot 3^5\cdot C_6\mathfrak{C}_{12}^2.$$

$$\frac{\partial(C_6,C_{12},C_{18})}{\partial(z_1,z_2,z_3)}=-432C_9\frac{\partial(C_6,C_9,C_{12})}{\partial(z_1,z_2,z_3)}=2^9\cdot 3^7\cdot C_9\mathfrak{C}_{12}^2.$$

$$\frac{\partial(C_9,C_{12},C_{18})}{\partial(z_1,z_2,z_3)}=\frac{3}{2}(C_{12}-C_6^2)\frac{\partial(C_6,C_9,C_{12})}{\partial(z_1,z_2,z_3)}=-2^4\cdot 3^5\cdot(C_{12}-C_6^2)\mathfrak{C}_{12}^2.$$

By

$$1728\mathfrak{C}_{12}^3=C_{18}^2-C_{12}^3,$$

we find that

$$3\cdot 1728\mathfrak{C}_{12}^2\frac{\partial\mathfrak{C}_{12}}{\partial z_j}=2C_{18}\frac{\partial C_{18}}{\partial z_j}-3C_{12}^2\frac{\partial C_{12}}{\partial z_j}, \quad (j=1,2,3).$$

This gives the other identities. □

Corollary 4.4. *The invariants C_6, C_9, C_{12}, $\mathfrak{C}_{12}^2$ and C_{18} can be represented by:*

$$\mathfrak{C}_{12}^2=-\frac{1}{2592}(C_6,C_9,C_{12}). \tag{4.36}$$

$$C_6=\frac{2}{3}\frac{(C_6,C_9,C_{18})}{(C_6,C_9,C_{12})}. \tag{4.37}$$

$$C_9=-\frac{1}{432}\frac{(C_6,C_{12},C_{18})}{(C_6,C_9,C_{12})}. \tag{4.38}$$

$$C_{12}=\frac{2}{3}\frac{(C_9,C_{12},C_{18})}{(C_6,C_9,C_{12})}+\frac{4}{9}\frac{(C_6,C_9,C_{18})^2}{(C_6,C_9,C_{12})^2}. \tag{4.39}$$

$$C_{18}=-\frac{(C_6,C_9,C_{12})(C_6,C_{12},\mathfrak{C}_{12})}{(C_6,C_{12},C_{18})}. \tag{4.40}$$

Proof. We find that

$$\frac{(C_6,C_9,C_{18})}{(C_6,C_9,C_{12})}=\frac{3}{2}C_6.$$

$$\frac{(C_6,C_{12},C_{18})}{(C_6,C_9,C_{12})}=-432C_9.$$

$$\frac{(C_9,C_{12},C_{18})}{(C_6,C_9,C_{12})}=\frac{3}{2}(C_{12}-C_6^2).$$

$$C_{18} = \frac{(C_6, C_{12}, \mathfrak{C}_{12})}{432C_9}.$$

□

Proposition 4.5. *The following algebraic equations of the sixth degree hold:*

$$\begin{aligned}
&27(C_6, C_9, \mathfrak{C}_{12})(C_6, C_9, C_{12})^4(C_6, C_{12}, C_{18}) \\
=&18(C_9, C_{12}, C_{18})^2(C_6, C_9, C_{12})^2(C_6, C_{12}, C_{18}) \\
&+ 8(C_6, C_9, C_{18})^4(C_6, C_{12}, C_{18}) \\
&+ 24(C_9, C_{12}, C_{18})(C_6, C_9, C_{18})^2(C_6, C_9, C_{12})(C_6, C_{12}, C_{18}) \\
&+ 27(C_6, C_9, C_{18})(C_6, C_{12}, \mathfrak{C}_{12})(C_6, C_9, C_{12})^4.
\end{aligned} \tag{4.41}$$

$$\begin{aligned}
&27(C_6, C_{18}, \mathfrak{C}_{12})(C_6, C_9, C_{12})^5 \\
&+8(C_6, C_{12}, C_{18})(C_6, C_9, C_{18})^4 \\
&+18(C_6, C_{12}, C_{18})(C_9, C_{12}, C_{18})^2(C_6, C_9, C_{12})^2 \\
&+24(C_6, C_{12}, C_{18})(C_9, C_{12}, C_{18})(C_6, C_9, C_{18})^2(C_6, C_9, C_{12}) = 0.
\end{aligned} \tag{4.42}$$

$$\begin{aligned}
&27(C_9, C_{18}, \mathfrak{C}_{12})(C_6, C_9, C_{12})^5 \\
&+18(C_9, C_{12}, C_{18})^3(C_6, C_9, C_{12})^2 \\
&+8(C_6, C_9, C_{18})^4(C_9, C_{12}, C_{18}) \\
&+24(C_9, C_{12}, C_{18})^2(C_6, C_9, C_{18})^2(C_6, C_9, C_{12}) = 0.
\end{aligned} \tag{4.43}$$

Proof. They are obtained from the identities:

$$(C_6, C_9, \mathfrak{C}_{12}) = \frac{3}{2}(C_{12}^2 - C_6 C_{18}).$$

$$(C_6, C_{18}, \mathfrak{C}_{12}) = 648 C_9 C_{12}^2.$$

$$(C_9, C_{18}, \mathfrak{C}_{12}) = \frac{9}{4} C_{12}^2 (C_6^2 - C_{12}).$$

□

Proposition 4.6. *The invariants C_6, C_9, C_{12}, $\mathfrak{C}_{12}$ and C_{18} satisfy the following duality formula:*

$$(C_9, C_{12}, \mathfrak{C}_{12})(C_6, C_{12}, C_{18}) = (C_6, C_{12}, \mathfrak{C}_{12})(C_9, C_{12}, C_{18}). \tag{4.44}$$

Proof. It is obtained from the equation

$$(C_9, C_{12}, \mathfrak{C}_{12}) = \frac{3}{2} C_{18}(C_6^2 - C_{12}).$$

□

Put

$$\begin{cases}
Y_1 = (C_6, C_{12}, \mathfrak{C}_{12}), \\
Y_2 = (C_9, C_{12}, \mathfrak{C}_{12}), \\
Y_3 = (C_6, C_{12}, C_{18}), \\
Y_4 = (C_9, C_{12}, C_{18}).
\end{cases} \tag{4.45}$$

$$\begin{cases} X_1 = (C_6, C_9, C_{12}), \\ X_2 = (C_6, C_9, \mathfrak{C}_{12}), \\ X_3 = (C_6, C_9, C_{18}), \\ X_4 = (C_6, C_{18}, \mathfrak{C}_{12}), \\ X_5 = (C_9, C_{18}, \mathfrak{C}_{12}). \end{cases} \tag{4.46}$$

Then Proposition 4.5 and Proposition 4.6 give that

$$18X_1^2Y_3Y_4^2 + 24X_1X_3^2Y_3Y_4 + 27X_1^4X_3Y_1 + (8X_3^4 - 27X_1^4X_2)Y_3 = 0, \tag{4.47}$$

$$18X_1^2Y_3Y_4^2 + 24X_1X_3^2Y_3Y_4 + 8X_3^4Y_3 + 27X_1^5X_4 = 0, \tag{4.48}$$

$$18X_1^2Y_4^3 + 24X_1X_3^2Y_4^2 + 8X_3^4Y_4 + 27X_1^5X_5 = 0, \tag{4.49}$$

$$Y_2Y_3 - Y_1Y_4 = 0. \tag{4.50}$$

(4.47) – (4.48) gives that

$$X_3Y_1 - X_2Y_3 = X_1X_4. \tag{4.51}$$

(4.47) $\times Y_4$ – (4.49) $\times Y_3$ gives that

$$X_3Y_1Y_4 - X_2Y_3Y_4 = X_1X_5Y_3. \tag{4.52}$$

By (4.51), we have

$$X_4Y_4 = X_5Y_3. \tag{4.53}$$

Thus, we get the following theorem:

Theorem 4.7 (Main Theorem 3). *The functions X_1, X_2, X_3, X_4, X_5 and Y_1, Y_2, Y_3, Y_4 satisfy the system of algebraic equations of the sixth degree:*

$$\begin{cases} 18X_1^2Y_4^3 + 24X_1X_3^2Y_4^2 + 8X_3^4Y_4 + 27X_1^5X_5 = 0, \\ Y_2Y_3 = Y_1Y_4, \\ X_4Y_4 = X_5Y_3, \\ X_3Y_1 - X_2Y_3 = X_1X_4. \end{cases} \tag{4.54}$$

The first equation in (4.52) is an algebraic hypersurface of the sixth degree in $\mathbb{C}^4$. The second and the third equations are the algebraic hypersurfaces of the second degree in $\mathbb{C}^4$. The last equation is an algebraic hypersurface of the second degree in $\mathbb{C}^6$.

Let

$$\mathcal{A} = \{(C_6, C_9, C_{12}, \mathfrak{C}_{12}, C_{18}) \in \mathbb{C}^5\} \tag{4.55}$$

be the configuration space of the Hessian invariants. Let

$$\mathcal{B} = \{(X_1, X_2, X_3, X_4, X_5, Y_1, Y_2, Y_3, Y_4) \in \mathbb{C}^9\} \tag{4.56}$$

be the dual configuration space of $\mathcal{A}$. Then

$$\dim\mathcal{A} = 3, \quad \dim\mathcal{B} = 5. \tag{4.57}$$

Let us consider the Hessian polyhedral equations:

$$Z_1 = 432\frac{C_9^2}{C_6^3}, \quad Z_2 = 3\frac{C_{12}}{C_6^2}. \tag{4.58}$$

Note that the relation

$$432C_9^2 = C_6^3 - 3C_6C_{12} + 2C_{18}$$

implies that

$$Z_1 + Z_2 - 1 = 2\frac{C_{18}}{C_6^3}. \tag{4.59}$$

The relation

$$1728\mathfrak{C}_{12}^3 = C_{18}^2 - C_{12}^3$$

gives that

$$12\frac{\mathfrak{C}_{12}}{C_6^2} = \sqrt[3]{\frac{1}{4}(Z_1 + Z_2 - 1)^2 - \frac{1}{27}Z_2^3}. \tag{4.60}$$

We find that

$$\begin{aligned}
&Z_1^2 : Z_2^3 : (Z_1 + Z_2 - 1)^2 : [27(Z_1 + Z_2 - 1)^2 - 4Z_2^3] : 1\\
=&432^2C_9^4 : 27C_{12}^3 : 4C_{18}^2 : 432^2\mathfrak{C}_{12}^3 : C_6^6\\
=&432^2(\xi^3 - \eta^3)^4(\eta^3 - 1)^4(1 - \xi^3)^4\\
&:27(\xi^3 + \eta^3 + 1)^3[(\xi^3 + \eta^3 + 1)^3 + 216\xi^3\eta^3]^3\\
&:4[(\xi^3 + \eta^3 + 1)^6 - 540\xi^3\eta^3(\xi^3 + \eta^3 + 1)^3 - 5832\xi^6\eta^6]^2\\
&:432^2\xi^3\eta^3[27\xi^3\eta^3 - (\xi^3 + \eta^3 + 1)^3]^3\\
&:[\xi^6 + \eta^6 + 1 - 10(\xi^3\eta^3 + \xi^3 + \eta^3)]^6.
\end{aligned} \tag{4.61}$$

In fact,

$$Z_1 = \frac{1}{128}\frac{X_1Y_3^2}{X_3^3}, \quad Z_2 = 3 + \frac{9}{2}\frac{X_1Y_4}{X_3^2}. \tag{4.62}$$

Let

$$F_1 = C_{18}, \quad F_2 = \mathfrak{C}_{12}, \quad F_3 = C_{12}, \quad F_4 = C_9, \quad F_5 = C_6 \tag{4.63}$$

and

$$W = \frac{F_1F_5}{F_3F_4}. \tag{4.64}$$

Then

$$F_1 = \frac{Z_1^3Z_2^6W^6}{2^7 \cdot 3^{15}(Z_1 + Z_2 - 1)^5}, \tag{4.65}$$

$$F_3 = \frac{Z_1^2Z_2^5W^4}{2^4 \cdot 3^{11}(Z_1 + Z_2 - 1)^4}, \tag{4.66}$$

$$F_4 = \frac{Z_1^2Z_2^3W^3}{2^5 \cdot 3^9(Z_1 + Z_2 - 1)^3}, \tag{4.67}$$

$$F_5 = \frac{Z_1 Z_2^2 W^2}{2^2 \cdot 3^5 (Z_1 + Z_2 - 1)^2}, \tag{4.68}$$

$$F_2 = \frac{Z_1^2 Z_2^4 W^4}{2^6 \cdot 3^{11} (Z_1 + Z_2 - 1)^4} \sqrt[3]{\frac{1}{4}(Z_1 + Z_2 - 1)^2 - \frac{1}{27} Z_2^3}. \tag{4.69}$$

Put

$$\tau_1 = \xi^3 + \eta^3, \quad \tau_2 = \xi\eta. \tag{4.70}$$

Then

$$\begin{cases} C_6 = \tau_1^2 - 12\tau_2^3 - 10\tau_1 + 1, \\ C_{12} = (\tau_1 + 1)[(\tau_1 + 1)^3 + 216\tau_2^3], \\ \mathfrak{C}_{12} = \tau_2[27\tau_2^3 - (\tau_1 + 1)^3], \\ C_{18} = (\tau_1 + 1)^6 - 540\tau_2^3(\tau_1 + 1)^3 - 5832\tau_2^6, \\ C_9^2 = (\tau_1^2 - 4\tau_2^3)(\tau_2^3 - \tau_1 + 1)^2. \end{cases} \tag{4.71}$$

Thus,

$$\begin{aligned} &Z_1^2 : Z_2^3 : (Z_1 + Z_2 - 1)^2 : [27(Z_1 + Z_2 - 1)^2 - 4Z_2^3] : 1 \\ =&432^2(\tau_1^2 - 4\tau_2^3)^2(\tau_2^3 - \tau_1 + 1)^4 \\ &:27(\tau_1 + 1)^3[(\tau_1 + 1)^3 + 216\tau_2^3]^3 \\ &:4[(\tau_1 + 1)^6 - 540\tau_2^3(\tau_1 + 1)^3 - 5832\tau_2^6]^2 \\ &:432^2\tau_2^3[27\tau_2^3 - (\tau_1 + 1)^3]^3 \\ &:(\tau_1^2 - 12\tau_2^3 - 10\tau_1 + 1)^6. \end{aligned} \tag{4.72}$$

5. Invariant theory for the system of algebraic equations

In the affine coordinates, we have

$$\begin{cases} A(\xi, \eta) = \left(\frac{\eta}{\xi}, \frac{1}{\xi}\right), \\ B(\xi, \eta) = \left(\frac{\xi}{\eta}, \frac{1}{\eta}\right), \\ C(\xi, \eta) = (\omega\xi, \omega^2\eta), \\ D(\xi, \eta) = (\omega^2\xi, \eta), \\ E(\xi, \eta) = \left(\frac{\xi + \eta + 1}{\xi + \omega^2\eta + \omega}, \frac{\xi + \omega\eta + \omega^2}{\xi + \omega^2\eta + \omega}\right). \end{cases} \tag{5.1}$$

Note that $E^2 = -B$, $B^2 = I$, $A^3 = I$. We find that

$$\langle A, B \rangle = \{A, A^2, B, AB, BA, I\}. \tag{5.2}$$

Here,

$$A^2 = \begin{pmatrix} 0 & 0 & 1 \\ 1 & 0 & 0 \\ 0 & 1 & 0 \end{pmatrix}, \quad AB = \begin{pmatrix} 0 & 0 & 1 \\ 0 & 1 & 0 \\ 1 & 0 & 0 \end{pmatrix}, \quad BA = \begin{pmatrix} 0 & 1 & 0 \\ 1 & 0 & 0 \\ 0 & 0 & 1 \end{pmatrix}. \tag{5.3}$$

There are three involutions:

$$B^2 = (AB)^2 = (BA)^2 = I. \tag{5.4}$$

(1) $B(\xi, \eta) = (\frac{\xi}{\eta}, \frac{1}{\eta})$.
Let $B(\xi, \eta) = (\xi, \eta)$, we find that

$$\xi = \xi\eta \quad \text{and} \quad \eta^2 = 1.$$

Hence,

$$z_1(z_2 - z_3) = 0 \quad \text{and} \quad (z_2 + z_3)(z_2 - z_3) = 0.$$

The common factor is $z_2 - z_3 = 0$.

(2) $AB(\xi, \eta) = (\frac{1}{\xi}, \frac{\eta}{\xi})$.
Let $AB(\xi, \eta) = (\xi, \eta)$, we find that

$$\xi^2 = 1 \quad \text{and} \quad \eta = \xi\eta.$$

Hence,

$$(z_3 + z_1)(z_3 - z_1) = 0 \quad \text{and} \quad z_2(z_3 - z_1) = 0.$$

The common factor is $z_3 - z_1 = 0$.

(3) $BA(\xi, \eta) = (\eta, \xi)$.
Let $BA(\xi, \eta) = (\xi, \eta)$, we find that

$$\eta = \xi \quad \text{and} \quad \xi = \eta.$$

Hence,

$$z_2 - z_1 = 0 \quad \text{and} \quad z_1 - z_2 = 0.$$

The common factor is $z_1 - z_2 = 0$.

Put

$$G(z_1, z_2, z_3) := (z_1 - z_2)(z_2 - z_3)(z_3 - z_1). \tag{5.5}$$

The Hessian of G is zero.

Note that

$$E(z_1, z_2, z_3) = \frac{1}{\sqrt{-3}}(z_1 + z_2 + z_3, z_1 + \omega z_2 + \omega^2 z_3, z_1 + \omega^2 z_2 + \omega z_3). \tag{5.6}$$

According to Maschke [52], put

$$\varphi = z_1 z_2 z_3, \quad \psi = z_1^3 + z_2^3 + z_3^3, \quad \chi = z_1^3 z_2^3 + z_2^3 z_3^3 + z_3^3 z_1^3. \tag{5.7}$$

We have

$$\psi(E(z_1, z_2, z_3)) = \frac{-1}{\sqrt{-3}}(\psi + 6\varphi). \tag{5.8}$$

$$\varphi(E(z_1, z_2, z_3)) = \frac{-1}{3\sqrt{-3}}(\psi - 3\varphi). \tag{5.9}$$

Let

$$H(z_1, z_2, z_3) := \frac{1}{3}[(z_1+z_2+z_3)^3+(z_1+\omega z_2+\omega^2 z_3)^3+(z_1+\omega^2 z_2+\omega z_3)^3]. \tag{5.10}$$

$$K(z_1, z_2, z_3) := (z_1 + z_2 + z_3)(z_1 + \omega z_2 + \omega^2 z_3)(z_1 + \omega^2 z_2 + \omega z_3). \tag{5.11}$$

Then

$$H = \psi + 6\varphi, \quad K = \psi - 3\varphi. \tag{5.12}$$

The Hessian of H:

$$\text{Hessian}(H) = \begin{vmatrix} \frac{\partial^2 H}{\partial z_1^2} & \frac{\partial^2 H}{\partial z_1 \partial z_2} & \frac{\partial^2 H}{\partial z_1 \partial z_3} \\ \frac{\partial^2 H}{\partial z_1 \partial z_2} & \frac{\partial^2 H}{\partial z_2^2} & \frac{\partial^2 H}{\partial z_2 \partial z_3} \\ \frac{\partial^2 H}{\partial z_1 \partial z_3} & \frac{\partial^2 H}{\partial z_2 \partial z_3} & \frac{\partial^2 H}{\partial z_3^2} \end{vmatrix} = -216K. \tag{5.13}$$

The Hessian of K:

$$\text{Hessian}(K) = \begin{vmatrix} \frac{\partial^2 K}{\partial z_1^2} & \frac{\partial^2 K}{\partial z_1 \partial z_2} & \frac{\partial^2 K}{\partial z_1 \partial z_3} \\ \frac{\partial^2 K}{\partial z_1 \partial z_2} & \frac{\partial^2 K}{\partial z_2^2} & \frac{\partial^2 K}{\partial z_2 \partial z_3} \\ \frac{\partial^2 K}{\partial z_1 \partial z_3} & \frac{\partial^2 K}{\partial z_2 \partial z_3} & \frac{\partial^2 K}{\partial z_3^2} \end{vmatrix} = -54K. \tag{5.14}$$

In fact,

$$\begin{cases} C_6 = \psi^2 - 12\chi, \\ C_{12} = \psi(\psi^3 + 216\varphi^3), \\ \mathfrak{C}_{12} = \varphi(27\varphi^3 - \psi^3), \\ C_{18} = \psi^6 - 540\varphi^3\psi^3 - 5832\varphi^6. \end{cases} \tag{5.15}$$

Theorem 5.1 (Main Theorem 4). *The invariants G, H and K satisfy the following algebraic equations, which are the form-theoretic resolvents (algebraic resolvents) of G, H, K:*

$$\begin{cases} 4G^3 + H^2 G - C_6 G - 4C_9 = 0, \\ H(H^3 + 8K^3) - 9C_{12} = 0, \\ K(K^3 - H^3) - 27\mathfrak{C}_{12} = 0. \end{cases} \tag{5.16}$$

Proof. We have

$$\begin{aligned} \frac{C_9}{G} =&(z_1^2 + z_1 z_2 + z_2^2)(z_2^2 + z_2 z_3 + z_3^2)(z_3^2 + z_3 z_1 + z_1^2) \\ =&3z_1^2 z_2^2 z_3^2 + z_1^4 z_2^2 + z_2^4 z_3^2 + z_3^4 z_1^2 + z_1^2 z_2^4 + z_2^2 z_3^4 + z_3^2 z_1^4 \\ &+ z_1^3 z_2^3 + z_2^3 z_3^3 + z_3^3 z_1^3 + z_1^4 z_2 z_3 + z_2^4 z_3 z_1 + z_3^4 z_1 z_2 \\ &+ 2z_1 z_2 z_3(z_1^2 z_2 + z_2^2 z_3 + z_3^2 z_1 + z_1 z_2^2 + z_2 z_3^2 + z_3 z_1^2). \end{aligned}$$

Set

$$\sigma_1 = z_1 + z_2 + z_3, \quad \sigma_2 = z_1z_2 + z_2z_3 + z_3z_1.$$

We have

$$z_1^2z_2 + z_2^2z_3 + z_3^2z_1 + z_1z_2^2 + z_2z_3^2 + z_3z_1^2 = \sigma_1\sigma_2 - 3\varphi.$$

$$z_1^4z_2^2 + z_2^4z_3^2 + z_3^4z_1^2 + z_1^2z_2^4 + z_2^2z_3^4 + z_3^2z_1^4 = (\sigma_1^2 - 2\sigma_2)(\sigma_2^2 - 2\varphi\sigma_1) - 3\varphi^2.$$

Thus,

$$\frac{C_9}{G} = 3\varphi^2 + \chi + \varphi\psi + (\sigma_1^2 - 2\sigma_2)(\sigma_2^2 - 2\varphi\sigma_1) - 3\varphi^2 + 2\varphi(\sigma_1\sigma_2 - 3\varphi).$$

Note that

$$\psi = \sigma_1(\sigma_1^2 - 3\sigma_2) + 3\varphi, \quad \chi = \sigma_2(\sigma_2^2 - 3\varphi\sigma_1) + 3\varphi^2.$$

This gives that

$$(\sigma_1^2 - 2\sigma_2)(\sigma_2^2 - 2\varphi\sigma_1) = -2\varphi(\psi - 3\varphi) - 2\varphi\sigma_1\sigma_2 + \sigma_2^2(\sigma_1^2 - 2\sigma_2).$$

Hence,

$$\frac{C_9}{G} = \chi - \varphi\psi + \sigma_2^2(\sigma_1^2 - 2\sigma_2).$$

In fact,

$$\sigma_2^3 = \chi - 3\varphi^2 + 3\varphi\sigma_1\sigma_2.$$

This implies that

$$\begin{aligned}\frac{C_9}{G} =& \chi - \varphi\psi + \sigma_1^2\sigma_2^2 - 2\sigma_2^3 \\ =& -\chi - \varphi\psi + 6\varphi^2 + \sigma_1^2\sigma_2^2 - 6\varphi\sigma_1\sigma_2.\end{aligned}$$

We have

$$G = (z_1z_2^2 + z_2z_3^2 + z_3z_1^2) - (z_1^2z_2 + z_2^2z_3 + z_3^2z_1).$$

Hence,

$$\begin{aligned}G^2 =& (z_1z_2^2 + z_2z_3^2 + z_3z_1^2 + z_1^2z_2 + z_2^2z_3 + z_3^2z_1)^2 \\ & - 4(z_1z_2^2 + z_2z_3^2 + z_3z_1^2)(z_1^2z_2 + z_2^2z_3 + z_3^2z_1) \\ =& (\sigma_1\sigma_2 - 3\varphi)^2 - 4(\chi + 3\varphi^2 + \varphi\psi) \\ =& \sigma_1^2\sigma_2^2 - 6\varphi\sigma_1\sigma_2 - 3\varphi^2 - 4\chi - 4\varphi\psi.\end{aligned}$$

Therefore,

$$\frac{C_9}{G} - G^2 = 3(\chi + \varphi\psi + 3\varphi^2).$$

We find that

$$4\left(\frac{C_9}{G} - G^2\right) + C_6 = (\psi + 6\varphi)^2 = H^2.$$

Hence,

$$4G^3 + H^2G - C_6G - 4C_9 = 0.$$

By

$$H = \psi + 6\varphi, \quad K = \psi - 3\varphi,$$

we have

$$\varphi = \frac{1}{9}(H - K), \quad \psi = \frac{1}{3}(H + 2K).$$

The relations

$$C_{12} = \psi(\psi^3 + 216\varphi^3), \quad \mathfrak{C}_{12} = \varphi(27\varphi^3 - \psi^3)$$

give that

$$H(H^3 + 8K^3) = 9C_{12}, \quad K(K^3 - H^3) = 27\mathfrak{C}_{12}.$$

□

Let

$$r_1 = \frac{G^2}{C_6}, \quad r_2 = \frac{H^2}{C_6}, \quad r_3 = \frac{K^2}{C_6}. \tag{5.17}$$

Proposition 5.2. *The following identities hold:*

$$\begin{cases} Z_1 = 27r_1(4r_1 + r_2 - 1)^2, \\ Z_2 = \frac{1}{3}r_2^2 + \frac{8}{3}r_3 \cdot \sqrt{r_2 r_3}, \\ \sqrt[3]{\frac{1}{4}(Z_1 + Z_2 - 1)^2 - \frac{1}{27}Z_2^3} = \frac{4}{9}(r_3^2 - r_2 \cdot \sqrt{r_2 r_3}). \end{cases} \tag{5.18}$$

Proof. By

$$4G^3 + H^2G - C_6G - 4C_9 = 0,$$

we have

$$C_9 = G\left(G^2 + \frac{1}{4}H^2 - \frac{1}{4}C_6\right).$$

Thus

$$\frac{C_9^2}{C_6^3} = \frac{G^2}{C_6}\left(\frac{G^2}{C_6} + \frac{1}{4}\frac{H^2}{C_6} - \frac{1}{4}\right)^2.$$

This gives that

$$\frac{1}{432}Z_1 = r_1\left(r_1 + \frac{1}{4}r_2 - \frac{1}{4}\right)^2.$$

Hence,

$$Z_1 = 27r_1(4r_1 + r_2 - 1)^2.$$

By

$$C_{12} = \frac{1}{9}H(H^3 + 8K^3),$$

we have

$$\frac{C_{12}}{C_6^2} = \frac{1}{9}\left(\frac{H^2}{C_6}\right)^{\frac{1}{2}}\left[\left(\frac{H^2}{C_6}\right)^{\frac{3}{2}} + 8\left(\frac{K^2}{C_6}\right)^{\frac{3}{2}}\right].$$

This gives that

$$Z_2 = \frac{1}{3}r_2^2 + \frac{8}{3}r_3 \cdot \sqrt{r_2 r_3}.$$

By

$$\mathfrak{C}_{12} = \frac{1}{27}K(K^3 - H^3),$$

we have

$$\frac{\mathfrak{C}_{12}}{C_6^2} = \frac{1}{27}\left(\frac{K^2}{C_6}\right)^{\frac{1}{2}}\left[\left(\frac{K^2}{C_6}\right)^{\frac{3}{2}} - \left(\frac{H^2}{C_6}\right)^{\frac{3}{2}}\right].$$

This gives that

$$\sqrt[3]{\frac{1}{4}(Z_1 + Z_2 - 1)^2 - \frac{1}{27}Z_2^3} = \frac{4}{9}(r_3^2 - r_2 \cdot \sqrt{r_2 r_3}).$$

□

This gives the following theorem:

Theorem 5.3 (Main Theorem 5). *The function-theoretic resolvents (analytic resolvents) of Z_1 and Z_2 are given by:*

$$\begin{aligned} &Z_1^2 : Z_2^3 : (Z_1 + Z_2 - 1)^2 : [27(Z_1 + Z_2 - 1)^2 - 4Z_2^3] : 1 \\ =&729^2 r_1^2 (4r_1 + r_2 - 1)^4 : 27(r_2^2 + 8r_3 \cdot \sqrt{r_2 r_3})^3 \\ &:[256(r_3^2 - r_2 \cdot \sqrt{r_2 r_3})^3 + 4(r_2^2 + 8r_3 \cdot \sqrt{r_2 r_3})^3] \\ &:6912(r_3^2 - r_2 \cdot \sqrt{r_2 r_3})^3 : 729. \end{aligned} \tag{5.19}$$

For the Hessian polyhedral equations, the corresponding J-invariants (J_1, J_2) satisfy:

$$\begin{aligned} &J_1^2 : J_2^3 : (J_1 + J_2 - 1)^2 : [27(J_1 + J_2 - 1)^2 - 4J_2^3] : 1 \\ =&729^2 v_1^2 (4v_1 + v_2 - 1)^4 : 27(v_2^2 + 8v_3 \cdot \sqrt{v_2 v_3})^3 \\ &:[256(v_3^2 - v_2 \cdot \sqrt{v_2 v_3})^3 + 4(v_2^2 + 8v_3 \cdot \sqrt{v_2 v_3})^3] \\ &:6912(v_3^2 - v_2 \cdot \sqrt{v_2 v_3})^3 : 729, \end{aligned} \tag{5.20}$$

where

$$v_1 = v_1(\omega_1, \omega_2), \quad v_2 = v_2(\omega_1, \omega_2), \quad v_3 = v_3(\omega_1, \omega_2)$$

are three Picard modular functions with $(\omega_1, \omega_2) \in \mathfrak{S}_2 = \{(z_1, z_2) \in \mathbb{C}^2 : z_1 + \overline{z_1} - z_2\overline{z_2} > 0\}$.

Put

$$u = \frac{C_6 G}{C_9}, \quad v = \frac{C_9 H}{C_{12}}, \quad w = \frac{C_9 K}{\mathfrak{C}_{12}}. \tag{5.21}$$

We will denote the next three equations as the resolvents of the u, v and w's.

Theorem 5.4 (Main Theorem 6). *The functions u, v and w satisfy the following algebraic equations, which are the analytic and algebraic resolvents of u, v and w:*

$$\begin{cases} Z_1^2u^3 + 5184Z_2^2uv^2 - 108Z_1u - 432Z_1 = 0, \\ 6912Z_2^3v^4 + 8[27(Z_1+Z_2-1)^2 - 4Z_2^3]vw^3 - 9Z_1^2 = 0, \\ [27(Z_1+Z_2-1)^2 - 4Z_2^3]w^4 - 6912Z_2^3v^3w - 27Z_1^2 = 0. \end{cases} \tag{5.22}$$

Proof. We have

$$G = \frac{C_9u}{C_6}, \quad H = \frac{C_{12}v}{C_9}, \quad K = \frac{\mathfrak{C}_{12}w}{C_9}.$$

By

$$4G^3 + H^2G - C_6G - 4C_9 = 0,$$

we have

$$4\frac{C_9^3u^3}{C_6^3} + \frac{C_{12}^2v^2}{C_9^2}\cdot\frac{C_9u}{C_6} - C_6\cdot\frac{C_9u}{C_6} - 4C_9 = 0,$$

i.e.,

$$4C_9^4u^3 + C_6^2C_{12}^2v^2u - C_6^3C_9^2u - 4C_6^3C_9^2 = 0.$$

Thus,

$$4\frac{C_9^4}{C_6^6}u^3 + \frac{C_{12}^2}{C_6^4}uv^2 - \frac{C_9^2}{C_6^3}u - 4\frac{C_9^2}{C_6^3} = 0.$$

This implies that

$$4\left(\frac{Z_1}{432}\right)^2u^3 + \left(\frac{Z_2}{3}\right)^2uv^2 - \frac{Z_1}{432}u - 4\frac{Z_1}{432} = 0.$$

Hence,

$$Z_1^2u^3 + 5184Z_2^2uv^2 - 108Z_1u - 432Z_1 = 0.$$

Note that

$$9C_{12} = H(H^3 + 8K^3) = \frac{C_{12}v}{C_9}\left(\frac{C_{12}^3v^3}{C_9^3} + 8\frac{\mathfrak{C}_{12}^3w^3}{C_9^3}\right).$$

i.e.,

$$9C_9^4 = C_{12}^3v^4 + 8\mathfrak{C}_{12}^3vw^3.$$

Thus,

$$9\left(\frac{C_9^2}{C_6^3}\right)^2 = \left(\frac{C_{12}}{C_6^2}\right)^3v^4 + 8\left(\frac{\mathfrak{C}_{12}}{C_6^2}\right)^3vw^3.$$

This implies that

$$9\left(\frac{Z_1}{432}\right)^2 = \left(\frac{Z_2}{3}\right)^3v^4 + 8\cdot\frac{1}{1728}\left[\frac{1}{4}(Z_1+Z_2-1)^2 - \frac{1}{27}Z_2^3\right]vw^3.$$

Hence,

$$9Z_1^2 = 6912Z_2^3v^4 + 8[27(Z_1 + Z_2 - 1)^2 - 4Z_2^3]vw^3.$$

Note that

$$27\mathfrak{C}_{12} = K(K^3 - H^3) = \frac{\mathfrak{C}_{12}w}{C_9}\left(\frac{\mathfrak{C}_{12}^3w^3}{C_9^3} - \frac{C_{12}^3v^3}{C_9^3}\right),$$

i.e.,

$$27C_9^4 = \mathfrak{C}_{12}^3w^4 - C_{12}^3v^3w.$$

Thus,

$$27\left(\frac{C_9^2}{C_6^3}\right)^2 = \left(\frac{\mathfrak{C}_{12}}{C_6^2}\right)^3 w^4 - \left(\frac{C_{12}}{C_6^2}\right)^3 v^3w.$$

This implies that

$$27\cdot\left(\frac{Z_1}{432}\right)^2 = \frac{1}{1728}\left[\frac{1}{4}(Z_1 + Z_2 - 1)^2 - \frac{1}{27}Z_2^3\right]w^4 - \left(\frac{Z_2}{3}\right)^3 v^3w.$$

Hence,

$$27Z_1^2 = [27(Z_1 + Z_2 - 1)^2 - 4Z_2^3]w^4 - 6912Z_2^3v^3w.$$

□

Theorem 5.5 (Main Theorem 7). *The functions u, v and w can be expressed as the algebraic functions of r_1, r_2 and r_3:*

$$\begin{cases} u = \dfrac{4}{4r_1 + r_2 - 1}, \\ v = \dfrac{9}{4}\dfrac{(4r_1 + r_2 - 1)\sqrt{r_1}}{r_2\sqrt{r_2} + 8r_3\sqrt{r_3}}, \\ w = \dfrac{27}{4}\dfrac{(4r_1 + r_2 - 1)\sqrt{r_1}}{r_3\sqrt{r_3} - r_2\sqrt{r_2}}. \end{cases} \tag{5.23}$$

Proof. We have

$$u^2 = \frac{C_6^2G^2}{C_9^2} = \frac{\frac{G^2}{C_6}}{\frac{C_9^2}{C_6^3}} = \frac{r_1}{\frac{1}{432}Z_1} = \frac{16}{(4r_1 + r_2 - 1)^2}.$$

So,

$$u = \frac{4}{4r_1 + r_2 - 1}.$$

On the other hand,

$$v^2 = \frac{C_9^2H^2}{C_{12}^2} = \frac{\frac{C_9^2}{C_6^3}}{(\frac{C_{12}}{C_6^2})^2}\cdot\frac{H^2}{C_6} = \frac{\frac{1}{432}Z_1}{(\frac{1}{3}Z_2)^2}r_2 = \frac{\frac{1}{16}r_1(4r_1 + r_2 - 1)^2}{(\frac{1}{9}r_2^2 + \frac{8}{9}r_3\sqrt{r_2r_3})^2}r_2.$$

So,

$$v = \frac{9(4r_1 + r_2 - 1)\sqrt{r_1}}{4r_2\sqrt{r_2} + 32r_3\sqrt{r_3}}.$$

Similarly,

$$\begin{aligned} w^2 =& \frac{C_9^2 K^2}{\mathfrak{C}_{12}^2} = \frac{\frac{C_9^2}{C_6^3}}{(\frac{\mathfrak{C}_{12}}{C_6^2})^2} \cdot \frac{K^2}{C_6} = \frac{\frac{1}{432}Z_1}{\frac{1}{144}[\frac{1}{4}(Z_1 + Z_2 - 1)^2 - \frac{1}{27}Z_2^3]^{\frac{2}{3}}} r_3 \\ =& \frac{\frac{1}{16}r_1(4r_1 + r_2 - 1)^2 r_3}{\frac{1}{144}(\frac{4}{9})^2(r_3^2 - r_2 \cdot \sqrt{r_2 r_3})^2}. \end{aligned}$$

So,

$$w = \frac{27}{4} \frac{(4r_1 + r_2 - 1)\sqrt{r_1}}{r_3\sqrt{r_3} - r_2\sqrt{r_2}}.$$

□

Set

$$M := M_\infty = 81\sqrt{-3}\varphi^3, \quad N := N_\infty = 3\sqrt{-3}\psi^3. \tag{5.24}$$

Then

$$M_0 := M(E(z_1, z_2, z_3)) = K^3, \quad N_0 := N(E(z_1, z_2, z_3)) = H^3. \tag{5.25}$$

The equations

$$H(H^3 + 8K^3) = 9C_{12}, \quad K(K^3 - H^3) = 27\mathfrak{C}_{12}$$

give that

$$N_0(N_0 + 8M_0)^3 = 9^3 C_{12}^3, \quad M_0(M_0 - N_0)^3 = 27^3 \mathfrak{C}_{12}^3. \tag{5.26}$$

Let

$$\begin{cases} M_i := M(ED^i(z_1, z_2, z_3)) = (\psi - 3\omega^{2i}\varphi)^3, \\ N_i := N(ED^i(z_1, z_2, z_3)) = (\psi + 6\omega^{2i}\varphi)^3. \end{cases} \tag{5.27}$$

We find that

$$M_\infty M_0 M_1 M_2 = -81\sqrt{-3}\mathfrak{C}_{12}^3, \quad N_\infty N_0 N_1 N_2 = 3\sqrt{-3}C_{12}^3. \tag{5.28}$$

Note that

$$M_\infty(D(z_1, z_2, z_3)) = M_\infty(z_1, z_2, z_3), \quad N_\infty(D(z_1, z_2, z_3)) = N_\infty(z_1, z_2, z_3). \tag{5.29}$$

$$M_1(E(z_1, z_2, z_3)) = M_2(z_1, z_2, z_3), \quad N_1(E(z_1, z_2, z_3)) = N_2(z_1, z_2, z_3). \tag{5.30}$$

$$M_2(E(z_1, z_2, z_3)) = -M_1(z_1, z_2, z_3), \quad N_2(E(z_1, z_2, z_3)) = -N_1(z_1, z_2, z_3). \tag{5.31}$$

$$M_\infty(E(z_1, z_2, z_3)) = M_0(z_1, z_2, z_3), \quad N_\infty(E(z_1, z_2, z_3)) = N_0(z_1, z_2, z_3). \tag{5.32}$$

$$M_0(E(z_1, z_2, z_3)) = -M_\infty(z_1, z_2, z_3), \quad N_0(E(z_1, z_2, z_3)) = -N_\infty(z_1, z_2, z_3). \tag{5.33}$$

Let

$$L(z_1, z_2, z_3) := (z_1 - z_2)^3 (z_2 - z_3)^3 (z_3 - z_1)^3. \tag{5.34}$$

Note that

$$AE = EC, \quad CE = EA^2, \quad AC = \omega CA. \tag{5.35}$$

Put

$$L_{i,j,k}(z_1, z_2, z_3) := L(E^i A^j C^k (z_1, z_2, z_3)). \tag{5.36}$$

We find that

$$\begin{cases} L_{0,0,0} = L_{0,1,0} = L_{0,2,0} = L_{2,0,0} = L_{2,1,0} = L_{2,2,0}, \\ L_{0,0,1} = L_{0,1,1} = L_{0,2,1} = L_{2,0,1} = L_{2,1,1} = L_{2,2,1}, \\ L_{0,0,2} = L_{0,1,2} = L_{0,2,2} = L_{2,0,2} = L_{2,1,2} = L_{2,2,2}, \end{cases} \tag{5.37}$$

and

$$\begin{cases} L_{1,0,0} = L_{1,0,1} = L_{1,0,2}, \\ L_{1,1,0} = L_{1,1,1} = L_{1,1,2}, \\ L_{1,2,0} = L_{1,2,1} = L_{1,2,2}. \end{cases} \tag{5.38}$$

Here,

$$\begin{cases} L_{1,0,0} = L(E(z_1, z_2, z_3)) = (z_2^3 - z_3^3)^3, \\ L_{1,1,0} = L(EA(z_1, z_2, z_3)) = (z_3^3 - z_1^3)^3, \\ L_{1,2,0} = L(EA^2(z_1, z_2, z_3)) = (z_1^3 - z_2^3)^3, \end{cases} \tag{5.39}$$

and

$$\begin{cases} L_{0,0,0} = L(z_1, z_2, z_3) = (z_1 - z_2)^3 (z_2 - z_3)^3 (z_3 - z_1)^3, \\ L_{0,0,1} = L(C(z_1, z_2, z_3)) = (z_1 - \omega z_2)^3 (\omega z_2 - \omega^2 z_3)^3 (\omega^2 z_3 - z_1)^3, \\ L_{0,0,2} = L(C^2(z_1, z_2, z_3)) = (z_1 - \omega^2 z_2)^3 (\omega^2 z_2 - \omega z_3)^3 (\omega z_3 - z_1)^3. \end{cases} \tag{5.40}$$

We find that

$$L_{1,0,0} L_{1,1,0} L_{1,2,0} = L_{0,0,0} L_{0,0,1} L_{0,0,2} = C_9^3. \tag{5.41}$$

In fact,

$$\begin{cases} L_{1,0,0}(E(z_1, z_2, z_3)) = L_{0,0,0}(z_1, z_2, z_3), \\ L_{1,0,0}(A(z_1, z_2, z_3)) = L_{1,1,0}(z_1, z_2, z_3), \\ L_{1,0,0}(C(z_1, z_2, z_3)) = L_{1,0,0}(z_1, z_2, z_3). \end{cases}$$

$$\begin{cases} L_{1,1,0}(E(z_1, z_2, z_3)) = L_{0,0,1}(z_1, z_2, z_3), \\ L_{1,1,0}(A(z_1, z_2, z_3)) = L_{1,2,0}(z_1, z_2, z_3), \\ L_{1,1,0}(C(z_1, z_2, z_3)) = L_{1,1,0}(z_1, z_2, z_3). \end{cases}$$

$$\begin{cases} L_{1,2,0}(E(z_1, z_2, z_3)) = L_{0,0,2}(z_1, z_2, z_3), \\ L_{1,2,0}(A(z_1, z_2, z_3)) = L_{1,0,0}(z_1, z_2, z_3), \\ L_{1,2,0}(C(z_1, z_2, z_3)) = L_{1,2,0}(z_1, z_2, z_3). \end{cases}$$

$$\begin{cases} L_{0,0,0}(E(z_1, z_2, z_3)) = L_{1,0,0}(z_1, z_2, z_3), \\ L_{0,0,0}(A(z_1, z_2, z_3)) = L_{0,0,0}(z_1, z_2, z_3), \\ L_{0,0,0}(C(z_1, z_2, z_3)) = L_{0,0,1}(z_1, z_2, z_3). \end{cases}$$

$$\begin{cases} L_{0,0,1}(E(z_1, z_2, z_3)) = L_{1,2,0}(z_1, z_2, z_3), \\ L_{0,0,1}(A(z_1, z_2, z_3)) = L_{0,0,1}(z_1, z_2, z_3), \\ L_{0,0,1}(C(z_1, z_2, z_3)) = L_{0,0,2}(z_1, z_2, z_3). \end{cases}$$

$$\begin{cases} L_{0,0,2}(E(z_1, z_2, z_3)) = L_{1,1,0}(z_1, z_2, z_3), \\ L_{0,0,2}(A(z_1, z_2, z_3)) = L_{0,0,2}(z_1, z_2, z_3), \\ L_{0,0,2}(C(z_1, z_2, z_3)) = L_{0,0,0}(z_1, z_2, z_3). \end{cases}$$

Therefore,

$$\begin{cases} L_{1,0,0}(E(z_1, z_2, z_3)) = L_{0,0,0}(z_1, z_2, z_3), \\ L_{0,0,0}(E(z_1, z_2, z_3)) = L_{1,0,0}(z_1, z_2, z_3). \end{cases} \tag{5.42}$$

$$\begin{cases} L_{1,1,0}(E(z_1, z_2, z_3)) = L_{0,0,1}(z_1, z_2, z_3), \\ L_{0,0,1}(E(z_1, z_2, z_3)) = L_{1,2,0}(z_1, z_2, z_3), \\ L_{1,2,0}(E(z_1, z_2, z_3)) = L_{0,0,2}(z_1, z_2, z_3), \\ L_{0,0,2}(E(z_1, z_2, z_3)) = L_{1,1,0}(z_1, z_2, z_3). \end{cases} \tag{5.43}$$

$$\begin{cases} L_{1,0,0}(A(z_1, z_2, z_3)) = L_{1,1,0}(z_1, z_2, z_3), \\ L_{1,1,0}(A(z_1, z_2, z_3)) = L_{1,2,0}(z_1, z_2, z_3), \\ L_{1,2,0}(A(z_1, z_2, z_3)) = L_{1,0,0}(z_1, z_2, z_3). \end{cases} \tag{5.44}$$

$$\begin{cases} L_{0,0,0}(A(z_1, z_2, z_3)) = L_{0,0,0}(z_1, z_2, z_3), \\ L_{0,0,1}(A(z_1, z_2, z_3)) = L_{0,0,1}(z_1, z_2, z_3), \\ L_{0,0,2}(A(z_1, z_2, z_3)) = L_{0,0,2}(z_1, z_2, z_3). \end{cases} \tag{5.45}$$

$$\begin{cases} L_{1,0,0}(C(z_1, z_2, z_3)) = L_{1,0,0}(z_1, z_2, z_3), \\ L_{1,1,0}(C(z_1, z_2, z_3)) = L_{1,1,0}(z_1, z_2, z_3), \\ L_{1,2,0}(C(z_1, z_2, z_3)) = L_{1,2,0}(z_1, z_2, z_3). \end{cases} \tag{5.46}$$

$$\begin{cases} L_{0,0,0}(C(z_1, z_2, z_3)) = L_{0,0,1}(z_1, z_2, z_3), \\ L_{0,0,1}(C(z_1, z_2, z_3)) = L_{0,0,2}(z_1, z_2, z_3), \\ L_{0,0,2}(C(z_1, z_2, z_3)) = L_{0,0,0}(z_1, z_2, z_3). \end{cases} \tag{5.47}$$

Put

$$\zeta_1 = \frac{M_0}{C_9}, \quad \zeta_2 = \frac{N_0}{C_9}. \tag{5.48}$$

Theorem 5.6 (Main Theorem 8). *The functions ζ_1 and ζ_2 satisfy the following equations:*

$$\begin{cases} Z_1^2 \zeta_2 (\zeta_2 + 8\zeta_1)^3 = 2^8 \cdot 3^9 Z_2^3, \\ Z_1^2 \zeta_1 (\zeta_1 - \zeta_2)^3 = 27^3 [27(Z_1 + Z_2 - 1)^2 - 4Z_2^3]. \end{cases} \tag{5.49}$$

Proof. We find that

$$\zeta_2(\zeta_2+8\zeta_1)^3 = 9^3\frac{C_{12}^3}{C_9^4} = 9^3\frac{(\frac{C_{12}}{C_6^2})^3}{(\frac{C_9^2}{C_6^3})^2} = 9^3\frac{(\frac{Z_2}{3})^3}{(\frac{Z_1}{432})^2} = 2^8\cdot 3^9\frac{Z_2^3}{Z_1^2},$$

and

$$\zeta_1(\zeta_1-\zeta_2)^3 = 27^3\frac{\mathfrak{C}_{12}^3}{C_9^4} = 27^3\frac{(\frac{\mathfrak{C}_{12}}{C_6^2})^3}{(\frac{C_9^2}{C_6^3})^2}$$

$$=27^3\cdot\frac{1}{1728}\left[\frac{1}{4}(Z_1+Z_2-1)^2-\frac{1}{27}Z_2^3\right]\frac{1}{(\frac{Z_1}{432})^2}.$$

Thus,

$$Z_1^2\zeta_2(\zeta_2+8\zeta_1)^3 = 2^8\cdot 3^9 Z_2^3,$$

and

$$Z_1^2\zeta_1(\zeta_1-\zeta_2)^3 = 27^3[27(Z_1+Z_2-1)^2-4Z_2^3].$$

□

Put

$$\xi_1 := \frac{M_0^{\frac{1}{3}}}{C_6^{\frac{1}{2}}},\quad \xi_2 := \frac{N_0^{\frac{1}{3}}}{C_6^{\frac{1}{2}}}. \tag{5.50}$$

Theorem 5.7 (Main Theorem 9). *The functions ξ_1 and ξ_2 satisfy the following equations*:

$$\begin{cases} \xi_2^3(\xi_2^3+8\xi_1^3)^3 = 27Z_2^3,\\ 256\xi_1^3(\xi_1^3-\xi_2^3)^3 = 27[27(Z_1+Z_2-1)^2-4Z_2^3]. \end{cases} \tag{5.51}$$

Proof. We find that

$$\xi_2^3(\xi_2^3+8\xi_1^3)^3 = 9^3\left(\frac{C_{12}}{C_6^2}\right)^3 = 27Z_2^3,$$

and

$$\xi_1^3(\xi_1^3-\xi_2^3)^3 = 27^3\left(\frac{\mathfrak{C}_{12}}{C_6^2}\right)^3 = \frac{27}{256}[27(Z_1+Z_2-1)^2-4Z_2^3].$$

□

Theorem 5.7 gives that

$$Z_1 = \frac{2}{27}[\xi_2^3(\xi_2^3+8\xi_1^3)^3+64\xi_1^3(\xi_1^3-\xi_2^3)^3]^{\frac{1}{2}} - \frac{1}{3}\xi_2(\xi_2^3+8\xi_1^3)+1. \tag{5.52}$$

Note that

$$\zeta_1^2 = \frac{\frac{M_0^2}{C_6^3}}{\frac{C_9^2}{C_6^3}} = \frac{432\xi_1^6}{Z_1},\quad \zeta_2^2 = \frac{\frac{N_0^2}{C_6^3}}{\frac{C_9^2}{C_6^3}} = \frac{432\xi_2^6}{Z_1}. \tag{5.53}$$

Hence, ξ_1, ξ_2 and ζ_1, ζ_2 satisfy the following relations:

$$\begin{cases} \frac{4}{729}\zeta_1^4[\xi_2^3(\xi_2^3+8\xi_1^3)^3+64\xi_1^3(\xi_1^3-\xi_2^3)^3]=[432\xi_1^6+\frac{1}{3}\zeta_1^2\xi_2(\xi_2^3+8\xi_1^3)-\zeta_1^2]^2, \\ \frac{4}{729}\zeta_2^4[\xi_2^3(\xi_2^3+8\xi_1^3)^3+64\xi_1^3(\xi_1^3-\xi_2^3)^3]=[432\xi_2^6+\frac{1}{3}\zeta_2^2\xi_2(\xi_2^3+8\xi_1^3)-\zeta_2^2]^2. \end{cases} \tag{5.54}$$

Theorem 5.8 (Main Theorem 10). *The function-theoretic resolvents (analytic resolvents) of Z_1 and Z_2 are also given by:*

$$\begin{aligned} &Z_1^2 : Z_2^3 : (Z_1+Z_2-1)^2 : [27(Z_1+Z_2-1)^2-4Z_2^3] : 1 \\ =&[2(\xi_2^3(\xi_2^3+8\xi_1^3)^3+64\xi_1^3(\xi_1^3-\xi_2^3)^3)^{\frac{1}{2}}-9\xi_2(\xi_2^3+8\xi_1^3)+27]^2 \\ :&27\xi_2^3(\xi_2^3+8\xi_1^3)^3 : 4[\xi_2^3(\xi_2^3+8\xi_1^3)^3+64\xi_1^3(\xi_1^3-\xi_2^3)^3] \\ :&6912\xi_1^3(\xi_1^3-\xi_2^3)^3 : 729. \end{aligned} \tag{5.55}$$

Proof. It is obtained by Theorem 5.7 and (5.52). □

Note that $CD = DC$, we have

$$C^iD^j(z_1,z_2,z_3)=(z_1,\omega^{i+j}z_2,\omega^{2i+j}z_3). \tag{5.56}$$

We find that

$$\begin{cases} C_6(C^iD^j(z_1,z_2,z_3))=C_6(z_1,z_2,z_3), \\ C_9(C^iD^j(z_1,z_2,z_3))=C_9(z_1,z_2,z_3), \\ C_{12}(C^iD^j(z_1,z_2,z_3))=C_{12}(z_1,z_2,z_3), \\ \mathfrak{C}_{12}(C^iD^j(z_1,z_2,z_3))=\omega^{2j}\mathfrak{C}_{12}(z_1,z_2,z_3). \end{cases} \tag{5.57}$$

Let

$$\begin{cases} G_0:=G(z_1,z_2,z_3)=(z_1-z_2)(z_2-z_3)(z_3-z_1), \\ G_1:=G(C(z_1,z_2,z_3))=(z_1-\omega z_2)(\omega z_2-\omega^2z_3)(\omega^2z_3-z_1), \\ G_2:=G(C^2(z_1,z_2,z_3))=(z_1-\omega^2z_2)(\omega^2z_2-\omega z_3)(\omega z_3-z_1). \end{cases} \tag{5.58}$$

A straightforward calculation gives that

$$G_0G_1G_2=C_9. \tag{5.59}$$

Hence,

$$u_0u_1u_2:=\frac{C_6G_0}{C_9}\cdot\frac{C_6G_1}{C_9}\cdot\frac{C_6G_2}{C_9}=\frac{C_6^3}{C_9^2}=\frac{432}{Z_1}. \tag{5.60}$$

On the other hand,

$$H(D^j(z_1,z_2,z_3))=\psi+6\omega^{2j}\varphi. \tag{5.61}$$

Put

$$H_0:=\psi+6\varphi, \quad H_1:=\psi+6\omega^2\varphi, \quad H_2:=\psi+6\omega\varphi. \tag{5.62}$$

Then

$$H_0H_1H_2 = \psi^3 + 216\varphi^3 = \frac{C_{12}}{\psi}. \tag{5.63}$$

Hence,

$$v_0v_1v_2 := \frac{C_9H_0}{C_{12}} \cdot \frac{C_9H_1}{C_{12}} \cdot \frac{C_9H_2}{C_{12}} = \frac{C_9^3}{C_{12}^2\psi}. \tag{5.64}$$

Similarly,

$$K(D^j(z_1, z_2, z_3)) = \psi - 3\omega^{2j}\varphi. \tag{5.65}$$

Put

$$K_0 := \psi - 3\varphi, \quad K_1 := \psi - 3\omega^2\varphi, \quad K_2 := \psi - 3\omega\varphi. \tag{5.66}$$

Then

$$K_0K_1K_2 = \psi^3 - 27\varphi^3 = -\frac{\mathfrak{C}_{12}}{\varphi}. \tag{5.67}$$

Hence,

$$w_0w_1w_2 := \frac{C_9K_0}{\mathfrak{C}_{12}} \cdot \frac{C_9K_1}{\omega^2\mathfrak{C}_{12}} \cdot \frac{C_9K_2}{\omega\mathfrak{C}_{12}} = -\frac{C_9^3}{\mathfrak{C}_{12}^2\varphi}. \tag{5.68}$$

We have

$$\sqrt[3]{M_\infty} = -3\sqrt{-3}\varphi, \quad \sqrt[3]{M_0} = \psi - 3\varphi, \quad \sqrt[3]{M_1} = \psi - 3\omega^2\varphi, \quad \sqrt[3]{M_2} = \psi - 3\omega\varphi. \tag{5.69}$$

Hence,

$$\begin{cases} \sqrt[3]{M_0} + \sqrt[3]{M_1} + \sqrt[3]{M_2} = 3\psi, \\ \sqrt{-3}\sqrt[3]{M_\infty} + \sqrt[3]{M_0} + \omega\sqrt[3]{M_1} + \omega^2\sqrt[3]{M_2} = 0, \\ \sqrt[3]{M_0} + \omega^2\sqrt[3]{M_1} + \omega\sqrt[3]{M_2} = 0. \end{cases} \tag{5.70}$$

Similarly,

$$\sqrt[3]{N_\infty} = -\sqrt{-3}\psi, \quad \sqrt[3]{N_0} = \psi + 6\varphi, \quad \sqrt[3]{N_1} = \psi + 6\omega^2\varphi, \quad \sqrt[3]{N_2} = \psi + 6\omega\varphi. \tag{5.71}$$

Hence,

$$\begin{cases} -\sqrt{-3}\sqrt[3]{N_\infty} + \sqrt[3]{N_0} + \sqrt[3]{N_1} + \sqrt[3]{N_2} = 0, \\ \sqrt[3]{N_0} + \omega\sqrt[3]{N_1} + \omega^2\sqrt[3]{N_2} = 18\varphi, \\ \sqrt[3]{N_0} + \omega^2\sqrt[3]{N_1} + \omega\sqrt[3]{N_2} = 0. \end{cases} \tag{5.72}$$

6. Ternary cubic forms associated to Hessian polyhedra

Put

$$X(z_1, z_2, z_3) = z_1^3, \quad Y(z_1, z_2, z_3) = z_2^3, \quad Z(z_1, z_2, z_3) = z_3^3. \tag{6.1}$$

$$Q_1(z_1, z_2, z_3) = z_1z_2^2 + z_2z_3^2 + z_3z_1^2, \quad Q_2(z_1, z_2, z_3) = z_1^2z_2 + z_2^2z_3 + z_3^2z_1. \tag{6.2}$$

Then

$$\sqrt[3]{L_{1,0,0}} = Y - Z, \quad \sqrt[3]{L_{1,1,0}} = Z - X, \quad \sqrt[3]{L_{1,2,0}} = X - Y. \tag{6.3}$$

$$\sqrt[3]{L_{0,0,0}} = Q_1 - Q_2, \quad \sqrt[3]{L_{0,0,1}} = \omega^2 Q_1 - \omega Q_2, \quad \sqrt[3]{L_{0,0,2}} = \omega Q_1 - \omega^2 Q_2. \tag{6.4}$$

The functions φ, ψ, χ, X, Y, Z and Q_1, Q_2 satisfy the relations:

$$\begin{cases} \varphi^3 = XYZ, \\ \psi = X + Y + Z, \\ \chi = XY + YZ + ZX, \\ \chi + 3\varphi^2 + \varphi\psi = Q_1 Q_2, \\ \psi\chi + 6\varphi\chi + 6\varphi^2\psi + 9\varphi^3 = Q_1^3 + Q_2^3, \\ (X - Y)(Y - Z)(Z - X) = Q_1^3 - Q_2^3. \end{cases} \tag{6.5}$$

The invariants C_6, C_9, C_{12}, $\mathfrak{C}_{12}$ and C_{18} can be expressed as the functions of X, Y and Z:

$$\begin{cases} C_6 = (X + Y + Z)^2 - 12(XY + YZ + ZX), \\ C_9 = (X - Y)(Y - Z)(Z - X), \\ C_{12} = (X + Y + Z)[(X + Y + Z)^3 + 216XYZ], \\ \mathfrak{C}_{12} = \sqrt[3]{XYZ}[27XYZ - (X + Y + Z)^3], \\ C_{18} = (X + Y + Z)^6 - 540XYZ(X + Y + Z)^3 - 5832X^2Y^2Z^2. \end{cases} \tag{6.6}$$

Let

$$\begin{cases} W_2 = (X + Y + Z)^2 - 12(XY + YZ + ZX), \\ W_3 = (X - Y)(Y - Z)(Z - X), \\ \mathfrak{W}_3 = XYZ - \varphi^3, \\ W_4 = (X + Y + Z)[(X + Y + Z)^3 + 216\varphi^3], \\ \mathfrak{W}_4 = \varphi[27\varphi^3 - (X + Y + Z)^3], \\ W_6 = (X + Y + Z)^6 - 540\varphi^3(X + Y + Z)^3 - 5832\varphi^6. \end{cases} \tag{6.7}$$

We will regard (φ, X, Y, Z) as a point in the complex projective space $\mathbb{CP}^3$.

We find that

$$1728\mathfrak{W}_4^3 = W_6^2 - W_4^3 \tag{6.8}$$

and

$$\frac{\partial(\mathfrak{W}_3, W_4, \mathfrak{W}_4, W_6)}{\partial(\varphi, X, Y, Z)} = 0. \tag{6.9}$$

For the ternary cubic form

$$F_1(x_1, x_2, x_3) = Xx_1^3 + Yx_2^3 + Zx_3^3 - 3\varphi x_1 x_2 x_3, \tag{6.10}$$

we have

$$\begin{cases} \sqrt[3]{M_\infty} = -3\sqrt{-3}\varphi, \\ \sqrt[3]{M_0} = -3\varphi + X + Y + Z, \\ \sqrt[3]{M_1} = -3\omega^2\varphi + X + Y + Z, \\ \sqrt[3]{M_2} = -3\omega\varphi + X + Y + Z. \end{cases} \tag{6.11}$$

Correspondingly,

$$\begin{cases} \sqrt[3]{M_\infty}: & (\sqrt{-3}, 0, 0, 0) \in \mathbb{CP}^3, \\ \sqrt[3]{M_0}: & (1, 1, 1, 1) \in \mathbb{CP}^3, \\ \sqrt[3]{M_1}: & (\omega^2, 1, 1, 1) \in \mathbb{CP}^3, \\ \sqrt[3]{M_2}: & (\omega, 1, 1, 1) \in \mathbb{CP}^3. \end{cases} \tag{6.12}$$

For the ternary cubic form

$$F_2(x_1, x_2, x_3) = Xx_1^3 + Yx_2^3 + Zx_3^3 + 6\varphi x_1 x_2 x_3, \tag{6.13}$$

we have

$$\begin{cases} \sqrt[3]{N_\infty} = -\sqrt{-3}(X + Y + Z), \\ \sqrt[3]{N_0} = 6\varphi + X + Y + Z, \\ \sqrt[3]{N_1} = 6\omega^2\varphi + X + Y + Z, \\ \sqrt[3]{N_2} = 6\omega\varphi + X + Y + Z. \end{cases} \tag{6.14}$$

Correspondingly,

$$\begin{cases} \sqrt[3]{N_\infty}: & (0, -\sqrt{-3}, -\sqrt{-3}, -\sqrt{-3}) \in \mathbb{CP}^3, \\ \sqrt[3]{N_0}: & (1, 1, 1, 1) \in \mathbb{CP}^3, \\ \sqrt[3]{N_1}: & (\omega^2, 1, 1, 1) \in \mathbb{CP}^3, \\ \sqrt[3]{N_2}: & (\omega, 1, 1, 1) \in \mathbb{CP}^3. \end{cases} \tag{6.15}$$

In fact, $W_2 = 0$ is a quadratic ruled surface, $W_3 = 0$ is a cubic ruled surface, $\mathfrak{W}_3 = 0$ is a cubic surface, $W_4 = 0$ is a quartic surface, $\mathfrak{W}_4 = 0$ is a quartic surface, $W_6 = 0$ is a surface of degree six.

The action of Hessian groups on $\mathbb{CP}^3$ has two orbits:

$$\{(\sqrt{-3}, 0, 0, 0), (1, 1, 1, 1), (\omega^2, 1, 1, 1), (\omega, 1, 1, 1)\}$$

and

$$\{(0, -\sqrt{-3}, -\sqrt{-3}, -\sqrt{-3}), (1, 1, 1, 1), (\omega^2, 1, 1, 1), (\omega, 1, 1, 1)\}.$$

Each orbit has four points. The invariant surface $\mathfrak{W}_3 = 0$ passes through the points $(1, 1, 1, 1)$, $(\omega^2, 1, 1, 1)$ and $(\omega, 1, 1, 1)$. The invariant surface $W_2 = 0$ passes through the point $(\sqrt{-3}, 0, 0, 0)$. The invariant surface $W_3 = 0$ passes through all the five points $(\sqrt{-3}, 0, 0, 0)$, $(1, 1, 1, 1)$, $(\omega^2, 1, 1, 1)$, $(\omega, 1, 1, 1)$ and $(0, -\sqrt{-3}, -\sqrt{-3}, -\sqrt{-3})$. The invariant surface $W_4 = 0$ passes through the point $(\sqrt{-3}, 0, 0, 0)$. The invariant surface $\mathfrak{W}_4 = 0$ passes through the points $(0, -\sqrt{-3}, -\sqrt{-3}, -\sqrt{-3})$, $(1, 1, 1, 1)$, $(\omega^2, 1, 1, 1)$ and $(\omega, 1, 1, 1)$. The invariant surface $W_6 = 0$ does not pass through the five points.

When we restrict the action of Hessian groups to the cubic surface $\mathfrak{W}_3 = 0$, we get the action of Hessian groups on $\mathbb{CP}^2$. We set

$$\begin{cases} W_2 = (X+Y+Z)^2 - 12(XY+YZ+ZX), \\ W_3 = (X-Y)(Y-Z)(Z-X), \\ W_4 = (X+Y+Z)[(X+Y+Z)^3 + 216XYZ], \\ W_6 = (X+Y+Z)^6 - 540XYZ(X+Y+Z)^3 - 5832X^2Y^2Z^2, \\ W_{12} = XYZ[27XYZ - (X+Y+Z)^3]^3. \end{cases} \tag{6.16}$$

The invariants W_2, W_3, W_4, W_6 and W_{12} satisfy the relations:

$$\begin{cases} 432W_3^2 = W_2^3 - 3W_2W_4 + 2W_6, \\ 1728W_{12} = W_6^2 - W_4^3. \end{cases} \tag{6.17}$$

We will study the algebraic curves in $\mathbb{CP}^2$:

$$W_2 = 0, \quad W_3 = 0, \quad W_4 = 0, \quad W_6 = 0, \quad W_{12} = 0.$$

$W_2 = 0$ is a conic. $W_3 = 0$ is the product of three lines: $X - Y = 0$, $Y - Z = 0$ or $Z - X = 0$. $W_4 = 0$, i.e., $X + Y + Z = 0$ or $(X + Y + Z)^3 + 216XYZ = 0$. $W_6 = 0$, i.e., $(X+Y+Z)^3 - (270+162\sqrt{3})XYZ = 0$ or $(X+Y+Z)^3 - (270 - 162\sqrt{3})XYZ = 0$. $W_{12} = 0$, i.e., $X = 0$, $Y = 0$, $Z = 0$ or $(X + Y + Z)^3 - 27XYZ = 0$.

The curves $W_2 = 0$, $W_4 = 0$, $W_6 = 0$, $W_{12} = 0$ intersect $W_3 = 0$ in 6, 12, 18, 15 points respectively.

$$\{W_2 = 0\} \cap \{W_3 = 0\} = \{(\alpha_i, 1, 1), (1, \alpha_i, 1), (1, 1, \alpha_i), i = 1, 2\},$$

where α_1 and α_2 are two roots of the quadratic equation:

$$\alpha^2 - 20\alpha - 8 = 0.$$

$$\begin{aligned} &\{W_4 = 0\} \cap \{W_3 = 0\} \\ =&\{(-2, 1, 1), (1, -2, 1), (1, 1, -2), (\beta_i, 1, 1), (1, \beta_i, 1), (1, 1, \beta_i), i = 1, 2, 3\}, \end{aligned}$$

where β_1, β_2 and β_3 are the roots of the cubic equation:

$$\beta^3 + 6\beta^2 + 228\beta + 8 = 0.$$

$$\{W_6 = 0\} \cap \{W_3 = 0\} = \{(\gamma_i, 1, 1), (1, \gamma_i, 1), (1, 1, \gamma_i), 1 \le i \le 6\},$$

where γ_i $(1 \le i \le 6)$ are the roots of the equation of the sixth degree:

$$[\gamma^3 + 6\gamma^2 - (258 + 162\sqrt{3})\gamma + 8][\gamma^3 + 6\gamma^2 - (258 - 162\sqrt{3})\gamma + 8] = 0.$$

$$\{W_{12}=0\}\cap\{W_3=0\}=\{(1,0,0),(0,1,0),(0,0,1),(1,1,0),(1,0,1),(0,1,1),\\ (\delta_i,1,1),(1,\delta_i,1),(1,1,\delta_i),i=1,2,3\},$$

where δ_1, δ_2 and δ_3 are the roots of the cubic equation:

$$\delta^3+6\delta^2-15\delta+8=0.$$

In fact, by $W_3=0$, we have

$$(X,Y,Z)=(X,X,Z),(X,Y,Y),(Z,Y,Z).$$

Without loss of generality, for $(X,Y,Z)=(X,Y,Y)$, we find that

$$\begin{cases} W_2=X^2-20XY-8Y^2,\\ W_4=(X+2Y)(X^3+6X^2Y+228XY^2+8Y^3),\\ W_6=[X^3+6X^2Y-(258+162\sqrt{3})XY^2+8Y^3]\\ \qquad\times[X^3+6X^2Y-(258-162\sqrt{3})XY^2+8Y^3],\\ W_{12}=-XY^2(X^3+6X^2Y-15XY^2+8Y^3)^3. \end{cases}$$

Note that

$$\mathbb{C}[W_2,W_3,W_4,W_6,W_{12}]/I\cong\mathbb{C}[W_2,W_3,W_4], \tag{6.18}$$

where the ideal

$$I=(W_2^3-3W_2W_4-432W_3^2+2W_6,W_6^2-W_4^3-1728W_{12}).$$

We find that

$$\begin{cases} \sqrt[3]{L_{1,0,0}}+\sqrt[3]{L_{1,1,0}}+\sqrt[3]{L_{1,2,0}}=0,\\ \sqrt[3]{L_{0,0,0}}+\sqrt[3]{L_{0,0,1}}+\sqrt[3]{L_{0,0,2}}=0. \end{cases} \tag{6.19}$$

Moreover,

$$\begin{cases} L_{1,0,0}+L_{1,1,0}+L_{1,2,0}=3C_9,\\ L_{0,0,0}+L_{0,0,1}+L_{0,0,2}=3C_9. \end{cases} \tag{6.20}$$

In fact,

$$\sqrt{-3}(M_0+M_1+M_2)=N_\infty-M_\infty,\quad \sqrt{-3}(N_0+N_1+N_2)=N_\infty+8M_\infty. \tag{6.21}$$

$$\begin{aligned}(L_{1,0,0}+L_{1,1,0}+L_{1,2,0})^3&=27L_{1,0,0}L_{1,1,0}L_{1,2,0},\\ (L_{0,0,0}+L_{0,0,1}+L_{0,0,2})^3&=27L_{0,0,0}L_{0,0,1}L_{0,0,2}.\end{aligned} \tag{6.22}$$

We have

$$\begin{cases} \varphi(A(z_1,z_2,z_3))=\varphi(z_1,z_2,z_3),\\ \psi(A(z_1,z_2,z_3))=\psi(z_1,z_2,z_3),\\ X(A(z_1,z_2,z_3))=Y(z_1,z_2,z_3),\\ Y(A(z_1,z_2,z_3))=Z(z_1,z_2,z_3),\\ Z(A(z_1,z_2,z_3))=X(z_1,z_2,z_3),\\ Q_1(A(z_1,z_2,z_3))=Q_1(z_1,z_2,z_3),\\ Q_2(A(z_1,z_2,z_3))=Q_2(z_1,z_2,z_3). \end{cases} \tag{6.23}$$

$$\begin{cases} \varphi(B(z_1, z_2, z_3)) = \varphi(z_1, z_2, z_3), \\ \psi(B(z_1, z_2, z_3)) = \psi(z_1, z_2, z_3), \\ X(B(z_1, z_2, z_3)) = X(z_1, z_2, z_3), \\ Y(B(z_1, z_2, z_3)) = Z(z_1, z_2, z_3), \\ Z(B(z_1, z_2, z_3)) = Y(z_1, z_2, z_3), \\ Q_1(B(z_1, z_2, z_3)) = Q_2(z_1, z_2, z_3), \\ Q_2(B(z_1, z_2, z_3)) = Q_1(z_1, z_2, z_3). \end{cases} \tag{6.24}$$

$$\begin{cases} \varphi(C(z_1, z_2, z_3)) = \varphi(z_1, z_2, z_3), \\ \psi(C(z_1, z_2, z_3)) = \psi(z_1, z_2, z_3), \\ X(C(z_1, z_2, z_3)) = X(z_1, z_2, z_3), \\ Y(C(z_1, z_2, z_3)) = Y(z_1, z_2, z_3), \\ Z(C(z_1, z_2, z_3)) = Z(z_1, z_2, z_3), \\ Q_1(C(z_1, z_2, z_3)) = \omega^2 Q_1(z_1, z_2, z_3), \\ Q_2(C(z_1, z_2, z_3)) = \omega Q_2(z_1, z_2, z_3). \end{cases} \tag{6.25}$$

$$\begin{cases} \varphi(D(z_1, z_2, z_3)) = \omega^2 \varphi(z_1, z_2, z_3), \\ \psi(D(z_1, z_2, z_3)) = \psi(z_1, z_2, z_3), \\ X(D(z_1, z_2, z_3)) = X(z_1, z_2, z_3), \\ Y(D(z_1, z_2, z_3)) = Y(z_1, z_2, z_3), \\ Z(D(z_1, z_2, z_3)) = Z(z_1, z_2, z_3). \end{cases} \tag{6.26}$$

$$\begin{cases} (\sqrt{-3})^3 \varphi(E(z_1, z_2, z_3)) = \psi - 3\varphi, \\ -\sqrt{-3}\psi(E(z_1, z_2, z_3)) = \psi + 6\varphi, \\ (\sqrt{-3})^3 X(E(z_1, z_2, z_3)) = \psi + 6\varphi + 3Q_1 + 3Q_2, \\ (\sqrt{-3})^3 Y(E(z_1, z_2, z_3)) = \psi + 6\varphi + 3\omega^2 Q_1 + 3\omega Q_2, \\ (\sqrt{-3})^3 Z(E(z_1, z_2, z_3)) = \psi + 6\varphi + 3\omega Q_1 + 3\omega^2 Q_2, \\ -\sqrt{-3} Q_1(E(z_1, z_2, z_3)) = X + \omega^2 Y + \omega Z, \\ -\sqrt{-3} Q_2(E(z_1, z_2, z_3)) = X + \omega Y + \omega^2 Z. \end{cases} \tag{6.27}$$

We have

$$\sum_{\nu=0}^{2} \omega^{\nu} M_{\nu} = -27\psi^2\varphi, \quad \sum_{\nu=0}^{2} \omega^{2\nu} M_{\nu} = 81\psi\varphi^2. \tag{6.28}$$

$$\sum_{\nu=0}^{2} \omega^{\nu} N_{\nu} = 54\psi^2\varphi, \quad \sum_{\nu=0}^{2} \omega^{2\nu} N_{\nu} = 324\psi\varphi^2. \tag{6.29}$$

$$\sum_{\nu=0}^{2} \omega^{\nu} L_{0,0,\nu} = -9Q_1^2 Q_2, \quad \sum_{\nu=0}^{2} \omega^{2\nu} L_{0,0,\nu} = 9Q_1 Q_2^2. \tag{6.30}$$

Hence,

$$\left(\sum_{\nu=0}^{2} \omega^{\nu} M_{\nu}\right) \cdot \left(\sum_{\nu=0}^{2} \omega^{2\nu} M_{\nu}\right) = 3 M_{\infty} N_{\infty}. \tag{6.31}$$

$$\left(\sum_{\nu=0}^{2} \omega^{\nu} N_{\nu}\right) \cdot \left(\sum_{\nu=0}^{2} \omega^{2\nu} N_{\nu}\right) = -24 M_{\infty} N_{\infty}. \tag{6.32}$$

$$\left(\sum_{\nu=0}^{2} \omega^{\nu} L_{0,0,\nu}\right) \cdot \left(\sum_{\nu=0}^{2} \omega^{2\nu} L_{0,0,\nu}\right) = -81(\chi + 3\varphi^2 + \varphi\psi)^3. \tag{6.33}$$

In fact,

$$\begin{cases} \sqrt[3]{M_{\infty}}\sqrt[3]{M_0}\sqrt[3]{M_1}\sqrt[3]{M_2} = 3\sqrt{-3}\mathfrak{C}_{12}, \\ \sqrt[3]{N_{\infty}}\sqrt[3]{N_0}\sqrt[3]{N_1}\sqrt[3]{N_2} = -\sqrt{-3}C_{12}, \\ \sqrt[3]{L_{1,0,0}}\sqrt[3]{L_{1,1,0}}\sqrt[3]{L_{1,2,0}} = C_9, \\ \sqrt[3]{L_{0,0,0}}\sqrt[3]{L_{0,0,1}}\sqrt[3]{L_{0,0,2}} = C_9. \end{cases} \tag{6.34}$$

The Hessian of C_6 is given by

$$\text{Hessian}(C_6) = \begin{vmatrix} \frac{\partial^2 C_6}{\partial z_1^2} & \frac{\partial^2 C_6}{\partial z_1 \partial z_2} & \frac{\partial^2 C_6}{\partial z_1 \partial z_3} \\ \frac{\partial^2 C_6}{\partial z_1 \partial z_2} & \frac{\partial^2 C_6}{\partial z_2^2} & \frac{\partial^2 C_6}{\partial z_2 \partial z_3} \\ \frac{\partial^2 C_6}{\partial z_1 \partial z_3} & \frac{\partial^2 C_6}{\partial z_2 \partial z_3} & \frac{\partial^2 C_6}{\partial z_3^2} \end{vmatrix} = -108000\mathfrak{C}_{12}. \tag{6.35}$$

The Hessian of $\mathfrak{C}_{12}$ with respect to φ and ψ:

$$\text{Hessian}(\mathfrak{C}_{12}) = \begin{vmatrix} \frac{\partial^2 \mathfrak{C}_{12}}{\partial \varphi^2} & \frac{\partial^2 \mathfrak{C}_{12}}{\partial \varphi \partial \psi} \\ \frac{\partial^2 \mathfrak{C}_{12}}{\partial \varphi \partial \psi} & \frac{\partial^2 \mathfrak{C}_{12}}{\partial \psi^2} \end{vmatrix} = -9C_{12}. \tag{6.36}$$

The Hessian of C_{12} with respect to φ and ψ:

$$\text{Hessian}(C_{12}) = \begin{vmatrix} \frac{\partial^2 C_{12}}{\partial \varphi^2} & \frac{\partial^2 C_{12}}{\partial \varphi \partial \psi} \\ \frac{\partial^2 C_{12}}{\partial \varphi \partial \psi} & \frac{\partial^2 C_{12}}{\partial \psi^2} \end{vmatrix} = -15552\mathfrak{C}_{12}. \tag{6.37}$$

The functional determinant of $\mathfrak{C}_{12}$ and C_{12} with respect to φ and ψ:

$$\frac{\partial(\mathfrak{C}_{12}, C_{12})}{\partial(\varphi, \psi)} = \begin{vmatrix} \frac{\partial \mathfrak{C}_{12}}{\partial \varphi} & \frac{\partial \mathfrak{C}_{12}}{\partial \psi} \\ \frac{\partial C_{12}}{\partial \varphi} & \frac{\partial C_{12}}{\partial \psi} \end{vmatrix} = -4C_{18}. \tag{6.38}$$

We have

$$\begin{aligned} &C_{12}^3 : C_{18}^2 : 1728\mathfrak{C}_{12}^3 \\ =&\psi^3(\psi^3 + 216\varphi^3)^3 : (\psi^6 - 540\varphi^3\psi^3 - 5832\varphi^6)^2 : 1728\varphi^3(27\varphi^3 - \psi^3)^3. \end{aligned} \tag{6.39}$$

Note that

$$H_1 := \text{Hessian}(Q_1) = \begin{vmatrix} \frac{\partial^2 Q_1}{\partial z_1^2} & \frac{\partial^2 Q_1}{\partial z_1 \partial z_2} & \frac{\partial^2 Q_1}{\partial z_1 \partial z_3} \\ \frac{\partial^2 Q_1}{\partial z_1 \partial z_2} & \frac{\partial^2 Q_1}{\partial z_2^2} & \frac{\partial^2 Q_1}{\partial z_2 \partial z_3} \\ \frac{\partial^2 Q_1}{\partial z_1 \partial z_3} & \frac{\partial^2 Q_1}{\partial z_2 \partial z_3} & \frac{\partial^2 Q_1}{\partial z_3^2} \end{vmatrix} = -8(\psi - 3\varphi). \tag{6.40}$$

$$H_2 := \text{Hessian}(Q_2) = \begin{vmatrix} \frac{\partial^2 Q_2}{\partial z_1^2} & \frac{\partial^2 Q_2}{\partial z_1 \partial z_2} & \frac{\partial^2 Q_2}{\partial z_1 \partial z_3} \\ \frac{\partial^2 Q_2}{\partial z_1 \partial z_2} & \frac{\partial^2 Q_2}{\partial z_2^2} & \frac{\partial^2 Q_2}{\partial z_2 \partial z_3} \\ \frac{\partial^2 Q_2}{\partial z_1 \partial z_3} & \frac{\partial^2 Q_2}{\partial z_2 \partial z_3} & \frac{\partial^2 Q_2}{\partial z_3^2} \end{vmatrix} = -8(\psi - 3\varphi). \tag{6.41}$$

$$H_3 := \text{Hessian}(\varphi) = 2\varphi. \tag{6.42}$$

$$H_4 := \text{Hessian}(\psi) = 216\varphi. \tag{6.43}$$

We have

$$J_1 := \begin{vmatrix} \frac{\partial^2 Q_1}{\partial z_1^2} & \frac{\partial^2 Q_1}{\partial z_1 \partial z_2} & \frac{\partial^2 Q_1}{\partial z_1 \partial z_3} & \frac{\partial H_1}{\partial z_1} \\ \frac{\partial^2 Q_1}{\partial z_1 \partial z_2} & \frac{\partial^2 Q_1}{\partial z_2^2} & \frac{\partial^2 Q_1}{\partial z_2 \partial z_3} & \frac{\partial H_1}{\partial z_2} \\ \frac{\partial^2 Q_1}{\partial z_1 \partial z_3} & \frac{\partial^2 Q_1}{\partial z_2 \partial z_3} & \frac{\partial^2 Q_1}{\partial z_3^2} & \frac{\partial H_1}{\partial z_3} \\ \frac{\partial H_1}{\partial z_1} & \frac{\partial H_1}{\partial z_2} & \frac{\partial H_1}{\partial z_3} & 0 \end{vmatrix} = 6912[3Q_1^2 - (\psi + 6\varphi)Q_2]. \tag{6.44}$$

$$J_2 := \begin{vmatrix} \frac{\partial^2 Q_2}{\partial z_1^2} & \frac{\partial^2 Q_2}{\partial z_1 \partial z_2} & \frac{\partial^2 Q_2}{\partial z_1 \partial z_3} & \frac{\partial H_2}{\partial z_1} \\ \frac{\partial^2 Q_2}{\partial z_1 \partial z_2} & \frac{\partial^2 Q_2}{\partial z_2^2} & \frac{\partial^2 Q_2}{\partial z_2 \partial z_3} & \frac{\partial H_2}{\partial z_2} \\ \frac{\partial^2 Q_2}{\partial z_1 \partial z_3} & \frac{\partial^2 Q_2}{\partial z_2 \partial z_3} & \frac{\partial^2 Q_2}{\partial z_3^2} & \frac{\partial H_2}{\partial z_3} \\ \frac{\partial H_2}{\partial z_1} & \frac{\partial H_2}{\partial z_2} & \frac{\partial H_2}{\partial z_3} & 0 \end{vmatrix} = 6912[3Q_2^2 - (\psi + 6\varphi)Q_1]. \tag{6.45}$$

$$J_3 := \begin{vmatrix} \frac{\partial^2 \varphi}{\partial z_1^2} & \frac{\partial^2 \varphi}{\partial z_1 \partial z_2} & \frac{\partial^2 \varphi}{\partial z_1 \partial z_3} & \frac{\partial H_3}{\partial z_1} \\ \frac{\partial^2 \varphi}{\partial z_1 \partial z_2} & \frac{\partial^2 \varphi}{\partial z_2^2} & \frac{\partial^2 \varphi}{\partial z_2 \partial z_3} & \frac{\partial H_3}{\partial z_2} \\ \frac{\partial^2 \varphi}{\partial z_1 \partial z_3} & \frac{\partial^2 \varphi}{\partial z_2 \partial z_3} & \frac{\partial^2 \varphi}{\partial z_3^2} & \frac{\partial H_3}{\partial z_3} \\ \frac{\partial H_3}{\partial z_1} & \frac{\partial H_3}{\partial z_2} & \frac{\partial H_3}{\partial z_3} & 0 \end{vmatrix} = -12\varphi^2. \tag{6.46}$$

$$J_4 := \begin{vmatrix} \frac{\partial^2 \psi}{\partial z_1^2} & \frac{\partial^2 \psi}{\partial z_1 \partial z_2} & \frac{\partial^2 \psi}{\partial z_1 \partial z_3} & \frac{\partial H_4}{\partial z_1} \\ \frac{\partial^2 \psi}{\partial z_1 \partial z_2} & \frac{\partial^2 \psi}{\partial z_2^2} & \frac{\partial^2 \psi}{\partial z_2 \partial z_3} & \frac{\partial H_4}{\partial z_2} \\ \frac{\partial^2 \psi}{\partial z_1 \partial z_3} & \frac{\partial^2 \psi}{\partial z_2 \partial z_3} & \frac{\partial^2 \psi}{\partial z_3^2} & \frac{\partial H_4}{\partial z_3} \\ \frac{\partial H_4}{\partial z_1} & \frac{\partial H_4}{\partial z_2} & \frac{\partial H_4}{\partial z_3} & 0 \end{vmatrix} = -36^4\chi. \tag{6.47}$$

$$K_1 := \begin{vmatrix} \frac{\partial Q_1}{\partial z_1} & \frac{\partial Q_1}{\partial z_2} & \frac{\partial Q_1}{\partial z_3} \\ \frac{\partial H_1}{\partial z_1} & \frac{\partial H_1}{\partial z_2} & \frac{\partial H_1}{\partial z_3} \\ \frac{\partial J_1}{\partial z_1} & \frac{\partial J_1}{\partial z_2} & \frac{\partial J_1}{\partial z_3} \end{vmatrix} = 6912 \cdot 24[(\psi + 6\varphi)^3 - 27Q_2^3]. \tag{6.48}$$

$$K_2 := \begin{vmatrix} \frac{\partial Q_2}{\partial z_1} & \frac{\partial Q_2}{\partial z_2} & \frac{\partial Q_2}{\partial z_3} \\ \frac{\partial H_2}{\partial z_1} & \frac{\partial H_2}{\partial z_2} & \frac{\partial H_2}{\partial z_3} \\ \frac{\partial J_2}{\partial z_1} & \frac{\partial J_2}{\partial z_2} & \frac{\partial J_2}{\partial z_3} \end{vmatrix} = -6912 \cdot 24[(\psi + 6\varphi)^3 - 27Q_1^3]. \tag{6.49}$$

$$K_3 := \begin{vmatrix} \frac{\partial \varphi}{\partial z_1} & \frac{\partial \varphi}{\partial z_2} & \frac{\partial \varphi}{\partial z_3} \\ \frac{\partial H_3}{\partial z_1} & \frac{\partial H_3}{\partial z_2} & \frac{\partial H_3}{\partial z_3} \\ \frac{\partial J_3}{\partial z_1} & \frac{\partial J_3}{\partial z_2} & \frac{\partial J_3}{\partial z_3} \end{vmatrix} = 0. \tag{6.50}$$

$$K_4 := \begin{vmatrix} \frac{\partial \psi}{\partial z_1} & \frac{\partial \psi}{\partial z_2} & \frac{\partial \psi}{\partial z_3} \\ \frac{\partial H_4}{\partial z_1} & \frac{\partial H_4}{\partial z_2} & \frac{\partial H_4}{\partial z_3} \\ \frac{\partial J_4}{\partial z_1} & \frac{\partial J_4}{\partial z_2} & \frac{\partial J_4}{\partial z_3} \end{vmatrix} = -36^5 \cdot 54(Q_1^3 - Q_2^3). \tag{6.51}$$

$$(G, H, K) := \begin{vmatrix} \frac{\partial G}{\partial z_1} & \frac{\partial G}{\partial z_2} & \frac{\partial G}{\partial z_3} \\ \frac{\partial H}{\partial z_1} & \frac{\partial H}{\partial z_2} & \frac{\partial H}{\partial z_3} \\ \frac{\partial K}{\partial z_1} & \frac{\partial K}{\partial z_2} & \frac{\partial K}{\partial z_3} \end{vmatrix} = 27[3(Q_1^2 + Q_2^2) - (\psi + 6\varphi)(Q_1 + Q_2)]. \tag{6.52}$$

Put

$$U_1 = z_1 z_2^2, \quad U_2 = z_2 z_3^2, \quad U_3 = z_3 z_1^2, \tag{6.53}$$

$$V_1 = z_1^2 z_2, \quad V_2 = z_2^2 z_3, \quad V_3 = z_3^2 z_1. \tag{6.54}$$

Then

$$U_1 V_1 = XY, \quad U_2 V_2 = YZ, \quad U_3 V_3 = ZX. \tag{6.55}$$

$$U_1 U_2 U_3 = V_1 V_2 V_3 = XYZ = \varphi^3. \tag{6.56}$$

$$Q_1 = U_1 + U_2 + U_3, \quad Q_2 = V_1 + V_2 + V_3, \quad \psi = X + Y + Z. \tag{6.57}$$

The ternary cubic form

$$\begin{aligned} F(x_1, x_2, x_3) =& X x_1^3 + Y x_2^3 + Z x_3^3 + 6\varphi x_1 x_2 x_3 \\ &+ 3(U_1 x_1 x_2^2 + U_2 x_2 x_3^2 + U_3 x_3 x_1^2) + 3(V_1 x_1^2 x_2 + V_2 x_2^2 x_3 + V_3 x_3^2 x_1) \\ =& (z_1 x_1 + z_2 x_2 + z_3 x_3)^3. \end{aligned} \tag{6.58}$$

We have

$$\begin{cases} U_1(A(z_1, z_2, z_3)) = U_2(z_1, z_2, z_3), \\ U_2(A(z_1, z_2, z_3)) = U_3(z_1, z_2, z_3), \\ U_3(A(z_1, z_2, z_3)) = U_1(z_1, z_2, z_3). \end{cases} \tag{6.59}$$

$$\begin{cases} V_1(A(z_1, z_2, z_3)) = V_2(z_1, z_2, z_3), \\ V_2(A(z_1, z_2, z_3)) = V_3(z_1, z_2, z_3), \\ V_3(A(z_1, z_2, z_3)) = V_1(z_1, z_2, z_3). \end{cases} \tag{6.60}$$

$$\begin{cases} U_1(B(z_1, z_2, z_3)) = V_3(z_1, z_2, z_3), \\ U_2(B(z_1, z_2, z_3)) = V_2(z_1, z_2, z_3), \\ U_3(B(z_1, z_2, z_3)) = V_1(z_1, z_2, z_3). \end{cases} \tag{6.61}$$

$$\begin{cases} V_1(B(z_1, z_2, z_3)) = U_3(z_1, z_2, z_3), \\ V_2(B(z_1, z_2, z_3)) = U_2(z_1, z_2, z_3), \\ V_3(B(z_1, z_2, z_3)) = U_1(z_1, z_2, z_3). \end{cases} \tag{6.62}$$

$$\begin{cases} U_1(C(z_1, z_2, z_3)) = \omega^2 U_1(z_1, z_2, z_3), \\ U_2(C(z_1, z_2, z_3)) = \omega^2 U_2(z_1, z_2, z_3), \\ U_3(C(z_1, z_2, z_3)) = \omega^2 U_3(z_1, z_2, z_3). \end{cases} \tag{6.63}$$

$$\begin{cases} V_1(C(z_1, z_2, z_3)) = \omega V_1(z_1, z_2, z_3), \\ V_2(C(z_1, z_2, z_3)) = \omega V_2(z_1, z_2, z_3), \\ V_3(C(z_1, z_2, z_3)) = \omega V_3(z_1, z_2, z_3). \end{cases} \tag{6.64}$$

$$\begin{cases} U_1(D(z_1, z_2, z_3)) = \omega^2 U_1(z_1, z_2, z_3), \\ U_2(D(z_1, z_2, z_3)) = U_2(z_1, z_2, z_3), \\ U_3(D(z_1, z_2, z_3)) = \omega U_3(z_1, z_2, z_3). \end{cases} \tag{6.65}$$

$$\begin{cases} V_1(D(z_1, z_2, z_3)) = \omega V_1(z_1, z_2, z_3), \\ V_2(D(z_1, z_2, z_3)) = V_2(z_1, z_2, z_3), \\ V_3(D(z_1, z_2, z_3)) = \omega^2 V_3(z_1, z_2, z_3). \end{cases} \tag{6.66}$$

$$\begin{cases} (\sqrt{-3})^3 U_1(E(z_1, z_2, z_3)) = X + \omega^2 Y + \omega Z \\ \qquad + (\omega - 1)U_1 + (1 - \omega^2)U_2 + (\omega^2 - \omega)U_3 \\ \qquad + (\omega - \omega^2)V_1 + (1 - \omega)V_2 + (\omega^2 - 1)V_3, \\ (\sqrt{-3})^3 U_2(E(z_1, z_2, z_3)) = X + \omega^2 Y + \omega Z \\ \qquad + (1 - \omega^2)U_1 + (\omega^2 - \omega)U_2 + (\omega - 1)U_3 \\ \qquad + (\omega^2 - 1)V_1 + (\omega - \omega^2)V_2 + (1 - \omega)V_3, \\ (\sqrt{-3})^3 U_3(E(z_1, z_2, z_3)) = X + \omega^2 Y + \omega Z \\ \qquad + (\omega^2 - \omega)U_1 + (\omega - 1)U_2 + (1 - \omega^2)U_3 \\ \qquad + (1 - \omega)V_1 + (\omega^2 - 1)V_2 + (\omega - \omega^2)V_3. \end{cases} \tag{6.67}$$

$$\begin{cases} (\sqrt{-3})^3 V_1(E(z_1, z_2, z_3)) = X + \omega Y + \omega^2 Z \\ \qquad + (\omega - \omega^2)U_1 + (\omega^2 - 1)U_2 + (1 - \omega)U_3 \\ \qquad + (1 - \omega^2)V_1 + (\omega - 1)V_2 + (\omega^2 - \omega)V_3, \\ (\sqrt{-3})^3 V_2(E(z_1, z_2, z_3)) = X + \omega Y + \omega^2 Z \\ \qquad + (1 - \omega)U_1 + (\omega - \omega^2)U_2 + (\omega^2 - 1)U_3 \\ \qquad + (\omega - 1)V_1 + (\omega^2 - \omega)V_2 + (1 - \omega^2)V_3, \\ (\sqrt{-3})^3 V_3(E(z_1, z_2, z_3)) = X + \omega Y + \omega^2 Z \\ \qquad + (\omega^2 - 1)U_1 + (1 - \omega)U_2 + (\omega - \omega^2)U_3 \\ \qquad + (\omega^2 - \omega)V_1 + (1 - \omega^2)V_2 + (\omega - 1)V_3. \end{cases} \tag{6.68}$$

7. Some rational invariants on $\mathbb{CP}^2$

For the Hessian polyhedral equations:

$$\begin{cases} 432\dfrac{C_9^2}{C_6^3} = R_1(x,y), \\ 3\dfrac{C_{12}}{C_6^2} = R_2(x,y), \end{cases} \tag{7.1}$$

there are some particular cases:

$$\begin{cases} 432\dfrac{C_9^2}{C_6^3} = \dfrac{a_1x + a_2y + a_3}{c_1x + c_2y + c_3}, \\ 3\dfrac{C_{12}}{C_6^2} = \dfrac{b_1x + b_2y + b_3}{c_1x + c_2y + c_3}, \end{cases} \tag{7.2}$$

where $\begin{pmatrix} a_1 & a_2 & a_3 \\ b_1 & b_2 & b_3 \\ c_1 & c_2 & c_3 \end{pmatrix} \in GL(3, \mathbb{C})$.

Set

$$x_1 = C_6, \quad x_2 = 12C_9, \quad x_3 = C_{12}, \quad x_4 = 12\mathfrak{C}_{12}, \quad x_5 = C_{18}.$$

Then

$$3x_2^2 = x_1^3 - 3x_1x_3 + 2x_5, \quad x_4^3 = x_5^2 - x_3^3.$$

Put

$$A = \mathbb{C}[x_1, x_2, x_3, x_4, x_5]/I, \tag{7.3}$$

where the ideal

$$I = (x_1^3 - 3x_1x_3 - 3x_2^2 + 2x_5, x_3^3 + x_4^3 - x_5^2). \tag{7.4}$$

We find that

$$A \cong \mathbb{C}[x_1, x_2, x_3, x_4]/(x_3^3 + x_4^3 - \frac{1}{4}(x_1^3 - 3x_1x_3 - 3x_2^2)^2). \tag{7.5}$$

The work of Belyi ([4]) represents one of the most recent advances in Galois theory, known as the theory of Dessins d'Enfants called by its founder Grothendieck (see [24]). The core notion of the theory is that of a Belyi function (see [4]). In the planar case, a Belyi function is a rational function defined on the Riemann complex sphere $\mathbb{CP}^1$ which has at most three critical values, namely, 0, 1 and ∞. A pre-image of the segment $[1, \infty]$ under such a function is a planar map. The coefficients of such a function are algebraic numbers. This leads to an action of the absolute Galois group on planar maps (see [49]).

According to [16], let us recall a classical result due to Klein concerning the classification of genus zero Galois coverings, and related to the classification of regular polyhedra ([42]). We will introduce a certain number of Galois genus zero coverings of the sphere, corresponding to the well-known dessins.

The first family corresponds to the dessins consisting of a star with e rays where e is a positive integer. A corresponding Belyi function is

$$y = f(x) = x^e$$

where e is the degree of the covering, totally ramified over 0 and ∞ and unramified elsewhere. We call these dessins C_e. Their topological Galois group is the cyclic group with e elements, C_e.

The second family corresponds to the polygon with $2e$ edges and admits the following Belyi function

$$-4y = x^e + x^{-e} - 2.$$

We call these dessins D_{2e}. Their topological Galois group is the dihedral group with $2e$ elements, D_{2e}.

The third family corresponds to the tetrahedron T of degree 12 with Galois group the alternating permutation group on 4 letters A_4. A corresponding Belyi function is given by

$$yx^3(x^3+8)^3 = 2^6(x^3-1)^3.$$

The fourth family corresponds to the octahedron O of degree 24 with Galois group the full symmetric group on 4 letters S_4. A corresponding Belyi function is given by

$$y(x^8+14x^4+1)^3 = 2^2 \cdot 3^3 \cdot x^4(x^4-1)^4.$$

The fifth family corresponds to the icosahedron I of degree 60 with Galois group the alternating permutation group on 5 letters A_5. A corresponding Belyi function is given by

$$y(x^{20}+228x^{15}+494x^{10}-228x^5+1)^3 = x^5(x^{10}-11x^5-1)^5.$$

Let us define three rational functions (invariants) associated to the complex projective plane $\mathbb{CP}^2$:

$$\begin{cases} f_1^{\text{Hessian}}(\xi,\eta) := 432\dfrac{C_9(\xi,\eta)^2}{C_6(\xi,\eta)^3}, \\ f_2^{\text{Hessian}}(\xi,\eta) := 3\dfrac{C_{12}(\xi,\eta)}{C_6(\xi,\eta)^2}, \\ f_3^{\text{Hessian}}(\xi,\eta) := \dfrac{C_{18}(\xi,\eta)^2}{1728\mathfrak{C}_{12}(\xi,\eta)^3}. \end{cases} \tag{7.6}$$

In fact,

$$\begin{cases} f_1^{\text{Hessian}}(\xi,\eta) = 432\dfrac{(\xi^3-1)^2(\eta^3-1)^2(\xi^3-\eta^3)^2}{(\xi^6-10\xi^3\eta^3+\eta^6-10\xi^3-10\eta^3+1)^3}, \\ f_2^{\text{Hessian}}(\xi,\eta) = 3\dfrac{(\xi^3+\eta^3+1)[(\eta^3+\eta^3+1)^3+216\xi^3\eta^3]}{(\xi^6-10\xi^3\eta^3+\eta^6-10\xi^3-10\eta^3+1)^2}, \\ f_3^{\text{Hessian}}(\xi,\eta) = \dfrac{[(\xi^3+\eta^3+1)^6-540\xi^3\eta^3(\xi^3+\eta^3+1)^3-5832\xi^6\eta^6]^2}{1728\xi^3\eta^3[27\xi^3\eta^3-(\xi^3+\eta^3+1)^3]^3}. \end{cases} \tag{7.7}$$

Let

$$\begin{cases} F_1^{\text{Hessian}}(\xi,\eta) := (f_1^{\text{Hessian}}(\xi,\eta), f_3^{\text{Hessian}}(\xi,\eta)) \\ \qquad = \left(\frac{432C_9(\xi,\eta)^2}{C_6(\xi,\eta)^3}, \frac{C_{18}(\xi,\eta)^2}{1728\mathfrak{C}_{12}(\xi,\eta)^3}\right), \\ F_2^{\text{Hessian}}(\xi,\eta) := (f_2^{\text{Hessian}}(\xi,\eta), f_3^{\text{Hessian}}(\xi,\eta)) \\ \qquad = \left(\frac{3C_{12}(\xi,\eta)}{C_6(\xi,\eta)^2}, \frac{C_{18}(\xi,\eta)^2}{1728\mathfrak{C}_{12}(\xi,\eta)^3}\right). \end{cases} \tag{7.8}$$

We have

$$(F_1^{\text{Hessian}})^{-1}(0,0) = \{(\xi,\eta) \in \mathbb{C}^2 : C_9(\xi,\eta) = 0, C_{18}(\xi,\eta) = 0\}.$$
$$(F_1^{\text{Hessian}})^{-1}(\infty,0) = \{(\xi,\eta) \in \mathbb{C}^2 : C_6(\xi,\eta) = 0, C_{18}(\xi,\eta) = 0\}.$$
$$(F_1^{\text{Hessian}})^{-1}(0,1) = \{(\xi,\eta) \in \mathbb{C}^2 : C_9(\xi,\eta) = 0, C_{12}(\xi,\eta) = 0\}.$$
$$(F_1^{\text{Hessian}})^{-1}(\infty,1) = \{(\xi,\eta) \in \mathbb{C}^2 : C_6(\xi,\eta) = 0, C_{12}(\xi,\eta) = 0\}.$$
$$(F_1^{\text{Hessian}})^{-1}(0,\infty) = \{(\xi,\eta) \in \mathbb{C}^2 : C_9(\xi,\eta) = 0, \mathfrak{C}_{12}(\xi,\eta) = 0\}.$$
$$(F_1^{\text{Hessian}})^{-1}(\infty,\infty) = \{(\xi,\eta) \in \mathbb{C}^2 : C_6(\xi,\eta) = 0, \mathfrak{C}_{12}(\xi,\eta) = 0\}.$$
$$(F_2^{\text{Hessian}})^{-1}(0,0) = \{(\xi,\eta) \in \mathbb{C}^2 : C_{12}(\xi,\eta) = 0, C_{18}(\xi,\eta) = 0\}.$$
$$(F_2^{\text{Hessian}})^{-1}(\infty,0) = \{(\xi,\eta) \in \mathbb{C}^2 : C_6(\xi,\eta) = 0, C_{18}(\xi,\eta) = 0\}.$$
$$(F_2^{\text{Hessian}})^{-1}(0,1) = \{(\xi,\eta) \in \mathbb{C}^2 : C_{12}(\xi,\eta) = 0\}.$$
$$(F_2^{\text{Hessian}})^{-1}(\infty,1) = \{(\xi,\eta) \in \mathbb{C}^2 : C_6(\xi,\eta) = 0, C_{12}(\xi,\eta) = 0\}.$$
$$(F_2^{\text{Hessian}})^{-1}(0,\infty) = \{(\xi,\eta) \in \mathbb{C}^2 : C_{12}(\xi,\eta) = 0, \mathfrak{C}_{12}(\xi,\eta) = 0\}.$$
$$(F_2^{\text{Hessian}})^{-1}(\infty,\infty) = \{(\xi,\eta) \in \mathbb{C}^2 : C_6(\xi,\eta) = 0, \mathfrak{C}_{12}(\xi,\eta) = 0\}.$$

Hence, the cardinal numbers are given by:

$$\#(F_1^{\text{Hessian}})^{-1}(0,0) = 9 \times 18 = 162.$$
$$\#(F_1^{\text{Hessian}})^{-1}(\infty,0) = 6 \times 18 = 108.$$
$$\#(F_1^{\text{Hessian}})^{-1}(0,1) = 9 \times 12 = 108.$$
$$\#(F_1^{\text{Hessian}})^{-1}(\infty,1) = 6 \times 12 = 72.$$
$$\#(F_1^{\text{Hessian}})^{-1}(0,\infty) = 9 \times 12 = 108.$$
$$\#(F_1^{\text{Hessian}})^{-1}(\infty,\infty) = 6 \times 12 = 72.$$
$$\#(F_2^{\text{Hessian}})^{-1}(0,0) = 12 \times 18 = 216.$$
$$\#(F_2^{\text{Hessian}})^{-1}(\infty,0) = 6 \times 18 = 108.$$
$$\#(F_2^{\text{Hessian}})^{-1}(0,1) = \infty.$$
$$\#(F_2^{\text{Hessian}})^{-1}(\infty,1) = 6 \times 12 = 72.$$
$$\#(F_2^{\text{Hessian}})^{-1}(0,\infty) = 12 \times 12 = 144.$$

$$\#(F_2^{\text{Hessian}})^{-1}(\infty, \infty) = 6 \times 12 = 72.$$

The function $F_1^{\text{Hessian}}(\xi, \eta)$ is ramified at the points $(0, 0)$, $(\infty, 0)$, $(0, 1)$, $(\infty, 1)$, $(0, \infty)$, (∞, ∞), with degrees 162, 108, 108, 72, 108, 72, respectively. Their least common multiple is 648. The function $F_2^{\text{Hessian}}(\xi, \eta)$ is ramified at the points $(0, 0)$, $(\infty, 0)$, $(0, 1)$, $(\infty, 1)$, $(0, \infty)$, (∞, ∞), with degrees 216, 108, ∞, 72, 144, 72, respectively. Their least common multiple is 1296. Thus, the rational invariants F_1^{Hessian} and F_2^{Hessian} correspond to the Hessian groups G_{648} and G_{1296}, respectively.

Let us consider the four lines:

$$\begin{cases} l_1 : f_1^{\text{Hessian}} = 0, \\ l_2 : f_2^{\text{Hessian}} = 0, \\ l_3 : f_1^{\text{Hessian}} = \infty, f_2^{\text{Hessian}} = \infty, \\ l_4 : f_1^{\text{Hessian}} + f_2^{\text{Hessian}} = 1. \end{cases}$$

In fact,

$$l_1 = \{(\xi, \eta) \in \mathbb{C}^2 : C_9(\xi, \eta) = 0\},$$

$$l_2 = \{(\xi, \eta) \in \mathbb{C}^2 : C_{12}(\xi, \eta) = 0\},$$

$$l_3 = \{(\xi, \eta) \in \mathbb{C}^2 : C_6(\xi, \eta) = 0\},$$

$$l_4 = \{(\xi, \eta) \in \mathbb{C}^2 : C_{18}(\xi, \eta) = 0\}.$$

This leads to the geometric object:

$$\mathbb{CP}^2 - \{l_1, l_2, l_3, l_4\}.$$

Let

$$T = \begin{pmatrix} -1 & 0 & 1 \\ & -1 & 1 \\ & & 1 \end{pmatrix}, \quad S_1 = \begin{pmatrix} 1 & 0 & 0 \\ 0 & 0 & 1 \\ 0 & 1 & 0 \end{pmatrix}, \quad S_2 = \begin{pmatrix} & & 1 \\ & 1 & \\ 1 & & \end{pmatrix}. \tag{7.9}$$

Then

$$T(x, y) = (1 - x, 1 - y). \tag{7.10}$$

$$\begin{cases} S_1(x, y) = \left(\dfrac{x}{y}, \dfrac{1}{y}\right), \\ S_1T(x, y) = \left(\dfrac{1 - x}{1 - y}, \dfrac{1}{1 - y}\right), \\ TS_1(x, y) = \left(\dfrac{y - x}{y}, \dfrac{y - 1}{y}\right), \\ S_1TS_1(x, y) = \left(\dfrac{y - x}{y - 1}, \dfrac{y}{y - 1}\right). \end{cases} \tag{7.11}$$

$$\begin{cases} S_2(x,y) = \left(\dfrac{1}{x}, \dfrac{y}{x}\right), \\ S_2T(x,y) = \left(\dfrac{1}{1-x}, \dfrac{1-y}{1-x}\right), \\ TS_2(x,y) = \left(\dfrac{x-1}{x}, \dfrac{x-y}{x}\right), \\ S_2TS_2(x,y) = \left(\dfrac{x}{x-1}, \dfrac{x-y}{x-1}\right). \end{cases} \tag{7.12}$$

For $(x, y) = (\infty, 0)$, we have

$$T(x,y) = (\infty, 1).$$

$$\begin{cases} S_1(x,y) = (\infty, \infty), \\ S_1T(x,y) = (\infty, 1), \\ TS_1(x,y) = (\infty, \infty), \\ S_1TS_1(x,y) = (\infty, 0). \end{cases}$$

$$\begin{cases} S_2(x,y) = (0, 0), \\ S_2T(x,y) = (0, 0), \\ TS_2(x,y) = (1, 1), \\ S_2TS_2(x,y) = (1, 1). \end{cases}$$

The orbit consists of five points:

$$\{(\infty, 0), (\infty, 1), (\infty, \infty), (0, 0), (1, 1)\}.$$

For $(x, y) = (\infty, 1)$, we have

$$T(x,y) = (\infty, 0).$$

$$\begin{cases} S_1(x,y) = (\infty, 1), \\ S_1T(x,y) = (\infty, \infty), \\ TS_1(x,y) = (\infty, 0), \\ S_1TS_1(x,y) = (\infty, \infty). \end{cases}$$

$$\begin{cases} S_2(x,y) = (0, 0), \\ S_2T(x,y) = (0, 0), \\ TS_2(x,y) = (1, 1), \\ S_2TS_2(x,y) = (1, 1). \end{cases}$$

The orbit consists of five points:

$$\{(\infty, 1), (\infty, 0), (\infty, \infty), (0, 0), (1, 1)\}.$$

For $(x, y) = (0, 1)$, we have

$$T(x,y) = (1, 0).$$

$$\begin{cases} S_1(x,y) = (0,1), \\ S_1T(x,y) = (\infty,\infty), \\ TS_1(x,y) = (1,0), \\ S_1TS_1(x,y) = (\infty,\infty). \end{cases}$$

$$\begin{cases} S_2(x,y) = (\infty,\infty), \\ S_2T(x,y) = (1,0), \\ TS_2(x,y) = (\infty,\infty), \\ S_2TS_2(x,y) = (0,1). \end{cases}$$

The orbit consists of three points:

$$\{(0,1),(1,0),(\infty,\infty)\}.$$

According to [63], let

$$G = \langle T_1, T_2, S, U_1, U_2 \rangle \tag{7.13}$$

where

$$T_1 = \begin{pmatrix} 1 & 1 & -\omega \\ & 1 & 1 \\ & & 1 \end{pmatrix}, \quad T_2 = \begin{pmatrix} 1 & \omega & -\omega \\ & 1 & \overline{\omega} \\ & & 1 \end{pmatrix}, \tag{7.14}$$

$$S = -\overline{\omega} J, \quad J = \begin{pmatrix} & & 1 \\ & -1 & \\ 1 & & \end{pmatrix}, \tag{7.15}$$

$$U_1 = \begin{pmatrix} 1 & & \\ & -\omega & \\ & & 1 \end{pmatrix}, \quad U_2 = \begin{pmatrix} -1 & & \\ & -\omega & \\ & & -1 \end{pmatrix}. \tag{7.16}$$

We find that

$$\begin{aligned} G &\cong \langle S, ST_1, ST_2, SU_1, SU_2 : S^2 = (ST_1)^4 = (ST_2)^4 = \omega I, \\ &\qquad\qquad (SU_1)^6 = (SU_2)^6 = I \rangle \\ &\cong \langle a_1, a_2, a_3, a_4, a_5 : a_1^2 = a_2^4 = a_3^4 = 1, a_4^6 = a_5^6 = 1 \rangle \\ &\cong \mathbb{Z}_2 * \mathbb{Z}_4 * \mathbb{Z}_4 * \mathbb{Z}_6 * \mathbb{Z}_6. \end{aligned} \tag{7.17}$$

Note that

$$\begin{cases} \det(\lambda I - S) = (\lambda + \overline{\omega})(\lambda - \overline{\omega})^2, \\ \det(\lambda I - ST_1) = (\lambda + \omega)(\lambda^2 + \overline{\omega}), \\ \det(\lambda I - ST_2) = (\lambda + \omega)(\lambda^2 + \overline{\omega}), \\ \det(\lambda I - SU_1) = (\lambda + 1)(\lambda + \overline{\omega})(\lambda - \overline{\omega}), \\ \det(\lambda I - SU_2) = (\lambda + 1)(\lambda + \overline{\omega})(\lambda - \overline{\omega}). \end{cases}$$

One of the most interesting problems in algebraic geometry is to obtain a generalization of the theory of elliptic modular functions to the case of higher genus. We know that in the elliptic case the invariant j plays a fundamental role. Igusa (see [30]) studied the j invariants for the case of genus two, i.e. the hyperelliptic curves. He gave the invariants J_2, J_4, J_6, J_8 and J_{10}.

For the Picard curve

$$y^3 = x(x-1)(x-k_1)(x-k_2), \tag{7.18}$$

put

$$\begin{cases} J_1 = J_1(k_1, k_2) = \dfrac{k_2^2(k_2-1)^2}{k_1^2(k_1-1)^2(k_1-k_2)^2}, \\ J_2 = J_2(k_1, k_2) = \dfrac{k_1^2(k_1-1)^2}{k_2^2(k_2-1)^2(k_2-k_1)^2}. \end{cases} \tag{7.19}$$

Let us study the following six Picard curves:

$$\begin{cases} y^3 = x(x-1)(x-k_1)(x-k_2), \\ y^3 = x(x-1)(x-k_2)(x-k_3), \\ y^3 = x(x-1)(x-k_3)(x-k_1), \\ y^3 = x(x-1)(x-k_0)(x-k_1), \\ y^3 = x(x-1)(x-k_0)(x-k_2), \\ y^3 = x(x-1)(x-k_0)(x-k_3), \end{cases} \tag{7.20}$$

with $k_0 = k_0(z_1, z_2)$, $k_1 = k_1(z_1, z_2)$, $k_2 = k_2(z_1, z_2)$ and $k_3 = k_3(z_1, z_2)$ (see [63]) where $(z_1, z_2) \in \mathfrak{S}_2 = \{(z_1, z_2) \in \mathbb{C}^2 : z_1 + \overline{z_1} - z_2\overline{z_2} > 0\}$. We are interested in the next six maps:

$$\begin{cases} (J_1(k_1(z_1,z_2), k_2(z_1,z_2)), J_2(k_1(z_1,z_2), k_2(z_1,z_2))) : \mathfrak{S}_2 \to \mathbb{CP}^2, \\ (J_1(k_2(z_1,z_2), k_3(z_1,z_2)), J_2(k_2(z_1,z_2), k_3(z_1,z_2))) : \mathfrak{S}_2 \to \mathbb{CP}^2, \\ (J_1(k_3(z_1,z_2), k_1(z_1,z_2)), J_2(k_3(z_1,z_2), k_1(z_1,z_2))) : \mathfrak{S}_2 \to \mathbb{CP}^2, \\ (J_1(k_0(z_1,z_2), k_1(z_1,z_2)), J_2(k_0(z_1,z_2), k_1(z_1,z_2))) : \mathfrak{S}_2 \to \mathbb{CP}^2, \\ (J_1(k_0(z_1,z_2), k_2(z_1,z_2)), J_2(k_0(z_1,z_2), k_2(z_1,z_2))) : \mathfrak{S}_2 \to \mathbb{CP}^2, \\ (J_1(k_0(z_1,z_2), k_3(z_1,z_2)), J_2(k_0(z_1,z_2), k_3(z_1,z_2))) : \mathfrak{S}_2 \to \mathbb{CP}^2. \end{cases} \tag{7.21}$$

In [63], we find that

(1) Picard curves and Picard integrals

$$s_1\left(\frac{2}{3}, \frac{2}{3}, -\frac{1}{3}; \lambda_1, \lambda_2\right), s_2\left(\frac{2}{3}, \frac{2}{3}, -\frac{1}{3}; \lambda_1, \lambda_2\right).$$

(2) Picard modular functions

$$s_1\left(\frac{3}{4}, \frac{1}{2}, -\frac{1}{4}; J_1, J_2\right), s_2\left(\frac{3}{4}, \frac{1}{2}, -\frac{1}{4}; J_1, J_2\right).$$

Let us recall the following facts:

(1) $GL(2)$: $\mathbb{H}/SL(2,\mathbb{Z})(2) \cong \mathbb{P}^1 - \{0,1,\infty\}$.

(2) $GL(3)$: $\mathfrak{S}_2/U(2,1;\mathcal{O}_K)(1-\omega) \cong \mathbb{P}^2 - \{\text{four points}\}$.

This leads us to study the special functions on the configuration

$$\begin{aligned}&\mathbb{P}^2 - \{\text{four points}\}\\ \cong&\mathbb{P}^2 - \{(1,0),(0,1),(\infty,\infty),(0,0)\} \cong \mathbb{P}^2 - \{(1,0),(0,1),(\infty,\infty),(1,1)\}.\end{aligned}$$

In fact, the fundamental modular function J (Hauptmodul) was first constructed by Dedekind in 1877 and independently by Klein one year later. In our case, we constructed two fundamental modular functions J_1 and J_2.

(1) $GL(2)$ (Dedekind, Klein):

$$J(1-J)\frac{d^2z}{dJ^2} + \left(\frac{2}{3} - \frac{7}{6}J\right)\frac{dz}{dJ} - \frac{1}{144}z = 0.$$

A solution is given by

$$z = F\left(\frac{1}{12}, \frac{1}{12}; \frac{2}{3}; J\right).$$

(2) $GL(3)$ (see [63]):

$$\begin{cases} J_1(1-J_1)\dfrac{\partial^2 z}{\partial J_1^2} + J_2(1-J_1)\dfrac{\partial^2 z}{\partial J_1 \partial J_2} + \left(1 - \dfrac{3}{2}J_1\right)\dfrac{\partial z}{\partial J_1} \\ -\dfrac{1}{4}J_2\dfrac{\partial z}{\partial J_2} - \dfrac{1}{16}z = 0, \\ J_2(1-J_2)\dfrac{\partial^2 z}{\partial J_2^2} + J_1(1-J_2)\dfrac{\partial^2 z}{\partial J_1 \partial J_2} + \left(1 - \dfrac{3}{2}J_2\right)\dfrac{\partial z}{\partial J_2} \\ -\dfrac{1}{4}J_1\dfrac{\partial z}{\partial J_1} - \dfrac{1}{16}z = 0. \end{cases} \tag{7.22}$$

A solution is given by

$$z = F_1\left(\frac{1}{4}; \frac{1}{4}, \frac{1}{4}; 1; J_1, J_2\right). \tag{7.23}$$

The following five values are interesting:

$$(J_1, J_2) = (1,0), \quad (0,1), \quad (\infty,\infty), \quad (0,0) \quad \text{or} \quad (1,1).$$

References

[1] P. Appell and J. Kampé de Fériet, *Fonctions Hypergéométriques et Hypersphériques, Polynomes d'Hermite*, Gauthier-Villars, 1926.

[2] M. F. Atiyah, The icosahedron, Math. Medley **18** (1990), 1-12.

[3] M. F. Atiyah and P. Sutcliffe, Polyhedra in physics, chemistry and geometry, Milan J. Math. **71** (2003), 33-58.

[4] G. V. Belyĭ, On Galois extensions of a maximal cyclotomic field, Math. USSR Izv. **14** (1980), 247-256.

[5] H. Burkhardt, Untersuchungen aus dem Gebiete der hyperelliptischen Modulfunctionen, II, Math. Ann. **38** (1891), 161-224.

[6] H. Burkhardt, Untersuchungen aus dem Gebiete der hyperelliptischen Modulfunctionen, III, Math. Ann. **41** (1893), 313-343.

[7] A. B. Coble, An application of the form-problems associated with certain Cremona groups to the solution of equations of higher degree, Trans. Amer. Math. Soc. **9** (1908), 396-424.

[8] A. B. Coble, An application of Moore's cross-ratio group to the solution of the sextic equation, Trans. Amer. Math. Soc. **12** (1911), 311-325.

[9] A. B. Coble, Point sets and allied Cremona groups, I, Trans. Amer. Math. Soc. **16** (1915), 155-198.

[10] A. B. Coble, Point sets and allied Cremona groups, II, Trans. Amer. Math. Soc. **17** (1916), 345-385.

[11] A. B. Coble, Point sets and allied Cremona groups, III, Trans. Amer. Math. Soc. **18** (1917), 331-372.

[12] A. M. Cohen, Finite complex reflection groups, Ann. Sci. Éc. Norm. Sup. 4e série **9** (1976), 379-436.

[13] P. B. Cohen and J. Wolfart, Algebraic Appell-Lauricella functions, Analysis **12** (1992), 359-376.

[14] P. B. Cohen and J. Wolfart, Fonctions hypergéométriques en plusieurs variables et espaces des modules de variétés abéliennes, Ann. Sci. Éc. Norm. Sup. **26** (1993), 665-690.

[15] J. H. Conway, R. T. Curtis, S. P. Norton, R. A. Parker and R. A. Wilson, *Atlas of Finite Groups, Maximal Subgroups and Ordinary Characters for Simple Groups*, Clarendon Press, Oxford, 1985.

[16] J.-M. Couveignes and L. Granboulan, Dessins from a geometric point of view, In: *The Grothendieck Theory of Dessins d'Enfants*, Edited by L. Schneps, 79-113, London Math. Soc. Lecture Note Series **200**, Cambridge University Press, 1994.

[17] H. S. M. Coxeter, *Regular Polytopes*, Second Edition, The MacMillan Company, New York, 1963.

[18] H. S. M. Coxeter, *Regular Complex Polytopes*, Cambridge University Press, 1974.

[19] H. S. M. Coxeter, Finite groups generated by unitary reflections, Abh. Math. Sem. Univ. Hamburg **31** (1967), 125-135.

[20] P. Deligne and G. D. Mostow, Monodromy of hypergeometric functions and non-lattice integral monodromy, Publ. Math. I.H.E.S. **63** (1986), 5-89.

[21] P. Deligne and G. D. Mostow, *Commensurabilities among Lattices in* $PU(1,n)$, Annals of Math. Studies **132**, Princeton University Press, 1993.

[22] R. Fricke, *Lehrbuch der Algebra*, Vol. II, Braunschweig, 1926.

[23] J. J. Gray, *Linear Differential Equations and Group Theory from Riemann to Poincaré*, Second Edition, Birkhäuser, 2000.

[24] A. Grothendieck, Esquisse d'un programme, In: *Geometric Galois Actions, 1. Around Grothendieck's Esquisse d'un Programme*, Edited by L. Schneps and P. Lochak, 5-48, London Math. Soc. Lecture Note Series **242**, Cambridge University Press, 1997.

[25] C. Hermite, *Œuvres de Charles Hermite*, Vol. II, E. Picard Ed., Gauthier-Villars, 1908.

[26] D. Hilbert, Mathematical problems, Bull. Amer. Math. Soc. **8** (1902), 437-479.

[27] F. Hirzebruch, The ring of Hilbert modular forms for real quadratic fields of small discriminant, In: *Modular Functions of One Variable VI, Proceedings, Bonn 1976*, Edited by J. P. Serre and D. B. Zagier, 287-323, Lecture Notes in Math. **627**, Springer-Verlag, 1977.

[28] R. P. Holzapfel, *Geometry and Arithmetic Around Euler Partial Differential Equations*, Dt. Verl. d. Wiss., Berlin/Reidel, Dordrecht 1986.

[29] B. Hunt, *The Geometry of some special Arithmetic Quotients*, Lecture Notes in Math. **1637**, Springer-Verlag, 1996.

[30] J. Igusa, Arithmetic variety of moduli for genus two, Ann. of Math. **72** (1960), 612-649.

[31] C. G. J. Jacobi, *Fundamenta nova theoriae functionum ellipticarum*, Cambridge University Press, 2012.

[32] F. Klein, Vergleichende Betrachtungen über neuere geometrische Forschungen (Das Erlanger Programm), Programm zum Eintritt in die philosophische Fakultät und den Senat der k. Friedrich-Alexanders-Universität zu Erlangen, Erlangen, A. Deichert, 1872, In: *Gesammelte Mathematische Abhandlungen*, Bd. I, Springer-Verlag, Berlin, 1921, 460-497.

[33] F. Klein, *Gesammelte Mathematische Abhandlungen*, Bd. II, Springer-Verlag, Berlin, 1922.

[34] F. Klein, Über binäre Formen mit linearen Transformationen in sich selbst, Math. Ann. **9** (1875), In: *Gesammelte Mathematische Abhandlungen*, Bd. II, Springer-Verlag, Berlin, 1922, 275-301.

[35] F. Klein, Über (algebraisch integrierbare) lineare Differentialgleichungen (Erster Aufsatz), Math. Ann. **11** (1876), In: *Gesammelte Mathematische Abhandlungen*, Bd. II, Springer-Verlag, Berlin, 1922, 302-306.

[36] F. Klein, Über (algebraisch integrierbare) lineare Differentialgleichungen (Zweiter Aufsatz), Math. Ann. **12** (1877), In: *Gesammelte Mathematische Abhandlungen*, Bd. II, Springer-Verlag, Berlin, 1922, 307-320.

[37] F. Klein, Weitere Untersuchungen über das Ikosaeder, Math. Ann. **12** (1877), In: *Gesammelte Mathematische Abhandlungen*, Bd. II, Springer-Verlag, Berlin, 1922, 321-384.

[38] F. Klein, Über die Auflösung gewisser Gleichungen vom siebenten und achten Grade, Math. Ann. **15** (1879), In: *Gesammelte Mathematische Abhandlungen*, Bd. II, Springer-Verlag, Berlin, 1922, 390-428.

[39] F. Klein, Zur Theorie der allgemeinen Gleichungen sechsten und siebenten Grades, Math. Ann. **28** (1886/87), In: *Gesammelte Mathematische Abhandlungen*, Bd. II, Springer-Verlag, Berlin, 1922, 439-472.

[40] F. Klein, Sur la résolution, par les fonctions hyperelliptiques, de l'équation du vingt-septième degré, de laquelle dépend la détermination des vingt-sept droites d'une surface cubique, J. de Math. pure et appl. **4** (1888), In: *Gesammelte Mathematische Abhandlungen*, Bd. II, Springer-Verlag, Berlin, 1922, 473-479.

[41] F. Klein, Über die Auflösung der allgemeinen Gleichungen fünften und sechsten Grades, Math. Ann. **61** (1905), In: *Gesammelte Mathematische Abhandlungen*, Bd. II, Springer-Verlag, Berlin, 1922, 481-504.

[42] F. Klein, *Lectures on the Icosahedron and the Solution of Equations of the Fifth Degree*, Translated by G. G. Morrice, second and revised edition, Dover Publications, Inc., 1956.

[43] F. Klein, *Vorlesungen über die Entwicklung der Mathematik im 19. Jahrhundert*, Berlin, 1926.

[44] F. Klein, *Vorlesungen über das Ikosaeder und die Auflösung der Gleichungen vom fünften Grade*, Reprint of the 1884 original. Edited, with an introduction and commentary by Peter Slodowy, Birkhäuser Verlag, Basel; B. G. Teubner, Stuttgart, 1993.

[45] F. Klein, Ueber die Transformation der elliptischen Functionen und die Auflösung der Gleichungen fünften Grades, Math. Ann. **14** (1879), 111-172.

[46] F. Klein, Ueber die Transformation siebenter Ordnung der elliptischen Functionen, Math. Ann. **14** (1879), 428-471.

[47] R. P. Langlands, Problems in the theory of automorphic forms, In: *Lectures in Modern Analysis and Applications*, III, 18-61, Lecture Notes in Math. **170**, Springer, Berlin, 1970.

[48] R. P. Langlands, Representation theory: its rise and its role in number theory, In: *Proceedings of the Gibbs Symposium* (New Haven, CT, 1989), 181-210, Amer. Math. Soc., Providence, RI, 1990.

[49] N. Magot and A. Zvonkin, Belyi functions for Archimedean solids, Discrete Math. **217** (2000), 249-271.

[50] Yu. I. Manin, *Cubic Forms, Algebra, Geometry, Arithmetic*, Translated from Russian by M. Hazewinkel, Second Edition, North-Holland, 1986.

[51] H. Maschke, Ueber die quaternäre, endliche, lineare Substitutionsgruppe der Borchardt'schen Moduln, Math. Ann. **30** (1887), 496-515.

[52] H. Maschke, Aufstellung des vollen Formensystems einer quaternären Gruppe von 51840 linearen Substitutionen, Math. Ann. **33** (1889), 317-344.

[53] G. A. Miller, H. F. Blichfeldt and L. E. Dickson, *Theory and Applications of Finite Groups*, Wiley, New York, 1916.

[54] G. D. Mostow, Generalized Picard lattices arising from half-integral conditions, Publ. Math. I.H.E.S. **63** (1986), 91-106.

[55] D. Mumford, *Tata Lectures on Theta II, Jacobian theta functions and differential equations*, Progress in Math. Vol. **43**, Birkhäuser, 1984.

[56] J.-P. Serre, Extensions icosaédriques, Séminaire de Théorie des Nombres de Bordeaux 1979/80, No. 19, In: *Œuvres Collected Papers*, Vol. III, 1972-1984, 550-554, Springer-Verlag, 1986.

[57] C. C. Shephard and J. A. Todd, Finite unitary reflection groups, Canad. J. Math. **6** (1954), 274-304.

[58] P. Slodowy, Das Ikosaeder und die Gleichungen fünften Grades, In: *Arithmetik und Geometrie*, 71-113, Math. Miniaturen, **3**, Birkhäuser, 1986.

[59] H. Umemura, Resolution of algebraic equations by theta constants, In: D. Mumford, *Tata Lectures on Theta II*, Progress in Math. Vol. **43**, Birkhäuser, 1984, 261-272.

[60] H. Weber, *Lehrbuch der Algebra*, Vol. II, Chelsea Publishing Company, New York, 1961.

[61] H. Weyl, *Symmetry*, Princeton University Press, 1952.

[62] A. Witting, Ueber Jacobi'sche Functionen k^{ter} Ordnung zweier Variabler, Math. Ann. **29** (1887), 157-170.

[63] L. Yang, Geometry and arithmetic associated with Appell hypergeometric functions, Part I of this monograph.

Part III. Galois representations arising from twenty-seven lines on a cubic surface and the arithmetic associated with Hessian polyhedra

Contents

1. Introduction

In his celebrated lecture: Mathematical problems (see [17]), which was delivered before the International Congress of Mathematicians at Paris in 1900, David Hilbert gave his famous 23 problems. The 12th problem is the extension of Kronecker's theorem on abelian fields to any algebraic realm of rationality. Hilbert said: "The extension of Kronecker's theorem to the case that, in place of the realm of rational numbers or of the imaginary quadratic field, any algebraic field whatever is laid down as realm of rationality, seems to me of the greatest importance. I regard this problem as one of the most profound and far-reaching in the theory of numbers and of functions. It will be seen that in the problem the three fundamental branches of mathematics, number theory, algebra and function theory, come into closest touch with one another, and I am certain that the theory of analytical functions of several variables in particular would be notably enriched if one should succeed in finding and discussing those functions which play the part for any algebraic number field corresponding to that of the exponential function in the field of rational numbers and of the elliptic modular functions in the imaginary quadratic number field."

In his paper [56], Weil said: " First of all, it will be necessary to extend the theory of abelian functions to non-abelian extensions of fields of algebraic functions. As I hope to show, such an extension is indeed possible. In any case, we face here a series of important and difficult problems, whose solution will perhaps require the efforts of more than one generation." Later on, Weil [58] pointed out: " We take up the non-commutative case."

According to Tate ([52], see also [30]), the biggest problem after Artin's reciprocity law was to extend it in some way to non-abelian extensions, or, from an analytic point of view, to identify his non-abelian L-series. The breakthrough came out of Langlands's discovery in the 1960's of the conjectural principle of functoriality, which included a formulation of a non-abelian reciprocity law as a special case. It turns out that in the non-abelian case, one should describe the representations of $\text{Gal}(\overline{K}/K)$ rather than $\text{Gal}(\overline{K}/K)$ itself. Thus, the global reciprocity law is formulated as a general conjectural correspondence between Galois representations and automorphic forms.

The Langlands program describes profound relations between motives (arithmetic data) and automorphic representations (analytic data). Wiles's spectacular work on the Shimura-Taniyama-Weil conjecture, which established the proof of Fermat's Last Theorem, can be regarded as confirmation of such a relation in the case of elliptic curves. In general, the arithmetic information wrapped up in motives comes from solutions of polynomial equations with rational coefficients. The analytic information from automorphic representations is backed up by the structure of Lie theory. Thus, the Langlands program translates large parts of algebraic geometry into the language of representation theory of Lie groups. The Langlands correspondence embodies a large part of number theory, arithmetic algebraic geometry and representation theory of Lie groups. The Artin conjecture on L-functions and the Ramanujan-Petersson conjecture would follow from the Langlands correspondence. Over number fields, the Langlands correspondence in its full generality seems still to be out of reach. Even its precise formulation is very involved.

The investigation of the relation between the arithmetic of Galois representations and the analytic behavior of corresponding L-series is one of the central topics in number theory and arithmetic geometry. One of the outstanding problems in this field is Artin's conjecture which predicts the holomorphy of the Artin L-series of nontrivial irreducible complex representations of the absolute Galois group of number fields. The idea of describing extensions of $\mathbb{Q}$ via the action of the Galois group $\text{Gal}(\overline{\mathbb{Q}}/\mathbb{Q})$ on certain groups and other algebraic objects is very fruitful. Many examples of constructions of abelian and non-abelian extensions of $\mathbb{Q}$ are based on this idea. A complete classification of all these extensions in terms of Galois representations and in terms of certain objects of analysis and algebraic geometry (automorphic forms and motives) is the central problem of the Langlands program (see [34]).

In fact, elliptic curves can be considered as a special kind of cubic hypersurfaces. A cubic hypersurface is a projective algebraic variety defined by a homogeneous equation $F_3(x_0, \cdots, x_n) = 0$ of degree three with coefficients in some ground field k. When $n = 2$, it gives a cubic curve, i.e., an elliptic curve. Now, we will study the case of $n = 3$, cubic surfaces.

Over an algebraically closed field k, every irreducible cubic surface is a rational surface. The class of a hyperplane section h of a surface F is the canonical class $-K_F$. Any smooth cubic surface can be obtained from the projective plane $\mathbb{P}^2$ by blowing up of 6 generic points. The appropriate birational mapping $\phi : \mathbb{P}^2 \to F$ is determined by the linear system of cubic curves passing through the 6 points. There are 27 straight lines on F, each of which is exceptional; they are the only exceptional curves on F. The configuration of these 27 lines is rich in symmetries. The automorphism group of the corresponding graph is isomorphic to the Weyl group of E_6. Cubic surfaces belong to the class of del Pezzo surfaces, i.e. two-dimensional Fano varieties. Over an algebraically non-closed field k, there are smooth cubic surfaces F which are not birationally isomorphic to $\mathbb{P}^2$ over k. Among these surfaces one finds surfaces possessing k-points, and these are unirational over k. The group of birational automorphisms of a minimal surface has been determined and an arithmetic theory of cubic surfaces has been developed in [32] by the non-associative structures, such as quasi-groups and Moufang loops.

In his paper [57], Weil showed that for nonsingular rational surfaces over an algebraic number field K, the Hasse-Weil L-function, except for a finite number of factors, is the same as a suitable Artin L-function for a certain extension of K. For instance, for a nonsingular cubic surface in the projective 3-space, one thus gets an L-function belonging to the extension of K determined by the 27 straight lines on the surface. The Galois group for this is $W(E_6)$, a group of order $2^7 \cdot 3^4 \cdot 5$ and has a simple subgroup of index 2. Therefore, the Hasse-Weil L-function for a given cubic surface is essentially an Artin L-function of a definitely non-abelian type for one dimensional representations, i.e., characters. Here is a rather unexpected connection between number theory and algebraic geometry.

According to [32], let k be a perfect field and let V run through the smooth cubic surfaces over k. we have the realization problems.

(1) Which subgroups of $W(E_6)$ are realized as an image of the Galois group by its representation on $\mathrm{Pic}(V \otimes \overline{k})$ for some V?

(2) Which extensions of the field k correspond with kernels of such representations?

In his monograph [32], Manin said: "Of course, the answer must depend heavily on k. The derived group $W'(E_6)$ is simple, and the whole of it obviously cannot be realized over a local base field which has only solvable extensions. On the other hand, $W(E_6)$ can be realized over some function field. The more interesting case of a number field is completely unknown. The search for a reasonable approach to problem (2) may involve a pattern for a non-commutative class field theory. Here, it appears that topological-analytic considerations and something like Hecke operators do not suffice."

In the present paper, we will show that three apparently disjoint objects: Galois representations arising from twenty-seven lines on a cubic surface (number theory and arithmetic algebraic geometry), Picard modular forms (automorphic forms), rigid Calabi-Yau threefolds and their arithmetic (Diophantine geometry) are intimately related to Hessian polyhedra and their invariants. We investigate the Artin L-functions for six dimensional representations associated to a kind of cubic surfaces. We study the linear representations of the Galois group $\mathrm{Gal}(\overline{\mathbb{Q}}/\mathbb{Q})$ which constitutes a primary object of number theory and find the correspondence between some transcendental functions and representations. Moreover, there exists the duality which connects them with each other. Our linear representations come from the Picard groups of cubic surfaces. We construct a Galois representation whose image is a proper subgroup of $W(E_6)$, the Weyl group of the exceptional Lie algebra E_6. We give a conjecture about the identification of two different kinds of L-functions which can be considered as a higher dimensional counterpart of the Langlands-Tunnell theorem. We obtain the structure of rational points in the rigid Calabi-Yau threefolds. More precisely, we will investigate the relations among invariant theory (algebra), Picard modular forms (analysis), Galois representations (arithmetic) and cubic surfaces (geometry and topology). Just as the applications of modular curves and elliptic modular functions to number theory, i.e., one uses $GL(2)$ to study $GL(1)$, we will apply Picard modular surfaces and Picard modular functions to number theory, i.e., we will use $GL(3)$ (or $U(3)$) to study $GL(1)$. As an application, we find an answer to Manin's problems described as above. Moreover, we give a conjecture which involves the noncommutative class field theory. It should be pointed out that our Galois representations arise from twenty-seven lines on a cubic surface, while the previous Galois representations either arise from torsion points of an elliptic curve or an abelian variety (e.g. the Tate modules), or arise from Mordell-Weil lattices of an elliptic curve, i.e., both of them come from elliptic curves (cubic curves).

Main Results.

We have the following dictionaries.

(1)

$GL(2)$

elliptic curves (cubic curves)

elliptic modular forms

regular polyhedra

algebraic solutions of hypergeometric differential equations

A_4, S_4, A_5

$PSL(2, \mathcal{O}_K)/PSL(2, \mathcal{O}_K)(\sqrt{5}) = A_5$

Hilbert modular surfaces over $K = \mathbb{Q}(\sqrt{5})$ (Hirzebruch)

diagonal cubic surface of Clebsch and Klein, F_1, 27 real lines

two dimensional real surface

(2)

$GL(3)$
cubic surfaces
Picard modular forms
Hessian polyhedra
algebraic solutions of Appell hypergeometric
partial differential equations
Hessian groups : G_{648}, G_{1296}
$U(2,1;\mathcal{O}_K)/U(2,1;\mathcal{O}_K)(1-\omega) = S_4$(see [22], [24])
Picard modular surfaces over $K = \mathbb{Q}(\sqrt{-3})$
our cubic surface, F_2, 15 real lines and 12 complex lines
two dimensional complex surface

Let us consider the following Galois representation:

$$\rho : \mathrm{Gal}(\overline{K}/K) \to \mathrm{Aut}(X),$$

where X is some geometric object.

(1) $X = V_n(k)$ is an n-dimensional linear space over a field k, such as $\mathbb{C}$, $\mathbb{R}$, $\mathbb{Q}_p$ and $\mathbb{F}_p$, $\mathrm{Aut}(V_n(k)) = GL(n,k)$(Lie groups, p-adic Lie groups, finite groups)

(2) $X = T_\ell(\mu)$ is the Tate module of the number field K, $\mathrm{Aut}(T_\ell(\mu)) = GL(1, \mathbb{Z}_\ell)$($\ell$-adic Lie groups)

(3) $X = T_\ell(E)$ is the Tate module of the elliptic curve E, $\mathrm{Aut}(T_\ell(E)) = GL(2, \mathbb{Z}_\ell)$($\ell$-adic Lie groups)

(4) $X = (E(k(C)), \langle , \rangle)$ is the Mordell-Weil lattice associated to a rational elliptic surface, where $k(C)$ is the function field of a smooth projective curve C over an algebraically closed field k, $\mathrm{Aut}(E(k(C)), \langle , \rangle) = \mathrm{Aut}(E_r)$, $r = 6, 7, 8$(finite groups)

(5) $X = \mathcal{L}$ is the configuration of 27 lines on the cubic surface S preserving the intersection behavior of the lines, $\mathrm{Aut}(\mathcal{L}) = W(E_6)$(finite groups)

From the point of view of finite simple groups, the significance of $W(E_6)$ comes from the following three pairs (A_5, S_5), (A_6, S_6), $(O_5(3), W(E_6))$ and Brauer's theorem (see [4]): If G is a simple group of an order $g = 5 \cdot 3^a \cdot 2^b > 5$, then G is isomorphic to one of the following three groups: the alternating group A_5 of order 60, the alternating group A_6 of order 360 or the orthogonal group $O_5(3)$ of order 25920. The first two pairs appeared in Klein's celebrated book [28]. In this paper, we will study the third one.

One of the significance of the group $W(E_6)$ is that it possesses in terms of different Chevalley groups (see [26], [10]):

$$G_{25,920} \cong PU(3,1;\mathbb{F}_4) \cong PSp(4,\mathbb{Z}/3\mathbb{Z}),$$

where $G_{25,920} \subset W(E_6)$ is the simple subgroup of index two consisting of all even elements. We have the action of the unitary reflection group of order 25,920: $G_{25,920}$ on $\mathbb{P}^4 = \{(Y_0, Y_1, Y_2, Y_3, Y_4)\}$. The following theorems say that the invariants of $G_{25,920}$ (see section two for the details):

$$\Phi, u, t, \Psi_1, C_6, C_9, \Phi_3, t_3, C_{12}, C_{18}$$

can be expressed as the polynomials of the invariants f_0, f_1, G, H, K and C_6, where

$$\Phi = Y_0^4 + 8Y_0Y_1^3,$$
$$u = Y_0\psi + 6Y_1\varphi,$$
$$t = Y_0^6 - 20Y_0^3Y_1^3 - 8Y_1^6,$$
$$\Psi_1 = \psi(-Y_0^3 + 4Y_1^3) + 18\varphi Y_0^2Y_1,$$
$$C_6 = Y_2^6 + Y_3^6 + Y_4^6 - 10(Y_2^3Y_3^3 + Y_3^3Y_4^3 + Y_4^3Y_2^3),$$
$$C_9 = (Y_2^3 - Y_3^3)(Y_3^3 - Y_4^3)(Y_4^3 - Y_2^3),$$
$$\Phi_3 = -\psi^3Y_0 + 18\varphi\psi^2Y_1 + 108\varphi^3Y_0,$$
$$t_3 = \psi^3(Y_0^3 + 8Y_1^3) - 54\varphi\psi^2Y_0^2Y_1 + 324\varphi^2\psi Y_0Y_1^2 + 216\varphi^3(Y_0^3 - Y_1^3),$$
$$C_{12} = (Y_2^3 + Y_3^3 + Y_4^3)[(Y_2^3 + Y_3^3 + Y_4^3)^3 + 216Y_2^3Y_3^3Y_4^3],$$
$$C_{18} = (Y_2^3 + Y_3^3 + Y_4^3)^6 - 540Y_2^3Y_3^3Y_4^3(Y_2^3 + Y_3^3 + Y_4^3)^3 - 5832Y_2^6Y_3^6Y_4^6.$$

The other three invariants are given by

$$\Psi = Y_0^3Y_1 - Y_1^4,$$

$$\Psi_2 = -\psi^2Y_1^2 - 3\varphi\psi Y_0^2 + 18\varphi^2Y_0Y_1,$$
$$\mathfrak{C}_{12} = Y_2Y_3Y_4[27Y_2^3Y_3^3Y_4^3 - (Y_2^3 + Y_3^3 + Y_4^3)^3].$$

Here, the invariants f_0, f_1, G, H, K are given by

$$f_0 = Y_0 + 2Y_1, \quad f_1 = Y_0 - Y_1. \tag{1.1}$$

$$G = (Y_2 - Y_3)(Y_3 - Y_4)(Y_4 - Y_2), \quad H = \psi + 6\varphi, \quad K = \psi - 3\varphi \tag{1.2}$$

with

$$\varphi = Y_2Y_3Y_4, \quad \psi = Y_2^3 + Y_3^3 + Y_4^3. \tag{1.3}$$

Recall that in [61], we have proved the following:

Theorem 1.1 (see [61], Main Theorem 4) (The resolution of algebraic system of equations associated with the Hessian polyhedra). *The invariants G, H and K*

satisfy the following algebraic equations, which are the form-theoretic resolvents (algebraic resolvents) of G, H, K:

$$\begin{cases} 4G^3 + H^2G - C_6G - 4C_9 = 0, \\ H(H^3 + 8K^3) - 9C_{12} = 0, \\ K(K^3 - H^3) - 27\mathfrak{C}_{12} = 0. \end{cases} \tag{1.4}$$

Consequently,

$$C_{18} = -\frac{1}{27}(H^6 - 20H^3K^3 - 8K^6). \tag{1.5}$$

Now, the next theorem can be considered as a generalization of the above theorem form the Hessian groups to the finite simple group $G_{25,920}$:

Theorem 1.2 (Main Theorem 1) (The resolution of algebraic system of equations associated with the simple group $G_{25,920}$). *The invariants* f_0, f_1, H *and* K *satisfy the following algebraic equations, which are the form-theoretic resolvents (algebraic resolvents) of* f_0, f_1, H *and* K:

$$u = \frac{1}{3}(f_0H + 2f_1K). \tag{1.6}$$

$$\begin{cases} \Phi = \frac{1}{9}f_0(f_0^3 + 8f_1^3), \\ \Psi = \frac{1}{9}f_1(f_0^3 - f_1^3), \\ t = -\frac{1}{27}(f_0^6 - 20f_0^3f_1^3 - 8f_1^6). \end{cases} \tag{1.7}$$

$$\begin{cases} \Psi_1 = \frac{1}{9}[(f_0^3 - 4f_1^3)H - 6f_0^2f_1K], \\ \Psi_2 = \frac{1}{9}(-f_0^2HK + 2f_0f_1K^2 - f_1^2H^2), \\ \Phi_3 = \frac{1}{9}[f_0(H^3 - 4K^3) - 6f_1H^2K], \\ t_3 = \frac{1}{27}[f_0^3(H^3 + 8K^3) - 18f_0^2f_1H^2K + 36f_0f_1^2HK^2 + 8f_1^3(H^3 - K^3)] \\ \quad = \frac{1}{27}[(f_0^3 + 8f_1^3)H^3 - 18f_0^2f_1H^2K + 36f_0f_1^2HK^2 + 8(f_0^3 - f_1^3)K^3]. \end{cases} \tag{1.8}$$

We find the following nonlinear duality:

$$f_0 \longleftrightarrow H, \quad f_1 \longleftrightarrow K. \tag{1.9}$$

Here, f_0 and f_1 are linear, H and K are cubic. The functions u and t_3 are invariant under the above nonlinear dual transformation. Under the nonlinear dual transformation (1.9),

$$C_{12} \longleftrightarrow \Phi, \quad \mathfrak{C}_{12} \longleftrightarrow -\frac{1}{3}\Psi, \quad C_{18} \longleftrightarrow t, \quad \Psi_1 \longleftrightarrow \Phi_3. \tag{1.10}$$

We will study the analogies between Hessian invariants and Picard modular forms, which can be considered as the higher dimensional counterpart of the analogies between regular polyhedral invariants and elliptic modular forms studied by Klein.

Conjecture 1.3 (Hessian invariants and Picard modular forms).

$$\mathbb{C}[z_1, z_2, z_3]^{Hessian\ groups} = \mathbb{C}[C_6, C_9, C_{12}]. \tag{1.11}$$

The map

$$\phi : \mathbb{C}[C_6, C_9, C_{12}] \to \mathbb{C}[G_2, G_3, G_4] \tag{1.12}$$

given by

$$\phi(h(C_6, C_9, C_{12})) = h(G_2, G_3, G_4) \tag{1.13}$$

is an algebra isomorphism.

Here,

$$\begin{cases} C_6(z_1, z_2, z_3) = z_1^6 + z_2^6 + z_3^6 - 10(z_1^3 z_2^3 + z_2^3 z_3^3 + z_3^3 z_1^3), \\ C_9(z_1, z_2, z_3) = (z_1^3 - z_2^3)(z_2^3 - z_3^3)(z_3^3 - z_1^3), \\ C_{12}(z_1, z_2, z_3) = (z_1^3 + z_2^3 + z_3^3)[(z_1^3 + z_2^3 + z_3^3)^3 + 216 z_1^3 z_2^3 z_3^3]. \end{cases}$$

G_2, G_3, G_4 are of weight 2, 3 and 4, respectively, which were introduced by Holzapfel in [22].

Now, we will give some further relations between the invariants G, H, K and C_6, C_9, C_{12}, C_{18}, $\mathfrak{C}_{12}$, which improve the results in [61]. Put

$$X(z_1, z_2, z_3) = z_1^3, \quad Y(z_1, z_2, z_3) = z_2^3,$$

$$Z(z_1, z_2, z_3) = z_3^3, \quad \varphi = z_1 z_2 z_3,$$

$$Q_1(z_1, z_2, z_3) = z_1 z_2^2 + z_2 z_3^2 + z_3 z_1^2, \quad Q_2(z_1, z_2, z_3) = z_1^2 z_2 + z_2^2 z_3 + z_3^2 z_1.$$

Theorem 1.4 (Main Theorem 2) (Invariants and their relations, I). *Let*

$$\begin{cases} W_2 = (X+Y+Z)^2 - 12(XY+YZ+ZX), \\ W_3 = (X-Y)(Y-Z)(Z-X), \\ \mathfrak{W}_3 = XYZ - \varphi^3, \\ W_4 = (X+Y+Z)[(X+Y+Z)^3 + 216XYZ], \\ \mathfrak{W}_4 = \varphi[27XYZ - (X+Y+Z)^3], \\ W_6 = (X+Y+Z)^6 - 540XYZ(X+Y+Z)^3 - 5832X^2Y^2Z^2, \\ \mathfrak{V}_3 = Q_1^3 + Q_2^3 - (X+Y+Z+6\varphi)(XY+YZ+ZX) \\ \qquad - 6\varphi^2(X+Y+Z) - 9\varphi^3, \\ \mathfrak{V}_2 = Q_1 Q_2 - (XY+YZ+ZX) - \varphi(X+Y+Z) - 3\varphi^2, \\ \mathfrak{U}_3 = Q_1^3 - Q_2^3 - (X-Y)(Y-Z)(Z-X). \end{cases} \tag{1.14}$$

They satisfy the following four relations:

$$\begin{cases} W_2^3 - 3W_2W_4 + 2W_6 = 432W_3^2, \\ 8U^3(W_6^2 - W_4^3 - 1728\mathfrak{W}_4^3) = 27\mathfrak{W}_3(W_4 - 9U^4)^3, \end{cases} \tag{1.15}$$

where

$$27U^8 - 18W_4U^4 - 8W_6U^2 - W_4^2 = 0, \tag{1.16}$$

and the Jacobian:

$$\frac{\partial(W_2, W_3, W_4, \mathfrak{W}_3)}{\partial(X, Y, Z, \varphi)} = 288\mathfrak{W}_4^2. \tag{1.17}$$

$$\frac{\partial(\mathfrak{V}_3, \mathfrak{V}_2)}{\partial(Q_1, Q_2)} = 3(\mathfrak{U}_3 + W_3). \tag{1.18}$$

We find the following correspondence:

$$(G_2, G_3, G_4, z^2) \longleftrightarrow (W_2, W_3, W_4, \mathfrak{W}_4).$$

Here,

$$\Delta^2 = 16G_2^4G_4 - 128G_2^2G_4^2 - 4G_2^3G_3^2 + 144G_2G_3^2G_4 - 27G_3^4 + 256G_4^3, \quad z^6 = \Delta^2.$$

$$\begin{aligned} 6912\mathfrak{W}_4^3 =& W_2^6 + 9W_2^2W_4^2 + 432^2W_3^4 - 4W_4^3 \\ &- 6W_2^4W_4 - 864W_2^3W_3^2 + 2592W_2W_3^2W_4. \end{aligned}$$

Theorem 1.5 (Main Theorem 3) (Invariants and their relations, II). *Let*

$$x = Q_1 - Q_2, \quad y = X + Y + Z + 6\varphi, \quad z = X + Y + Z - 3\varphi. \tag{1.19}$$

Then x, y and z satisfy the following three equations:

$$\begin{aligned} &2^4 \cdot 3^8\mathfrak{W}_3^2(4\mathfrak{U}_3 - 4x^3 - xy^2 + W_2x + 4W_3)^3 \\ =&2^{10} \cdot 3^{11}\mathfrak{W}_3^2\mathfrak{V}_2^3(\mathfrak{U}_3 + W_3) - (4\mathfrak{U}_3 - 4x^3 - xy^2 + W_2x + 4W_3)\mathfrak{V}_2^2 \\ &\times \{2^8 \cdot 3^{11}\mathfrak{W}_3^2\mathfrak{V}_2 + [(9W_4 - y(y^3 + 8z^3))^2 - 2^6 \cdot 3^{10}W_2\mathfrak{W}_3^2] \\ &+ 2^5[27\mathfrak{W}_4 - z(z^3 - y^3)][9W_4 - y(y^3 + 8z^3)] + 2^8[27\mathfrak{W}_4 - z(z^3 - y^3)]^2\}. \end{aligned}$$

$$\begin{aligned} &[9W_4 - y(y^3 + 8z^3)]^4 \\ &+ 2^{12}[9W_4 - y(y^3 + 8z^3)]\{3^{18}\mathfrak{W}_3^4 + [27\mathfrak{W}_4 - z(z^3 - y^3)]^3\} = 2^{12} \cdot 3^{20}\mathfrak{W}_3^4W_4. \end{aligned}$$

$$\begin{aligned} &[9W_4 - y(y^3 + 8z^3)]^6 \\ &- 2^{11} \cdot 5[9W_4 - y(y^3 + 8z^3)]^3\{3^{18}\mathfrak{W}_3^4 + [27\mathfrak{W}_4 - z(z^3 - y^3)]^3\} \\ &- 2^{21}\{3^{18}\mathfrak{W}_3^4 + [27\mathfrak{W}_4 - z(z^3 - y^3)]^3\}^2 = 2^{18} \cdot 3^{30}\mathfrak{W}_3^6W_6. \end{aligned}$$

We have the following table which describes the relation between Hessian polyhedra, invariant theory and Picard modular forms.

Hessian polyhedra $\longleftrightarrow$ Picard modular functions

$$(C_6, C_9, C_{12}) \longleftrightarrow (G_2, G_3, G_4), \quad \text{weight } 2,\ 3,\ 4$$

$$(G, H, K) \longleftrightarrow (\xi_1, \xi_2, \xi_3), \quad \text{weight } 1$$

$$(\varphi, X, Y, Z, Q_1, Q_2) \longleftrightarrow (\zeta_1, \zeta_2, \zeta_3, \zeta_4, \zeta_5, \zeta_6), \quad \text{weight } 1$$

It is known that the Hessian group of order 216 is generated by the five generators (see [61]):

$$A = \begin{pmatrix} 0 & 1 & 0 \\ 0 & 0 & 1 \\ 1 & 0 & 0 \end{pmatrix}, \quad B = \begin{pmatrix} 1 & 0 & 0 \\ 0 & 0 & 1 \\ 0 & 1 & 0 \end{pmatrix},$$

$$C = \begin{pmatrix} 1 & 0 & 0 \\ 0 & \omega & 0 \\ 0 & 0 & \omega^2 \end{pmatrix}, \quad D = \begin{pmatrix} 1 & 0 & 0 \\ 0 & \omega & 0 \\ 0 & 0 & \omega \end{pmatrix},$$

$$E = \frac{1}{\sqrt{-3}} \begin{pmatrix} 1 & 1 & 1 \\ 1 & \omega & \omega^2 \\ 1 & \omega^2 & \omega \end{pmatrix}.$$

The actions of A, B, C and E on $(X, Y, Z, \varphi, Q_1, Q_2)$ are given as follows:

$$A(X, Y, Z, \varphi, Q_1, Q_2) = \begin{pmatrix} & 1 & & & & \\ & & 1 & & & \\ 1 & & & & & \\ & & & 1 & & \\ & & & & 1 & \\ & & & & & 1 \end{pmatrix} \begin{pmatrix} X \\ Y \\ Z \\ \varphi \\ Q_1 \\ Q_2 \end{pmatrix}, \tag{1.20}$$

$$B(X, Y, Z, \varphi, Q_1, Q_2) = \begin{pmatrix} 1 & & & & & \\ & & 1 & & & \\ & 1 & & & & \\ & & & 1 & & \\ & & & & & 1 \\ & & & & 1 & \end{pmatrix} \begin{pmatrix} X \\ Y \\ Z \\ \varphi \\ Q_1 \\ Q_2 \end{pmatrix}, \tag{1.21}$$

$$C(X, Y, Z, \varphi, Q_1, Q_2) = \begin{pmatrix} 1 & & & & & \\ & 1 & & & & \\ & & 1 & & & \\ & & & 1 & & \\ & & & & \overline{\omega} & \\ & & & & & \omega \end{pmatrix} \begin{pmatrix} X \\ Y \\ Z \\ \varphi \\ Q_1 \\ Q_2 \end{pmatrix}, \tag{1.22}$$

$$(\sqrt{-3})^3 E(X, Y, Z, \varphi, Q_1, Q_2)$$
$$= \begin{pmatrix} 1 & 1 & 1 & 6 & 3 & 3 \\ 1 & 1 & 1 & 6 & 3\overline{\omega} & 3\omega \\ 1 & 1 & 1 & 6 & 3\omega & 3\overline{\omega} \\ 1 & 1 & 1 & -3 & 0 & 0 \\ 3 & 3\overline{\omega} & 3\omega & 0 & 0 & 0 \\ 3 & 3\omega & 3\overline{\omega} & 0 & 0 & 0 \end{pmatrix} \begin{pmatrix} X \\ Y \\ Z \\ \varphi \\ Q_1 \\ Q_2 \end{pmatrix}, \tag{1.23}$$

Proposition 1.6. *The space curves* $W_2 = \mathfrak{W}_3 = \mathfrak{V}_3 = \mathfrak{V}_2 = 0$, $W_3 = \mathfrak{W}_3 = \mathfrak{V}_3 = \mathfrak{V}_2 = 0$, $W_4 = \mathfrak{W}_3 = \mathfrak{V}_3 = \mathfrak{V}_2 = 0$, $W_6 = \mathfrak{W}_3 = \mathfrak{V}_3 = \mathfrak{V}_2 = 0$ *and* $\mathfrak{W}_4 = \mathfrak{W}_3 = \mathfrak{V}_3 = \mathfrak{V}_2 = 0$ *are invariant curves on the invariant surface* $\mathfrak{W}_3 = \mathfrak{V}_3 = \mathfrak{V}_2 = 0$ *under the action of the subgroup of Hessian group generated by* A, B, C *and* E.

Put

$$g_1 = E(x), g_2 = EC(x), g_3 = EC^2(x), g_4 = x, g_5 = C(x), g_6 = C^2(x). \tag{1.24}$$

We find that

$$S: \begin{cases} g_1 + g_2 + g_3 = 0, \\ g_4 + g_5 + g_6 = 0, \\ g_1^3 + g_2^3 + g_3^3 = g_4^3 + g_5^3 + g_6^3, \end{cases} \tag{1.25}$$

which is a cubic surface.

The study of the general cubic surface dates from 1849, in which year the 27 lines were found by Cayley and Salmon; the discovery of the Sylvester pentahedron followed two years later, and the theory attained a remarkable degree of elegance and completeness with the introduction of the plane representation, made independently by Clebsch and Cremona in 1866 (see [44]). Nevertheless, even today the study of what Cayley has called 'the complicated and many-sided symmetry' of the relations between the 27 lines can not be said to have been exhausted, especially Schläfli's well-known notation for the lines readily lend itself to the description and deeper investigation of that symmetry (see e.g. [45] and [46]).

As far as the equation defining a cubic surface is concerned, we have the following result, claimed by Sylvester (1851) and Steiner, and proved by Clebsch (1861). It is the so-called pentahedral form of the cubic.

Theorem 1.7 (see [26]). *A general cubic form* $F(x_0 : x_1 : x_2 : x_3)$ *can be put, in a unique way, in the form*

$$a_1 y_1^3 + a_2 y_2^3 + a_3 y_3^3 + a_4 y_4^3 + a_5 y_5^3 = 0,$$

where the coordinates y_i *are linear in* $(x_0 : x_1 : x_2 : x_3)$ *and satisfy*

$$y_1 + y_2 + y_3 + y_4 + y_5 = 0.$$

A second special form of the equation, the hexahedral form, is given by the polar hexagons:

Theorem 1.8 (see [26]). *A general cubic form $F(x_0 : x_1 : x_2 : x_3)$ can be put in a four-dimensional family of forms:*

$$z_1^3 + z_2^3 + z_3^3 + z_4^3 + z_5^3 + z_6^3 = 0,$$

where the z_i are linear in the x_i.

In fact, the pentahedral form of the cubic is closed related to the Clebsch-Klein diagonal cubic surface. In 1873 Klein proved that the famous diagonal surface of Clebsch, which is the surface in $\mathbb{P}^4(\mathbb{C})$ with equations

$$x_0^3 + x_1^3 + x_2^3 + x_3^3 + x_4^3 = 0, \quad x_0 + x_1 + x_2 + x_3 + x_4 = 0,$$

can be obtained from $\mathbb{P}^2(\mathbb{C})$ by blowing up 6 points in $\mathbb{P}^2(\mathbb{C})$ in a special position, namely the 6 points in $\mathbb{P}^2(\mathbb{R}) = S^2/\{\pm 1\}$ corresponding to the 12 vertices of an icosahedron inscribed in S^2. Hirzebruch blew up 10 more points, namely those corresponding to the 20 vertices of the dual dodecahedron. The resulting surface Y can also be obtained from the Clebsch diagonal surface by blowing up 10 Eckhardt points, that is points, where 3 of the 27 lines on the surface meet. Let $\mathcal{O}_K \subset \mathbb{Q}(\sqrt{5})$ be the ring of integers and $\Gamma \subset SL(2, \mathcal{O}_K)$ the congruence subgroup mod 2. Hirzebruch proved that the icosahedral surface Y is the minimal resolution of $\overline{\mathbb{H}^2/\Gamma}$.

The significance of our cubic surface is that it is the hypersurface of the hexahedral form of the cubic.

There are 27 lines on the cubic surface S:

$$l_1 : \quad g_1 = g_2 + g_3 = g_4 = g_5 + g_6 = 0,$$

$$l_2 : \quad g_1 = g_2 + g_3 = g_5 = g_4 + g_6 = 0,$$

$$l_3 : \quad g_1 = g_2 + g_3 = g_6 = g_4 + g_5 = 0,$$

$$l_4 : \quad g_2 = g_1 + g_3 = g_4 = g_5 + g_6 = 0,$$

$$l_5 : \quad g_2 = g_1 + g_3 = g_5 = g_4 + g_6 = 0,$$

$$l_6 : \quad g_2 = g_1 + g_3 = g_6 = g_4 + g_5 = 0,$$

$$l_7 : \quad g_3 = g_1 + g_2 = g_4 = g_5 + g_6 = 0,$$

$$l_8 : \quad g_3 = g_1 + g_2 = g_5 = g_4 + g_6 = 0,$$

$$l_9 : \quad g_3 = g_1 + g_2 = g_6 = g_4 + g_5 = 0,$$

and

$$l_{j,1} : \quad g_1 = \omega^j g_4, \quad g_2 = \omega^j g_5, \quad g_3 = \omega^j g_6, \quad g_1 + g_2 + g_3 = 0,$$

$$l_{j,2} : \quad g_1 = \omega^j g_4, \quad g_2 = \omega^j g_6, \quad g_3 = \omega^j g_5, \quad g_1 + g_2 + g_3 = 0,$$

$$l_{j,3} : \quad g_1 = \omega^j g_5, \quad g_2 = \omega^j g_4, \quad g_3 = \omega^j g_6, \quad g_1 + g_2 + g_3 = 0,$$

$$l_{j,4} : \quad g_1 = \omega^j g_5, \quad g_2 = \omega^j g_6, \quad g_3 = \omega^j g_4, \quad g_1 + g_2 + g_3 = 0,$$

$$l_{j,5}: \quad g_1 = \omega^j g_6, \quad g_2 = \omega^j g_4, \quad g_3 = \omega^j g_5, \quad g_1 + g_2 + g_3 = 0,$$

$$l_{j,6}: \quad g_1 = \omega^j g_6, \quad g_2 = \omega^j g_5, \quad g_3 = \omega^j g_4, \quad g_1 + g_2 + g_3 = 0,$$

where $j \equiv 0, 1, 2 \pmod 3$.

Let us consider the configuration of the 27 lines (forgetting the cubic surface). That means we consider simply the set of 27 lines together with the incidence relations they satisfy. There is a group of automorphisms of the configuration meaning the permutations of the set of 27 lines, preserving the incidence relations.

There are three difficulties: 1. The realization of 27 lines on the cubic surface by algebraic equations. 2. Under the action of the subgroup of $W(E_6)$, 27 lines on the cubic surface may not lie on the same cubic surface. 3. Under the action of the subgroup of $W(E_6)$, the incidence relations of 27 lines may not preserve.

The double six is given by

$$N = \begin{pmatrix} a_1 & a_2 & a_3 & a_4 & a_5 & a_6 \\ b_1 & b_2 & b_3 & b_4 & b_5 & b_6 \end{pmatrix} = \begin{pmatrix} l_{1,1} & l_{2,2} & l_{2,3} & l_{1,4} & l_{1,5} & l_{2,6} \\ l_{2,1} & l_{1,2} & l_{1,3} & l_{2,4} & l_{2,5} & l_{1,6} \end{pmatrix}. \tag{1.26}$$

We find that

$$c_{12} = l_1, \quad c_{13} = l_9, \quad c_{14} = l_{0,5}, \quad c_{15} = l_{0,4}, \quad c_{16} = l_5. \tag{1.27}$$

$$c_{23} = l_{0,6}, \quad c_{24} = l_6, \quad c_{25} = l_8, \quad c_{26} = l_{0,3}. \tag{1.28}$$

$$c_{34} = l_2, \quad c_{35} = l_4, \quad c_{36} = l_{0,2}. \tag{1.29}$$

$$c_{45} = l_{0,1}, \quad c_{46} = l_7, \quad c_{56} = l_3. \tag{1.30}$$

We find that the action of the group H generated by A, B, C and E on the 27 lines of our cubic surface S gives the permutations of the lines which preserve the intersection behavior of the lines.

Theorem 1.9 (Main Theorem 4). *The group H generated by A, B, C and E is a subgroup of the group of automorphisms of the configuration of* 27 *lines on the cubic surface: $Aut(\mathcal{L}) = W(E_6)$.*

This gives the Galois representation:

$$\rho : G_{\mathbb{Q}} = \mathrm{Gal}(\overline{\mathbb{Q}}/\mathbb{Q}) \to \mathrm{Aut}(\mathcal{L}) = W(E_6). \tag{1.31}$$

Here,

$$\mathrm{Im}(\rho) = H = \langle A, B, C, E \rangle, \quad \ker(\rho) = \mathrm{Gal}(\overline{\mathbb{Q}}/\mathcal{K}). \tag{1.32}$$

We have

$$\mathrm{Im}(\rho) \cong \mathrm{Gal}(\overline{\mathbb{Q}}/\mathbb{Q})/\ker(\rho) = \mathrm{Gal}(\overline{\mathbb{Q}}/\mathbb{Q})/\mathrm{Gal}(\overline{\mathbb{Q}}/\mathcal{K}) \cong \mathrm{Gal}(\mathcal{K}/\mathbb{Q}). \tag{1.33}$$

Hence,

$$\mathrm{Gal}(\mathcal{K}/\mathbb{Q}) \cong H. \tag{1.34}$$

Conjecture 1.10 (Main Conjecture). *Let $\rho : Gal(\overline{\mathbb{Q}}/\mathbb{Q}) \to W(E_6)$ be a Galois representation. When the image of ρ: $Im(\rho) \cong H \leq W(E_6)$, if ρ is odd, i.e.,*

$\det \rho(\sigma_c) = -1$, where $\sigma_c \in Gal(\overline{\mathbb{Q}}/\mathbb{Q})$ is the complex conjugate, then there exists a Picard modular form f of weight one such that

$$L(\rho, s) = L(f, s)$$

up to finitely many Euler factors, where $L(\rho, s)$ is the Artin L-function associated to the Galois representation ρ and $L(f, s)$ is the automorphic L-function associated to the Picard modular form f. If ρ is even, i.e., $\det \rho(\sigma_c) = 1$, then there exists an automorphic form f associated to $U(2,1)$: $\Delta f = s(s-2)f$ with $s = 1$ and (see [59])

$$\Delta = (z_1 + \overline{z_1} - z_2\overline{z_2}) \left[(z_1 + \overline{z_1})\frac{\partial^2}{\partial z_1 \partial \overline{z_1}} + \frac{\partial^2}{\partial z_2 \partial \overline{z_2}} + z_2 \frac{\partial^2}{\partial \overline{z_1} \partial z_2} + \overline{z_2} \frac{\partial^2}{\partial z_1 \partial \overline{z_2}}\right],$$
$$(z_1, z_2) \in \mathfrak{S}_2 = \{(z_1, z_2) \in \mathbb{C}^2 : z_1 + \overline{z_1} - z_2\overline{z_2} > 0\},$$

such that

$$L(\rho, s) = L(f, s)$$

up to finitely many Euler factors, where $L(\rho, s)$ is the Artin L-function associated to the Galois representation ρ and $L(f, s)$ is the automorphic L-function associated to the automorphic form f.

Note that the group $W(E_6)$ is combinatorially defined, while the subgroup H is defined on algebraic variety. Thus we give a connection between combinatoric and algebraic geometry. Furthermore, the Galois representation ρ : $\text{Gal}(\overline{\mathbb{Q}}/\mathbb{Q}) \to H$ comes from number theory, the Picard modular forms come from analysis and representation theory of Lie groups. Here, Galois symmetry and geometric symmetry meet together. Therefore, our theorem and conjecture gives a connection which involves combinatoric, algebraic geometry, number theory, analysis and representation theory. More precisely, we find that Hessian polyhedra and their invariants play a role of bridge which connects Galois representations arising from 27 lines on a cubic surface (arithmetic) with Picard modular forms (analysis). In fact, there are two aspects about Hessian polyhedra. One is Hessian group which is related to analysis, the other is the system of invariants which is related to algebra. Thus, Hessian polyhedra give applications to number theory and algebraic geometry, especially arithmetic algebraic geometry and noncommutative class field theory.

Theorem 1.11 (Main Theorem 5). *The variety associated with Hessian polyhedra*

$$E: \quad y^2 = x^3 - 27C_{12}x + 54C_{18} \tag{1.35}$$

can be expressed as the fiber product of two isogenous, semi-stable, rational elliptic modular surfaces

$$E_{1,t}: \quad (t-1)(z_1^3 + z_2^3 + z_3^3) - 3(t+2)z_1z_2z_3 = 0$$

and

$$E_{2,t}: \quad y^2 = x^3 - 3t(t^3 + 8)x - 2(t^6 - 20t^3 - 8).$$

Hence, E is a rigid Calabi-Yau threefold. There are only finitely many rational points in $E_{1,t}$ and $E_{2,t}$. However, there exist infinitely many nontrivial rational points in E. More precisely, $P = (3C_6, 108C_9) \in E$ is of infinite order.

The present paper consists of five sections. In section two, we study Hessian polyhedra and cubic forms associated to the finite simple group $G_{25,920}$. In section three, we study Hessian polyhedra and Picard modular forms. In section four, we investigate Hessian polyhedra and Galois representations arising from 27 lines on a cubic surface. In the last section, we study Hessian polyhedra and the arithmetic of rigid Calabi-Yau threefolds.

Acknowledgement. The author thanks John McKay for his stimulating discussion. The author also thanks the referees very much for their helpful comments.

2. Hessian polyhedra and cubic forms associated to $G_{25,920}$

Let us recall some basic facts about the unitary reflection groups of order 25,920 (see [26]). We have actions of $G_{25,920}$ on $\mathbb{P}^3$ and on $\mathbb{P}^4$, both of which in fact are generated by unitary reflections.

Burkhardt (see [7] and [8]) determined the invariants of $G_{25,920}$ acting on $\mathbb{P}^4$. Let the coordinates be given as

$$Y_0 = Y_{00}, Y_1 = Y_{10}, Y_2 = Y_{01}, Y_3 = Y_{11}, Y_4 = Y_{12}$$

where the $Y_{\alpha\beta}$ are the theta functions of

$$\mathbb{P}^4 = \{Y_{\alpha\beta} = \frac{1}{2}(X_{\alpha\beta} + X_{-\alpha-\beta})\}.$$

Here

$$X_{\alpha\beta} = \Theta \begin{bmatrix} 0 & 0 \\ \frac{\alpha}{3} & \frac{\beta}{3} \end{bmatrix} (\tau, z), \quad \tau \in \mathbb{S}_2, z \in \mathbb{C}^2, \alpha, \beta \in \mathbb{Z}/3\mathbb{Z}.$$

Using the isomorphism $G_{25,920} \cong PSp(4, \mathbb{Z}/3\mathbb{Z})$, it suffices to use generators of the symplectic group $PSp(4, \mathbb{Z}/3\mathbb{Z})$ to get generators of the action of $G_{25,920}$ on $\mathbb{P}^4$. However, Burkhardt used instead the corresponding hyperelliptic curves, and a certain Weierstrass form for them to describe the level 3 structure. The generators of the group $G_{25,920}$ are transformations B, C, D and S_2, which act on $\mathbb{P}^4 = \{(Y_0, Y_1, Y_2, Y_3, Y_4)\}$ as follows:

$$\begin{aligned} &B(Y_0, Y_1, Y_2, Y_3, Y_4) \\ =&\frac{1}{\sqrt{-3}}(Y_0 + 2Y_1, Y_0 - Y_1, Y_2 + Y_3 + Y_4, Y_2 + \omega Y_3 + \omega^2 Y_4, Y_2 + \omega^2 Y_3 + \omega Y_4), \end{aligned}$$

$$\begin{aligned} C(Y_0, Y_1, Y_2, Y_3, Y_4) &= (Y_0, Y_1, Y_4, Y_2, Y_3), \\ D(Y_0, Y_1, Y_2, Y_3, Y_4) &= (-Y_0, -Y_2, -Y_1, -Y_3, -Y_4), \\ S_2(Y_0, Y_1, Y_2, Y_3, Y_4) &= (Y_0, \omega^2 Y_1, Y_2, \omega^2 Y_3, \omega^2 Y_4). \end{aligned}$$

In the form of matrices, we have

$$B = \frac{1}{\sqrt{-3}} \begin{pmatrix} 1 & 2 & & & \\ 1 & -1 & & & \\ & & 1 & 1 & 1 \\ & & 1 & \omega & \omega^2 \\ & & 1 & \omega^2 & \omega \end{pmatrix}, \quad C = \begin{pmatrix} 1 & & & & \\ & 1 & & & \\ & & 0 & 0 & 1 \\ & & 1 & 0 & 0 \\ & & 0 & 1 & 0 \end{pmatrix},$$

$$D = \begin{pmatrix} -1 & & & & \\ & 0 & -1 & & \\ & -1 & 0 & & \\ & & & -1 & \\ & & & & -1 \end{pmatrix}, \quad S_2 = \begin{pmatrix} 1 & & & & \\ & \omega^2 & & & \\ & & 1 & & \\ & & & \omega^2 & \\ & & & & \omega^2 \end{pmatrix}.$$

Note that

$$B^4 = C^3 = D^2 = S_2^3 = I$$

and

$$B^2 = -\begin{pmatrix} 1 & & & & \\ & 1 & & & \\ & & 1 & 0 & 0 \\ & & 0 & 0 & 1 \\ & & 0 & 1 & 0 \end{pmatrix}.$$

In fact, $B, C, D, S_2 \in SL(5, \mathbb{C})$.

Consider the hyperplane $\mathcal{S}$ given by $\mathcal{S} = \{Y_0 = 0\}$ and the plane $\mathcal{J} = \{Y_0 = Y_1 = 0\}$. The stabilizer of each is a subgroup of order 648, but these two subgroups of order 648 are not conjugate to each other; indeed,

$$N(\mathcal{S}) = \langle C, D, S_2 \rangle, \quad N(\mathcal{J}) = \langle B, C, S_2 \rangle.$$

Hence, $G_{25,920}$ is generated by the subgroup of order 648 acting on $\mathcal{J}$, generated by B, C and S_2, and by the centralizer of $\mathcal{J}$. Burkhardt shows this to be a homogenous tetrahedral group, and its invariants are:

$$\begin{cases} \Phi = Y_0^4 + 8Y_0Y_1^3, \\ \Psi = Y_0^3Y_1 - Y_1^4, \\ t = Y_0^6 - 20Y_0^3Y_1^3 - 8Y_1^6, \end{cases} \tag{2.1}$$

which satisfy the relation:

$$\Phi^3 - 64\Psi^3 = t^2. \tag{2.2}$$

Furthermore, set

$$\begin{cases} \varphi = Y_2Y_3Y_4, \\ \psi = Y_2^3 + Y_3^3 + Y_4^3, \\ u = Y_0\psi + 6Y_1\varphi. \end{cases} \tag{2.3}$$

All invariants of $G_{25,920}$ can be written as linear combinations of the following:

$$\Phi, u, t, \Psi_1, C_6, C_9, \Phi_3, t_3, C_{12}, C_{18},$$

where

$$\begin{cases} \Psi_1 = \psi(-Y_0^3 + 4Y_1^3) + 18\varphi Y_0^2 Y_1, \\ \Psi_2 = -\psi^2 Y_1^2 - 3\varphi\psi Y_0^2 + 18\varphi^2 Y_0 Y_1, \\ \Phi_3 = -\psi^3 Y_0 + 18\varphi\psi^2 Y_1 + 108\varphi^3 Y_0, \\ t_3 = \psi^3(Y_0^3 + 8Y_1^3) - 54\varphi\psi^2 Y_0^2 Y_1 + 324\varphi^2\psi Y_0 Y_1^2 + 216\varphi^3(Y_0^3 - Y_1^3), \end{cases} \tag{2.4}$$

and

$$\begin{cases} C_6 = Y_2^6 + Y_3^6 + Y_4^6 - 10(Y_2^3 Y_3^3 + Y_3^3 Y_4^3 + Y_4^3 Y_2^3), \\ C_9 = (Y_2^3 - Y_3^3)(Y_3^3 - Y_4^3)(Y_4^3 - Y_2^3), \\ C_{12} = (Y_2^3 + Y_3^3 + Y_4^3)[(Y_2^3 + Y_3^3 + Y_4^3)^3 + 216 Y_2^3 Y_3^3 Y_4^3], \\ \mathfrak{C}_{12} = Y_2 Y_3 Y_4 [27 Y_2^3 Y_3^3 Y_4^3 - (Y_2^3 + Y_3^3 + Y_4^3)^3], \\ C_{18} = (Y_2^3 + Y_3^3 + Y_4^3)^6 - 540 Y_2^3 Y_3^3 Y_4^3 (Y_2^3 + Y_3^3 + Y_4^3)^3 - 5832 Y_2^6 Y_3^6 Y_4^6. \end{cases} \tag{2.5}$$

They satisfy the following relations:

$$\begin{cases} \Psi_1^2 = 16\Psi\Psi_2 + u^2\Phi, \\ \Psi_1^3 = 2\Phi^2\Phi_3 - tt_3 - 3u^2\Phi\Psi_1 - u^3 t, \end{cases} \tag{2.6}$$

and

$$\begin{cases} 432 C_9^2 = C_6^3 - 3C_6 C_{12} + 2C_{18}, \\ 1728 \mathfrak{C}_{12}^3 = C_{18}^2 - C_{12}^3. \end{cases} \tag{2.7}$$

Burkhardt calculates the following expressions (see [7], [8], [26]). There are invariants of degrees 4, 6, 10, 12, 18 and 45. The invariant of degree 45 is just the product of the 45 reflection hyperplanes defining the arrangement in $\mathbb{P}^4$. The others are given by:

$$J_4 = \Phi + 8u,$$

$$J_6 = t + 20\Psi_1 - 8C_6,$$

$$J_{10} = \frac{1}{24}(\Phi\Psi_1 + ut + 2\Phi C_6 + 2u\Psi_1 - 2\Phi_3 - 2uC_6),$$

$$J_{12} = \frac{1}{24}(3t\Psi_1 + 3u\Phi^2 + 19\Psi_1^2 - 9u^2\Phi - 10C_6 t - 11t_3 + 9u^3 - 2C_6\Psi_1 - 4C_{12} + 4C_6^2),$$

$$\begin{aligned} J_{18} = &\frac{1}{864}(72t\Psi\Psi_2 + 9u\Phi^2\Psi_1 + 9u^2\Phi t + 288C_6\Psi^3 \\ &+ 4\Phi^2\Phi_3 - 18tt_3 - 42u^2\Phi\Psi_1 - 20u^3 t - 18C_6 t\Psi_1 \\ &- 18C_6\Phi^2 u + 84C_{12}t - 72u\Phi\Phi_3 + 162u^3\Psi_1 \\ &- 240C_6\Psi\Psi_2 + 12C_6 u^2\Phi - 6C_6^2 t + 24\Psi_1 C_{12} - 36u^2\Phi_3 \\ &- 6C_6 t_3 - 18C_6 u^3 - 12C_6^2\Psi_1 - 4C_{18} + 6C_6 C_{12} - 2C_6^3). \end{aligned}$$

There is a relation between these, which takes the form $J_{45}^2 =$ rational expression in the other invariants. There is also an invariant of degree 40: the product of the 40 Steiner primes. Denote this by J_{40}, the following relation holds:

$$3^{33} F_{40} = [J_4^2(2^9 J_{12} - J_4^3) - 3 \cdot 2^{18} J_{10}^2]^2 - 2^{19}[J_4 J_6 - 3 \cdot 2^8 J_{10}][J_4^3 J_{18} - 2^{11} J_{10}^3].$$

Now, we put

$$f_0 = Y_0 + 2Y_1, \quad f_1 = Y_0 - Y_1. \tag{2.8}$$

Then

$$Y_0 = \frac{1}{3}(f_0 + 2f_1), \quad Y_1 = \frac{1}{3}(f_0 - f_1).$$

Recall that (see [61])

$$H = \psi + 6\varphi, \quad K = \psi - 3\varphi. \tag{2.9}$$

Thus,

$$\psi = \frac{1}{3}(H + 2K), \quad \varphi = \frac{1}{9}(H - K).$$

We have the following theorem which says that the above invariants of $G_{25,920}$ can be expressed as the polynomials of the invariants f_0, f_1, H and K.

Theorem 2.1 (Main Theorem 1). *The invariants f_0, f_1, H and K satisfy the following algebraic equations, which are the form-theoretic resolvents (algebraic resolvents) of f_0, f_1, H and K:*

$$u = \frac{1}{3}(f_0 H + 2f_1 K). \tag{2.10}$$

$$\begin{cases} \Phi = \dfrac{1}{9} f_0(f_0^3 + 8f_1^3), \\ \Psi = \dfrac{1}{9} f_1(f_0^3 - f_1^3), \\ t = -\dfrac{1}{27}(f_0^6 - 20 f_0^3 f_1^3 - 8 f_1^6). \end{cases} \tag{2.11}$$

$$\begin{cases} \Psi_1 = \dfrac{1}{9}[(f_0^3 - 4f_1^3)H - 6f_0^2 f_1 K], \\ \Psi_2 = \dfrac{1}{9}(-f_0^2 HK + 2f_0 f_1 K^2 - f_1^2 H^2), \\ \Phi_3 = \dfrac{1}{9}[f_0(H^3 - 4K^3) - 6f_1 H^2 K], \\ t_3 = \dfrac{1}{27}[f_0^3(H^3 + 8K^3) - 18 f_0^2 f_1 H^2 K + 36 f_0 f_1^2 HK^2 + 8f_1^3(H^3 - K^3)] \\ \quad = \dfrac{1}{27}[(f_0^3 + 8f_1^3)H^3 - 18 f_0^2 f_1 H^2 K + 36 f_0 f_1^2 HK^2 + 8(f_0^3 - f_1^3)K^3]. \end{cases} \tag{2.12}$$

Proof. We have

$$\begin{aligned} u &= \frac{1}{9}[(f_0+2f_1)(H+2K)+2(f_0-f_1)(H-K)] \\ &= \frac{1}{3}(f_0H+2f_1K). \end{aligned}$$

$$\begin{aligned} \Phi &= \frac{1}{81}(f_0+2f_1)[(f_0+2f_1)^3+8(f_0-f_1)^3] \\ &= \frac{1}{9}(f_0+2f_1)f_0(f_0^2-2f_0f_1+4f_1^2) \\ &= \frac{1}{9}f_0(f_0^3+8f_1^3). \end{aligned}$$

$$\begin{aligned} \Psi &= \frac{1}{81}(f_0-f_1)[(f_0+2f_1)^3-(f_0-f_1)^3] \\ &= \frac{1}{9}(f_0-f_1)f_1(f_0^2+f_0f_1+f_1^2) \\ &= \frac{1}{9}f_1(f_0^3-f_1^3). \end{aligned}$$

$$\begin{aligned} t &= \frac{1}{729}[(f_0+2f_1)^6-20(f_0+2f_1)^3(f_0-f_1)^3-8(f_0-f_1)^6] \\ &= -\frac{1}{27}(f_0^6-20f_0^3f_1^3-8f_1^6). \end{aligned}$$

$$\begin{aligned} \Psi_1 =& \frac{1}{9}\psi(f_0^3-6f_0^2f_1-4f_1^3)+\frac{2}{3}\varphi(f_0^3+3f_0^2f_1-4f_1^3) \\ =& \frac{1}{9}f_0^3(\psi+6\varphi)-\frac{2}{3}f_0^2f_1(\psi-3\varphi)-\frac{4}{9}f_1^3(\psi+6\varphi) \\ =& \frac{1}{9}(f_0^3H-6f_0^2f_1K-4f_1^3H). \end{aligned}$$

$$\begin{aligned} \Psi_2 =& \frac{1}{9}[-\psi^2(f_0^2-2f_0f_1+f_1^2)-3\varphi\psi(f_0^2+4f_0f_1+4f_1^2) \\ &+18\varphi^2(f_0^2+f_0f_1-2f_1^2)] \\ =& -\frac{1}{9}f_0^2(\psi+6\varphi)(\psi-3\varphi)+\frac{2}{9}f_0f_1(\psi-3\varphi)^2-\frac{1}{9}f_1^2(\psi+6\varphi)^2 \\ =& \frac{1}{9}(-f_0^2HK+2f_0f_1K^2-f_1^2H^2). \end{aligned}$$

$$\Phi_3 = \frac{1}{3}f_0(-\psi^3+18\varphi\psi^2+108\varphi^3)+\frac{2}{3}f_1(-\psi^3-9\varphi\psi^2+108\varphi^3).$$

Here,

$$\begin{aligned} -\psi^3+18\varphi\psi^2+108\varphi^3 &= \frac{1}{3}(H^3-4K^3), \\ -\psi^3-9\varphi\psi^2+108\varphi^3 &= -H^2K. \end{aligned}$$

Hence,

$$\Phi_3 = \frac{1}{9}[f_0(H^3-4K^3)-6f_1H^2K].$$

$$\begin{aligned}t_3 =&\frac{1}{3}\psi^3(f_0^3-2f_0^2f_1+4f_0f_1^2)-2\varphi\psi^2(f_0^3+3f_0^2f_1-4f_1^3)\\&+12\varphi^2\psi(f_0^3-3f_0f_1^2+2f_1^3)+72\varphi^3(f_0^2f_1+f_0f_1^2+f_1^3)\\=&\frac{1}{3}f_0^3(\psi^3-6\varphi\psi^2+36\varphi^2\psi)-\frac{2}{3}f_0^2f_1(\psi^3+9\varphi\psi^2-108\varphi^3)\\&+\frac{4}{3}f_0f_1^2(\psi^3-27\varphi^2\psi+54\varphi^3)+8f_1^3(\varphi\psi^2+3\varphi^2\psi+9\varphi^3).\end{aligned}$$

Here,

$$\begin{aligned}\psi^3-6\varphi\psi^2+36\varphi^2\psi&=\frac{1}{9}(H^3+8K^3),\\\psi^3+9\varphi\psi^2-108\varphi^3&=H^2K,\\\psi^3-27\varphi^2\psi+54\varphi^3&=HK^2,\\\varphi\psi^2+3\varphi^2\psi+9\varphi^3&=\frac{1}{27}(H^3-K^3).\end{aligned}$$

Hence,

$$t_3=\frac{1}{27}[f_0^3(H^3+8K^3)-18f_0^2f_1H^2K+36f_0f_1^2HK^2+8f_1^3(H^3-K^3)].$$

□

Note that (see [61])

$$\begin{cases}C_{12}=\frac{1}{9}H(H^3+8K^3),\\\mathfrak{C}_{12}=\frac{1}{27}K(K^3-H^3),\\C_{18}=-\frac{1}{27}(H^6-20H^3K^3-8K^6),\end{cases}\tag{2.13}$$

we find the following nonlinear duality:

$$f_0\longleftrightarrow H,\quad f_1\longleftrightarrow K.\tag{2.14}$$

Here, f_0 and f_1 are linear, H and K are cubic. The functions u and t_3 are invariant under the above dual transformation. Under the dual transformation (2.14),

$$C_{12}\longleftrightarrow\Phi,\quad\mathfrak{C}_{12}\longleftrightarrow-\frac{1}{3}\Psi,\quad C_{18}\longleftrightarrow t,\quad\Psi_1\longleftrightarrow\Phi_3.\tag{2.15}$$

Recall that (see [61])

$$Z_1=432\frac{C_9^2}{C_6^3},\quad Z_2=3\frac{C_{12}}{C_6^2}.$$

Now, we put

$$Z_3=\frac{\Phi^3}{64\Psi^3},\quad Z_4=\frac{\Psi_1^2}{16\Psi\Psi_2},\quad Z_5=\frac{2\Phi\Phi_3}{3u^2\Psi_1},\quad Z_6=\frac{\Psi_1^2}{3u^2\Phi},\quad Z_7=\frac{ut}{3\Phi\Psi_1}.\tag{2.16}$$

Then

$$\frac{t^2}{64\Psi^3}=Z_3-1,\qquad \frac{u^2\Phi}{16\Psi\Psi_2}=Z_4-1,\qquad \frac{tt_3}{3u^2\Phi\Psi_1}=Z_5-Z_6-Z_7-1. \tag{2.17}$$

Recall that (see [61])

$$r_1=\frac{G^2}{C_6},\quad r_2=\frac{H^2}{C_6},\quad r_3=\frac{K^2}{C_6},$$

where $G=(Y_2-Y_3)(Y_3-Y_4)(Y_4-Y_2)$. Put

$$r_4=\frac{f_1}{f_0}. \tag{2.18}$$

By Theorem 2.1, We find that

$$Z_3=\frac{1}{64}\frac{(1+8r_4^3)^3}{r_4^3(1-r_4^3)^3}. \tag{2.19}$$

$$Z_4=-\frac{1}{16}\frac{[(1-4r_4^3)\sqrt{r_2}-6r_4\sqrt{r_3}]^2}{r_4(1-r_4^3)(\sqrt{r_2r_3}-2r_4r_3+r_4^2r_2)}. \tag{2.20}$$

$$Z_5=\frac{2}{3}\frac{(1+8r_4^3)(r_2\sqrt{r_2}-4r_3\sqrt{r_3}-6r_4r_2\sqrt{r_3})}{(\sqrt{r_2}+2r_4\sqrt{r_3})^2[(1-4r_4^3)\sqrt{r_2}-6r_4\sqrt{r_3}]}. \tag{2.21}$$

$$Z_6=\frac{1}{3}\frac{[(1-4r_4^3)\sqrt{r_2}-6r_4\sqrt{r_3}]^2}{(\sqrt{r_2}+2r_4\sqrt{r_3})^2(1+8r_4^3)}. \tag{2.22}$$

$$Z_7=-\frac{1}{3}\frac{(\sqrt{r_2}+2r_4\sqrt{r_3})(1-20r_4^3-8r_4^6)}{(1+8r_4^3)[(1-4r_4^3)\sqrt{r_2}-6r_4\sqrt{r_3}]}. \tag{2.23}$$

We have

$$\begin{cases}\sqrt{-3}f_0(B(Y_0,Y_1,Y_2,Y_3,Y_4))=f_0+2f_1,\\ \sqrt{-3}f_1(B(Y_0,Y_1,Y_2,Y_3,Y_4))=f_0-f_1.\end{cases} \tag{2.24}$$

$$\begin{cases}f_0(C(Y_0,Y_1,Y_2,Y_3,Y_4))=f_0,\\ f_1(C(Y_0,Y_1,Y_2,Y_3,Y_4))=f_1.\end{cases} \tag{2.25}$$

Note that

$$CD=\begin{pmatrix}-1&&&&\\&0&-1&0&0\\&0&0&0&-1\\&-1&0&0&0\\&0&0&-1&0\end{pmatrix},\quad DC=\begin{pmatrix}-1&&&&\\&0&0&0&-1\\&-1&0&0&0\\&0&-1&0&0\\&0&0&-1&0\end{pmatrix}.$$

$$C^2D=\begin{pmatrix}-1&&&&\\&0&-1&0&0\\&0&0&-1&0\\&0&0&0&-1\\&-1&0&0&0\end{pmatrix},\quad DC^2=\begin{pmatrix}-1&&&&\\&0&0&-1&0\\&-1&0&0&0\\&0&0&0&-1\\&0&-1&0&0\end{pmatrix}.$$

$$(CD)^2 = (DC^2)^2 = \begin{pmatrix} 1 & & & & \\ & 0 & 0 & 0 & 1 \\ & 0 & 0 & 1 & 0 \\ & 0 & 1 & 0 & 0 \\ & 1 & 0 & 0 & 0 \end{pmatrix}.$$

$$(DC)^2 = (C^2D)^2 = \begin{pmatrix} 1 & & & & \\ & 0 & 0 & 1 & 0 \\ & 0 & 0 & 0 & 1 \\ & 1 & 0 & 0 & 0 \\ & 0 & 1 & 0 & 0 \end{pmatrix}.$$

The above matrices are the permutation matrices.

On the other hand, Maschke determined the invariants of $G_{25,920}$ acting on $\mathbb{P}^3 = \{(z_0, z_1, z_2, z_3)\}$. According to Maschke [35], $G_{25,920}$ is generated by the following elements: M, N, P, Q and R, where

$$M = FAFC^2EFA^2C, \quad N = AEA^2,$$

$$P = FAFECFA^2CJ, \quad Q = BEJ^3, \quad R = F.$$

Here,

$$A(z_0, z_1, z_2, z_3) = (z_0, z_2, z_3, z_1),$$

$$B(z_0, z_1, z_2, z_3) = (-z_0, z_1, z_3, z_2),$$

$$C(z_0, z_1, z_2, z_3) = (z_0, z_1, \omega z_2, \omega^2 z_3),$$

$$D(z_0, z_1, z_2, z_3) = (\omega z_0, z_1, \omega z_2, \omega z_3),$$

$$E(z_0, z_1, z_2, z_3)$$
$$= \left(z_0, \frac{1}{\sqrt{-3}}(z_1 + z_2 + z_3), \frac{1}{\sqrt{-3}}(z_1 + \omega z_2 + \omega^2 z_3), \frac{1}{\sqrt{-3}}(z_1 + \omega^2 z_2 + \omega z_3)\right),$$

$$F(z_0, z_1, z_2, z_3) = (-z_2, z_1, -z_0, -z_3),$$

$$J(z_0, z_1, z_2, z_3) = (iz_0, iz_1, iz_2, iz_3).$$

In the form of matrices, we have

$$A = \begin{pmatrix} 1 & & & \\ & & 1 & \\ & & & 1 \\ & 1 & & \end{pmatrix}, \quad B = \begin{pmatrix} -1 & & & \\ & 1 & & \\ & & & 1 \\ & & 1 & \end{pmatrix},$$

$$C = \begin{pmatrix} 1 & & & \\ & 1 & & \\ & & \omega & \\ & & & \omega^2 \end{pmatrix}, \quad D = \begin{pmatrix} \omega & & & \\ & 1 & & \\ & & \omega & \\ & & & \omega \end{pmatrix},$$

$$E = \frac{1}{\sqrt{-3}} \begin{pmatrix} \sqrt{-3} & & & \\ & 1 & 1 & 1 \\ & 1 & \omega & \omega^2 \\ & 1 & \omega^2 & \omega \end{pmatrix}, \quad F = \begin{pmatrix} & & -1 & \\ & 1 & & \\ -1 & & & \\ & & & -1 \end{pmatrix},$$

and $J = iI$. Note that $A, B, C, D, E, F, J \in SU(4)$.

We find that

$$B^2 = F^2 = (AB)^2 = (BA)^2 = (BFB)^2 = I.$$

$$(ABF)^2 = (BFA)^2 = (ABF \cdot BFA)^2 = -I.$$

$$BFB = FBF, \quad ABF = -FAB.$$

Here,

$$AB = \begin{pmatrix} -1 & & & \\ & 0 & 0 & 1 \\ & 0 & 1 & 0 \\ & 1 & 0 & 0 \end{pmatrix}, \quad BA = \begin{pmatrix} -1 & & & \\ & 0 & 1 & 0 \\ & 1 & 0 & 0 \\ & 0 & 0 & 1 \end{pmatrix},$$

$$ABF = \begin{pmatrix} 0 & 0 & 1 & 0 \\ 0 & 0 & 0 & -1 \\ -1 & 0 & 0 & 0 \\ 0 & 1 & 0 & 0 \end{pmatrix}, \quad BFA = \begin{pmatrix} 0 & 0 & 0 & 1 \\ 0 & 0 & 1 & 0 \\ 0 & -1 & 0 & 0 \\ -1 & 0 & 0 & 0 \end{pmatrix},$$

$$BFB = \begin{pmatrix} 0 & 0 & 0 & 1 \\ 0 & 1 & 0 & 0 \\ 0 & 0 & -1 & 0 \\ 1 & 0 & 0 & 0 \end{pmatrix}, \quad ABF \cdot BFA = \begin{pmatrix} 0 & -1 & 0 & 0 \\ 1 & 0 & 0 & 0 \\ 0 & 0 & 0 & -1 \\ 0 & 0 & 1 & 0 \end{pmatrix}.$$

They play the similar role of the permutation matrices.

3. Hessian polyhedra and Picard modular forms

In his celebrated lecture [28], Klein set

$$\mathbf{A}_0 = z_1 z_2, \quad \mathbf{A}_1 = z_1^2, \quad \mathbf{A}_2 = -z_2^2$$

and

$$A = \mathbf{A}_0^2 + \mathbf{A}_1 \mathbf{A}_2.$$

The invariants associated to the icosahedron: f, H and T (see [28] or [61]) can be expressed as the functions of $\mathbf{A}_0$, $\mathbf{A}_1$ and $\mathbf{A}_2$:

$$B = 8\mathbf{A}_0^4 \mathbf{A}_1 \mathbf{A}_2 - 2\mathbf{A}_0^2 \mathbf{A}_1^2 \mathbf{A}_2^2 + \mathbf{A}_1^3 \mathbf{A}_2^3 - \mathbf{A}_0(\mathbf{A}_1^5 + \mathbf{A}_2^5),$$

$$\begin{aligned} C = {} & 320\mathbf{A}_0^6 \mathbf{A}_1^2 \mathbf{A}_2^2 - 160\mathbf{A}_0^4 \mathbf{A}_1^3 \mathbf{A}_2^3 + 20\mathbf{A}_0^2 \mathbf{A}_1^4 \mathbf{A}_2^4 + 6\mathbf{A}_1^5 \mathbf{A}_2^5 \\ & - 4\mathbf{A}_0(\mathbf{A}_1^5 + \mathbf{A}_2^5)(32\mathbf{A}_0^4 - 20\mathbf{A}_0^2 \mathbf{A}_1 \mathbf{A}_2 + 5\mathbf{A}_1^2 \mathbf{A}_2^2) + \mathbf{A}_1^{10} + \mathbf{A}_2^{10}, \end{aligned}$$

$$
\begin{aligned}
D = (\mathbf{A}_1^5 - \mathbf{A}_2^5)[&-1024\mathbf{A}_0^{10} + 3840\mathbf{A}_0^8\mathbf{A}_1\mathbf{A}_2 - 3840\mathbf{A}_0^6\mathbf{A}_1^2\mathbf{A}_2^2 \\
&+ 1200\mathbf{A}_0^4\mathbf{A}_1^3\mathbf{A}_2^3 - 100\mathbf{A}_0^2\mathbf{A}_1^4\mathbf{A}_2^4 + \mathbf{A}_1^{10} + \mathbf{A}_2^{10} + 2\mathbf{A}_1^5\mathbf{A}_2^5 \\
&+ \mathbf{A}_0(\mathbf{A}_1^5 + \mathbf{A}_2^5)(352\mathbf{A}_0^4 - 160\mathbf{A}_0^2\mathbf{A}_1\mathbf{A}_2 + 10\mathbf{A}_1^2\mathbf{A}_2^2)].
\end{aligned}
$$

They satisfy the identity

$$D^2 = -1728B^5 + C^3 + 720ACB^3 - 80A^2C^2B + 64A^3(5B^2 - AC)^2.$$

Moreover, the Jacobian of A, B and C satisfies the relation (see [14]):

$$\frac{\partial(A,B,C)}{\partial(\mathbf{A}_0,\mathbf{A}_1,\mathbf{A}_2)} = \begin{vmatrix} \frac{\partial A}{\partial \mathbf{A}_0} & \frac{\partial A}{\partial \mathbf{A}_1} & \frac{\partial A}{\partial \mathbf{A}_2} \\ \frac{\partial B}{\partial \mathbf{A}_0} & \frac{\partial B}{\partial \mathbf{A}_1} & \frac{\partial B}{\partial \mathbf{A}_2} \\ \frac{\partial C}{\partial \mathbf{A}_0} & \frac{\partial C}{\partial \mathbf{A}_1} & \frac{\partial C}{\partial \mathbf{A}_2} \end{vmatrix} = -10D.$$

The general Jacobian equation of the sixth degree is the following:

$$(z-A)^6 - 4A(z-A)^5 + 10B(z-A)^3 - C(z-A) + (5B^2 - AC) = 0.$$

Let $z_0, z_1, z_2, z_3, z_4, z_\infty$ be the six roots, put

$$y = (z_\infty - z_0)(z_1 - z_4)(z_2 - z_3).$$

Brioschi remarked that the square root of this expression is already rational in the $\mathbb{A}$'s, and gives rise to an equation of the fifth degree. Let x be this square root, Brioschi found for the five values of which x is susceptible the following formula:

$$
\begin{aligned}
x_\nu = &- \epsilon^\nu \mathbf{A}_1(4\mathbf{A}_0^2 - \mathbf{A}_1\mathbf{A}_2) + \epsilon^{2\nu}(2\mathbf{A}_0\mathbf{A}_1^2 - \mathbf{A}_2^3) \\
&+ \epsilon^{3\nu}(-2\mathbf{A}_0\mathbf{A}_2^2 + \mathbf{A}_1^3) + \epsilon^{4\nu}\mathbf{A}_2(4\mathbf{A}_0^2 - \mathbf{A}_1\mathbf{A}_2),
\end{aligned}
$$

with $\epsilon = \exp(2\pi i/5)$, while for the corresponding equation of the fifth degree he found this:

$$x^5 + 10Bx^3 + 5(9B^2 - AC)x - D = 0.$$

In fact, the Jacobian equation of the sixth degree with $A = 0$ is none other than that simplest resolvent of the sixth degree which Klein established in the case of the icosahedron: put $\mathbf{A}_0 = z_1 z_2$, $\mathbf{A}_1 = z_1^2$, $\mathbf{A}_2 = -z_2^2$ and correspondingly:

$$B = -f, \quad C = -H, \quad D = T.$$

The resolvent of the fifth degree is transformed into the following:

$$x^5 + 10Bx^3 + 45B^2x - D = 0.$$

The actions of the icosahedral group on $\mathbf{A}_0$, $\mathbf{A}_1$ and $\mathbf{A}_2$ are given by

$$S: \quad \mathbf{A}_0' = \mathbf{A}_0, \quad \mathbf{A}_1' = \epsilon^4\mathbf{A}_1, \quad \mathbf{A}_2' = \epsilon\mathbf{A}_2;$$

$$T: \begin{cases} \sqrt{5}\mathbf{A}_0' = \mathbf{A}_0 + \mathbf{A}_1 + \mathbf{A}_2, \\ \sqrt{5}\mathbf{A}_1' = 2\mathbf{A}_0 + (\epsilon^2 + \epsilon^3)\mathbf{A}_1 + (\epsilon + \epsilon^4)\mathbf{A}_2, \\ \sqrt{5}\mathbf{A}_2' = 2\mathbf{A}_0 + (\epsilon + \epsilon^4)\mathbf{A}_1 + (\epsilon^2 + \epsilon^3)\mathbf{A}_2; \end{cases}$$

$$U: \quad \mathbf{A}_0' = -\mathbf{A}_0, \quad \mathbf{A}_1' = -\mathbf{A}_2, \quad \mathbf{A}_2' = -\mathbf{A}_1.$$

This leads to the investigation of the binary quadratic forms:

$$\mathbf{A}_1 z_1^2 + 2\mathbf{A}_0 z_1 z_2 - \mathbf{A}_2 z_2^2$$

and Hilbert modular surfaces (see [20]). Klein (see [28]) set

$$\delta_\nu = (\epsilon^{4\nu}\mathbf{A}_1 - \epsilon^\nu \mathbf{A}_2)[(1+\sqrt{5})\mathbf{A}_0 + \epsilon^{4\nu}\mathbf{A}_1 + \epsilon^\nu \mathbf{A}_2][(1-\sqrt{5})\mathbf{A}_0 + \epsilon^{4\nu}\mathbf{A}_1 + \epsilon^\nu \mathbf{A}_2],$$

where $\epsilon = \exp(2\pi i/5)$, $\nu = 0, 1, 2, 3, 4$ and found that

$$\begin{cases} \delta_0 + \delta_1 + \delta_2 + \delta_3 + \delta_4 = 0, \\ \delta_0^3 + \delta_1^3 + \delta_2^3 + \delta_3^3 + \delta_4^3 = 0. \end{cases}$$

Moreover,

$$S(\delta_0) = \delta_1, S(\delta_1) = \delta_2, S(\delta_2) = \delta_3, S(\delta_3) = \delta_4, S(\delta_4) = \delta_0.$$

$$U(\delta_0) = \delta_0, U(\delta_1) = \delta_4, U(\delta_2) = \delta_3, U(\delta_3) = \delta_2, U(\delta_4) = \delta_1.$$

$T(\delta_0) = \delta_0$. However, there does not exist the analogous transformation formulas for $T(\delta_i)$ $(i = 1, 2, 3, 4)$.

It is well-known that the group A_5 is isomorphic to the finite group I of those elements of $SO(3)$ which carry a given icosahedron centered at the origin of the standard Euclidean space $\mathbb{R}^3$ to itself. The group I operates linearly on $\mathbb{R}^3$ (standard coordinates x_0, x_1, x_2) and thus also on $\mathbb{RP}^2$ and $\mathbb{CP}^2$. We are concerned with the action on $\mathbb{CP}^2$ (see [20] and [21]). A curve in $\mathbb{CP}^2$ which is mapped to itself by all elements of I is given by a homogeneous polynomial in x_0, x_1, x_2 which is I-invariant up to constant factors and hence I-invariant, because I is a simple group. The graded ring of all I-invariant polynomials in x_0, x_1, x_2 is generated by homogeneous polynomials A, B, C, D of degrees 2, 6, 10, 15 with $A = x_0^2 + x_1^2 + x_2^2$. The action of I on $\mathbb{CP}^2$ has exactly one minimal orbit where "minimal" means that the number of points in the orbit is minimal. This orbit has six points, they are called poles. These are the points of $\mathbb{RP}^2 \subset \mathbb{CP}^2$ which are represented by the six axes through the vertices of the icosahedron. Klein uses coordinates

$$A_0 = x_0, \quad A_1 = x_1 + ix_2, \quad A_2 = x_1 - ix_2$$

and puts the icosahedron in such a position that the six poles are given by

$$(A_0, A_1, A_2) = \left(\frac{\sqrt{5}}{2}, 0, 0\right), \quad \left(\frac{1}{2}, \varepsilon^\nu, \varepsilon^{-\nu}\right)$$

with $\varepsilon = \exp(2\pi i/5)$ and $0 \leq \nu \leq 4$.

The invariant curve $A = 0$ does not pass through the poles. There is exactly one invariant curve $B = 0$ of degree 6 which passes through the poles, exactly one invariant curve $C = 0$ of degree 10 which has higher multiplicity than the curve $B = 0$ in the poles and exactly one invariant curve $D = 0$ of degree 15. In fact, $B = 0$ has an ordinary double point (multiplicity 2) in each pole, $C = 0$ has a double cusp (multiplicity 4) in each pole and $D = 0$ is the union of the 15 lines connecting the six poles. Klein gives formulas for the homogeneous polynomials A, B, C, D (determined up to constant factors). They generate the ring of all I-invariant polynomials. According to Klein the ring of I-invariant polynomials is given as follows:

$$\mathbb{C}[A_0, A_1, A_2]^I = \mathbb{C}[A, B, C, D]/(R(A, B, C, D) = 0).$$

The relation $R(A, B, C, D) = 0$ is of degree 30.

The equations for B and C show that the two tangents of $B = 0$ in the pole $(\sqrt{5}/2, 0, 0)$ are given by $A_1 = 0$, $A_2 = 0$. They coincide with the tangents of $C = 0$ in that pole. Therefore the curves $B = 0$ and $C = 0$ have in each pole the intersection multiplicity 10. Thus they intersect only in the poles.

When we restrict the action of I to the conic $A = 0$, we get the well-known action of I on $\mathbb{CP}^1$. The curves $B = 0$, $C = 0$, $D = 0$ intersect $A = 0$ transversally in 12, 20, 30 points respectively.

Put

$$\Gamma(\sqrt{5}) = \left\{ \begin{pmatrix} \alpha & \beta \\ \gamma & \delta \end{pmatrix} \in SL(2, \mathcal{O}_K) : \alpha \equiv \delta \equiv 1 (\text{mod } \sqrt{5}), \beta \equiv \gamma \equiv 0 (\text{mod } \sqrt{5}) \right\},$$

where $K = \mathbb{Q}(\sqrt{5})$.

Theorem 3.1 (Hirzebruch) (see [20] and [13]). *The ring of symmetric Hilbert modular forms for $\Gamma(\sqrt{5})$ is equal to $\mathbb{C}[A_0, A_1, A_2]$ where A_0, A_1, A_2 have weight 1.*

Corollary 3.2 (see [20] and [13]). *The ring of symmetric Hilbert modular forms for $SL(2, \mathcal{O}_K)$ of even weight is equal to $\mathbb{C}[A, B, C]$.*

Let

$$\Gamma(2) = \left\{ \begin{pmatrix} \alpha & \beta \\ \gamma & \delta \end{pmatrix} \in SL(2, \mathcal{O}_K) : \alpha \equiv \delta \equiv 1 (\text{mod } 2), \beta \equiv \gamma \equiv 0 (\text{mod } 2) \right\},$$

where $K = \mathbb{Q}(\sqrt{5})$.

Theorem 3.3 (see [19]). *The graded ring of Hilbert modular forms of even weight relative to the group $\Gamma(2)$ is the ring of polynomials in the five Eisenstein series $E_0, \cdots, E_4$, with the relations*

$$\sigma_2(E_0, \cdots, E_4) = 0, \quad \sigma_4(E_0, \cdots, E_4) = 0$$

and their consequences, but no other relations. Here, σ_j is the j-th elementary symmetric function of $E_0, \cdots, E_4$.

Corollary 3.4 (see [19]). *The ring $M(SL(2, \mathcal{O}_K))$ is generated by the four modular forms:*

$$\sigma_1 = \sum_{0 \le i \le 4} E_i, \quad \sigma_3 = \sum_{0 \le i < j < k \le 4} E_i E_j E_k,$$

$$\sigma_5 = \prod_{0\leq i\leq 4} E_i, \quad \Delta = \prod_{0\leq i<j\leq 4} (E_i - E_j)$$

of weight 2, 6, 10, *and* 20, *respectively, subject to the single relation:*

$$\begin{aligned}\Delta^2 =&\sigma_5(3125\sigma_5^3 + 2000\sigma_5^2\sigma_3\sigma_1^2 + 256\sigma_5^2\sigma_1^5 \\ &- 900\sigma_5\sigma_3^3\sigma_1 - 128\sigma_5\sigma_3^2\sigma_1^4 + 16\sigma_3^4\sigma_1^3 + 108\sigma_3^5).\end{aligned}$$

Moreover, Hirzebruch found the connection between lattices, theta functions, coding theory and Hilbert modular surface associated with $\mathbb{Q}(\sqrt{5})$ (see [18], pp.796-798). Let p be an odd prime and $\zeta = e^{2\pi i/p}$. The field $K = \mathbb{Q}(\zeta)$ contains the totally real field $k = \mathbb{Q}(\zeta + \zeta^{-1})$. For the rings of integers $\mathcal{O}_K$ and $\mathcal{O}_k$ we have

$$\mathcal{O}_K = \mathbb{Z}\zeta + \cdots + \mathbb{Z}\zeta^{p-1},$$

$$\mathcal{O}_k = \mathbb{Z}(\zeta + \zeta^{-1}) + \cdots + \mathbb{Z}(\zeta^{\frac{p-1}{2}} + \zeta^{\frac{p+1}{2}}).$$

The Galois group G of k over $\mathbb{Q}$ is cyclic of order $\frac{p-1}{2}$. Let $\mathrm{Tr} : k \to \mathbb{Q}$ be the trace, i.e.

$$\mathrm{Tr}(y) = \sum_{\sigma\in G} \sigma(y), \quad \text{for} \quad y \in k.$$

We consider the following quadratic form Q over the $\mathbb{Z}$-lattice $\mathcal{O}_K$

$$Q : \mathcal{O}_K \to \mathbb{Q}, \quad Q(x) = 2\mathrm{Tr}\left(\frac{x\overline{x}}{p}\right).$$

It has discriminant $\frac{1}{p}$.

The ideal $(1 - \zeta) = \mathfrak{p}$ in $\mathcal{O}_K$ has norm p. Let $\varrho : \mathcal{O}_K \to \mathbb{F}_p = GF(p)$ be the corresponding homomorphism. We also consider $\varrho : \mathcal{O}_K^n \to \mathbb{F}_p^n$. For a self-dual code $C \subset \mathbb{F}_p^n$ (n even) we define the lattice $\Lambda = \varrho^{-1}(C)$. If we restrict Q to Λ, then we obtain an integral even unimodular form on the lattice Λ. It is well-known that $n(p-1) \equiv 0 \pmod 8$.

The theta functions in the Hilbert modular sense for the totally real field k depend on $\frac{p-1}{2}$ variables $z_\sigma \in \mathbb{H}$ (upper half plane) indexed by the elements of the Galois group G of k over $\mathbb{Q}$. For $y \in k$

$$\mathrm{Tr}(zy) \triangleq \sum_{\sigma\in G} z_\sigma y_\sigma \quad (y_\sigma = \sigma(y)).$$

The fundamental theta functions are

$$\theta_j(z) = \sum_{x\in\mathfrak{p}+j} e^{2\pi i\mathrm{Tr}\left(z\frac{x\overline{x}}{p}\right)} \quad \text{for} \quad j = 0, 1, \cdots, \frac{p-1}{2}.$$

The θ_j are Hilbert modular forms of weight 1 for the group

$$SL_2(\mathcal{O}_k, \widetilde{\mathfrak{p}}) = \{A | A \in SL_2(\mathcal{O}_k) \quad \text{and} \quad A \equiv 1 \bmod \widetilde{\mathfrak{p}}\}$$

where $\widetilde{\mathfrak{p}}$ is the prime ideal in k lying over p

$$\widetilde{\mathfrak{p}} = ((\zeta - \overline{\zeta})^2), \quad \widetilde{\mathfrak{p}}^{\frac{p-1}{2}} = (p).$$

In fact, the group $SL_2(\mathbb{F}_p)$ acts on the $\frac{p+1}{2}$-dimensional vector space over $\mathbb{C}$ generated by the θ_j. For $p \equiv 1 \bmod 4$ this is an action of $PSL_2(\mathbb{F}_p)$.

The lattice Λ gives a theta function

$$\theta_C(z) = \sum_{x \in \Lambda} e^{2\pi i \mathrm{Tr}\left(z \frac{x\overline{x}}{p}\right)}.$$

Consider the Lee weight enumerator $L_C(X_0, \cdots, X_{\frac{p-1}{2}})$.

Theorem 3.5 (see [18], p.797).

$$\theta_C = L_C(\theta_0, \cdots, \theta_{\frac{p-1}{2}}).$$

θ_C is a Hilbert modular form of weight n for the group $SL_2(\mathcal{O}_k)$. The polynomial $L_C(X_0, \cdots, X_{\frac{p-1}{2}})$ is an invariant polynomial for the above mentioned representation of dimension $\frac{p+1}{2}$ of the group $SL_2(\mathbb{F}_p)$. All Hilbert modular forms θ_C, θ_j occurring are symmetric, i.e. invariant under the action of the Galois group G on the variables z_σ.

In his paper on $\mathbb{Q}(\sqrt{5})$, which is k for $p = 5$, Hirzebruch proved that the ring of symmetric Hilbert modular forms for $SL_2(\mathcal{O}_k, \widetilde{\mathfrak{p}})$, $\widetilde{\mathfrak{p}} = (\sqrt{5})$, equals $\mathbb{C}[A_0, A_1, A_2]$ where A_0, A_1, A_2 have weight 1. He proved that the ring of symmetric Hilbert modular forms for $SL_2(\mathcal{O}_k)$ of even weight equals $\mathbb{C}[A, B, C]$ where A, B, C are the Klein invariants of degrees 2, 6, 10. Moreover, he found that

$$A_0 = \theta_0, \quad A_1 = 2\theta_1, \quad A_2 = 2\theta_2.$$

Together with Theorem 3.5, this explains the above mentioned connection.

For $p = 3$ (where $k = \mathbb{Q}$). In this case C is a ternary code. According to [13], the Lee weight enumerator of the code C satisfies the following identity:

$$W_C(\theta_0, \theta_1) = \theta_C.$$

We consider the functions θ_0 and θ_1. The lattice Λ is isomorphic to A_2. The lattice A_2 is the $\mathbb{Z}$-module $\mathbb{Z}^2$ with the quadratic form

$$(x, y)^2 = 2x^2 - 2xy + 2y^2.$$

Therefore,

$$\theta_0(z) = \sum_{(x,y) \in \mathbb{Z}^2} q^{x^2 - xy + y^2}, \quad \theta_1(z) = q^{\frac{1}{3}} \sum_{(x,y) \in \mathbb{Z}^2} q^{x^2 - xy + y^2 + x - y} \tag{3.1}$$

with $q = e^{2\pi i z}$. The functions θ_0 and θ_1 are modular forms of weight 1 for the group

$$\Gamma(3) = \left\{ \begin{pmatrix} a & b \\ c & d \end{pmatrix} \in SL_2(\mathbb{Z}) : a \equiv d \equiv 1 (\bmod 3), b \equiv c \equiv 0 (\bmod 3) \right\}.$$

We have

$$\theta_0\left(-\frac{1}{z}\right)=\frac{z}{\sqrt{-3}}(\theta_0(z)+2\theta_1(z)),\quad \theta_1\left(-\frac{1}{z}\right)=\frac{z}{\sqrt{-3}}(\theta_0(z)-\theta_1(z)).$$

$$\theta_0(z+1)=\theta_0(z),\quad \theta_1(z+1)=\omega\theta_1(z),\quad \omega=e^{\frac{2\pi i}{3}}.$$

The Eisenstein series E_4 is the theta function of the unimodular E_8-lattice and

$$E_4=\theta_0^4+8\theta_0\theta_1^3,\quad E_6=\theta_0^6-20\theta_0^3\theta_1^3-8\theta_1^6. \tag{3.2}$$

Theorem 3.6 (see [13]). *The algebra of all modular forms for the group* $\Gamma(3)$ *is isomorphic to the polynomial algebra* $\mathbb{C}[\theta_0,\theta_1]$.

The group $SL_2(\mathbb{F}_3)$ acts on the polynomial algebra $\mathbb{C}[\theta_0,\theta_1]$. The ring of invariant polynomials under this action is denoted by $\mathbb{C}[\theta_0,\theta_1]^{SL_2(\mathbb{F}_3)}$.

Theorem 3.7 (see [13]).

$$\mathbb{C}[\theta_0,\theta_1]^{SL_2(\mathbb{F}_3)}=\mathbb{C}[E_4,E_6].$$

The Hamming weight enumerator $L_C=H_C=H_C(\theta_0,\theta_1)$ is a polynomial in the modular forms E_4 and E_6^2. For the ternary Golay code of length 12,

$$H_C=\frac{5}{8}E_4^3+\frac{3}{8}E_6^2.$$

According to [11], the Hamming weight enumerator classifies codewords according to the number of nonzero coordinates. More detailed information is supplied by the complete weight enumerator (c.w.e.), which gives the number of codewords of each composition. For example the c.w.e. of a ternary code C is

$$c.w.e._C(x,y,z)=\sum_{u\in C}x^{n_0(u)}y^{n_1(u)}z^{n_{-1}(u)},$$

where $n_i(u)$ is the number of times $i\in\mathbb{F}_3$ occurs in u. There is also a version of the MacWilliams identity for c.w.e.'s.

Theorem 3.8 (Gleason) (see [11]). *If C is a Type III code then $W_C(x,y)$ is invariant under a group G_3 of order* 48, *and belongs to* $\mathbb{C}[\psi_4,\xi_{12}]$, *where*

$$\psi_4=x^4+8xy^3$$

is the weight enumerator of $\mathcal{C}_4$ *and*

$$\xi_{12}=y^3(x^3-y^3)^3.$$

Equivalently, $W_C(x,y)$ can be written as a polynomial in the weight enumerators of $\mathcal{C}_4$ and $\mathcal{C}_{12}$.

To describe the complete weight enumerator, we first define some further polynomials. Let

$$a=x^3+y^3+z^3,\quad p=3xyz,$$

$$b = x^3y^3 + y^3z^3 + z^3x^3,$$
$$\beta_6 = a^2 - 12b = x^6 + y^6 + z^6 - 10(x^3y^3 + y^3z^3 + z^3x^3),$$
$$\pi_9 = (x^3 - y^3)(y^3 - z^3)(z^3 - x^3),$$
$$\alpha_{12} = a(a^3 + 8p^3).$$

Theorem 3.9 (see [11]). *If C is a Type III code which contains the all-ones vector then the complete weight enumerator is invariant under a group G_4 of order* 2592, *and belongs to*

$$R \oplus \beta_6\pi_9^2 R,$$

where

$$R = \mathbb{C}[\beta_6^2, \alpha_{12}, \pi_9^4].$$

In other words the complete weight enumerator can be written uniquely as a polynomial in β_6^2, α_{12} and π_9^4, plus $\beta_6\pi_9^2$ times another such polynomial.

Here, G_3 is generated by $\frac{1}{\sqrt{3}}\begin{pmatrix}1 & 2\\ 1 & -1\end{pmatrix}$, $\begin{pmatrix}1 & 0\\ 0 & \omega\end{pmatrix}$ and is the reflection group 3[6]2. G_4 is generated by all permutations, $\mathrm{diag}(1, 1, \omega)$ and

$$\frac{1}{\sqrt{3}}\begin{pmatrix}1 & 1 & 1\\ 1 & \omega & \overline{\omega}\\ 1 & \overline{\omega} & \omega\end{pmatrix},$$

has a center $Z(G_4)$ of order 12, and $G_4/Z(G_4)$ is the Hessian group of order 216.

It is well-known that a complete system of invariants for the Hessian groups (the corresponding geometric objects are Hessian polyhedra) has degrees 6, 9, 12, 12 and 18 and can be given explicitly by the following forms:

$$\begin{cases} C_6(z_1, z_2, z_3) = z_1^6 + z_2^6 + z_3^6 - 10(z_1^3z_2^3 + z_2^3z_3^3 + z_3^3z_1^3),\\ C_9(z_1, z_2, z_3) = (z_1^3 - z_2^3)(z_2^3 - z_3^3)(z_3^3 - z_1^3),\\ C_{12}(z_1, z_2, z_3) = (z_1^3 + z_2^3 + z_3^3)[(z_1^3 + z_2^3 + z_3^3)^3 + 216z_1^3z_2^3z_3^3],\\ \mathfrak{C}_{12}(z_1, z_2, z_3) = z_1z_2z_3[27z_1^3z_2^3z_3^3 - (z_1^3 + z_2^3 + z_3^3)^3],\\ C_{18}(z_1, z_2, z_3) = (z_1^3 + z_2^3 + z_3^3)^6 - 540z_1^3z_2^3z_3^3(z_1^3 + z_2^3 + z_3^3)^3 - 5832z_1^6z_2^6z_3^6. \end{cases} \tag{3.3}$$

They satisfy the following relations:

$$\begin{cases} 432C_9^2 = C_6^3 - 3C_6C_{12} + 2C_{18},\\ 1728\mathfrak{C}_{12}^3 = C_{18}^2 - C_{12}^3. \end{cases} \tag{3.4}$$

For the Hessian group 3[3]3[3]3, of order $6 \times 9 \times 12$, the three forms are C_6, C_9, C_{12}. For the closely related Hessian group 2[4]3[3]3, of order $6 \times 12 \times 18$, they are C_6, C_{12}, C_9^2 (see [12]).

Note that

$$\beta_6 = C_6, \quad \pi_9 = C_9, \quad \alpha_{12} = C_{12}, \quad R = \mathbb{C}[C_6^2, C_{12}, C_9^4] \tag{3.5}$$

under the identification

$$(x, y, z) = (z_1, z_2, z_3).$$

On the other hand,

$$\psi_4 = 9C_{12}, \quad \xi_{12} = -27^3\mathfrak{C}_{12}^3 \tag{3.6}$$

under the identification

$$(x, y) = (H, K).$$

Thus, we find a connection between Hessian invariants (Hessian groups) and coding theory.

Two theories apparently far apart, that of weight polynomials of codes over the field $\mathbb{F}_2$ and that of theta functions of lattices over $\mathbb{Z}$, are too similar to be strangers to each other. The most striking similarity is no doubt the following: The weight polynomials of certain codes $\cdots$ generate the graded algebra of polynomials in two variables fixed by the operation of a subgroup H of $GL(2, \mathbb{C})$; the theta functions of even unimodular lattices are modular forms which generate an algebra visibly isomorphic to the preceding (see [6]). Since the times of Klein the analogies between invariant theory and modular forms have emerged.

Denote $E_2(z)$, $E_4(z)$, $E_6(z)$ the Eisenstein series of order 2, 4, 6 and the cusp form $\Delta(z)$ of weight 12 as

$$E_2(z) = 1 - 24\sum_{m=1}^{\infty}\sigma_1(m)q^m,$$

$$E_4(z) = 1 + 240\sum_{m=1}^{\infty}\sigma_3(m)q^m,$$

$$E_6(z) = 1 - 504\sum_{m=1}^{\infty}\sigma_5(m)q^m,$$

$$\Delta(z) = \frac{1}{1728}(E_4^3 - E_6^2) = q\prod_{m=1}^{\infty}(1 - q^m)^{24},$$

where $q = \exp(2\pi i z)$, and the sum of rth power of divisor function is $\sigma_r(m) := \sum_{d|m} d^r$. As is well known E_4, E_6 are modular forms of weight 4 and 6 but E_2 is not. The following result is due to Hecke.

Theorem 3.10 (see [9]). *The algebra of modular forms of weight multiple of* 4 *is* $\mathbb{C}[E_4, \Delta]$. *The algebra of modular forms of even weight is* $\mathbb{C}[E_4, E_6]$.

Let $H_2 = \langle M_2, N_2\rangle$, $M_2 = \frac{1}{\sqrt{2}}\begin{pmatrix}1 & 1\\ 1 & -1\end{pmatrix}$, $N_2 = \begin{pmatrix}1 & 0\\ 0 & i\end{pmatrix}$. H_2 is of order 192. There is a subgroup $G_2 \leq H_2$ of index 2 defined as the kernel of the following linear character defined on the generators of H_2 as

$$\chi(M_2) = -1, \quad \chi(N_2) = 1.$$

Define the following invariants for the group H_2 of degree 8 and 24, respectively:

$$\psi_8 = x^8 + 14x^4y^4 + y^8,$$

$$\nu_{24} = x^4y^4(x^4 - y^4)^4.$$

Then

$$\mathbb{C}[x,y]^{H_2} = \mathbb{C}[\psi_8, \nu_{24}],$$
$$\mathbb{C}[x,y]^{G_2} = \mathbb{C}[\psi_8, k_{12}],$$

where

$$k_{12} = x^{12} - 33(x^8y^4 + x^4y^8) + y^{12}.$$

Theorem 3.11 (see [9]). *The map*

$$\phi_1 : \mathbb{C}[\psi_8, \nu_{24}] \to \mathbb{C}[E_4, \Delta]$$

defined by

$$\phi_1(h(\psi_8, \nu_{24})) = h(E_4, \Delta)$$

is an algebra isomorphism. The map

$$\phi_2 : \mathbb{C}[\psi_8, k_{12}] \to \mathbb{C}[E_4, E_6]$$

defined by

$$\phi_2(h(\psi_8, k_{12})) = h(E_4, E_6)$$

is an algebra isomorphism.

Let $G_3 = \langle M_3, N_3 \rangle$, $M_3 = \frac{1}{\sqrt{3}}\begin{pmatrix} 1 & 2 \\ 1 & -1 \end{pmatrix}$, $N_3 = \begin{pmatrix} 1 & 0 \\ 0 & \omega \end{pmatrix}$, $\omega = \exp(2\pi i/3)$. Then $G_3 \cong SL_2(\mathbb{F}_3)$. Primary invariants for G_3 are

$$\psi_4 = x^4 + 8xy^3,$$

$$k_6 = x^6 - 20x^3y^3 - 8y^6.$$
$$\mathbb{C}[x,y]^{G_3} = \mathbb{C}[\psi_4, k_6].$$

Theorem 3.12 (see [9]). *The map*

$$\phi_3 : \mathbb{C}[\psi_4, k_6] \to \mathbb{C}[E_4, E_6]$$

given by

$$\phi_3(h(\psi_4, k_6)) = h(E_4, E_6)$$

is an algebra isomorphism.

In our case, we have the following:

Theorem 3.13.

$$\mathbb{C}[H, K]^{G_3} = \mathbb{C}[C_{12}, C_{18}].$$

The map

$$\varphi : \mathbb{C}[C_{12}, C_{18}] \to \mathbb{C}[E_4, E_6]$$

given by

$$\varphi(h(C_{12}, C_{18})) = h(E_4, E_6)$$

is an algebra isomorphism.

Consequently,

$$\varphi(h(C_{12}, -\mathfrak{C}_{12}^3)) = h(E_4, \Delta).$$

Now, we will study the analogies between Hessian invariants and Picard modular forms, which can be considered as the higher dimensional counterpart of the analogies between regular polyhedral invariants and elliptic modular forms studied by Klein. Let us recall some basic facts about Picard modular forms. In his monograph [22], Holzapfel proved the following results:

Theorem 3.14 (see [22]).

(1) The monodromy group of the hypergeometric differential equation system considered by Picard is the principal congruence subgroup $U(2,1;\mathbb{Z}[\omega])(1-\omega)$ of the full Eisenstein lattice of the ball $U(2,1;\mathbb{Z}[\omega])$ with respect to the principal ideal $\mathbb{Z}[\omega]\cdot(1-\omega)$, $\omega = \exp(2\pi i/3)$.

(2) The rings of automorphic forms of $U(2,1;\mathbb{Z}[\omega])(1-\omega)$ and $U(2,1;\mathbb{Z}[\omega])$ (of Nebentypus χ) are polynomial rings

$$\bigoplus_{m=0}^{\infty}[U(2,1;\mathbb{Z}[\omega])(1-\omega), m]_{\chi} = \mathbb{C}[\xi_1, \xi_2, \xi_3],$$

$$\bigoplus_{m=0}^{\infty}[U(2,1;\mathbb{Z}[\omega]), m]_{\chi} = \mathbb{C}[G_2, G_3, G_4],$$

with ξ_1, ξ_2, ξ_3 of weight 1, G_k of weight k, $k = 2, 3, 4$.

A key result is the following one:

$$\bigoplus_{m=0}^{\infty}[SU(2,1;\mathbb{Z}[\omega])(1-\omega), m] = \mathbb{C}[\xi_1, \xi_2, \xi_3, \zeta],$$

with ξ_k as above for $k = 1, 2, 3$, ζ of weight 2,

$$\zeta^3 = (\xi_3^2 - \xi_2^2)(\xi_3^2 - \xi_1^2)(\xi_2^2 - \xi_1^2).$$

(3) $\mathbb{B}^2/U(2,1;\mathbb{Z}[\omega])(1-\omega) \cong \mathbb{P}^2 - \{$four points in general position$\}$.

(4) If F_1, F_2, F_3 are three fundamental solutions of Picard's system of differential equations, then there is a basis ξ_1', ξ_2', ξ_3' of the space of automorphic forms of weight 1 with respect to a group isomorphic to $U(2,1;\mathbb{Z}[\omega])(1-\omega)$, such that $(\xi_1' : \xi_2' : \xi_3')$ is the inverse of the multi-valued map $(F_1 : F_2 : F_3)$.

(5) Let $\mathbb{B}^2/\widehat{U(2,1;\mathbb{Z}[\omega])} = \overline{\mathbb{B}^2}/U(2,1;\mathbb{Z}[\omega])$ be the Baily-Borel compactification, where $\overline{\mathbb{B}^2} = \mathbb{B}^2 \cup \partial_{U(2,1;\mathbb{Z}[\omega])}\mathbb{B}^2$, $\partial_{U(2,1;\mathbb{Z}[\omega])}\mathbb{B}^2 \subset \partial\mathbb{B}^2$ the set of $U(2,1;\mathbb{Z}[\omega])$-cusps. Then we have the following commutative diagram:

$$\begin{array}{ccc}
P & \longrightarrow & C(P): Y^3 = X^4 + G_2(P)X^2 + G_3(P)X + G_4(P) \\
\overline{\mathbb{B}^2} & \longrightarrow & \{\text{Picard curves}\}/\text{projective equivalence} \\
\downarrow & & \downarrow \\
\overline{\mathbb{B}^2}/U(2,1;\mathbb{Z}[\omega]) & \xrightarrow{=} & \mathbb{B}^2/\widehat{U(2,1;\mathbb{Z}[\omega])}
\end{array}$$

($Y^3 = X^4$ is excluded from the set of Picard curves. On $\mathbb{B}^2 - U(2,1;\mathbb{Z}[\omega])\cdot D \subset \overline{\mathbb{B}^2}$, $D = \{(z,0) : |z| < 1\}$ a sub-disc of $\mathbb{B}^2$, projective equivalence can be replaced by isomorphism.) The image of $\mathbb{B}^2 - U(2,1;\mathbb{Z}[\omega])\cdot D$ is the set of isomorphism classes of all smooth Picard curves.

(6) *The images of the $U(2,1;\mathbb{Z}[\omega])$-fixed point set of $\mathbb{B}^2$ is the set of classes of Picard curves with larger automorphism groups (larger than $\mathbb{Z}/3\mathbb{Z}$).*

(7) *The Lagrange resolvent map, which associates to each polynomial of degree four a polynomial of degree three, admits by a change of fibres procedure over the moduli space of Picard curves (change Picard curves to elliptic curves) a geometric interpretation as a morphism $\mathbb{B}^2/U(2,1;\mathbb{Z}[\omega]) \to \mathbb{H}/SL(2,\mathbb{Z})$. There is a lift along the automorphic form (quotient) morphisms $\mathbb{B}^2 \to \mathbb{B}^2/U(2,1;\mathbb{Z}[\omega])$, $\mathbb{H} \to \mathbb{H}/SL(2,\mathbb{Z})$.*

Moreover, Picard modular forms can be considered as theta constants (see [23]). Theta functions $\vartheta\begin{bmatrix} a \\ b \end{bmatrix}$ with characteristics $a, b \in \mathbb{Q}^g$ are holomorphic functions on $\mathbb{C}^g \times \mathbb{H}_g$, $\mathbb{H}_g$ the generalized Siegel upper half plane uniformizing the moduli space of principally polarized abelian varieties of dimension g. Explicitly the theta functions

$$\vartheta\begin{bmatrix} a \\ b \end{bmatrix} : \mathbb{C}^g \times \mathbb{H}_g \to \mathbb{C}$$

are defined by

$$\vartheta\begin{bmatrix} a \\ b \end{bmatrix}(z,\Omega) = \sum_{n\in\mathbb{Z}^g} \exp\left[\pi i^t(n+a)\Omega(n+a) + 2\pi i^t(n+a)(z+b)\right].$$

The restrictions $\vartheta|0 \times \mathbb{H}_g$,

$$\theta\begin{bmatrix} a \\ b \end{bmatrix}(\Omega) = \vartheta\begin{bmatrix} a \\ b \end{bmatrix}(0,\Omega)$$

are called theta constants (with characteristics a, b).

Theorem 3.15 (Feustel, Shiga) (see [23]). *Let $\theta_i(\Omega) = \vartheta_i(0,\Omega)$, $i = 0,1,2$, be the theta constants on $\mathbb{H}_3$ restricting the theta functions*

$$\vartheta_k(z,\Omega) = \vartheta\begin{bmatrix} 0 & \frac{1}{6} & 0 \\ \frac{k}{3} & \frac{1}{6} & \frac{k}{3} \end{bmatrix}(z,\Omega), \quad k = 0,1,2, \quad z \in \mathbb{C}^3.$$

Set

$$Th_1 = \theta_0^3 + \theta_1^3 + \theta_2^3, \quad Th_2 = -3\theta_0^3 + \theta_1^3 + \theta_2^3,$$
$$Th_3 = \theta_0^3 - 3\theta_1^3 + \theta_2^3, \quad Th_4 = \theta_0^3 + \theta_1^3 - 3\theta_2^3,$$

and

$$th_i(\tau) = Th_i(*\tau), \quad i = 1,2,3,4, \quad \tau \in \mathbb{B}^2,$$

where the embedding $* : \mathbb{B}^2 \hookrightarrow \mathbb{H}_3$. *Then the functions* $th_i(\tau)$ *are the normalized Picard modular forms satisfying*

$$t_1 + t_2 + t_3 + t_4 = 0, \quad t_1,\ t_2,\ t_3 \text{ are linearly independent,}$$

and the functional equations

$$\gamma^*(t_i) = (\det \gamma)^2 \cdot sgn(\overline{\gamma}) \cdot j_\gamma \cdot t_{\overline{\gamma}(i)} \quad \text{for } i = 1, 2, 3, 4.$$

The analogies between elliptic modular forms and Picard modular forms are given as follows:

(1)

$$\bigoplus_{m=0}^{\infty} [SL(2,\mathbb{Z}), m] = \mathbb{C}[g_2, g_3],$$

$$\bigoplus_{m=0}^{\infty} [SL(2,\mathbb{Z})(2), m] = \mathbb{C}[\varepsilon_1, \varepsilon_2],$$

where g_i is of weight i, $i = 2,\ 3$, ε_1, ε_2 are of weight 1.

(2)

$$\bigoplus_{m=0}^{\infty} [U(2,1;\mathcal{O}_K), m]_\chi = \mathbb{C}[G_2, G_3, G_4],$$

$$\bigoplus_{m=0}^{\infty} [U(2,1;\mathcal{O}_K(1-\omega)), m]_\chi = \mathbb{C}[\xi_1, \xi_2, \xi_3],$$

where G_k is of weight k, $k = 2,\ 3,\ 4$, ξ_1, ξ_2, ξ_3 are of weight 1.

This leads us to give the following conjecture, which is the higher dimensional counterpart of Theorem 3.11 and Theorem 3.12.

Conjecture 3.16.

$$\mathbb{C}[z_1, z_2, z_3]^{Hessian\ groups} = \mathbb{C}[C_6, C_9, C_{12}]. \tag{3.7}$$

The map

$$\phi : \mathbb{C}[C_6, C_9, C_{12}] \to \mathbb{C}[G_2, G_3, G_4] \tag{3.8}$$

given by

$$\phi(h(C_6, C_9, C_{12})) = h(G_2, G_3, G_4) \tag{3.9}$$

is an algebra isomorphism.

We have the following correspondences

$$(C_6, C_9, C_{12}) \longleftrightarrow (G_2, G_3, G_4), \tag{3.10}$$

$$(G, H, K) \longleftrightarrow (\xi_1, \xi_2, \xi_3). \tag{3.11}$$

$$(C_{12}, C_{18}) \longleftrightarrow (E_4, E_6) = (g_2, g_3), \tag{3.12}$$

$$(H, K) \longleftrightarrow (\varepsilon_1, \varepsilon_2). \tag{3.13}$$

Note that

$$C_{12} = C_{12}(z_1, z_2, z_3) \longleftrightarrow G_4, \quad \text{three-variable function.}$$

$$C_{12} = C_{12}(H, K) \longleftrightarrow g_2, \quad \text{two-variable function.}$$

(1) $GL(2)$: ε_1, ε_2: theta functions of weight one, elliptic modular functions

(2) $GL(3)$: ξ_1, ξ_2, ξ_3: theta functions of weight one, Picard modular functions

$$GL(2): \quad (x, y) \longleftrightarrow (\theta_0(z), \theta_1(z)), z \in \mathbb{H}.$$

$$E_4(z) = \theta_0(z)^4 + 8\theta_0(z)\theta_1(z)^3, \quad E_6(z) = \theta_0(z)^6 - 20\theta_0(z)^3\theta_1(z)^3 - 8\theta_1(z)^6.$$

$\theta_0(z)$, $\theta_1(z)$ are of weight 1.

$$U(2,1): \quad (z_1, z_2, z_3) \longleftrightarrow (\vartheta_0(w_1, w_2), \vartheta_1(w_1, w_2), \vartheta_2(w_1, w_2)), (w_1, w_2) \in \mathfrak{S}_2, \tag{3.14}$$

where $\mathfrak{S}_2 = \{(z_1, z_2) \in \mathbb{C}^2 : z_1 + \overline{z_1} > z_2\overline{z_2}\}$. $\vartheta_0(w_1, w_2)$, $\vartheta_1(w_1, w_2)$ and $\vartheta_2(w_1, w_2)$ are of weight $\frac{1}{3}$. By

$$9C_{12} = H(H^3 + 8K^3), \quad -27C_{18} = H^6 - 20H^3K^3 - 8K^6,$$

we have

$$(H, K) \longleftrightarrow (\theta_0(z), \theta_1(z)),$$

$$(9C_{12}, -27C_{18}) \longleftrightarrow (E_4(z), E_6(z)).$$

Note that

$$H = z_1^3 + z_2^3 + z_3^3 + 6z_1z_2z_3, \quad K = z_1^3 + z_2^3 + z_3^3 - 3z_1z_2z_3.$$

Thus,

$$\begin{aligned}\theta_0(z) =& \vartheta_0(w_1, w_2)^3 + \vartheta_1(w_1, w_2)^3 + \vartheta_2(w_1, w_2)^3 \\ &+ 6\vartheta_0(w_1, w_2)\vartheta_1(w_1, w_2)\vartheta_2(w_1, w_2),\end{aligned}$$

$$\begin{aligned}\theta_1(z) =& \vartheta_0(w_1, w_2)^3 + \vartheta_1(w_1, w_2)^3 + \vartheta_2(w_1, w_2)^3 \\ &- 3\vartheta_0(w_1, w_2)\vartheta_1(w_1, w_2)\vartheta_2(w_1, w_2).\end{aligned}$$

Hence,

$$\vartheta_0(w_1, w_2)^3 + \vartheta_1(w_1, w_2)^3 + \vartheta_2(w_1, w_2)^3 = \frac{1}{3}[\theta_0(z) + 2\theta_1(z)], \tag{3.15}$$

$$\vartheta_0(w_1, w_2)\vartheta_1(w_1, w_2)\vartheta_2(w_1, w_2) = \frac{1}{9}[\theta_0(z) - \theta_1(z)]. \tag{3.16}$$

The group $PSL(2,\mathbb{Z})/\Gamma(3)$ is the tetrahedral group. The theta functions $\theta_0(z)$, $\theta_1(z)$ define a mapping

$$\frac{\theta_0}{\theta_1} : \overline{\mathbb{H}/\Gamma(3)} \to \mathbb{P}^1(\mathbb{C}).$$

This mapping is a bijection. The group $PSL(2,\mathbb{Z})/\Gamma(3)$ acts on $\overline{\mathbb{H}/\Gamma(3)}$. Under the above mapping this action corresponds to the action of the tetrahedral group on a tetrahedron lying inside the Riemann sphere $\mathbb{P}^1(\mathbb{C})$ (see [13]).

In our case, the theta functions $\vartheta_0(w_1,w_2)$, $\vartheta_1(w_1,w_2)$ and $\vartheta_2(w_1,w_2)$ define a mapping

$$\left(\frac{\vartheta_1}{\vartheta_0}, \frac{\vartheta_2}{\vartheta_0}\right) : \overline{\mathfrak{S}_2/U(2,1;\mathcal{O}_K)(1-\omega)} \to \mathbb{P}^2(\mathbb{C}),$$

where $K = \mathbb{Q}(\sqrt{-3})$ and $\mathcal{O}_K = \mathbb{Z}[\omega]$.

Let $\mathfrak{S}_2$ be the Siegel domain

$$\mathfrak{S}_2 = \{(z_1,z_2) \in \mathbb{C}^2 : z_1 + \overline{z_1} > |z_2|^2\}.$$

The imaginary quadratic field $K = \mathbb{Q}(\sqrt{-3})$ is called the field of Eisenstein numbers. Its ring of integers

$$\mathcal{O} = \mathcal{O}_K = \{a/2 + b\sqrt{-3}/2 : a,b \in \mathbb{Z}, a \equiv b(\text{mod}2)\} = \mathbb{Z} + \mathbb{Z}[\omega]$$

with $\omega = \frac{-1+\sqrt{-3}}{2}$ is called the ring of Eisenstein integers. We define the Cayley transform $C : \mathbb{B}^2 := \{(w_1,w_2) \in \mathbb{C}^2 : |w_1|^2 + |w_2|^2 < 1\} \to \mathfrak{S}_2$, where

$$C = \begin{pmatrix} -\overline{\omega} & & -\omega \\ & 1 & \\ -1 & & 1 \end{pmatrix} \in GL(3,\mathcal{O}_K).$$

It is known that

$$U(2,1) = \{g \in GL(3,\mathbb{C}) : gI_{2,1}g^* = I_{2,1}\},$$

where $I_{2,1} = \text{diag}\{1,1,-1\}$.

Let

$$G = CU(2,1)C^{-1}, \quad K = C(U(2)\times U(1))C^{-1}, \quad G(\mathcal{O}) = CU(2,1;\mathcal{O}_K)C^{-1},$$

then $G(\mathcal{O})$ is an arithmetic subgroup of G and $S := G(\mathcal{O})\backslash G/K$ is called a Picard modular surface. Denote $J = \begin{pmatrix} 0 & 0 & 1 \\ 0 & -1 & 0 \\ 1 & 0 & 0 \end{pmatrix}$, then $CI_{2,1}C^* = J$. It follows that

$$G = \{g \in GL(3,\mathbb{C}) : gJg^* = J\}.$$

This is the other realization of $U(2,1)$. For simplicity, from now on, we use the same symbol $U(2,1)$ to denote as G. In fact, $G = \{g \in GL(3,\mathbb{C}) : g^*Jg = J\}$.

The principal congruence subgroup $U(2,1;\mathcal{O}_K)(1-\omega)$ of $U(2,1;\mathcal{O}_K)$ is the monodromy group of the Appell hypergeometric partial differential equations with parameters $(a,b,b',c)=(\frac{1}{3},\frac{1}{3},\frac{1}{3},\frac{1}{3})$. Moreover, $U(2,1;\mathcal{O}_K)(1-\omega)$ is the modular group of the algebraic family of projective plane curves C of equation type

$$C: \quad y^3 = p_4(x), \quad \deg p_4(x) = 4$$

(Picard curves) endowed with a fixed order of the ramification points of the 3-sheeted Galois covering $C \to \mathbb{P}^1$ induced by the projection $(x,y) \mapsto x$. More precisely, the noncompact surface $\mathfrak{S}_2/U(2,1;\mathcal{O}_K)(1-\omega)$ classifies the isomorphy classes of these curves endowed with four ordered points. $\mathfrak{S}_2/U(2,1;\mathcal{O}_K)(1-\omega)$ is isomorphic to $\mathbb{P}^2 - \{4 \text{ points}\}$. The four points come from four cusps lying on $\partial\mathfrak{S}_2$. $SU(2,1;\mathcal{O}_K)(1-\omega)$ has up to $SU(2,1;\mathcal{O}_K)(1-\omega)$-equivalence exactly four cusps and three elliptic points $\widetilde{S_0}$, $\widetilde{S_1}$, $\widetilde{S_2} \in \mathfrak{S}_2$ with stationary groups of order 3. For the smooth compactification $\overline{\mathfrak{S}_2/SU(2,1;\mathcal{O}_K)(1-\omega)}$ of $\mathfrak{S}_2/SU(2,1;\mathcal{O}_K)(1-\omega)$ one needs four elliptic curves E_0, E_1, E_2, E_3. So $\overline{\mathfrak{S}_2/SU(2,1;\mathcal{O}_K)(1-\omega)}$ or $(\overline{\mathfrak{S}_2/SU(2,1;\mathcal{O}_K)(1-\omega)}, E_0+E_1+E_2+E_3)$ is an elliptic bounded surface (see [25]).

According to Picard (see [40] or [22]), the five generators of the principal congruence subgroup $U(2,1;\mathcal{O}_K)(1-\omega)$ are given as follows:

$$g_1 = \begin{pmatrix} 1 & \overline{\omega}-\omega & 1-\overline{\omega} \\ 0 & \overline{\omega} & 1-\omega \\ 0 & 0 & 1 \end{pmatrix}, \quad g_2 = \begin{pmatrix} 1 & \overline{\omega}-1 & 1-\overline{\omega} \\ 0 & \overline{\omega} & 1-\overline{\omega} \\ 0 & 0 & 1 \end{pmatrix}, \quad g_3 = \begin{pmatrix} 1 & 0 & 0 \\ 0 & \omega & 0 \\ 0 & 0 & 1 \end{pmatrix},$$

$$g_4 = \begin{pmatrix} \omega & 0 & 1-\overline{\omega} \\ 0 & 1 & 0 \\ 1-\overline{\omega} & 0 & -2\omega \end{pmatrix}, \quad g_5 = \begin{pmatrix} 1 & 0 & 0 \\ \overline{\omega}-\omega & \overline{\omega} & 0 \\ 1-\overline{\omega} & 1-\omega & 1 \end{pmatrix}.$$

In fact (see [60]),

$$\begin{cases} g_1 = U_1^{-4} T_1^{-1} T_2^{-2}, \\ g_2 = U_1^{-4} T_1^{-2} T_2^{-1}, \\ g_3 = U_1^4, \\ g_4 = U_2^3 S^3 [T_1, T_2] S^3 (S^4 U_2)^{-1} [T_1, T_2], \\ g_5 = S^3 U_1^{-4} T_1^{-1} T_2 S^3. \end{cases}$$

Here,

$$T_1 = \begin{pmatrix} 1 & 1 & -\omega \\ & 1 & 1 \\ & & 1 \end{pmatrix}, \quad T_2 = \begin{pmatrix} 1 & \omega & -\omega \\ & 1 & \overline{\omega} \\ & & 1 \end{pmatrix}, \quad S = -\overline{\omega} J = \begin{pmatrix} & & -\overline{\omega} \\ & \overline{\omega} & \\ -\overline{\omega} & & \end{pmatrix},$$

$$U_1 = \begin{pmatrix} 1 & & \\ & -\omega & \\ & & 1 \end{pmatrix}, \quad U_2 = \begin{pmatrix} -1 & & \\ & -\omega & \\ & & -1 \end{pmatrix}.$$

$$[T_1, T_2] := T_1 T_2 T_1^{-1} T_2^{-1} = \begin{pmatrix} 1 & 0 & \overline{\omega} - \omega \\ & 1 & 0 \\ & & 1 \end{pmatrix}.$$

Note that $g_3^3 = g_4^3 = I$ and $g_3 g_4 = g_4 g_3$. Hence,

$$\langle g_3, g_4 \rangle \cong \mathbb{Z}/3\mathbb{Z} \times \mathbb{Z}/3\mathbb{Z}.$$

In [61], we set

$$\varphi = z_1 z_2 z_3, \quad \psi = z_1^3 + z_2^3 + z_3^3, \quad \chi = z_1^3 z_2^3 + z_2^3 z_3^3 + z_3^3 z_1^3.$$

Put

$$X(z_1, z_2, z_3) = z_1^3, \quad Y(z_1, z_2, z_3) = z_2^3, \quad Z(z_1, z_2, z_3) = z_3^3. \tag{3.17}$$

$$Q_1(z_1, z_2, z_3) = z_1 z_2^2 + z_2 z_3^2 + z_3 z_1^2, \quad Q_2(z_1, z_2, z_3) = z_1^2 z_2 + z_2^2 z_3 + z_3^2 z_1. \tag{3.18}$$

The functions φ, ψ, χ, X, Y, Z and Q_1, Q_2 satisfy the relations:

$$\begin{cases} \varphi^3 = XYZ, \\ \psi = X + Y + Z, \\ \chi = XY + YZ + ZX, \\ \chi + 3\varphi^2 + \varphi\psi = Q_1 Q_2, \\ \psi\chi + 6\varphi\chi + 6\varphi^2\psi + 9\varphi^3 = Q_1^3 + Q_2^3, \\ (X - Y)(Y - Z)(Z - X) = Q_1^3 - Q_2^3. \end{cases} \tag{3.19}$$

Let

$$\begin{cases} W_2 = (X + Y + Z)^2 - 12(XY + YZ + ZX), \\ W_3 = (X - Y)(Y - Z)(Z - X), \\ \mathfrak{W}_3 = XYZ - \varphi^3, \\ W_4 = (X + Y + Z)[(X + Y + Z)^3 + 216XYZ], \\ \mathfrak{W}_4 = \varphi[27XYZ - (X + Y + Z)^3], \\ W_6 = (X + Y + Z)^6 - 540XYZ(X + Y + Z)^3 - 5832X^2Y^2Z^2, \\ \mathfrak{V}_3 = Q_1^3 + Q_2^3 - (X + Y + Z + 6\varphi)(XY + YZ + ZX) \\ \qquad - 6\varphi^2(X + Y + Z) - 9\varphi^3, \\ \mathfrak{V}_2 = Q_1 Q_2 - (XY + YZ + ZX) - \varphi(X + Y + Z) - 3\varphi^2, \\ \mathfrak{U}_3 = Q_1^3 - Q_2^3 - (X - Y)(Y - Z)(Z - X). \end{cases} \tag{3.20}$$

Theorem 3.17. *The invariants W_2, W_3, $\mathfrak{W}_3$, W_4, $\mathfrak{W}_4$ and W_6 satisfy the following algebraic relations:*

$$\begin{cases} W_2^3 - 3W_2 W_4 + 2W_6 = 432 W_3^2, \\ 8U^3(W_6^2 - W_4^3 - 1728\mathfrak{W}_4^3) = 27\mathfrak{W}_3(W_4 - 9U^4)^3, \end{cases} \tag{3.21}$$

where

$$27U^8 - 18W_4U^4 - 8W_6U^2 - W_4^2 = 0. \tag{3.22}$$

Proof. By the definition of W_2, W_3, W_4 and W_6, we have

$$W_2^3 - 3W_2W_4 + 2W_6 = 432W_3^2.$$

We find that

$$W_6^2 - W_4^3 - 1728\mathfrak{W}_4^3 = 1728\mathfrak{W}_3[27XYZ - (X+Y+Z)^3]^3. \tag{3.23}$$

Put

$$U = X + Y + Z, \quad V = XYZ. \tag{3.24}$$

Then

$$\begin{cases} U(U^3 + 216V) = W_4, \\ U^6 - 540U^3V - 5832V^2 = W_6. \end{cases} \tag{3.25}$$

By the first equation in (3.25), we have

$$V = \frac{W_4 - U^4}{216U}. \tag{3.26}$$

Substituting the above expression (3.26) in the second equation in (3.25), we obtain that

$$27U^8 - 18W_4U^4 - 8W_6U^2 - W_4^2 = 0.$$

Set $t = U^2$. Let

$$t^4 - \frac{2}{3}W_4t^2 - \frac{8}{27}W_6t - \frac{1}{27}W_4^2 = (t^2 + kt + l)(t^2 - kt + m). \tag{3.27}$$

Then

$$\begin{cases} l + m - k^2 = -\frac{2}{3}W_4, \\ k(m - l) = -\frac{8}{27}W_6, \\ lm = -\frac{1}{27}W_4^2. \end{cases} \tag{3.28}$$

By the first two equations in (3.28), we have

$$\begin{cases} l = \frac{1}{2k}(k^3 - \frac{2}{3}W_4k + \frac{8}{27}W_6), \\ m = \frac{1}{2k}(k^3 - \frac{2}{3}W_4k - \frac{8}{27}W_6). \end{cases} \tag{3.29}$$

Substituting these two expressions in the third equation in (3.28), we have

$$k^6 - \frac{4}{3}W_4k^4 + \frac{16}{27}W_4^2k^2 - \frac{64}{729}W_6^2 = 0, \tag{3.30}$$

i.e.,

$$(k^2 - \frac{4}{9}W_4)^3 = \frac{64}{729}(W_6^2 - W_4^3).$$

Hence,

$$k = \pm\frac{2}{3}\sqrt{W_4 + \sqrt[3]{W_6^2 - W_4^3}}. \tag{3.31}$$

Consequently,

$$l = -\frac{1}{9}W_4 + \frac{2}{9}\sqrt[3]{W_6^2 - W_4^3} \pm \frac{2}{9}\frac{W_6}{\sqrt{W_4 + \sqrt[3]{W_6^2 - W_4^3}}}. \tag{3.32}$$

$$m = -\frac{1}{9}W_4 + \frac{2}{9}\sqrt[3]{W_6^2 - W_4^3} \mp \frac{2}{9}\frac{W_6}{\sqrt{W_4 + \sqrt[3]{W_6^2 - W_4^3}}}. \tag{3.33}$$

Thus,

$$t = \frac{1}{2}(-k \pm \sqrt{k^2 - 4l}), \quad \text{or} \quad t = \frac{1}{2}(k \pm \sqrt{k^2 - 4m}). \tag{3.34}$$

We have

$$W_6^2 - W_4^3 - 1728\mathfrak{W}_4^3 = 1728\mathfrak{W}_3(27V - U^3)^3 = \frac{27}{8}\mathfrak{W}_3\frac{(W_4 - 9U^4)^3}{U^3}.$$

Therefore,

$$8U^3(W_6^2 - W_4^3 - 1728\mathfrak{W}_4^3) = 27\mathfrak{W}_3(W_4 - 9U^4)^3.$$

□

Theorem 3.18. *The invariants W_2, W_3, $\mathfrak{W}_3$, W_4, $\mathfrak{W}_4$, $\mathfrak{U}_3$, $\mathfrak{V}_2$ and $\mathfrak{V}_3$ satisfy the following differential relations:*

$$\frac{\partial(W_2, W_3, W_4, \mathfrak{W}_3)}{\partial(X, Y, Z, \varphi)} = 288\mathfrak{W}_4^2. \tag{3.35}$$

$$\frac{\partial(\mathfrak{V}_3, \mathfrak{V}_2)}{\partial(Q_1, Q_2)} = 3(\mathfrak{U}_3 + W_3). \tag{3.36}$$

Proof. We have

$$\begin{cases} \frac{\partial W_2}{\partial X} = 2X - 10Y - 10Z, \\ \frac{\partial W_2}{\partial Y} = 2Y - 10Z - 10X, \\ \frac{\partial W_2}{\partial Z} = 2Z - 10X - 10Y, \\ \frac{\partial W_2}{\partial \varphi} = 0. \end{cases}$$

$$\begin{cases} \dfrac{\partial W_3}{\partial X} = (Y-Z)(Y+Z-2X), \\ \dfrac{\partial W_3}{\partial Y} = (Z-X)(Z+X-2Y), \\ \dfrac{\partial W_3}{\partial Z} = (X-Y)(X+Y-2Z), \\ \dfrac{\partial W_3}{\partial \varphi} = 0. \end{cases}$$

$$\begin{cases} \dfrac{\partial W_4}{\partial X} = 4(X+Y+Z)^3 + 216XYZ + 216(X+Y+Z)YZ, \\ \dfrac{\partial W_4}{\partial Y} = 4(X+Y+Z)^3 + 216XYZ + 216(X+Y+Z)ZX, \\ \dfrac{\partial W_4}{\partial Z} = 4(X+Y+Z)^3 + 216XYZ + 216(X+Y+Z)XY, \\ \dfrac{\partial W_4}{\partial \varphi} = 0. \end{cases}$$

$$\begin{cases} \dfrac{\partial \mathfrak{W}_3}{\partial X} = YZ, \\ \dfrac{\partial \mathfrak{W}_3}{\partial Y} = ZX, \\ \dfrac{\partial \mathfrak{W}_3}{\partial Z} = XY, \\ \dfrac{\partial \mathfrak{W}_3}{\partial \varphi} = -3\varphi^2. \end{cases}$$

Hence,

$$\frac{\partial(W_2, W_3, W_4, \mathfrak{W}_3)}{\partial(X, Y, Z, \varphi)} = -24\varphi^2 \Delta,$$

where

$$\begin{aligned} \Delta =& [(X+Y+Z)^3 + 54YZ(2X+Y+Z)][-(X+Y+Z)^3 \\ & - 9X(X+Y+Z)^2 + 27X^2(Y+Z) + 27X(Y^2+Z^2)] \\ & + [(X+Y+Z)^3 + 54ZX(2Y+Z+X)][-(X+Y+Z)^3 \\ & - 9Y(X+Y+Z)^2 + 27Y^2(Z+X) + 27Y(Z^2+X^2)] \\ & + [(X+Y+Z)^3 + 54XY(2Z+X+Y)][-(X+Y+Z)^3 \\ & - 9Z(X+Y+Z)^2 + 27Z^2(X+Y) + 27Z(X^2+Y^2)]. \end{aligned}$$

By the proof of Proposition 4.3 in [61], we have

$$\Delta = -12[27XYZ - (X+Y+Z)^3]^2.$$

Thus,

$$\frac{\partial(W_2, W_3, W_4, \mathfrak{W}_3)}{\partial(X, Y, Z, \varphi)} = 288\varphi^2[27XYZ - (X+Y+Z)^3]^2 = 288\mathfrak{W}_4^2.$$

On the other hand, we find that

$$\frac{\partial(\mathfrak{V}_3, \mathfrak{V}_2)}{\partial(Q_1, Q_2)} = 3(Q_1^3 - Q_2^3) = 3(\mathfrak{U}_3 + W_3).$$

□

Combine Theorem 3.17 with Theorem 3.18, we get the proof of Theorem 1.4 (Main Theorem 2).

According to [22], we denote the elementary symmetric functions of X_1, X_2, X_3, X_4 by

$$\begin{aligned}
\Sigma_1 &= -(X_1 + X_2 + X_3 + X_4),\\
\Sigma_2 &= X_1X_2 + X_1X_3 + X_1X_4 + X_2X_3 + X_2X_4 + X_3X_4,\\
\Sigma_3 &= -(X_1X_2X_3 + X_1X_2X_4 + X_1X_3X_4 + X_2X_3X_4),\\
\Sigma_4 &= X_1X_2X_3X_4,
\end{aligned}$$

and

$$\Delta = \prod_{1 \leq i < j \leq 4} (X_i - X_j).$$

Then

$$27\Delta^2 = 4(\Sigma_2^2 - 3\Sigma_1\Sigma_3 + 12\Sigma_4)^3 - (27\Sigma_1^2\Sigma_4 + 27\Sigma_3^2 + 2\Sigma_2^3 - 72\Sigma_2\Sigma_4 + 9\Sigma_1\Sigma_2\Sigma_3)^2.$$

Their images in $\mathbb{C}[x_1, x_2, x_3, x_4] = \mathbb{C}[X_1, X_2, X_3, X_4]/(X_1 + X_2 + X_3 + X_4)$ are denoted by $G_1 = 0$, G_2, G_3 and G_4, respectively. Hence,

$$\Delta^2 = 16G_2^4G_4 - 128G_2^2G_4^2 - 4G_2^3G_3^2 + 144G_2G_3^2G_4 - 27G_3^4 + 256G_4^3.$$

Proposition 3.19 (see [22]). *There are two isomorphisms of graded rings:*

$$\bigoplus_{m=0}^{\infty} [U(2,1;\mathcal{O}_K), m]_\chi \cong \mathbb{C}[G_2, G_3, G_4]$$

and

$$\bigoplus_{m=0}^{\infty} [SU(2,1;\mathcal{O}_K), m]_\chi \cong \mathbb{C}[G_2, G_3, G_4, z^2].$$

The fundamental relation between the generators is $z^6 = \Delta^2$.

We find the following correspondence:

$$(G_2, G_3, G_4, z^2) \longleftrightarrow (W_2, W_3, W_4, \mathfrak{W}_4).$$

Here, W_2, W_3, W_4 and $\mathfrak{W}_4$ satisfy the single relation:

$$\begin{aligned}
6912\mathfrak{W}_4^3 =& W_2^6 + 9W_2^2W_4^2 + 432^2W_3^4 - 4W_4^3 - 6W_2^4W_4\\
&- 864W_2^3W_3^2 + 2592W_2W_3^2W_4.
\end{aligned}$$

The equations $W_4 = 0$ and $\mathfrak{W}_4 = 0$ define two surfaces of order 4. The intersection of the cubic surface $\mathfrak{W}_3 = 0$ with the quartic surface $W_4 = 0$ gives four elliptic curves

$$\begin{cases} \varphi^3 + 6\varphi XY + (X+Y)XY = 0, \\ \varphi^3 + 6\omega\varphi XY + (X+Y)XY = 0, \\ \varphi^3 + 6\overline{\omega}\varphi XY + (X+Y)XY = 0, \\ \varphi^3 + (X+Y)XY = 0. \end{cases}$$

The intersection of the cubic surface $\mathfrak{W}_3 = 0$ with the quartic surface $\mathfrak{W}_4 = 0$ gives the other four elliptic curves

$$\begin{cases} \varphi^3 - 3\varphi XY + (X+Y)XY = 0, \\ \varphi^3 - 3\omega\varphi XY + (X+Y)XY = 0, \\ \varphi^3 - 3\overline{\omega}\varphi XY + (X+Y)XY = 0, \\ XYZ = 0. \end{cases}$$

The equation $W_2 = 0$ defines a quadric surface, the equation $W_3 = 0$ is the union of three planes. The intersection of $W_3 = 0$ with $\mathfrak{W}_3 = 0$ gives three elliptic curves

$$Y^2Z - \varphi^3 = 0, \quad Z^2X - \varphi^3 = 0, \quad X^2Y - \varphi^3 = 0.$$

The intersection of $W_2 = 0$ with $\mathfrak{W}_3 = 0$ gives an algebraic curve of degree six

$$\varphi^6 - 10XY(X+Y)\varphi^3 + (X^4Y^2 - 10X^3Y^3 + X^2Y^4) = 0.$$

It is known that the Hessian group of order 216 is generated by the five generators (see [61]):

$$A = \begin{pmatrix} 0 & 1 & 0 \\ 0 & 0 & 1 \\ 1 & 0 & 0 \end{pmatrix}, \quad B = \begin{pmatrix} 1 & 0 & 0 \\ 0 & 0 & 1 \\ 0 & 1 & 0 \end{pmatrix},$$

$$C = \begin{pmatrix} 1 & 0 & 0 \\ 0 & \omega & 0 \\ 0 & 0 & \omega^2 \end{pmatrix}, \quad D = \begin{pmatrix} 1 & 0 & 0 \\ 0 & \omega & 0 \\ 0 & 0 & \omega \end{pmatrix},$$

$$E = \frac{1}{\sqrt{-3}} \begin{pmatrix} 1 & 1 & 1 \\ 1 & \omega & \omega^2 \\ 1 & \omega^2 & \omega \end{pmatrix}.$$

We have

$$\begin{cases} \varphi(A(z_1, z_2, z_3)) = \varphi, \\ X(A(z_1, z_2, z_3)) = Y, \\ Y(A(z_1, z_2, z_3)) = Z, \\ Z(A(z_1, z_2, z_3)) = X, \\ Q_1(A(z_1, z_2, z_3)) = Q_1, \\ Q_2(A(z_1, z_2, z_3)) = Q_2. \end{cases}$$

$$\begin{cases} \varphi(B(z_1, z_2, z_3)) = \varphi, \\ X(B(z_1, z_2, z_3)) = X, \\ Y(B(z_1, z_2, z_3)) = Z, \\ Z(B(z_1, z_2, z_3)) = Y, \\ Q_1(B(z_1, z_2, z_3)) = Q_2, \\ Q_2(B(z_1, z_2, z_3)) = Q_1. \end{cases}$$

$$\begin{cases} \varphi(C(z_1, z_2, z_3)) = \varphi, \\ X(C(z_1, z_2, z_3)) = X, \\ Y(C(z_1, z_2, z_3)) = Y, \\ Z(C(z_1, z_2, z_3)) = Z, \\ Q_1(C(z_1, z_2, z_3)) = \overline{\omega} Q_1, \\ Q_2(C(z_1, z_2, z_3)) = \omega Q_2. \end{cases}$$

$$\begin{cases} \varphi(D(z_1, z_2, z_3)) = \overline{\omega}\varphi, \\ X(D(z_1, z_2, z_3)) = X, \\ Y(D(z_1, z_2, z_3)) = Y, \\ Z(D(z_1, z_2, z_3)) = Z, \\ Q_1(D(z_1, z_2, z_3)) = \overline{\omega} z_1 z_2^2 + z_2 z_3^2 + \omega z_3 z_1^2, \\ Q_2(D(z_1, z_2, z_3)) = \omega z_1^2 z_2 + z_2^2 z_3 + \overline{\omega} z_3^2 z_1. \end{cases}$$

$$\begin{cases} (\sqrt{-3})^3 \varphi(E(z_1, z_2, z_3)) = X + Y + Z - 3\varphi, \\ (\sqrt{-3})^3 X(E(z_1, z_2, z_3)) = X + Y + Z + 6\varphi + 3Q_1 + 3Q_2, \\ (\sqrt{-3})^3 Y(E(z_1, z_2, z_3)) = X + Y + Z + 6\varphi + 3\overline{\omega} Q_1 + 3\omega Q_2, \\ (\sqrt{-3})^3 Z(E(z_1, z_2, z_3)) = X + Y + Z + 6\varphi + 3\omega Q_1 + 3\overline{\omega} Q_2, \\ (\sqrt{-3})^3 Q_1(E(z_1, z_2, z_3)) = 3X + 3\overline{\omega} Y + 3\omega Z, \\ (\sqrt{-3})^3 Q_2(E(z_1, z_2, z_3)) = 3X + 3\omega Y + 3\overline{\omega} Z. \end{cases}$$

In the form of matrices, we have

$$A(X, Y, Z, \varphi, Q_1, Q_2) = \begin{pmatrix} & 1 & & & & \\ & & 1 & & & \\ 1 & & & & & \\ & & & 1 & & \\ & & & & 1 & \\ & & & & & 1 \end{pmatrix} \begin{pmatrix} X \\ Y \\ Z \\ \varphi \\ Q_1 \\ Q_2 \end{pmatrix}, \tag{3.37}$$

$$
\begin{aligned}
&B(X, Y, Z, \varphi, Q_1, Q_2) \\
&= \begin{pmatrix} 1 & & & & & \\ & & 1 & & & \\ & 1 & & & & \\ & & & 1 & & \\ & & & & & 1 \\ & & & & 1 & \end{pmatrix} \begin{pmatrix} X \\ Y \\ Z \\ \varphi \\ Q_1 \\ Q_2 \end{pmatrix},
\end{aligned} \tag{3.38}
$$

$$
\begin{aligned}
&C(X, Y, Z, \varphi, Q_1, Q_2) \\
&= \begin{pmatrix} 1 & & & & & \\ & 1 & & & & \\ & & 1 & & & \\ & & & 1 & & \\ & & & & \overline{\omega} & \\ & & & & & \omega \end{pmatrix} \begin{pmatrix} X \\ Y \\ Z \\ \varphi \\ Q_1 \\ Q_2 \end{pmatrix},
\end{aligned} \tag{3.39}
$$

$$
\begin{aligned}
&(\sqrt{-3})^3 E(X, Y, Z, \varphi, Q_1, Q_2) \\
&= \begin{pmatrix} 1 & 1 & 1 & 6 & 3 & 3 \\ 1 & 1 & 1 & 6 & 3\overline{\omega} & 3\omega \\ 1 & 1 & 1 & 6 & 3\omega & 3\overline{\omega} \\ 1 & 1 & 1 & -3 & 0 & 0 \\ 3 & 3\overline{\omega} & 3\omega & 0 & 0 & 0 \\ 3 & 3\omega & 3\overline{\omega} & 0 & 0 & 0 \end{pmatrix} \begin{pmatrix} X \\ Y \\ Z \\ \varphi \\ Q_1 \\ Q_2 \end{pmatrix},
\end{aligned} \tag{3.40}
$$

We find that

$$A^3 = B^2 = C^3 = E^4 = I. \tag{3.41}$$

Theorem 3.20 (see [61], Main Theorem 4). *The invariants G, H and K satisfy the following algebraic equations, which are the form-theoretic resolvents (algebraic resolvents) of G, H, K:*

$$
\begin{cases} 4G^3 + H^2 G - C_6 G - 4C_9 = 0, \\ H(H^3 + 8K^3) - 9C_{12} = 0, \\ K(K^3 - H^3) - 27\mathfrak{C}_{12} = 0. \end{cases}
$$

Consequently,

$$C_{18} = -\frac{1}{27}(H^6 - 20H^3K^3 - 8K^6).$$

Note that

$$W_2 = C_6, W_3 = C_9, W_4 = C_{12}, \mathfrak{W}_4 = \mathfrak{C}_{12}, W_6 = C_{18}.$$

Set $(x, y, z) = (G, H, K)$. Then

$$x = Q_1 - Q_2, \quad y = X + Y + Z + 6\varphi, \quad z = X + Y + Z - 3\varphi. \tag{3.42}$$

We have

$$\frac{1}{9}y(y^3+8z^3)=(X+Y+Z)[(X+Y+Z)^3+216\varphi^3]=W_4-216(X+Y+Z)\mathfrak{W}_3. \tag{3.43}$$

$$\frac{1}{27}z(z^3-y^3)=\varphi[27\varphi^3-(X+Y+Z)^3]=\mathfrak{W}_4-27\varphi\mathfrak{W}_3. \tag{3.44}$$

Thus,

$$X+Y+Z=\frac{9W_4-y(y^3+8z^3)}{1944\mathfrak{W}_3}.$$

$$\varphi=\frac{27\mathfrak{W}_4-z(z^3-y^3)}{729\mathfrak{W}_3}.$$

$$XYZ=\mathfrak{W}_3+\varphi^3=\mathfrak{W}_3+\frac{[27\mathfrak{W}_4-z(z^3-y^3)]^3}{729^3\mathfrak{W}_3^3}.$$

Note that $U=X+Y+Z$ and $V=XYZ$ satisfy the equations:

$$U(U^3+216V)=W_4, \quad U^6-540U^3V-5832V^2=W_6.$$

We obtain the following relations:

$$\begin{aligned}&[9W_4-y(y^3+8z^3)]^4\\&+2^{12}[9W_4-y(y^3+8z^3)]\{3^{18}\mathfrak{W}_3^4+[27\mathfrak{W}_4-z(z^3-y^3)]^3\}=2^{12}\cdot 3^{20}\mathfrak{W}_3^4W_4.\end{aligned}$$

$$\begin{aligned}&[9W_4-y(y^3+8z^3)]^6\\&-2^{11}\cdot 5[9W_4-y(y^3+8z^3)]^3\{3^{18}\mathfrak{W}_3^4+[27\mathfrak{W}_4-z(z^3-y^3)]^3\}\\&-2^{21}\{3^{18}\mathfrak{W}_3^4+[27\mathfrak{W}_4-z(z^3-y^3)]^3\}^2=2^{18}\cdot 3^{30}\mathfrak{W}_3^6W_6.\end{aligned}$$

Similarly, we have

$$4x^3+xy^2-W_2x-4W_3=4\mathfrak{U}_3-12(Q_1-Q_2)\mathfrak{V}_2.$$

Hence,

$$Q_1-Q_2=\frac{4\mathfrak{U}_3-4x^3-xy^2+W_2x+4W_3}{12\mathfrak{V}_2}.$$

By the identity:

$$(Q_1-Q_2)^3=Q_1^3-Q_2^3-3Q_1Q_2(Q_1-Q_2),$$

where

$$Q_1^3-Q_2^3=\mathfrak{U}_3+W_3,$$

$$Q_1Q_2=\mathfrak{V}_2+(XY+YZ+ZX)+\varphi(X+Y+Z)+3\varphi^2.$$

Note that

$$\begin{aligned}XY+YZ+ZX&=\frac{1}{12}[(X+Y+Z)^2-W_2]\\&=\frac{1}{2^8\cdot 3^{11}\mathfrak{W}_3^2}[(9W_4-y(y^3+8z^3))^2-2^6\cdot 3^{10}W_2\mathfrak{W}_3^2].\end{aligned}$$

We obtain the following relation:

$$\begin{aligned}&2^4\cdot 3^8\mathfrak{W}_3^2(4\mathfrak{U}_3-4x^3-xy^2+W_2x+4W_3)^3\\=&2^{10}\cdot 3^{11}\mathfrak{W}_3^2\mathfrak{V}_2^3(\mathfrak{U}_3+W_3)-(4\mathfrak{U}_3-4x^3-xy^2+W_2x+4W_3)\mathfrak{V}_2^2\\&\times\{2^8\cdot 3^{11}\mathfrak{W}_3^2\mathfrak{V}_2+[(9W_4-y(y^3+8z^3))^2-2^6\cdot 3^{10}W_2\mathfrak{W}_3^2]\\&+2^5[27\mathfrak{W}_4-z(z^3-y^3)][9W_4-y(y^3+8z^3)]+2^8[27\mathfrak{W}_4-z(z^3-y^3)]^2\}.\end{aligned}$$

Therefore, we finish the proof of Theorem 1.5 (Main Theorem 3). □

Proposition 3.21. *The space curves $W_2=\mathfrak{W}_3=\mathfrak{V}_3=\mathfrak{V}_2=0$, $W_3=\mathfrak{W}_3=\mathfrak{V}_3=\mathfrak{V}_2=0$, $W_4=\mathfrak{W}_3=\mathfrak{V}_3=\mathfrak{V}_2=0$, $W_6=\mathfrak{W}_3=\mathfrak{V}_3=\mathfrak{V}_2=0$ and $\mathfrak{W}_4=\mathfrak{W}_3=\mathfrak{V}_3=\mathfrak{V}_2=0$ are invariant curves on the invariant surface $\mathfrak{W}_3=\mathfrak{V}_3=\mathfrak{V}_2=0$ under the action of the subgroup of Hessian group generated by A, B, C and E.*

Proof. Note that on the surface $\mathfrak{W}_3=\mathfrak{V}_3=\mathfrak{V}_2=0$, $XYZ=\varphi^3$,

$$Q_1Q_2=XY+YZ+ZX+\varphi(X+Y+Z)+3\varphi^2,$$

$$Q_1^3+Q_2^3=(X+Y+Z+6\varphi)(XY+YZ+ZX)+6\varphi^2(X+Y+Z)+9\varphi^3.$$

Hence,

$$\begin{aligned}&(\sqrt{-3})^9\mathfrak{W}_3(E(X,Y,Z,\varphi,Q_1,Q_2))\\=&(X+Y+Z+6\varphi)^3+27Q_1^3+27Q_2^3-27(X+Y+Z+6\varphi)Q_1Q_2\\&-(X+Y+Z-3\varphi)^3\\=&0.\end{aligned}$$

Similarly,

$$\begin{aligned}&(\sqrt{-3})^9\mathfrak{V}_3(E(X,Y,Z,\varphi,Q_1,Q_2))\\=&54(X^3+Y^3+Z^3+6XYZ)\\&-81(X^2Y+Y^2Z+Z^2X+XY^2+YZ^2+ZX^2)\\&-9(X+Y+Z)[3(X+Y+Z+6\varphi)^2-27Q_1Q_2]\\&-18(X+Y+Z-3\varphi)^2(X+Y+Z+6\varphi)-9(X+Y+Z-3\varphi)^3\\=&243(X+Y+Z)[Q_1Q_2-(X+Y+Z)\varphi-3\varphi^2]-729\varphi^3\\&-243(X^2Y+Y^2Z+Z^2X+XY^2+YZ^2+ZX^2)\\=&729(XYZ-\varphi^3)\\=&0.\end{aligned}$$

$$\begin{aligned}&(\sqrt{-3})^6\mathfrak{V}_2(E(X,Y,Z,\varphi,Q_1,Q_2))\\=&9(X^2+Y^2+Z^2-XY-YZ-ZX)-3(X+Y+Z+6\varphi)^2+27Q_1Q_2\\&-3(X+Y+Z+6\varphi)(X+Y+Z-3\varphi)-3(X+Y+Z-3\varphi)^2\\=&27[Q_1Q_2-(XY+YZ+ZX)-\varphi(X+Y+Z)-3\varphi^2]\\=&0.\end{aligned}$$

On the other hand, we have

$$\begin{cases} \mathfrak{W}_3(A(X,Y,Z,\varphi,Q_1,Q_2)) = \mathfrak{W}_3, \\ \mathfrak{W}_3(B(X,Y,Z,\varphi,Q_1,Q_2)) = \mathfrak{W}_3, \\ \mathfrak{W}_3(C(X,Y,Z,\varphi,Q_1,Q_2)) = \mathfrak{W}_3. \end{cases}$$

$$\begin{cases} \mathfrak{V}_3(A(X,Y,Z,\varphi,Q_1,Q_2)) = \mathfrak{V}_3, \\ \mathfrak{V}_3(B(X,Y,Z,\varphi,Q_1,Q_2)) = \mathfrak{V}_3, \\ \mathfrak{V}_3(C(X,Y,Z,\varphi,Q_1,Q_2)) = \mathfrak{V}_3. \end{cases}$$

$$\begin{cases} \mathfrak{V}_2(A(X,Y,Z,\varphi,Q_1,Q_2)) = \mathfrak{V}_2, \\ \mathfrak{V}_2(B(X,Y,Z,\varphi,Q_1,Q_2)) = \mathfrak{V}_2, \\ \mathfrak{V}_2(C(X,Y,Z,\varphi,Q_1,Q_2)) = \mathfrak{V}_2. \end{cases}$$

Hence, $\mathfrak{W}_3 = \mathfrak{V}_3 = \mathfrak{V}_2 = 0$ is an invariant surface. On this surface, we have

$$Q_1^3 - Q_2^3 = \pm(X-Y)(Y-Z)(Z-X).$$

Hence,

$$(\sqrt{-3})^9 W_3(E(X,Y,Z,\varphi,Q_1,Q_2)) = 81\sqrt{-3}(Q_1^3 - Q_2^3) = \pm 81\sqrt{-3}W_3.$$

Furthermore,

$$\begin{cases} W_3(A(X,Y,Z,\varphi,Q_1,Q_2)) = (Y-Z)(Z-X)(X-Y) = W_3, \\ W_3(B(X,Y,Z,\varphi,Q_1,Q_2)) = (X-Z)(Z-Y)(Y-X) = -W_3, \\ W_3(C(X,Y,Z,\varphi,Q_1,Q_2)) = (X-Y)(Y-Z)(Z-X) = W_3. \end{cases}$$

This implies that $W_3 = \mathfrak{W}_3 = \mathfrak{V}_3 = \mathfrak{V}_2 = 0$ is an invariant curve on the above surface.

We have

$$\begin{aligned} &(\sqrt{-3})^6 W_2(E(X,Y,Z,\varphi,Q_1,Q_2)) \\ =& -27(X+Y+Z+6\varphi)^2 + 324Q_1Q_2 \\ =& -27[(X+Y+Z)^2 - 12(XY+YZ+ZX)] \\ =& -27W_2. \end{aligned}$$

Moreover,

$$\begin{cases} W_2(A(X,Y,Z,\varphi,Q_1,Q_2)) = W_2, \\ W_2(B(X,Y,Z,\varphi,Q_1,Q_2)) = W_2, \\ W_2(C(X,Y,Z,\varphi,Q_1,Q_2)) = W_2. \end{cases}$$

This implies that $W_2 = \mathfrak{W}_3 = \mathfrak{V}_3 = \mathfrak{V}_2 = 0$ is an invariant curve on the above surface.

We have

$$\begin{aligned}
&(\sqrt{-3})^{12}W_4(E(X,Y,Z,\varphi,Q_1,Q_2))\\
=&729(X+Y+Z+6\varphi)[(X+Y+Z+6\varphi)^3+24(Q_1^3+Q_2^3)\\
&-24(X+Y+Z+6\varphi)Q_1Q_2]\\
=&729(X+Y+Z+6\varphi)(X+Y+Z)[(X+Y+Z)^2-6(X+Y+Z)\varphi+36\varphi^2]\\
=&729(X+Y+Z)[(X+Y+Z)^3+216\varphi^3]\\
=&729(X+Y+Z)[(X+Y+Z)^3+216XYZ]\\
=&729W_4.
\end{aligned}$$

Moreover,

$$\begin{cases}
W_4(A(X,Y,Z,\varphi,Q_1,Q_2))=W_4,\\
W_4(B(X,Y,Z,\varphi,Q_1,Q_2))=W_4,\\
W_4(C(X,Y,Z,\varphi,Q_1,Q_2))=W_4.
\end{cases}$$

This implies that $W_4=\mathfrak{W}_3=\mathfrak{V}_3=\mathfrak{V}_2=0$ is an invariant curve on the above surface.

On the above invariant surface one has

$$W_2^3-3W_2W_4+2W_6=432W_3^2,\quad W_6^2-W_4^3=1728\mathfrak{W}_4^3.$$

Thus, $W_6=\mathfrak{W}_3=\mathfrak{V}_3=\mathfrak{V}_2=0$ and $\mathfrak{W}_4=\mathfrak{W}_3=\mathfrak{V}_3=\mathfrak{V}_2=0$ are invariant curves on the above surface.

□

4. Hessian polyhedra and Galois representations associated with cubic surfaces

Manin's book [32] (see also [33]) is devoted to the arithmetic of smooth cubic surfaces. We give an account of some of the results of this book that are characteristic of cubic surfaces only. A cubic surface X has a very specific property: every point $x\in X$ generates a birational automorphism t_x of X by reflection in x. This enables us to describe the group of birational automorphisms of X. On the other hand, it allows us to introduce on $X(k)$ the structure of the ternary relation of collinearity of points. We recall that a symmetric quasigroup C is a set with a binary composition $(x,y)\mapsto x\circ y$ such that the ternary relation $x\circ y=z$ is symmetric under all permutations of x, y and z, and such that $x\circ y=y\circ x$ and $x\circ(x\circ y)=y$ for any pair $x,y\in C$. Suppose that an equivalence relation S is given on $X(k)$, we say that S is admissible if collinearity of points induces the structure of a symmetric quasigroup on $X(k)/S$. We say that a symmetric quasigroup C is abelian if for any element u the composition law $xy=u\circ(x\circ y)$ turns C into an abelian group. If any three elements of C generate an abelian subquasigroup, we say that C is a CH-quasigroup. We fix an element u in C and set $xy=u\circ(x\circ y)$, this composition law defines on C the structure of a commutative Moufang loop (CML), that is, it is commutative, has a unit element u and inverse, and moreover satisfies the three weak associativity properties: $x(xy) = x^2y$, $(xy)(xz) = x^2(yz)$ and $x(y(xz)) = (x^2y)z$.

Theorem 4.1 (see [32], [33]). *Let S be an admissible equivalence relation on a smooth cubic surface X. Then $X(k)/S$ is a CH-quasigroup, and in the corresponding CML the relation $x^6 = 1$ is satisfied identically; this CML is finite if and only if it is finitely generated, and in this case its order is $2^a \cdot 3^b$.*

In our case, let $H = \langle A, B, C, E\rangle$. Note that the order of H is $2^3 \cdot 3^2$.

Let k be a perfect field, and let V run through the smooth cubic surfaces over k. The classes of the lines on $V \otimes \overline{k}$ generate the group $N(V) = \mathrm{Pic}(V \otimes \overline{k})$ and the action of the Galois group $G = \mathrm{Gal}(\overline{k}/k)$ on $N(V)$ preserves symmetry and it implicitly contains an extremely large amount of information on the arithmetic and geometry of V.

Let k be a global field. Let χ denote the character of the representation of G on $N(V)$, and let $L(s, \chi, k)$ be the Artin L-function of the field k corresponding to the character χ.

Theorem 4.2 (Weil) (see [32]). *The Hasse-Weil zeta function of the surface V coincides (up to a finite number of Euler factors) with the product*

$$\zeta(s,k)\zeta(s-2,k)L(s-1,\chi,k).$$

Definition 4.3 (see [32]). Let $r \geq 1$ be an integer. We consider a composed object $\{N_r, \omega_r, (,)\}$, where

(1) $N_r = \mathbb{Z}^{r+1} = \bigoplus_{i=0}^{r} \mathbb{Z}l_i$; (l_i) a chosen basis.

(2) $\omega_r = (-3, 1, \cdots, 1) \in N_r$.

(3) $(,)$ is a bilinear form $N_r \times N_r \to \mathbb{Z}$ given by the formula

$$(l_0, l_0) = 1, \quad (l_i, l_i) = -1 \quad \text{if } i \geq 1, \quad (l_i, l_j) = 0 \quad \text{if } i \neq j.$$

(4) $R_r := \{l \in N_r : \quad (l, \omega_r) = 0, \quad (l, l) = -2\}$.

(5) $I_r := \{l \in N_r : \quad (l, \omega_r) = (l, l) = -1\}$.

Theorem 4.4 (see [32]). *Let V be a smooth cubic surface over a perfect field k.*

(1) The triple $\{N(V), \omega_V, \text{intersection number}\}$ is isomorphic to $\{N_6, \omega_6, (,)\}$ described in Definition 4.3.

(2) The Galois group G acts on $N(V)$ and preserves ω_V and the intersection number.

(3) The set of classes of lines on $V \otimes \overline{k}$ goes over into I_r under the isomorphism (1).

Theorem 4.5 (see [32]). *Let $3 \leq r \leq 8$.*

(1) The scalar product (with opposite sign) on $\mathbb{R}\otimes_{\mathbb{Z}} N_r \cong \mathbb{R}^{r+1}$ induces on the orthogonal complement of ω_r the structure of a Euclidean space. The set R_r is a root system in it of type $A_1 \times A_2$, A_4, D_5, E_6, E_7, E_8, respectively (the sum of the indices is equal to r).

(2) The following groups coincide:

(a) the group of automorphisms of the lattice N_r preserving ω_r and the scalar product;

(b) the group of permutations of the vectors from I_r preserving their pairwise scalar products;

(c) the Weyl group $W(R_r)$ of the system R_r generated by the reflections with respect to the roots.

In fact, Fano varieties in dimension two are called Del Pezzo surfaces. Over an algebraically closed field, these are: $\mathbb{P}^2$, $\mathbb{P}^1 \times \mathbb{P}^1$ and S_d, where S_d is the blow up of $\mathbb{P}^2$ at $9-d$ points, and the degree $d = 1, \cdots, 8$. However, starting with $r = 9$, the system R_r becomes infinite. Let V be a Del Pezzo surface of degree $1 \leq d \leq 7$, and let $f : V \to \mathbb{P}^2$ be its representation in the form of a monoidal transformation of the plane with as centre the union of $r = 9 - d$ points $x_1, \cdots, x_r$. It is shown that (see [32]) the map $D \mapsto$ (class of $\mathcal{O}_V(D)$) $\in \text{Pic}(V)$ establishes a one-one onto correspondence between exceptional curves on V and exceptional classes in the Picard group. These classes generate the Picard group. When $r = 6$, the number of exceptional curves on V is 27.

The significance of cubic surfaces comes from the following: the higher the degree of a rational surface, the smaller the rank of its Picard module, and therefore the less freedom there is for the action of the Galois group. Del Pezzo surfaces of degree three are cubic surfaces in $\mathbb{P}^3$ and the Weyl group $W(E_6)$ has enough large order. We have the following theorem (see [33]): Any minimal rational surface X over k of degree $d \leq 4$ is k-birationally nontrivial. The triple $\{N(V), \omega_V, \text{intersection number on } N(V)\}$ can be constructed for every surface V. The action of the Galois group and the group of automorphisms of V on this triple yields essential invariants of the surface.

Note that when $r = 6$, we have $|I_6| = 27$, $|R_6| = 72$, $|W(E_6)| = 2^7 \cdot 3^4 \cdot 5$, $|S_6| = 720$, $|W(E_6)/S_6| = 72$. On the other hand, in our case, for $H = \langle A, B, C, E \rangle$, we have $|H| = 72$.

According to [26], the modern point of view is to consider cubic surfaces as del Pezzo surfaces of degree 3. The combinatorics of the 27 lines is then encoded in the Picard group of the del Pezzo surface. In fact, the complement in $\text{Pic}(S)$ to the hyperplane section, call it $\text{Pic}^0(S)$, is isomorphic to the root lattice of E_6. The equation of the surface is given by the embedding of S by means of the linear system of elliptic curves through the six given points.

Fix six points in $\mathbb{P}^2$, say $x = (p_1, \cdots, p_6)$, such that the p_i are in general position, i.e., no three lie on a line, and not all six lie on a conic. Let $\widehat{\mathbb{P}^2_x}$ denote the blow up of $\mathbb{P}^2$ at all six points, $\varrho_x : \widehat{\mathbb{P}^2_x} \to \mathbb{P}^2$. Consider the following curves as classes in $\text{Pic}(\widehat{\mathbb{P}^2_x})$:

(1) $a_1, \cdots, a_6$, the exceptional divisors over $(p_1, \cdots, p_6)$;

(2) $b_1, \cdots, b_6$, b_i the proper transform of the conic q_i passing through all points p_j, $j \neq i$;

(3) c_{ik}, the proper transform of the line $\overline{p_i p_k}$.

If we consider the surface $\widehat{\mathbb{P}^2_x}$, we have $H^2(\widehat{\mathbb{P}^2_x}, \mathbb{Z}) = [l]\mathbb{Z} \oplus_i \mathbb{Z}a_i$. Let Q be the intersection form on $H^2(\widehat{\mathbb{P}^2_x}, \mathbb{Z})$, then the classes a_i, b_i, c_{ij} fulfill $Q(a_i, a_i) = Q(b_i, b_i) = Q(c_{ij}, c_{ij}) = -1$. In a well-known manner one takes a rank 6 subset, which is isomorphic to the root lattice of type E_6.

Consider the orthocomplement of the canonical class on $\widehat{\mathbb{P}^2_x}$, and denote this by $\mathrm{Pic}^0(\widehat{\mathbb{P}^2_x})$. Recall that the anti-canonical class is $3l + \sum_{i=1}^{6} a_i$, and that the anti-canonical embedding of $\widehat{\mathbb{P}^2_x}$ is as a cubic surface. Consequently we may view $\mathrm{Pic}^0(\widehat{\mathbb{P}^2_x})$ as the orthocomplement of the hyperplane section class of $\mathrm{Pic}(S_x)$, where S_x is the cubic surface which is the anti-canonical embedding. The following elements λ with $Q(\lambda, \lambda) = -2$ form a basis of $\mathrm{Pic}^0(S_x)$:

$$\begin{aligned}
\alpha_0 &= l - a_1 - a_2 - a_3, \\
\alpha_1 &= a_1 - a_2, \\
\alpha_2 &= a_2 - a_3, \\
\alpha_3 &= a_3 - a_4, \\
\alpha_4 &= a_4 - a_5, \\
\alpha_5 &= a_5 - a_6.
\end{aligned}$$

These also form a base of a root system of type E_6, by taking $\alpha_1, \cdots, \alpha_5$ as the sub-root system of type A_5. Since the classes a_i, b_i, c_{ij} are exceptional, they all represent elements of $\mathrm{Pic}^0(S_x)$.

According to [32], we have the realization problems.

(1) Which subgroups of $W(E_6)$ are realized as an image of the Galois group by its representation on $\mathrm{Pic}(V \otimes \overline{k})$ for some V?

(2) Which extensions of the field k correspond with kernels of such representations?

Shioda (see [49] and [50]) considered Galois representations and algebraic equations arising from Mordell-Weil lattices. Let $E = E_\lambda$ be the elliptic curve, defined by one of the following equations, over $K_0 = k_0(t)$ where $k_0 = \mathbb{Q}(\lambda) = \mathbb{Q}(p_i, q_j)$.

$$(E_8) \quad y^2 = x^3 + x\left(\sum_{i=0}^{3} p_i t^i\right) + \left(\sum_{i=0}^{3} q_i t^i + t^5\right)$$
$$\lambda = (p_0, p_1, p_2, p_3, q_0, q_1, q_2, q_3) \in \mathbb{A}^8$$

$$(E_7) \quad y^2 = x^3 + x(p_0 + p_1 t + t^3) + \left(\sum_{i=0}^{4} q_i t^i \right)$$

$$\lambda = (p_0, p_1, q_0, q_1, q_2, q_3, q_4) \in \mathbb{A}^7$$

$$(E_6) \quad y^2 = x^3 + x \left(\sum_{i=0}^{2} p_i t^i \right) + \left(\sum_{i=0}^{2} q_i t^i + t^4 \right)$$

$$\lambda = (p_0, p_1, p_2, q_0, q_1, q_2) \in \mathbb{A}^6$$

Let $f : S_\lambda \to \mathbb{P}^1$ be the associated elliptic surface. The fibre $f^{-1}(\infty)$ is an additive singular fibre of type II, III or IV according to the case (E_r) for $r = 8, 7$ or 6. Assume that λ satisfies the condition:

every fibre of f over $t \neq \infty$ is irreducible.

Then the Mordell-Weil lattice $E(k(t))$ is isomorphic to E_r^*, the dual lattice of E_r, where k is the algebraic closure of k_0, and we get the Galois representation

$$\varrho = \varrho_\lambda : \mathrm{Gal}(k/k_0) \to \mathrm{Aut}(E_r^*) = \mathrm{Aut}(E_r).$$

Recall that $\mathrm{Aut}(E_r) = W(E_r)$ for $r = 8$ or 7, and $W(E_r)\{\pm 1\}$ for $r = 6$, where $W(E_r)$ denotes the Weyl group of type E_r. In any case, we have $\mathrm{Im}(\varrho_\lambda) \subset W(E_r)$.

We assume that λ is generic, i.e. p_i, q_j are algebraically independent over $\mathbb{Q}$. Then the above condition holds.

Theorem 4.6 (Shioda). *Let λ be generic over $\mathbb{Q}$. Then*

(i) the image of the Galois representation ϱ_λ is the full Weyl group $W(E_r)$: $Im(\varrho_\lambda) = W(E_r)$.

(ii) Let $\mathcal{K}_\lambda/k_0$ be the Galois extension corresponding to $Ker(\varrho_\lambda)$, and let $\{P_1, \cdots, P_r\}$ be a basis of $E(k(t)) \simeq E_r^*$ consisting of minimal vectors. Further let $u_i = sp'_\infty(P_i) \in \mathcal{K}_\lambda \subset k$. Then $u_1, \cdots, u_r$ are algebraically independent over $\mathbb{Q}$, and we have

$$\mathcal{K}_\lambda = k_0(u_1, \cdots, u_r) = \mathbb{Q}(u_1, \cdots, u_r),$$

$$Gal(\mathbb{Q}(u_1, \cdots, u_r)/\mathbb{Q}(p_i, q_j)) = W(E_r).$$

(iii) $W(E_r)$ acts on the vector space $\mathbb{Q}u_1 \oplus \cdots \oplus \mathbb{Q}u_r$ and hence on the polynomial ring $\mathbb{Q}[u_1, \cdots, u_r]$, and the ring of the invariants is:

$$\mathbb{Q}[u_1, \cdots, u_r]^{W(E_r)} = \mathbb{Q}[p_i, q_j].$$

In particular, p_i and q_j form the fundamental invariants of the Weyl group $W(E_r)$ and we can explicitly write

$$p_i \text{ or } q_j = J_d(u_1, \cdots, u_r),$$

where J_d denotes a $W(E_r)$-invariant of degree d with $d \in \{2, 5, 6, 8, 9, 12\}$, $\{2, 6, 8, 10, 12, 14, 18\}$ or $\{2, 8, 12, 14, 18, 20, 24, 30\}$ for $r = 6$, 7 or 8.

Theorem 4.7 (Shioda). *Every Galois extension of* $\mathbb{Q}$ *with Galois group* $W(E_r)$ *is obtained as* $\mathcal{K}_\lambda$ *for some* $\lambda \in \Lambda$. *In other words, every such extension arises from the Mordell-Weil lattice of the elliptic curve* $E_\lambda/\mathbb{Q}(t)$ *for some* $\lambda \in \mathbb{Q}^r$.

In the equation (E_6), let $y = y' \pm t^2$ and $(x : y' : t : 1) = (X : Y : Z : W)$. Then we have a cubic surface $V = V_\lambda^{\pm}$ in $\mathbb{P}^3$:

$$Y^2W \pm 2YZ^2 = X^3 + X(p_0W^2 + p_1ZW + p_2Z^2) + q_0W^3 + q_1ZW^2 + q_2Z^2W.$$

It is smooth if and only if the above condition about λ holds. Under this assumption, the narrow Mordell-Weil lattice $E_\lambda(k(t))^0 \simeq E_6$ is isomorphic to the primitive part of $NS(V)$. The 27 minimal sections of the form $P = (at + b, t^2 + dt + e)$ in the Mordell-Weil lattice $E_\lambda(k(t))$ are transformed into the 27 lines on $V = V_\lambda^+$ defined by the equation

$$X = aZ + bW, \quad Y = dZ + eW.$$

(Similarly, the 27 sections $-P$ are mapped to the 27 lines on V_λ^-.)

Therefore, Shioda answered the problems raised by Manin when the image is surjective, i.e. $\text{Im}(\varrho) = W(E_6)$.

The Galois representations arising from Mordell-Weil lattices

$$\varrho : G = \text{Gal}(k/k_0) \to \text{Aut}(E(k(C)), \langle , \rangle) = \text{a finite group},$$

where $k(C)$ is the function field of a smooth projective curve C over an algebraically closed field k, are quite different from those arising from the torsion points of an elliptic curve or an abelian variety (e.g. the Tate modules), because we are dealing with points of infinite order here.

In fact, Mordell-Weil lattices E_6, E_7 and E_8 come from Klein singularities E_6, E_7 and E_8 which are intimately related to tetrahedron, octahedron and icosahedron. In contrast to Shioda's approach, we study Galois representations and the algebraic equations arising from Hessian polyhedra. It should be pointed out that Shioda used elliptic cubic surfaces, which are not the cubic surfaces of general type. However, we will use the general type cubic surfaces of type F_2.

For $(z_1, z_2, z_3) \in \mathbb{P}^2$, X, Y, Z, φ, Q_1 and Q_2 are cubic polynomials of z_1, z_2 and z_3. We will study the invariants x, y and z as functions of X, Y, Z, φ, Q_1 and Q_2.

We have

$$A(x) = x, \quad B(x) = -x, \quad B^2(x) = x.$$

$$C(x) = \overline{\omega}Q_1 - \omega Q_2, \quad C^2(x) = \omega Q_1 - \overline{\omega}Q_2, \quad C^3(x) = x.$$

$$E(x) = Y - Z, \quad E^2(x) = x.$$

$$EC(x) = X - Y, \quad EC^2(x) = Z - X.$$

$$E^2C(x) = \omega Q_1 - \overline{\omega}Q_2, \quad E^2C^2(x) = \overline{\omega}Q_1 - \omega Q_2.$$

$$A(y) = y, \quad B(y) = y, \quad C(y) = y.$$

$$E(y) = \frac{\sqrt{-3}}{3}(y + 2z), \quad E^2(y) = -y, \quad E^3(y) = -\frac{\sqrt{-3}}{3}(y + 2z), \quad E^4(y) = y.$$

$$A(z)=z, \quad B(z)=z, \quad C(z)=z.$$

$$E(z)=\frac{\sqrt{-3}}{3}(y-z), \quad E^2(z)=-z, \quad E^3(z)=-\frac{\sqrt{-3}}{3}(y-z), \quad E^4(z)=z.$$

Put

$$g_1=E(x), g_2=EC(x), g_3=EC^2(x), g_4=x, g_5=C(x), g_6=C^2(x). \tag{4.1}$$

We find that

$$g_1+g_2+g_3=0, \quad g_4+g_5+g_6=0. \tag{4.2}$$

$$g_1^3+g_2^3+g_3^3=3W_3=3C_9, \quad g_4^3+g_5^3+g_6^3=3(Q_1^3-Q_2^3)=3C_9. \tag{4.3}$$

Hence, we obtain the following variety:

$$S: \begin{cases} g_1+g_2+g_3=0, \\ g_4+g_5+g_6=0, \\ g_1^3+g_2^3+g_3^3=g_4^3+g_5^3+g_6^3, \end{cases} \tag{4.4}$$

which is a cubic surface in $\mathbb{P}^5$. Moreover, we get the following map:

$$\mathbb{P}^2 \to \mathbb{P}^5$$
$$(z_1,z_2,z_3) \mapsto (g_1,g_2,g_3,g_4,g_5,g_6)$$

by which the projective plane $\mathbb{CP}^2$ is mapped biholomorphically to the cubic surface $S \subset \mathbb{CP}^5$.

We find that

$$\begin{cases} E(g_1)=g_4, \\ E(g_2)=g_6, \\ E(g_3)=g_5, \end{cases} \quad \begin{cases} E(g_4)=g_1, \\ E(g_5)=g_2, \\ E(g_6)=g_3. \end{cases} \tag{4.5}$$

$$\begin{cases} A(g_1)=g_3, \\ A(g_2)=g_1, \\ A(g_3)=g_2, \end{cases} \quad \begin{cases} A(g_4)=g_4, \\ A(g_5)=g_5, \\ A(g_6)=g_6. \end{cases} \tag{4.6}$$

$$\begin{cases} B(g_1)=-g_1, \\ B(g_2)=-g_3, \\ B(g_3)=-g_2, \end{cases} \quad \begin{cases} B(g_4)=-g_4, \\ B(g_5)=-g_6, \\ B(g_6)=-g_5. \end{cases} \tag{4.7}$$

$$\begin{cases} C(g_1)=g_1, \\ C(g_2)=g_2, \\ C(g_3)=g_3, \end{cases} \quad \begin{cases} C(g_4)=g_5, \\ C(g_5)=g_6, \\ C(g_6)=g_4. \end{cases} \tag{4.8}$$

Hence, the cubic surface S is invariant under the action of the subgroup generated by A, B, C and E.

There are 27 lines on the cubic surface S:

$$l_1: \quad g_1=g_2+g_3=g_4=g_5+g_6=0,$$

$$
\begin{aligned}
l_2: &\quad g_1 = g_2 + g_3 = g_5 = g_4 + g_6 = 0,\\
l_3: &\quad g_1 = g_2 + g_3 = g_6 = g_4 + g_5 = 0,\\
l_4: &\quad g_2 = g_1 + g_3 = g_4 = g_5 + g_6 = 0,\\
l_5: &\quad g_2 = g_1 + g_3 = g_5 = g_4 + g_6 = 0,\\
l_6: &\quad g_2 = g_1 + g_3 = g_6 = g_4 + g_5 = 0,\\
l_7: &\quad g_3 = g_1 + g_2 = g_4 = g_5 + g_6 = 0,\\
l_8: &\quad g_3 = g_1 + g_2 = g_5 = g_4 + g_6 = 0,\\
l_9: &\quad g_3 = g_1 + g_2 = g_6 = g_4 + g_5 = 0,
\end{aligned}
$$

and

$$
\begin{aligned}
l_{j,1}: &\quad g_1 = \omega^j g_4, \quad g_2 = \omega^j g_5, \quad g_3 = \omega^j g_6, \quad g_1 + g_2 + g_3 = 0,\\
l_{j,2}: &\quad g_1 = \omega^j g_4, \quad g_2 = \omega^j g_6, \quad g_3 = \omega^j g_5, \quad g_1 + g_2 + g_3 = 0,\\
l_{j,3}: &\quad g_1 = \omega^j g_5, \quad g_2 = \omega^j g_4, \quad g_3 = \omega^j g_6, \quad g_1 + g_2 + g_3 = 0,\\
l_{j,4}: &\quad g_1 = \omega^j g_5, \quad g_2 = \omega^j g_6, \quad g_3 = \omega^j g_4, \quad g_1 + g_2 + g_3 = 0,\\
l_{j,5}: &\quad g_1 = \omega^j g_6, \quad g_2 = \omega^j g_4, \quad g_3 = \omega^j g_5, \quad g_1 + g_2 + g_3 = 0,\\
l_{j,6}: &\quad g_1 = \omega^j g_6, \quad g_2 = \omega^j g_5, \quad g_3 = \omega^j g_4, \quad g_1 + g_2 + g_3 = 0,
\end{aligned}
$$

where $j \equiv 0, 1, 2 \pmod 3$.

We find that

$$
\begin{aligned}
&E(l_1) = l_1, \quad A(l_1) = l_7, \quad B(l_1) = l_1, \quad C(l_1) = l_2.\\
&E(l_2) = l_4, \quad A(l_2) = l_8, \quad B(l_2) = l_3, \quad C(l_2) = l_3.\\
&E(l_3) = l_7, \quad A(l_3) = l_9, \quad B(l_3) = l_2, \quad C(l_3) = l_1.\\
&E(l_4) = l_3, \quad A(l_4) = l_1, \quad B(l_4) = l_7, \quad C(l_4) = l_5.\\
&E(l_5) = l_6, \quad A(l_5) = l_2, \quad B(l_5) = l_9, \quad C(l_5) = l_6.\\
&E(l_6) = l_9, \quad A(l_6) = l_3, \quad B(l_6) = l_8, \quad C(l_6) = l_4.\\
&E(l_7) = l_2, \quad A(l_7) = l_4, \quad B(l_7) = l_4, \quad C(l_7) = l_8.\\
&E(l_8) = l_5, \quad A(l_8) = l_5, \quad B(l_8) = l_6, \quad C(l_8) = l_9.\\
&E(l_9) = l_8, \quad A(l_9) = l_6, \quad B(l_9) = l_5, \quad C(l_9) = l_7.\\
&E(l_{j,1}) = l_{2j,2}, \quad A(l_{j,1}) = l_{j,4}, \quad B(l_{j,1}) = l_{j,1}, \quad C(l_{j,1}) = l_{j,4}.\\
&E(l_{j,2}) = l_{2j,1}, \quad A(l_{j,2}) = l_{j,6}, \quad B(l_{j,2}) = l_{j,2}, \quad C(l_{j,2}) = l_{j,3}.\\
&E(l_{j,3}) = l_{2j,5}, \quad A(l_{j,3}) = l_{j,2}, \quad B(l_{j,3}) = l_{j,6}, \quad C(l_{j,3}) = l_{j,6}.\\
&E(l_{j,4}) = l_{2j,3}, \quad A(l_{j,4}) = l_{j,5}, \quad B(l_{j,4}) = l_{j,5}, \quad C(l_{j,4}) = l_{j,5}.\\
&E(l_{j,5}) = l_{2j,6}, \quad A(l_{j,5}) = l_{j,1}, \quad B(l_{j,5}) = l_{j,4}, \quad C(l_{j,5}) = l_{j,1}.
\end{aligned}
$$

$$E(l_{j,6}) = l_{2j,4}, \quad A(l_{j,6}) = l_{j,3}, \quad B(l_{j,6}) = l_{j,3}, \quad C(l_{j,6}) = l_{j,2}.$$

It is well-known that the Segre cubic threefold is given by (see [26])

$$\mathcal{S}_3 : \quad \begin{cases} x_0 + x_1 + x_2 + x_3 + x_4 + x_5 = 0, \\ x_0^3 + x_1^3 + x_2^3 + x_3^3 + x_4^3 + x_5^3 = 0. \end{cases}$$

The hyperplane sections $\{x_i = 0\}$ of $\mathcal{S}_3$ are Clebsch diagonal cubic surfaces with equation

$$S_3 : \quad \begin{cases} x_0 + x_1 + x_2 + x_3 + x_4 = 0, \\ x_0^3 + x_1^3 + x_2^3 + x_3^3 + x_4^3 = 0. \end{cases}$$

The relation between S_3 and the icosahedral group was studied by Hirzebruch. It turns out that S_3 is A_5-equivariantly birational to the Hilbert modular surface for $\mathcal{O}_K$ of level $\sqrt{5}$ where $K = \mathbb{Q}(\sqrt{5})$ (see [26]).

We find that the hyperplane sections $\{x_i + x_j + x_k = 0\}$ of $\mathcal{S}_3$ are our cubic surfaces with equation

$$S : \quad \begin{cases} g_1 + g_2 + g_3 = 0, \\ (-g_4) + (-g_5) + (-g_6) = 0, \\ g_1^3 + g_2^3 + g_3^3 + (-g_4)^3 + (-g_5)^3 + (-g_6)^3 = 0. \end{cases}$$

Our cubic surface can be expressed as

$$S : \quad x_1^2 x_2 + x_1 x_2^2 + x_3^2 x_4 + x_3 x_4^2 = 0. \tag{4.9}$$

The Hessian variety of S (up to a constant) is given by

$$(x_1^2 + x_1 x_2 + x_2^2)(x_3^2 + x_3 x_4 + x_4^2) = 0, \tag{4.10}$$

which are four planes.

Let us consider the group of permutations (see [26]), $\text{Aut}(\mathcal{L})$, of the 27 lines, by which we mean the permutations of the lines preserving the intersection behavior of the lines. For this it is useful to consider the famous double sixes and the notation for the 27 lines introduced by Schläfli. A double six is an array

$$N = \begin{bmatrix} a_1 & a_2 & a_3 & a_4 & a_5 & a_6 \\ b_1 & b_2 & b_3 & b_4 & b_5 & b_6 \end{bmatrix}$$

of 12 of the 27 lines with the property that two of these 12 meet if and only if they are in different rows and columns. (This notation distinguishes this particular set of 12, although any such double six is equivalent to it under $\text{Aut}(\mathcal{L})$.) The other lines are given by the $\binom{6}{2} = 15$ $c_{ij} = a_i b_j \cap a_j b_i$, where $a_i b_j$ denotes the tritangent spanned by those two lines. There are 36 double sixes, namely the N above, 15 N_{ij} and 20 N_{ijk}:

$$N_{ij} = \begin{bmatrix} a_i & b_i & c_{jk} & c_{jl} & c_{jm} & c_{jn} \\ a_j & b_j & c_{ik} & c_{il} & c_{im} & c_{in} \end{bmatrix},$$

$$N_{ijk} = \begin{bmatrix} a_i & a_j & a_k & c_{mn} & c_{ln} & c_{lm} \\ c_{jk} & c_{ik} & c_{ij} & b_l & b_m & b_n \end{bmatrix}.$$

Since a double six describes the intersection behavior of the lines, we see that S_6 (the symmetric group on six letters) acts by permutations on a double six and a $\mathbb{Z}_2$ acts by exchanging rows. Since there are 36 double sixes, we have $|\text{Aut}(\mathcal{L})| = |S_6| \cdot 2 \cdot 36 = 51840$. In fact $\text{Aut}(\mathcal{L})$ is the Weyl group $W(E_6)$, and the 36 double sixes correspond to the positive roots. The 27 lines correspond to the 27 fundamental weights of E_6, and many other sets of objects (lines, tritangents, etc.) correspond to natural sets of objects (roots, weights, etc.) of E_6.

According to [44], a nonsingular cubic surface can only be of one of the five types: F_1, F_2, F_3, F_4 and F_5. Our cubic surface is of F_2 which has 15 real lines and 12 complex lines of the 2nd kind.

The 15 real lines of a cubic surface F_2 are of two different sorts: 6 of them, $l_{0,1}, \cdots, l_{0,6}$, constituting 2 complementary triplets, are elliptic; the 9 others $l_1, \cdots, l_9$ are hyperbolic of the 1st kind and constitute a Steiner set. The remaining lines $l_{1,1}, \cdots, l_{1,6}, l_{2,1}, \cdots, l_{2,6}$ of F_2 form a self-conjugate double-six, which we call 'of the 9th kind', in which each pair of corresponding lines is a pair of conjugate complex lines of the 2nd kind; hence the σ-transformation inherent to it changes each line of F_2 in its conjugate.

F_2 has 15 other self-conjugate double-sixes, which are the 15 double-sixes permutable with the one just considered. Each sextuplet of such a double-six is consequently self-conjugate, and consists of 4 real lines and 2 conjugate complex lines of the 2nd kind; but, while in 9 of these 15 double-sixes (which we call 'of the 4th kind') the 4 real lines of each sextuplet are 2 elliptic and 2 hyperbolic, in the 6 others (which we call 'of the 5th kind', and which constitute 2 permutable triads of associate double-sixes) the 4 real lines of each sextuplet are 1 elliptic and 3 hyperbolic. If ω is any one of the 15 double-sixes considered, its 8 real lines are the only real lines of F_2 skew to a well-determined real line of F_2 which is hyperbolic or elliptic according as ω is of the 4th or 5th kind.

In fact, the Clebsch diagonal cubic surface is of F_1 which has its 27 lines all real: 12 of them are elliptic, and 15 are hyperbolic of the 1st kind.

Let Γ_i be the group of the lines of a real cubic surface F_i and $\mathfrak{G}$ be the group of the 27 lines of a cubic surface.

Segre [44] proved that the group Γ_1 inherent to a cubic surface F_1 is icosahedral, and can be defined as the group of the 60 substitutions of $\mathfrak{G}$ which transform into themselves each of the 2 sextuplets of elliptic lines and also the system of 5 principal tritangent planes of F_1, inducing among these planes a substitution of even class.

The group Γ_2 inherent to a cubic surface F_2 is of order 36; it is simply isomorphic with the direct product of 2 symmetric groups of degree 3, its transformations being uniquely defined by the property of performing an arbitrary substitution among the 3 elliptic left-handed lines of F_2 and an arbitrary substitution among the 3 elliptic right-handed lines of F_2. The group operates transitively among the 9 pairs of complementary triplets of the 1st kind of F_2, any 2 such pairs being transformed one into the other by 4 transformations of Γ_2; in particular, the subgroup of Γ_2 transforming into itself one of these pairs is trirectangular, and coincides with the subgroup of Γ_2 which leaves unchanged each of the 2 elliptic lines of the pair.

In our case, we find that the set of 15 real lines, i.e., $\{l_1, \cdots, l_9\}$ and $\{l_{0,1}, \cdots, l_{0,6}\}$ are invariant under the actions of A, B, C and E, respectively. For the complex lines, we have

$$E(l_{1,1}) = l_{2,2}, E(l_{1,2}) = l_{2,1}, E(l_{1,3}) = l_{2,5},$$

$$E(l_{1,4}) = l_{2,3}, E(l_{1,5}) = l_{2,6}, E(l_{1,6}) = l_{2,4},$$

$$E(l_{2,1}) = l_{1,2}, E(l_{2,2}) = l_{1,1}, E(l_{2,3}) = l_{1,5},$$

$$E(l_{2,4}) = l_{1,3}, E(l_{2,5}) = l_{1,6}, E(l_{2,6}) = l_{1,4}.$$

We find that

$$\begin{cases} l_{1,1} \cap l_{1,2} = \overline{l_{2,1} \cap l_{2,2}} = (-2\omega, \omega, \omega, -2, 1, 1), \\ l_{1,1} \cap l_{1,3} = \overline{l_{2,1} \cap l_{2,3}} = (\omega, \omega, -2\omega, 1, 1, -2), \\ l_{1,1} \cap l_{1,4} = \overline{l_{2,1} \cap l_{2,4}} = \emptyset, \\ l_{1,1} \cap l_{1,5} = \overline{l_{2,1} \cap l_{2,5}} = \emptyset, \\ l_{1,1} \cap l_{1,6} = \overline{l_{2,1} \cap l_{2,6}} = (\omega, -2\omega, \omega, 1, -2, 1). \end{cases}$$

$$\begin{cases} l_{1,2} \cap l_{1,3} = \overline{l_{2,2} \cap l_{2,3}} = \emptyset, \\ l_{1,2} \cap l_{1,4} = \overline{l_{2,2} \cap l_{2,4}} = (\omega, -2\omega, \omega, 1, 1, -2), \\ l_{1,2} \cap l_{1,5} = \overline{l_{2,2} \cap l_{2,5}} = (\omega, \omega, -2\omega, 1, -2, 1), \\ l_{1,2} \cap l_{1,6} = \overline{l_{2,2} \cap l_{2,6}} = \emptyset. \end{cases}$$

$$\begin{cases} l_{1,3} \cap l_{1,4} = \overline{l_{2,3} \cap l_{2,4}} = (-2\omega, \omega, \omega, 1, -2, 1), \\ l_{1,3} \cap l_{1,5} = \overline{l_{2,3} \cap l_{2,5}} = (\omega, -2\omega, \omega, -2, 1, 1), \\ l_{1,3} \cap l_{1,6} = \overline{l_{2,3} \cap l_{2,6}} = \emptyset. \end{cases}$$

$$\begin{cases} l_{1,4} \cap l_{1,5} = \overline{l_{2,4} \cap l_{2,5}} = \emptyset, \\ l_{1,4} \cap l_{1,6} = \overline{l_{2,4} \cap l_{2,6}} = (\omega, \omega, -2\omega, -2, 1, 1). \end{cases}$$

$$l_{1,5} \cap l_{1,6} = \overline{l_{2,5} \cap l_{2,6}} = (-2\omega, \omega, \omega, 1, 1, -2).$$

$$\begin{cases} l_{1,1} \cap l_{2,1} = \emptyset, \\ l_{1,1} \cap l_{2,2} = \emptyset, \\ l_{1,1} \cap l_{2,3} = \emptyset, \\ l_{1,1} \cap l_{2,4} = (1, \overline{\omega}, \omega, \overline{\omega}, \omega, 1), \\ l_{1,1} \cap l_{2,5} = (\overline{\omega}, 1, \omega, \omega, \overline{\omega}, 1), \\ l_{1,1} \cap l_{2,6} = \emptyset. \end{cases}$$

$$\begin{cases} l_{1,2} \cap l_{2,1} = \emptyset, \\ l_{1,2} \cap l_{2,2} = \emptyset, \\ l_{1,2} \cap l_{2,3} = (1, \omega, \overline{\omega}, \overline{\omega}, \omega, 1), \\ l_{1,2} \cap l_{2,4} = \emptyset, \\ l_{1,2} \cap l_{2,5} = \emptyset, \\ l_{1,2} \cap l_{2,6} = (\overline{\omega}, \omega, 1, \omega, \overline{\omega}, 1). \end{cases}$$

$$\begin{cases} l_{1,3} \cap l_{2,1} = \emptyset, \\ l_{1,3} \cap l_{2,2} = (1, \overline{\omega}, \omega, \omega, \overline{\omega}, 1), \\ l_{1,3} \cap l_{2,3} = \emptyset, \\ l_{1,3} \cap l_{2,4} = \emptyset, \\ l_{1,3} \cap l_{2,5} = \emptyset, \\ l_{1,3} \cap l_{2,6} = (\overline{\omega}, 1, \omega, \overline{\omega}, \omega, 1). \end{cases}$$

$$\begin{cases} l_{1,4} \cap l_{2,1} = (1, \omega, \overline{\omega}, \omega, \overline{\omega}, 1), \\ l_{1,4} \cap l_{2,2} = \emptyset, \\ l_{1,4} \cap l_{2,3} = \emptyset, \\ l_{1,4} \cap l_{2,4} = \emptyset, \\ l_{1,4} \cap l_{2,5} = (\overline{\omega}, \omega, 1, \overline{\omega}, \omega, 1), \\ l_{1,4} \cap l_{2,6} = \emptyset. \end{cases}$$

$$\begin{cases} l_{1,5} \cap l_{2,1} = (\omega, 1, \overline{\omega}, \overline{\omega}, \omega, 1), \\ l_{1,5} \cap l_{2,2} = \emptyset, \\ l_{1,5} \cap l_{2,3} = \emptyset, \\ l_{1,5} \cap l_{2,4} = (\omega, \overline{\omega}, 1, \omega, \overline{\omega}, 1), \\ l_{1,5} \cap l_{2,5} = \emptyset, \\ l_{1,5} \cap l_{2,6} = \emptyset. \end{cases}$$

$$\begin{cases} l_{1,6} \cap l_{2,1} = \emptyset, \\ l_{1,6} \cap l_{2,2} = (\omega, \overline{\omega}, 1, \overline{\omega}, \omega, 1), \\ l_{1,6} \cap l_{2,3} = (\omega, 1, \overline{\omega}, \omega, \overline{\omega}, 1), \\ l_{1,6} \cap l_{2,4} = \emptyset, \\ l_{1,6} \cap l_{2,5} = \emptyset, \\ l_{1,6} \cap l_{2,6} = \emptyset. \end{cases}$$

The double six is given by

$$N = \begin{pmatrix} a_1 & a_2 & a_3 & a_4 & a_5 & a_6 \\ b_1 & b_2 & b_3 & b_4 & b_5 & b_6 \end{pmatrix} = \begin{pmatrix} l_{1,1} & l_{2,2} & l_{2,3} & l_{1,4} & l_{1,5} & l_{2,6} \\ l_{2,1} & l_{1,2} & l_{1,3} & l_{2,4} & l_{2,5} & l_{1,6} \end{pmatrix}. \tag{4.11}$$

We find that

$$l_{1,1} \cap l_{1,2} = \overline{l_{2,1} \cap l_{2,2}} = (-2\omega, \omega, \omega, -2, 1, 1),$$
$$l_1 \cap l_{1,1} = \overline{l_1 \cap l_{2,1}} = (0, -\omega, \omega, 0, -1, 1),$$
$$l_1 \cap l_{1,2} = \overline{l_1 \cap l_{2,2}} = (0, \omega, -\omega, 0, -1, 1).$$

Note that $\overline{l_1} = l_1$. Hence, $l_1 \cap l_{1,1} \in l_1$, $l_1 \cap l_{2,1} \in l_1$, $l_1 \cap l_{1,2} \in l_1$, $l_1 \cap l_{2,2} \in l_1$. Thus, $c_{12} = l_1$.

By the same method as above, we obtain that

$$c_{12} = l_1, \quad c_{13} = l_9, \quad c_{14} = l_{0,5}, \quad c_{15} = l_{0,4}, \quad c_{16} = l_5. \tag{4.12}$$

$$c_{23} = l_{0,6}, \quad c_{24} = l_6, \quad c_{25} = l_8, \quad c_{26} = l_{0,3}. \tag{4.13}$$

$$c_{34} = l_2, \quad c_{35} = l_4, \quad c_{36} = l_{0,2}. \tag{4.14}$$

$$c_{45} = l_{0,1}, \quad c_{46} = l_7, \quad c_{56} = l_3. \tag{4.15}$$

Therefore, we have

$$E(a_1) = a_2, \quad A(a_1) = a_4, \quad B(a_1) = a_1, \quad C(a_1) = a_4.$$
$$E(a_2) = a_1, \quad A(a_2) = a_6, \quad B(a_2) = a_2, \quad C(a_2) = a_3.$$
$$E(a_3) = a_5, \quad A(a_3) = a_2, \quad B(a_3) = a_6, \quad C(a_3) = a_6.$$
$$E(a_4) = a_3, \quad A(a_4) = a_5, \quad B(a_4) = a_5, \quad C(a_4) = a_5.$$
$$E(a_5) = a_6, \quad A(a_5) = a_1, \quad B(a_5) = a_4, \quad C(a_5) = a_1.$$
$$E(a_6) = a_4, \quad A(a_6) = a_3, \quad B(a_6) = a_3, \quad C(a_6) = a_2.$$
$$E(b_1) = b_2, \quad A(b_1) = b_4, \quad B(b_1) = b_1, \quad C(b_1) = b_4.$$
$$E(b_2) = b_1, \quad A(b_2) = b_6, \quad B(b_2) = b_2, \quad C(b_2) = b_3.$$
$$E(b_3) = b_5, \quad A(b_3) = b_2, \quad B(b_3) = b_6, \quad C(b_3) = b_6.$$
$$E(b_4) = b_3, \quad A(b_4) = b_5, \quad B(b_4) = b_5, \quad C(b_4) = b_5.$$
$$E(b_5) = b_6, \quad A(b_5) = b_1, \quad B(b_5) = b_4, \quad C(b_5) = b_1.$$
$$E(b_6) = b_4, \quad A(b_6) = b_3, \quad B(b_6) = b_3, \quad C(b_6) = b_2.$$
$$E(c_{12}) = c_{12}, \quad A(c_{12}) = c_{46}, \quad B(c_{12}) = c_{12}, \quad C(c_{12}) = c_{34}.$$
$$E(c_{13}) = c_{25}, \quad A(c_{13}) = c_{24}, \quad B(c_{13}) = c_{16}, \quad C(c_{13}) = c_{46}.$$
$$E(c_{14}) = c_{23}, \quad A(c_{14}) = c_{45}, \quad B(c_{14}) = c_{15}, \quad C(c_{14}) = c_{45}.$$
$$E(c_{15}) = c_{26}, \quad A(c_{15}) = c_{14}, \quad B(c_{15}) = c_{14}, \quad C(c_{15}) = c_{14}.$$
$$E(c_{16}) = c_{24}, \quad A(c_{16}) = c_{34}, \quad B(c_{16}) = c_{13}, \quad C(c_{16}) = c_{24}.$$
$$E(c_{23}) = c_{15}, \quad A(c_{23}) = c_{26}, \quad B(c_{23}) = c_{26}, \quad C(c_{23}) = c_{36}.$$
$$E(c_{24}) = c_{13}, \quad A(c_{24}) = c_{56}, \quad B(c_{24}) = c_{25}, \quad C(c_{24}) = c_{35}.$$

$$E(c_{25}) = c_{16}, \quad A(c_{25}) = c_{16}, \quad B(c_{25}) = c_{24}, \quad C(c_{25}) = c_{13}.$$
$$E(c_{26}) = c_{14}, \quad A(c_{26}) = c_{36}, \quad B(c_{26}) = c_{23}, \quad C(c_{26}) = c_{23}.$$
$$E(c_{34}) = c_{35}, \quad A(c_{34}) = c_{25}, \quad B(c_{34}) = c_{56}, \quad C(c_{34}) = c_{56}.$$
$$E(c_{35}) = c_{56}, \quad A(c_{35}) = c_{12}, \quad B(c_{35}) = c_{46}, \quad C(c_{35}) = c_{16}.$$
$$E(c_{36}) = c_{45}, \quad A(c_{36}) = c_{23}, \quad B(c_{36}) = c_{36}, \quad C(c_{36}) = c_{26}.$$
$$E(c_{45}) = c_{36}, \quad A(c_{45}) = c_{15}, \quad B(c_{45}) = c_{45}, \quad C(c_{45}) = c_{15}.$$
$$E(c_{46}) = c_{34}, \quad A(c_{46}) = c_{35}, \quad B(c_{46}) = c_{35}, \quad C(c_{46}) = c_{25}.$$
$$E(c_{56}) = c_{46}, \quad A(c_{56}) = c_{13}, \quad B(c_{56}) = c_{34}, \quad C(c_{56}) = c_{12}.$$

For

$$N = \begin{pmatrix} a_1 & a_2 & a_3 & a_4 & a_5 & a_6 \\ b_1 & b_2 & b_3 & b_4 & b_5 & b_6 \end{pmatrix},$$

we have

$$E(N) = \begin{pmatrix} a_2 & a_1 & a_5 & a_3 & a_6 & a_4 \\ b_2 & b_1 & b_5 & b_3 & b_6 & b_4 \end{pmatrix}.$$

$$A(N) = \begin{pmatrix} a_4 & a_6 & a_2 & a_5 & a_1 & a_3 \\ b_4 & b_6 & b_2 & b_5 & b_1 & b_3 \end{pmatrix}.$$

$$B(N) = \begin{pmatrix} a_1 & a_2 & a_6 & a_5 & a_4 & a_3 \\ b_1 & b_2 & b_6 & b_5 & b_4 & b_3 \end{pmatrix}.$$

$$C(N) = \begin{pmatrix} a_4 & a_3 & a_6 & a_5 & a_1 & a_2 \\ b_4 & b_3 & b_6 & b_5 & b_1 & b_2 \end{pmatrix}.$$

For

$$N_{12} = \begin{pmatrix} a_1 & b_1 & c_{23} & c_{24} & c_{25} & c_{26} \\ a_2 & b_2 & c_{13} & c_{14} & c_{15} & c_{16} \end{pmatrix},$$

we have

$$E(N_{12}) = \begin{pmatrix} a_2 & b_2 & c_{15} & c_{13} & c_{16} & c_{14} \\ a_1 & b_1 & c_{25} & c_{23} & c_{26} & c_{24} \end{pmatrix}.$$

$$A(N_{12}) = \begin{pmatrix} a_4 & b_4 & c_{26} & c_{56} & c_{16} & c_{36} \\ a_6 & b_6 & c_{24} & c_{54} & c_{14} & c_{34} \end{pmatrix}.$$

$$B(N_{12}) = \begin{pmatrix} a_1 & b_1 & c_{26} & c_{25} & c_{24} & c_{23} \\ a_2 & b_2 & c_{16} & c_{15} & c_{14} & c_{13} \end{pmatrix}.$$

$$C(N_{12}) = \begin{pmatrix} a_4 & b_4 & c_{36} & c_{35} & c_{31} & c_{32} \\ a_3 & b_3 & c_{46} & c_{45} & c_{41} & c_{42} \end{pmatrix}.$$

For

$$N_{123} = \begin{pmatrix} a_1 & a_2 & a_3 & c_{56} & c_{46} & c_{45} \\ c_{23} & c_{13} & c_{12} & b_4 & b_5 & b_6 \end{pmatrix},$$

we have

$$E(N_{123}) = \begin{pmatrix} a_2 & a_1 & a_5 & c_{46} & c_{34} & c_{36} \\ c_{15} & c_{25} & c_{12} & b_3 & b_6 & b_4 \end{pmatrix}.$$

$$A(N_{123}) = \begin{pmatrix} a_4 & a_6 & a_2 & c_{13} & c_{35} & c_{15} \\ c_{26} & c_{24} & c_{46} & b_5 & b_1 & b_3 \end{pmatrix}.$$

$$B(N_{123}) = \begin{pmatrix} a_1 & a_2 & a_6 & c_{34} & c_{35} & c_{45} \\ c_{26} & c_{16} & c_{12} & b_5 & b_4 & b_3 \end{pmatrix}.$$

$$C(N_{123}) = \begin{pmatrix} a_4 & a_3 & a_6 & c_{12} & c_{25} & c_{15} \\ c_{36} & c_{46} & c_{34} & b_5 & b_1 & b_2 \end{pmatrix}.$$

The other N_{ij} and N_{ijk} can be calculated similarly. We find that the action of the group H generated by A, B, C and E on the 27 lines gives the permutations of the lines which preserve the intersection behavior of the lines. Hence, we conclude that the group H is a subgroup of $\text{Aut}(\mathcal{L}) = W(E_6)$.

Theorem 4.8 (Main Theorem 4). *The group H generated by A, B, C and E is a subgroup of $Aut(\mathcal{L}) = W(E_6)$.*

For the complex conjugate $\sigma_c \in \text{Gal}(\overline{\mathbb{Q}}/\mathbb{Q})$, we have

$$\sigma_c(N) = \begin{pmatrix} b_1 & b_2 & b_3 & b_4 & b_5 & b_6 \\ a_1 & a_2 & a_3 & a_4 & a_5 & a_6 \end{pmatrix}.$$

$$\sigma_c(N_{12}) = \begin{pmatrix} b_1 & a_1 & c_{23} & c_{24} & c_{25} & c_{26} \\ b_2 & a_2 & c_{13} & c_{14} & c_{15} & c_{16} \end{pmatrix}.$$

$$\sigma_c(N_{123}) = \begin{pmatrix} b_1 & b_2 & b_3 & c_{56} & c_{46} & c_{45} \\ c_{23} & c_{13} & c_{12} & a_4 & a_5 & a_6 \end{pmatrix}.$$

This leads us to study the representation:

$$\rho : G_{\mathbb{Q}} = \text{Gal}(\overline{\mathbb{Q}}/\mathbb{Q}) \to \text{Aut}(\mathcal{L}), \tag{4.16}$$

i.e.,

$$\rho : \text{Gal}(\overline{\mathbb{Q}}/\mathbb{Q}) \to W(E_6). \tag{4.17}$$

Here,

$$\text{Im}(\rho) = H = \langle A, B, C, E \rangle, \quad \ker(\rho) = \text{Gal}(\overline{\mathbb{Q}}/\mathcal{K}). \tag{4.18}$$

We have

$$\text{Im}(\rho) \cong \text{Gal}(\overline{\mathbb{Q}}/\mathbb{Q})/\ker(\rho) = \text{Gal}(\overline{\mathbb{Q}}/\mathbb{Q})/\text{Gal}(\overline{\mathbb{Q}}/\mathcal{K}) \cong \text{Gal}(\mathcal{K}/\mathbb{Q}). \tag{4.19}$$

Hence,

$$\text{Gal}(\mathcal{K}/\mathbb{Q}) \cong H. \tag{4.20}$$

The (geometrical, analytical and algebraical) properties of the Hessian group $G = \langle A, B, C, D, E \rangle$ have been studied in our previous paper [61]. Note that $A^*A = B^*B = C^*C = I$, but $E^*E \neq I$. We have $\text{Tr}(A) = 3$, $\det(A) = 1$. $\text{Tr}(B) = 2$, $\det(B) = 1$. $\text{Tr}(C) = 3$, $\det(C) = 1$. $\text{Tr}(E) = 0$.

$$\det(E) = \frac{1}{(\sqrt{-3})^{18}} \det \begin{pmatrix} 1 & 1 & 1 & 9 & 3 & 3 \\ 1 & 1 & 1 & 9 & 3\overline{\omega} & 3\omega \\ 1 & 1 & 1 & 9 & 3\omega & 3\overline{\omega} \\ 1 & 1 & 1 & 0 & 0 & 0 \\ 3 & 3\overline{\omega} & 3\omega & 0 & 0 & 0 \\ 3 & 3\omega & 3\overline{\omega} & 0 & 0 & 0 \end{pmatrix} = -1.$$

Note that

$$E^2 = \begin{pmatrix} -1 & & & & & \\ & 0 & -1 & & & \\ & -1 & 0 & & & \\ & & & -1 & & \\ & & & & 0 & -1 \\ & & & & -1 & 0 \end{pmatrix}.$$

In fact, the elements of H have integer determinant and trace.

Now let us recall the definition of Artin L-functions (see [1], [15]). Suppose K is a number field, and E is a finite Galois extension of K with Galois group $G = \mathrm{Gal}(E/K)$. By a representation of G we understand a homomorphism σ of G into $GL(V)$, the group of invertible linear transformations of a complex vector space of dimension n. Given a finite place $\mathfrak{p}$ of K, we say a prime $\mathfrak{P}$ of $\mathcal{O}_E$ lies over $\mathfrak{p}$ if $\mathfrak{P}$ appears in the factorization of $\mathfrak{p}\mathcal{O}_E$ into prime ideals of $\mathcal{O}_E$. Given such a pair $\mathfrak{P}/\mathfrak{p}$, we have:

(1) The decomposition group $D_{\mathfrak{P}} = \{g \in G : g(\mathfrak{P}) = \mathfrak{P}\}$;

(2) The inertia subgroup $I_{\mathfrak{P}} = \{g \in D_{\mathfrak{P}} : g(x) = x (\mathrm{mod}\ \mathfrak{P}) \text{ for all } x \in \mathcal{O}_E\}$;

(3) The Frobenius automorphism $\mathrm{Fr}_{\mathfrak{P}}$ generating the cyclic group $D_{\mathfrak{P}}/I_{\mathfrak{P}}$ which is isomorphic to $\mathrm{Gal}(\mathcal{O}_E/\mathfrak{P} : \mathcal{O}_K/\mathfrak{p})$.

Let $V_{\mathfrak{P}}$ denote the subspace of V formed by vectors invariant by $\sigma(I_{\mathfrak{P}})$. Then $\sigma(\mathrm{Fr}_{\mathfrak{P}})$ is defined on $V_{\mathfrak{P}}$, and the Euler factor

$$L_{\mathfrak{p}}(\sigma, s) = [\det(I - \sigma(\mathrm{Fr}_{\mathfrak{P}}) \cdot N\mathfrak{p}^{-s}|_{V_{\mathfrak{P}}})]^{-1}$$

depends only on $\mathfrak{p}$, not $\mathfrak{P}$. The Artin L-function attached to σ is given by

$$L(\sigma, s) = L_{E/K}(\sigma, s) = \prod_{\mathfrak{p}} L_{\mathfrak{p}}(\sigma, s)$$

which is convergent for $\mathrm{Re}(s) > 1$.

In [2], Artin studied representations with a finite image $\mathrm{Im}(\rho)$ (see also [34]). In this case the representation ρ is always rational for some number field and is semisimple. Hence ρ is uniquely determined (up to equivalence) by its character χ_ρ. If ρ is an arbitrary rational representation then the function $L(\rho, s)$ is uniquely defined by its character χ_ρ. In view of this fact one often uses the notation $L(\chi_\rho, s) = L(\rho, s)$.

For arbitrary n, there is the following remarkable Reciprocity Conjecture:

Conjecture 4.9 (Langlands). *Suppose E is a finite Galois extension of F with Galois group $G = Gal(E/F)$, and $\sigma : G \to GL_n(\mathbb{C})$ is an irreducible representation of G. Then there exists an automorphic cuspidal representation π_σ on GL_n over F such that $L(s, \pi_\sigma) = L(s, \sigma)$.*

The truth of this conjecture implies the truth of Artin conjecture on the entirety of the Artin L-function $L(s, \sigma)$.

Conjecture 4.10 (Main Conjecture). *Let $\rho : Gal(\overline{\mathbb{Q}}/\mathbb{Q}) \to W(E_6)$ be a Galois representation. When the image of ρ: $Im(\rho) \cong H \leq W(E_6)$, if ρ is odd, i.e., $\det \rho(\sigma_c) = -1$, where $\sigma_c \in Gal(\overline{\mathbb{Q}}/\mathbb{Q})$ is the complex conjugate, then there exists a Picard modular form f of weight one such that*

$$L(\rho, s) = L(f, s)$$

up to finitely many Euler factors, where $L(\rho, s)$ is the Artin L-function associated to the Galois representation ρ and $L(f, s)$ is the automorphic L-function associated to the Picard modular form f. If ρ is even, i.e., $\det \rho(\sigma_c) = 1$, then there exists an automorphic form f associated to $U(2,1) : \Delta f = s(s-2)f$ with $s = 1$ and (see [59])

$$\Delta = (z_1 + \overline{z_1} - z_2\overline{z_2}) \left[(z_1 + \overline{z_1}) \frac{\partial^2}{\partial z_1 \partial \overline{z_1}} + \frac{\partial^2}{\partial z_2 \partial \overline{z_2}} + z_2 \frac{\partial^2}{\partial \overline{z_1} \partial z_2} + \overline{z_2} \frac{\partial^2}{\partial z_1 \partial \overline{z_2}} \right],$$

$$(z_1, z_2) \in \mathfrak{S}_2 = \{(z_1, z_2) \in \mathbb{C}^2 : z_1 + \overline{z_1} - z_2\overline{z_2} > 0\},$$

such that

$$L(\rho, s) = L(f, s)$$

up to finitely many Euler factors, where $L(\rho, s)$ is the Artin L-function associated to the Galois representation ρ and $L(f, s)$ is the automorphic L-function associated to the automorphic form f.

Here, we find that three topics: L-functions, Galois representations and automorphic forms or, equivalently, representations, which figure prominently in the modern higher arithmetic, appear naturally in our conjecture. The L-functions are attached to both the Galois representations and the automorphic representations and are the link that joins them. Our conjecture, like Langlands reciprocity law and Artin's abelian reciprocity law, gives a correspondence independent of any particular geometric construction.

(1) tetrahedron, A_4 (Langlands) (see [31]), elliptic modular forms of weight 1.

(2) octahedron, S_4 (Tunnell), elliptic modular forms of weight 1.

(3) icosahedron, A_5 (Artin conjecture), elliptic modular forms of weight 1.

(4) Hessian polyhedra, H (our conjecture), Picard modular forms of weight 1.

Note that the group $W(E_6)$ is combinatorially defined, while the subgroup H is defined on algebraic variety. Thus we give a connection between combinatoric and algebraic geometry. Furthermore, the Galois representation ρ : $\mathrm{Gal}(\overline{\mathbb{Q}}/\mathbb{Q}) \to H$ comes from number theory, the Picard modular forms come from analysis and representation theory of Lie groups. Here, Galois symmetry and geometric symmetry meet together. Therefore, our theorem and conjecture gives a connection which involves combinatoric, algebraic geometry, number theory, analysis and representation theory.

5. Hessian polyhedra and the arithmetic of rigid Calabi-Yau threefolds

A complete system of invariants for the Hessian groups (the corresponding geometric objects are Hessian polyhedra) has degrees 6, 9, 12, 12 and 18 and can be given explicitly by the following forms:

$$\begin{cases} C_6(z_1, z_2, z_3) = z_1^6 + z_2^6 + z_3^6 - 10(z_1^3 z_2^3 + z_2^3 z_3^3 + z_3^3 z_1^3), \\ C_9(z_1, z_2, z_3) = (z_1^3 - z_2^3)(z_2^3 - z_3^3)(z_3^3 - z_1^3), \\ C_{12}(z_1, z_2, z_3) = (z_1^3 + z_2^3 + z_3^3)[(z_1^3 + z_2^3 + z_3^3)^3 + 216 z_1^3 z_2^3 z_3^3], \\ \mathfrak{C}_{12}(z_1, z_2, z_3) = z_1 z_2 z_3 [27 z_1^3 z_2^3 z_3^3 - (z_1^3 + z_2^3 + z_3^3)^3], \\ C_{18}(z_1, z_2, z_3) = (z_1^3 + z_2^3 + z_3^3)^6 - 540 z_1^3 z_2^3 z_3^3 (z_1^3 + z_2^3 + z_3^3)^3 - 5832 z_1^6 z_2^6 z_3^6. \end{cases}$$

They satisfy the following relations:

$$\begin{cases} 432 C_9^2 = C_6^3 - 3 C_6 C_{12} + 2 C_{18}, \\ 1728 \mathfrak{C}_{12}^3 = C_{18}^2 - C_{12}^3. \end{cases}$$

The first relation can be considered as an elliptic curve with two parameters C_{12} and C_{18}:

$$432 C_9^2 = C_6^3 - 3 C_{12} C_6 + 2 C_{18}. \tag{5.1}$$

The second one can be considered as an elliptic curve with a parameter $\mathfrak{C}_{12}$:

$$C_{18}^2 = C_{12}^3 + 1728 \mathfrak{C}_{12}^3. \tag{5.2}$$

When $C_9 = 0$, the equation (5.1) reduces to

$$C_6^3 - 3 C_6 C_{12} + 2 C_{18} = 0.$$

In fact, Wirtinger studied the algebraic function z of two variables x, y defined by the equation (see [5])

$$z^3 + 3xz + 2y = 0.$$

In modern terms: the projection of the surface X with this equation to the (x, y)-plane is the semiuniversal unfolding of the 0-dimensional A_2-type singularity $z^3 = 0$. The discriminant curve $D \subset \mathbb{C}^2$ has the equation

$$x^3 + y^2 = 0, \quad \text{i.e.,} \quad 1728 \mathfrak{C}_{12}^3 = 0.$$

The fundamental group of $\mathfrak{C}^2 - D$ operates on the fibre over the base point by the monodromy representation. Wirtinger calculates $\pi_1(\mathbb{C}^2 - D)$ and finds a presentation with two generators and one relation $sts = tst$. In modern terms: π_1 is the braid group on 3 strings. The monodromy representation is the canonical homomorphism of this group to the symmetric group S_3. In his computation of $\pi_1(\mathbb{C}^2 - D)$, Wirtinger used an idea of Heegaard. Heegaard reduced the complex geometry of an algebroid covering $(X, x) \to (\mathbb{C}^2, 0)$ with a singularity $(D, 0)$ of the discriminant to a situation of 3-dimensional topology. He considered a small 4-ball $B \subset \mathbb{C}^2$ centered at 0 with boundary $\partial B = S^3$, a 3-sphere. The intersection $L = D \cap S^3$ is a knot or link in S^3. In Wirtinger's example it is the trefoil knot. This established a link between the geometry of singularities of complex surfaces and 3-dimensional topology.

Put

$$x = 3C_6, \quad y = 108C_9$$

in the elliptic curve (5.1), we have

$$E: \quad y^2 = x^3 - 27C_{12}x + 54C_{18}. \tag{5.3}$$

The j-invariant is given by

$$j = -\frac{C_{12}^3}{\mathfrak{C}_{12}^3}. \tag{5.4}$$

According to [61], put

$$\varphi = z_1 z_2 z_3, \quad \psi = z_1^3 + z_2^3 + z_3^3, \quad \chi = z_1^3 z_2^3 + z_2^3 z_3^3 + z_3^3 z_1^3.$$

$$G = (z_1 - z_2)(z_2 - z_3)(z_3 - z_1).$$

$$H = \psi + 6\varphi, \quad K = \psi - 3\varphi.$$

$$\begin{cases} C_6 = \psi^2 - 12\chi, \\ C_{12} = \psi(\psi^3 + 216\varphi^3), \\ \mathfrak{C}_{12} = \varphi(27\varphi^3 - \psi^3), \\ C_{18} = \psi^6 - 540\varphi^3\psi^3 - 5832\varphi^6. \end{cases}$$

Theorem 5.1 (see [61], Main Theorem 4). *The invariants G, H and K satisfy the following algebraic equations, which are the form-theoretic resolvents (algebraic resolvents) of G, H, K:*

$$\begin{cases} 4G^3 + H^2G - C_6G - 4C_9 = 0, \\ H(H^3 + 8K^3) - 9C_{12} = 0, \\ K(K^3 - H^3) - 27\mathfrak{C}_{12} = 0. \end{cases}$$

We have

$$C_{12} = \frac{1}{9}H(H^3 + 8K^3), \quad \mathfrak{C}_{12} = \frac{1}{27}K(K^3 - H^3).$$

Hence,

$$j = -27\frac{H^3(H^3 + 8K^3)^3}{K^3(K^3 - H^3)^3}. \tag{5.5}$$

Put

$$t = \frac{H}{K}.$$

Then

$$j = -27\frac{t^3(t^3+8)^3}{(1-t^3)^3}. \tag{5.6}$$

Note that

$$\varphi = \frac{1}{9}(H-K), \quad \psi = \frac{1}{3}(H+2K).$$

Hence,

$$C_{18} = -\frac{1}{27}(H^6 - 20H^3K^3 - 8K^6). \tag{5.7}$$

Now, we get

$$E: \quad y^2 = x^3 - 3H(H^3+8K^3)x - 2(H^6 - 20H^3K^3 - 8K^6).$$

Put

$$X = \frac{x}{K^2}, \quad Y = \frac{y}{K^3}.$$

We have

$$E_{2,t}: \quad Y^2 = X^3 - 3t(t^3+8)X - 2(t^6 - 20t^3 - 8). \tag{5.8}$$

For the elliptic curve

$$E_{2,t}/\mathbb{Q}(t): \quad y^2 = x^3 - 3t(t^3+8)x - 2(t^6 - 20t^3 - 8), \tag{5.9}$$

we find that

$$P_2 = (3t^2, 4(t^3-1)) \in E_{2,t}. \tag{5.10}$$

Moreover,

$$[2]P_2 = (3t^2, -4(t^3-1)), \quad -P_2 = (3t^2, -4(t^3-1)). \tag{5.11}$$

Hence, $[2]P_2 = -P_2$, i.e., $[3]P_2 = O$. So, P_2 is a 3-division point in $E_{2,t}$.
The discriminant of the elliptic curve (5.9) is given by

$$\Delta = 2^{12}\cdot 3^3 \cdot (t-1)^3(t^2+t+1)^3. \tag{5.12}$$

The bad fibers occur at

$$t = 1, \omega, \overline{\omega}, \infty.$$

They are of types (I_3, I_3, I_3, I_3) by [16], p.336. Correspondingly,

$$\begin{cases} t = 1, \text{i.e.}, \varphi = 0, \text{i.e.}, z_1z_2z_3 = 0, \\ t = \omega, \text{i.e.}, \psi - 3\overline{\omega}\varphi = 0, \text{i.e.}, \\ \qquad (z_1+z_2+\overline{\omega}z_3)(\omega z_1+z_2+\omega z_3)(\overline{\omega}z_1+z_2+z_3) = 0, \\ t = \overline{\omega}, \text{i.e.}, \psi - 3\omega\varphi = 0, \text{i.e.}, \\ \qquad (z_1+\omega z_2+z_3)(\omega z_1+z_2+z_3)(\overline{\omega}z_1+\overline{\omega}z_2+z_3) = 0, \\ t = \infty, \text{i.e.}, \psi - 3\varphi = 0, \text{i.e.}, \\ \qquad (z_1+z_2+z_3)(z_1+\omega z_2+\overline{\omega}z_3)(z_1+\overline{\omega}z_2+\omega z_3) = 0, \end{cases}$$

where $\omega = \exp(2\pi i/3)$. They are the 12 lines of the Hessian arrangement on $\mathbb{P}^2$.

Therefore, for the elliptic curve (5.3), we have

$$P_1 = (3C_6, 108C_9) \in E, \tag{5.13}$$

$$P_2 = (3H^2, 4(H^3 - K^3)) = (3(\psi + 6\varphi)^2, 108\varphi(\psi^2 + 3\psi\varphi + 9\varphi^2)) \in E. \tag{5.14}$$

Note that

$$t = \frac{H}{K} = \frac{\psi + 6\varphi}{\psi - 3\varphi}.$$

This gives the following identity:

$$E_{1,t}: \quad (t-1)(z_1^3 + z_2^3 + z_3^3) - 3(t+2)z_1 z_2 z_3 = 0, \tag{5.15}$$

which is a family of elliptic curves over $\mathbb{P}^1$. In the singular fiber: $t = 1$, $E_{1,t}$ degenerates into $z_1 z_2 z_3 = 0$. In the singular fiber: $t = \infty$, $E_{1,t}$ degenerates into $(z_1 + z_2 + z_3)(z_1 + \omega z_2 + \omega^2 z_3)(z_1 + \omega^2 z_2 + \omega z_3) = 0$. In the singular fiber: $t = \overline{\omega}$, $E_{1,t}$ degenerates into $(z_1 + \omega z_2 + z_3)(\omega z_1 + z_2 + z_3)(\overline{\omega} z_1 + \overline{\omega} z_2 + z_3) = 0$. In the singular fiber: $t = \omega$, $E_{1,t}$ degenerates into $(z_1 + z_2 + \overline{\omega} z_3)(\omega z_1 + z_2 + \omega z_3)(\overline{\omega} z_1 + z_2 + z_3) = 0$. They are the 12 lines of the Hessian arrangement on $\mathbb{P}^2$. Now, E decomposes into the pair $(E_{1,t}, E_{2,t})$.

By the linear transformation $t = 9r + 1$, $E_{1,t}$ becomes one of the Beauville's families of elliptic curves, which are modular elliptic surfaces for some conjugate of the given group (see [3] and [55]). The equation in $\mathbb{P}^2$ is given by

$$III: \quad z_1^3 + z_2^3 + z_3^3 = \left(\frac{1}{r} + 3\right) z_1 z_2 z_3.$$

The group $\Gamma = \Gamma(3)$. The number of components of singular fibers is $(3, 3, 3, 3)$. The Picard-Fuchs equation for this family of elliptic curves is given by

$$III: \quad r(27r^2 + 9r + 1)f'' + (9r+1)^2 f' + 3(9r+1)f = 0.$$

The solution of this equation in terms of modular forms is

$$III: \quad r = \eta(3\tau)^3 \eta\left(\frac{1}{3}\tau\right)^{-3}, \quad f = \eta\left(\frac{1}{3}\tau\right)^3 \eta(\tau)^{-1}.$$

In our present case, the Picard-Fuchs equation becomes

$$(t^3 - 1)\frac{d^2 f}{dt^2} + 3t^2 \frac{df}{dt} + tf = 0. \tag{5.16}$$

In fact, for $\overline{C} = \mathbb{P}^1$, $C = \mathbb{P}^1 - \{1, \infty, \text{roots of} \quad t^2 + t + 1 = 0\}$ and the family $f : X \to C$ is the modular family for the group $\Gamma_1(3)$. The Picard-Fuchs equation (for the holomorphic 1-forms on X_t) is (see [39])

$$(t-1)(t^2 + t + 1)\frac{d^2 f}{dt^2} + 3t^2 \frac{df}{dt} + tf = 0.$$

The solutions are given by

$$t = 9\frac{\eta(3\tau)^3}{\eta\left(\frac{1}{3}\tau\right)^3} + 1, \quad f = \frac{\eta\left(\frac{1}{3}\tau\right)^3}{\eta(\tau)}. \tag{5.17}$$

$\Gamma(3)$ is a genus zero principal congruence subgroup of $PSL(2,\mathbb{Z})$, the index of $\Gamma(3)$ in $PSL(2,\mathbb{Z})$ is $\mu = 12$. As a rational function of the Hauptmodul ρ, the j-function can be expressed as (see [36])

$$j(\rho) = \frac{\rho^3(\rho+6)^3(\rho^2-6\rho+36)^3}{(\rho-3)^3(\rho^2+3\rho+9)^3},$$

which is also the j-invariant of the semi-stable arithmetic elliptic surface over $\mathbb{P}^1$. In the elliptic curve $E_{1,t}$, set

$$\rho = \frac{z_1^3+z_2^3+z_3^3}{z_1z_2z_3} = \frac{3(t+2)}{t-1}. \tag{5.18}$$

We find that

$$j(\rho) = 27\frac{t^3(t^3+8)^3}{(t^3-1)^3}, \tag{5.19}$$

which is the j-invariant of the elliptic curve $E_{2,t}$.

Let us recall the following well-known results (see [48]): let k be a field of characteristic $\neq 3$ containing 3 cubic roots of unity. Consider the elliptic curve

$$E_3: \quad x^3+y^3+z^3-3\mu xyz = 0$$

defined over $K_3 = k(\mu)$, μ being a variable over k. Then E_3 has exactly 9 K_3-rational points (i.e. base points of the pencil) and they are of order 3.

Proposition 5.2 (Deuring Normal Form) (see [51]). *Let E/K be an elliptic curve over a field with char$(K) \neq 3$. Then E has a Weierstrass equation over $\overline{K}$ of the form*

$$E_\alpha: \quad y^2+\alpha xy+y = x^3, \quad \alpha \in \overline{K}, \alpha^3 \neq 27.$$

This equation has discriminant and j-invariant

$$\Delta = \alpha^3-27, \quad j = \alpha^3(\alpha^3-24)^3/(\alpha^3-27).$$

We will give some basic facts about families of elliptic curves with points of order 3: the Hessian family (see [27]).

Definition 5.3. The Hessian family of elliptic curves $E_\alpha: y^2+\alpha xy+y = x^3$ is defined for any field of characteristic different from 3 with j-invariant $j(\alpha) = j(E_\alpha)$ of E_α given by

$$j(\alpha) = \frac{\alpha^3(\alpha^3-24)^3}{\alpha^3-27}.$$

The curve E_α is nonsingular for $\alpha^3 \neq 27$, that is, α is not in $3\mu_3$, where μ_3 is the group of third roots of unity. Over the line k minus 3 points, $k - 3\mu_3$, the family E_α consists of elliptic curves with a constant section $(0,0)$ of order 3 where $2(0,0) = (0,-1)$. The Hessian family E_α has three singular fibers which are nodal cubics at the points of $3\mu_3 = \{3, 3\omega, 3\omega^2\}$, where $\omega^2 + \omega + 1 = 0$.

There are other versions of the Hessian family which in homogeneous coordinates take the form

$$H_\mu : \quad u^3 + v^3 + w^3 = 3\mu uvw,$$

or in affine coordinates with $w = -1$, it has the form

$$u^3 + v^3 = 1 - 3\mu uv.$$

If we set $y = -v^3$ and $x = -uv$, we obtain $x^3/y - y = 1 + 3\mu x$, or

$$E_{3\mu} : \quad y^2 + 3\mu xy + y = x^3.$$

This change of variable defines what is called a 3-isogeny of H_μ onto $E_{3\mu}$. There are nine cross-sections of the family H_μ given by

$$\begin{array}{ccc} (0,-1,1), & (0,-\omega,1), & (0,-\omega^2,1), \\ (1,0,-1), & (\omega,0,-\omega^2), & (\omega^2,0,-\omega), \\ (-1,1,0), & (-1,\omega^2,0), & (-1,\omega,0). \end{array}$$

The family H_μ is nonsingular over the line minus μ_3, and choosing, for example, $0 = (-1,1,0)$, one can show that these nine points form the subgroup of 3-division points of the family H_μ.

For $E_{3t} : y^2 + 3txy + y = x^3$, we have

$$j_1 = \frac{27t^3(9t^3-8)^3}{t^3-1}.$$

For $E_{2,t} : y^2 = x^3 - 3t(t^3+8)x - 2(t^6 - 20t^3 - 8)$, we have

$$j_2 = \frac{27t^3(t^3+8)^3}{(t^3-1)^3}.$$

Let $r = t^3 - 1$, then

$$\begin{cases} j_1 = j_1(r) = \dfrac{27}{r}(r+1)(9r+1)^3, \\ j_2 = j_2(r) = \dfrac{27}{r^3}(r+1)(r+9)^3. \end{cases}$$

Note that

$$j_1\left(\frac{1}{r}\right) = j_2(r).$$

This gives a one-to-one correspondence between two families of elliptic curves E_{3t} and $E_{2,t}$.

We will make use of certain elliptic modular surfaces associated to torsion free congruence subgroups of $SL(2,\mathbb{Z})$. Each of these has four singular fibers, all of which are of type I_b for various b. This is actually a complete list of rational elliptic surfaces with exactly four semi-stable singular fibers (see [3] and [16] for more details). Combine with the results in [3] and [16], we find that two families of elliptic curves $E_{1,t}$ and $E_{2,t}$ are isogenous. According to [48], Theorem 5.2, the points of $E_{1,t}$ and $E_{2,t}$ over $\mathbb{Q}(t)$ are of finite order because the group of sections of any elliptic modular surface is finite. In fact, $E_{1,t}$ is defined over $\mathbb{Q}(t)$, $E_{2,t}$ is defined over $\mathbb{Q}[t]$.

Isogenous elliptic curves are related by the modular polynomials (see [29]). For elliptic curves E and F with j-invariants j_E and j_F respectively, there is an isogeny $\lambda : E \to F$ with kernel isomorphic to the cyclic group of order n if and only if

$$\Phi_n(j_E, j_F) = 0,$$

where $\Phi_n(x,y)$ is the modular polynomial of order n. The roots of $\Phi_n(X, j(\tau)) = 0$ are $j(\alpha\tau)$ where $\alpha = \begin{pmatrix} a & b \\ 0 & d \end{pmatrix}$, $a,b,d \in \mathbb{Z}$ with $(a,b,d) = 1$, $ad = n$ and $0 \le b < d$. Such α are called primitive.

According to [49], suppose that the minimal Weierstrass equation of E over $K = k(t)$ is given by

$$y^2 + a_1(t)xy + a_3(t)y = x^3 + a_2(t)x^2 + a_4(t)x + a_6(t), \quad a_i(t) \in k[t].$$

Then the associated elliptic surface $f : S \to \mathbb{P}^1$ is a rational surface if and only if

$$\deg a_i(t) \le i \quad (\text{all } i) \text{ and } \quad \mathrm{Sing}(f) \ne \emptyset.$$

In case $\mathrm{char}(k) \ne 2, 3$, we can take the equation in a more familiar form:

$$y^2 = x^3 + p(t)x + q(t), \quad p(t), q(t) \in k[t].$$

Then the above condition is equivalent to $\deg p(t) \le 4$, $\deg q(t) \le 6$ and $\Delta = 4p(t)^3 + 27q(t)^2$ is not a constant, i.e. $\Delta \notin k$.

For $E_{2,t}$, we have $p(t) = -3t(t^3+8)$, $q(t) = -2(t^6 - 20t^3 - 8)$ and $\Delta = -2^8 \cdot 3^3(t^3-1)^3$. Hence, $E_{2,t}$ is a rational elliptic surface.

We need the following theorems due to Shioda (see [48]).

Proposition 5.4 (see [48], Proposition 4.2). *The elliptic modular surface B_Γ has t_1 singular fibers of types I_b $(b \ge 1)$, t_2 singular fibers of types I_b^* $(b \ge 1)$ and s singular fibers of type IV^*, where t_1, t_2 and s are respectively the number of cusps of the first kind, the number of cusps of the second kind and the number of elliptic points for Γ.*

Theorem 5.5 (see [48], Theorem 5.2). *Let B_Γ be the elliptic modular surface attached to Γ.*

(1) *If Γ has torsion (i.e. $s > 0$), then the group of sections $\mathcal{S}(B_\Gamma)$ is either trivial or a cyclic group of order 3.*

(2) If Γ has a cusp of the second kind (i.e. $t_2 > 0$), then the group of sections $\mathcal{S}(B_\Gamma)$ is either trivial or isomorphic to one of the groups

$$\mathbb{Z}/(2), \quad \mathbb{Z}/(4) \quad \text{or} \quad \mathbb{Z}/(2) \times \mathbb{Z}/(2).$$

(3) If Γ is torsion-free and all cusps are of the first kind, then the group of sections $\mathcal{S}(B_\Gamma)$ is isomorphic to a subgroup of $\mathbb{Z}/(m) \times \mathbb{Z}/(m)$, where m denotes the least common multiple of b_i's $(1 \leq i \leq t_1)$. Here we suppose that the singular fibers of B_Γ are of types I_{b_i} $(1 \leq i \leq t_1)$.

Combine with the above results, we find that the Mordell-Weil groups for $E_{1,t}$ and $E_{2,t}$ are finite groups. More precisely, the Mordell-Weil groups of $E_{1,t}$ and $E_{2,t}$ over $\mathbb{Q}(t)$ are isomorphic to the subgroups of $\mathbb{Z}/3\mathbb{Z} \times \mathbb{Z}/3\mathbb{Z}$.

Let $Y \to \mathbb{P}^1$, $Y' \to \mathbb{P}^1$ be two rational elliptic surfaces. Under some mild technical conditions on the singular fibers, the fiber product $Y \times_{\mathbb{P}^1} Y'$ admits a nice resolution W which is a projective simply connected threefold with trivial canonical bundle. It can be showed that some of them are rigid, i.e., admit no nontrivial deformation (see [42]).

According to [42], let $r : Y \to \mathbb{P}^1$ and $r' : Y' \to \mathbb{P}^1$ denote two relatively minimal, rational, elliptic surfaces with sections. Write S (resp. S') for the images of the singular fibers of Y (resp. Y') in $\mathbb{P}^1$. The fiber product $p : W = Y \times_{\mathbb{P}^1} Y' \to \mathbb{P}^1$ is nonsingular except at points in the fibers over $S'' = S \cap S'$. In order that the singularities of W be no worse than ordinary double points, we shall assume that the singular fibers of r and r' above the points in S'' are either irreducible nodal rational curves or cycles of smooth rational curves. In the notation of Kodaira such fibers are of type I_b, where $b > 0$ denotes the number of irreducible components. Such fibers are also called semi-stable.

The dualizing sheaf of W is most readily computed by regarding W as the hypersurface in $Y \times Y'$ obtained by pulling back the diagonal in $\mathbb{P}^1 \times \mathbb{P}^1$ via the map $r \times r'$. For a relatively minimal, regular elliptic surface, $r : Y \to \mathbb{P}^1$, the canonical sheaf is given by

$$\omega_Y \simeq r^* \mathcal{O}_{\mathbb{P}^1}(p_g(Y) - 1) \otimes \mathcal{O}_Y\Big(\sum_i (m_i - 1) F_i\Big),$$

where the second term is the contribution of the multiple fibers. An easy computation using the adjunction formula reveals that the dualizing sheaf ω_W is trivial exactly when $p_g(Y) = p_g(Y') = 0$ and there are no multiple fibers. Of course these conditions are fulfilled when both Y and Y' are rational and have sections. However it follows from Castelnuovo's rationality criterion and the vanishing of the Tate-Shafarevich group for rational elliptic surfaces with sections that this is the only case when the conditions for triviality of ω_W are fulfilled. In this case any small resolution has trivial canonical bundle.

Table 1. Certain rational elliptic modular surfaces (see [42])

Level	Congruence subgroup	Number of components in singular fibers
3	$\Gamma(3)$	$3,3,3,3$
4	$\Gamma_1(4)\cap\Gamma(2)$	$4,4,2,2$
5	$\Gamma_1(5)$	$5,5,1,1$
6	$\Gamma_1(6)$	$6,3,2,1$
8	$\Gamma_0(8)\cap\Gamma_1(4)$	$8,2,1,1$
9	$\Gamma_0(9)\cap\Gamma_1(3)$	$9,1,1,1$

Proposition 5.6 (see [42]). *Let $r: Y\to\mathbb{P}^1$ and $r': Y'\to\mathbb{P}^1$ denote relatively minimal rational elliptic surfaces with section. Suppose that the fibers of r and r' above all points in S'' are semi-stable. Then the fiber product $Y\times_{\mathbb{P}^1}Y'$ has only ordinary double point singularities. The projective variety, $\widetilde{W}$, obtained by blowing up the nodes has no infinitesimal deformations, exactly when all fibers of r and r' are semi-stable and one is in one of the following four cases, each of which actually occurs:*

(1) Y and Y' appear in Table 1 and are isogenous. In particular $S=S'=S''$ has 4 elements.

(2) Y and Y' appear in Table 1, each has at least one I_1 fiber, and the map r' has been modified by an automorphism of $\mathbb{P}^1$ so that $\sharp(S'')=3$. Furthermore the singular fibers of r and r' which do not lie above points of S'' are of type I_1.

(3) Y and Y' are not isogenous and $S=S'=S''$ has 5 elements.

(4) Y appears in Table 1, $S=S''$, $\sharp(S')=5$, and the singular fiber of r' which does not map to S'' has type I_1.

Therefore, we find that the fiber product $E=E_{1,t}\times_{\mathbb{P}^1}E_{2,t}$ is a rigid Calabi-Yau threefold.

The fiber product can be considered as a kind of topological surgery. We will study the structure of rational points under this topological surgery. From the point of view of Diophantine geometry, fiber products can be used to generate some new rational points in Calabi-Yau threefolds. For example, for the self-fiber product

$$E_{1,t}\times_{\mathbb{P}^1}E_{1,t}:\quad (x^3+y^3+z^3)rst=(r^3+s^3+t^3)xyz,$$

there are some trivial new rational points, such as $x=r$, $y=s$, $z=t$ and so on. It will be interesting to find some nontrivial new rational points in the fiber products. We will prove that there are infinitely many nontrivial new rational points in the fiber product $E=E_{1,t}\times_{\mathbb{P}^1}E_{2,t}$.

In actually dealing with specific elliptic curves over $\mathbb{Q}$ we find the following theorem due to Lutz and Nagell very useful and it gives a manageable method for determining $E_{\text{tor}}(\mathbb{Q})$.

Theorem 5.7 (Lutz-Nagell) (see [51]). *Let $E/\mathbb{Q}$ be an elliptic curve with Weierstrass equation*

$$y^2=x^3+Ax+B,\quad A,B\in\mathbb{Z}.$$

Suppose $P \in E(\mathbb{Q})$ is a non-zero torsion point. Then

(1) $x(P), y(P) \in \mathbb{Z}$.

(2) *Either* $[2]P = O$, *or else* $y(P)^2$ *divides* $4A^3 + 27B^2$.

For the elliptic curve (5.3), we know that $P = (3C_6, 108C_9) \in E$. Here, $A = -27C_{12}$ and $B = 54C_{18}$. Note that $A, B \in \mathbb{Z}$ when $(z_1, z_2, z_3) \in \mathbb{Z}^3$. In fact, E can be defined over $\mathbb{Q}[z_1, z_2, z_3]$. Suppose that P is a torsion point in E, then P is a torsion point in E with $(z_1, z_2, z_3) \in \mathbb{Z}^3$. We find that

$$[2]P = \left(-6C_6 + \frac{(C_6^2 - C_{12})^2}{64C_9^2}, -108C_9 + \frac{9C_6(C_6^2 - C_{12})}{8C_9} - \frac{(C_6^2 - C_{12})^3}{512C_9^3}\right). \tag{5.20}$$

If $[2]P = O$, then $C_9 = 0$ for every $(z_1, z_2, z_3) \in \mathbb{Z}^3$, which is impossible! Thus, $[2]P \neq O$. By Lutz-Nagell theorem, we have that $108^2C_9^2$ divides $2^8 \cdot 3^{12}\mathfrak{C}_{12}^3$, i.e., C_9^2 divides $2^4 \cdot 3^6\mathfrak{C}_{12}^3$ for every $(z_1, z_2, z_3) \in \mathbb{Z}^3$, which is also impossible! Therefore, P is of infinite order.

Now, we obtain the following theorem:

Theorem 5.8 (Main Theorem 5). *The variety associated with Hessian polyhedra*

$$E: \quad y^2 = x^3 - 27C_{12}x + 54C_{18} \tag{5.21}$$

can be expressed as the fiber product of two isogenous, semi-stable, rational elliptic modular surfaces

$$E_{1,t}: \quad (t-1)(z_1^3 + z_2^3 + z_3^3) - 3(t+2)z_1z_2z_3 = 0$$

and

$$E_{2,t}: \quad y^2 = x^3 - 3t(t^3 + 8)x - 2(t^6 - 20t^3 - 8).$$

Hence, E is a rigid Calabi-Yau threefold. There are only finitely many rational points in $E_{1,t}$ and $E_{2,t}$. However, there exist infinitely many nontrivial rational points in E. More precisely, $P = (3C_6, 108C_9) \in E$ is of infinite order.

The developments in physics stimulated the interest of mathematicians in Calabi-Yau varieties. There are some important conjectures for these special varieties. One of these is the modularity conjecture for Calabi-Yau threefolds defined over $\mathbb{Q}$ (see [37] and [62]).

Let us recall some basic facts about Calabi-Yau varieties (see [62]).

Definition 5.9. Let X be a smooth complex projective variety of dimension d. X is called a Calabi-Yau variety if

(1) $H^i(X, \mathcal{O}_X) = 0$ for every i, $0 < i < d$.

(2) The canonical bundle of X is trivial.

We introduce the Hodge cohomology groups and the Hodge numbers

$$H^{i,j}(X) := H^j(X, \Omega_X^i), \quad h^{i,j}(X) := \dim H^{i,j}(X).$$

By the duality induced from complex conjugation, $h^{i,j}(X) = h^{j,i}(X)$, and the Serre duality on the Hodge cohomology groups asserts that $h^{i,j}(X) = h^{d-j,d-i}(X)$. Also there is the Hodge decomposition

$$H^k(X,\mathbb{C}) = \bigoplus_{i+j=k} H^{i,j}(X).$$

The k-th Betti number $b_k(X) := \dim H^k(X,\mathbb{C})$ and the Euler characteristic is

$$\chi(X) = \sum_{k=0}^{2d} (-1)^i b_k(X).$$

The condition (1) is that $h^{i,0}(X) = 0$ for every i, $0 < i < d$, and the condition (2) implies that the geometric genus $p_g := h^{d,0}(X) = 1$.

The dimension three objects of our interest here are Calabi-Yau threefolds. Note that Calabi-Yau threefolds are Kähler manifolds, so that $h^{1,1} > 0$. The Hodge diamond of a Calabi-Yau threefold is given by

$$\begin{array}{ccccccc} & & & 1 & & & \\ & & 0 & & 0 & & \\ & 0 & & h^{1,1} & & 0 & \\ 1 & & h^{2,1} & & h^{1,2} & & 1 \\ & 0 & & h^{2,2} & & 0 & \\ & & 0 & & 0 & & \\ & & & 1 & & & \end{array}$$

The Betti numbers are

$$b_0 = b_6 = 1, \quad b_1 = b_5 = 0, \quad b_2 = b_4 = h^{1,1}, \quad b_3 = 2(1 + h^{2,1}).$$

The Euler characteristic is $\chi = 2(h^{1,1} - h^{2,1})$.

Definition 5.10. A Calabi-Yau threefold X is said to be rigid if $h^{2,1} = 0$

The name comes from the fact that this condition implies $H^1(X, T_X) = 0$, and hence that X has no infinitesimal complex deformations. In this case, the middle cohomology

$$H^3(X) = H^{3,0}(X) \oplus H^{0,3}(X) \cong \mathbb{C}^2$$

is two-dimensional.

Now we will formulate the modularity conjecture for rigid Calabi-Yau threefolds over $\mathbb{Q}$ (see [62]). Let X be a rigid Calabi-Yau threefold over $\mathbb{Q}$. We know that an integral model for X always exists. Let p be a good rational prime. Then the characteristic polynomial of the endomorphism Frob_p^* on $H^3_{\text{ét}}(\overline{X}, \mathbb{Q}_\ell)$ is

$$P_3^p(T) := \det(1 - \mathrm{Frob}_p^* T | H^3_{\text{ét}}(\overline{X}, \mathbb{Q}_\ell)).$$

Then it is given by

$$P_3^p(T) = 1 - t_3(p)T + p^3T^2 \in \mathbb{Z}[T]$$

with $t_3(p) \in \mathbb{Z}$, $|t_3(p)| \leq 2p^{3/2}$. Furthermore, by the Lefschetz fixed point formula, $t_3(p)$ can be expressed in terms of the number of rational points on X over $\mathbb{F}_p$. Let $t_i(p) := \mathrm{tr}(\mathrm{Frob}_p^* | H^i_{\text{ét}}(\overline{X}, \mathbb{Q}_\ell))$ denote the trace of the endomorphism Frob_p^* on $H^i_{\text{ét}}(\overline{X}, \mathbb{Q}_\ell)$. Then the Lefschetz fixed point formula asserts that

$$\# X(\mathbb{F}_p) = \sum_{i=0}^{6} (-1)^i t_i(p).$$

For a rigid Calabi-Yau threefold, this formula gives rise to the identity:

$$t_3(p) = 1 + p^3 + (1+p) t_2(p) - \# X(\mathbb{F}_p)$$

where $t_2(p) \leq p h^{1,1}$ and the equality holds when all cycles generating $H^{1,1}(X)$ are defined over $\mathbb{Q}$. The L-series of X is given by putting together all the local data:

$$L(X_{\mathbb{Q}}, s) := L(H^3_{\text{ét}}(\overline{X}, \mathbb{Q}_\ell), s) = (*) \prod_{p:\text{good}} \frac{1}{1 - t_3(p) p^{-s} + p^{3-2s}}$$

where $(*)$ is the factor corresponding to bad primes.

Now, we need the modularity theorem for rigid Calabi-Yau threefolds over $\mathbb{Q}$.

Theorem 5.11 (see [62] and [63]). *Let X be a rigid Calabi-Yau threefold defined over $\mathbb{Q}$. Then there is a cusp form f of weight 4 on some $\Gamma_0(N)$ where N is divisible only by bad primes, such that the two 2-dimensional ℓ-adic Galois representations $\rho_{X,\ell}$ and $\rho_{f,\ell}$ associated to X and f, respectively, have the isomorphic semi-simplifications ($\rho_{X,\ell} \sim \rho_{f,\ell}$). Consequently,*

$$L(X, s) = L(f, s).$$

More precisely, if we write $f(q) = \sum_{m=1}^{\infty} a_f(m) q^m$ with $q = e^{2\pi i z}$, then

$$t_3(p) = a_f(p)$$

for all primes not dividing N.

The fiber products of rational elliptic surfaces give some examples of rigid Calabi-Yau threefolds and their modularity. It is known that (see [3], [42] and [43]) there are exactly six torsion free genus zero congruence subgroups of $PSL(2, \mathbb{Z})$, and the list is complete. They are $\Gamma(3)$, $\Gamma_1(4) \cap \Gamma(2)$, $\Gamma_1(5)$, $\Gamma_1(6)$, $\Gamma_0(8) \cap \Gamma_1(4)$ and $\Gamma_0(9) \cap \Gamma_1(3)$ in the Table 1. The modularity of the rigid Calabi-Yau threefolds defined over $\mathbb{Q}$ corresponding to these subgroups are given as follows:

Theorem 5.12 (see [41], [54]). *Let Γ be one of the six groups in the Table 1. Then the rigid Calabi-Yau threefolds defined over $\mathbb{Q}$ associated to Γ are all modular. More precisely, let $\widetilde{S^2_\Gamma} := \widetilde{S_\Gamma \times_{C_\Gamma} S_\Gamma}$ be a rigid Calabi-Yau threefold defined over $\mathbb{Q}$ associated to Γ. Then there is a cusp form of weight 4 on Γ such that $L(\widetilde{S^2_\Gamma}, s) = L(f, s)$.*

In particular, when $\Gamma = \Gamma(3)$, the singular fibers are of types (I_3, I_3, I_3, I_3). The Hodge number $h^{1,1}(X) = 36$. The cusp form of weight 4 is given by $\eta(q^3)^8$. Here $\eta(q)$ is Dedekind eta-function $\eta(q) = q^{1/24}\prod_{n=1}^{\infty}(1-q^n)$ with $q = e^{2\pi i z}$, $z \in \mathbb{H}$. The equation for the rigid Calabi-Yau threefold is

$$(x^3 + y^3 + z^3)rst = (r^3 + s^3 + t^3)xyz.$$

Note that in the above construction, we only use the self-fiber products of semi-stable, rational elliptic modular surfaces. It will be interesting to study the fiber products of two isogenous, semi-stable, rational elliptic modular surfaces, such as our fiber product $E_{1,t} \times_{\mathbb{P}^1} E_{2,t}$. A natural question is: which modular form corresponds to this fiber product?

According to Vafa (see [53]), sometimes the physics of the singularities are unconventional. For example when a 4-cycle (say a $\mathbb{CP}^2$) shrinks in a Calabi-Yau threefold, it gives rise to very interesting unconventional new physical theories which were not anticipated! This is thus a great source of insight into new physics. In particular what types of singularities occur as well as what are the ways to resolve them will be of extreme importance for unravelling aspects of this new physics. It is tempting to speculate that these singularities may also lead to new invariants for four manifolds.

For the rigid Calabi-Yau threefold E in (5.21) where $C_{12} = C_{12}(z_1, z_2, z_3)$ and $C_{18} = C_{18}(z_1, z_2, z_3)$ with $(z_1, z_2, z_3) \in \mathbb{CP}^2$, we construct infinitely many subvarieties of E:

$$\begin{aligned} Y_m = \{(z_1, z_2, z_3, x([m]P), y([m]P)) :& (z_1, z_2, z_3) \in \mathbb{CP}^2, \\ & P = (3C_6, 108C_9) \in E, m \in \mathbb{Z}\}. \end{aligned} \tag{5.22}$$

It will be interesting to study the singularity when Y_m shrinks in E. This singularity maybe involves vanishing cycles and Picard-Lefschetz theory.

Note that the Calabi-Yau threefold E in (5.21) can be considered as an elliptic curve defined over $\mathbb{Q}(T_1, T_2)$:

$$E/\mathbb{Q}(T_1, T_2): \quad y^2 = x^3 - 27C_{12}x + 54C_{18}, \tag{5.23}$$

where $T_1 = z_1/z_3$, $T_2 = z_2/z_3$ and

$$\begin{cases} C_{12} = C_{12}(T_1, T_2) = (T_1^3 + T_2^3 + 1)[(T_1^3 + T_2^3 + 1)^3 + 216T_1^3T_2^3], \\ \mathfrak{C}_{12} = \mathfrak{C}_{12}(T_1, T_2) = T_1T_2[27T_1^3T_2^3 - (T_1^3 + T_2^3 + 1)^3], \\ C_{18} = C_{18}(T_1, T_2) = (T_1^3 + T_2^3 + 1)^6 - 540T_1^3T_2^3(T_1^3 + T_2^3 + 1)^3 - 5832T_1^6T_2^6. \end{cases} \tag{5.24}$$

Let k be a number field and A an abelian variety over $K = k(T_1, \cdots, T_n)$ where the T_i are indeterminates over k. By a theorem of Néron (see [47]), $A(K)$ is a finitely generated group. In our case, $E(\mathbb{Q}(T_1, T_2))$ is finitely generated by Néron's theorem. The j-invariant

$$j = j(T_1, T_2) = -\frac{C_{12}(T_1, T_2)^3}{\mathfrak{C}_{12}(T_1, T_2)^3} \notin \mathbb{Q}. \tag{5.25}$$

In fact, E is an elliptic threefold (see [38] for more details).

References

[1] E. Artin, Über eine neue Art von L-Reihen, Hamb. Abh. **3** (1923), 89-108, In: *The Collected Papers of Emil Artin*, Edited by S. Lang and J. Tate, 105-124, Addison-Wesley Publishing Company, 1965.

[2] E. Artin, Zur Theorie der L-Reihen mit allgemeinen Gruppencharakteren, Hamb. Abh. **8** (1930), 292-306, In: *The Collected Papers of Emil Artin*, Edited by S. Lang and J. Tate, 165-179, Addison-Wesley Publishing Company, 1965.

[3] A. Beauville, Les familles stables de courbes elliptiques sur $\mathbb{P}^1$ admettant quatre fibres singulières, C. R. Acad. Sci. Paris Sér. I Math. **294** (1982), 657-660.

[4] R. Brauer, On simple groups of order $5 \cdot 3^a \cdot 2^b$, In: *Collected Papers*, Vol. II, Edited by P. Fong and W. J. Wong, 421-470, The MIT Press, 1970.

[5] E. Brieskorn, Singularities in the work of Friedrich Hirzebruch, Surveys in Differential Geometry **VII** (2000), 17-60.

[6] M. Broué and M. Enguehard, Polynômes des poids de certains codes et fonctions thêta de certains réseaux, Ann. Sci. École Norm. Sup. (4) **5** (1972), 157-181.

[7] H. Burkhardt, Untersuchungen aus dem Gebiete der hyperelliptischen Modulfunctionen, II, Math. Ann. **38** (1891), 161-224.

[8] H. Burkhardt, Untersuchungen aus dem Gebiete der hyperelliptischen Modulfunctionen, III, Math. Ann. **41** (1893), 313-343.

[9] Y. Choie, B. Mourrain and P. Solé, Rankin-Cohen brackets and invariant theory, J. Algebraic Combinatorics **13** (2001), 5-13.

[10] J. H. Conway, R. T. Curtis, S. P. Norton, R. A. Parker and R. A. Wilson, *Atlas of Finite Groups, Maximal Subgroups and Ordinary Characters for Simple Groups*, Clarendon Press, Oxford, 1985.

[11] J. H. Conway and N. J. A. Sloane, *Sphere Packings, Lattices and Groups*, Second Edition, Springer-Verlag, 1993.

[12] H. S. M. Coxeter, *Regular Complex Polytopes*, Cambridge University Press, 1974.

[13] W. Ebeling, *Lattices and Codes, A Course Partially Based on Lectures by F. Hirzebruch*, 2nd Edition, Advanced Lectures in Mathematics, Vieweg, 2002.

[14] R. Fricke, *Lehrbuch der Algebra*, Bd. II, Braunschweig, 1926.

[15] S. Gelbart, An elementary introduction to the Langlands program, Bull. Amer. Math. Soc. **10** (1984), 177-219.

[16] S. Herfurtner, Elliptic surfaces with four singular fibres, Math. Ann. **291** (1991), 319-342.

[17] D. Hilbert, Mathematical problems, Bull. Amer. Math. Soc. **8** (1902), 437-479.

[18] F. Hirzebruch, *Gesammelte Abhandlungen*, Bd. II, 1963-1987, Springer-Verlag, 1987.

[19] F. Hirzebruch, Hilbert's modular group of the field $\mathbb{Q}(\sqrt{5})$ and the cubic diagonal surface of Clebsch and Klein, Russ. Math. Surv. **31**:5 (1976), 96-110, In: *Gesammelte Abhandlungen*, Bd. II, 394-408, Springer-Verlag, 1987.

[20] F. Hirzebruch, The ring of Hilbert modular forms for real quadratic fields of small discriminant, In: *Modular Functions of One Variable VI, Proceedings, Bonn 1976*, Edited by J. P. Serre and D. B. Zagier, 287-323, Lecture Notes in Math. **627**, Springer-Verlag, 1977, In: *Gesammelte Abhandlungen*, Bd. II, 501-536, Springer-Verlag, 1987.

[21] F. Hirzebruch, The icosahedron, Raymond and Beverly Sackler Distinguished Lectures in Mathematics, Tel Aviv University, 1981, In: *Gesammelte Abhandlungen*, Bd. II, 656-661, Springer-Verlag, 1987.

[22] R. P. Holzapfel, *Geometry and Arithmetic Around Euler Partial Differential Equations*, Dt. Verl. d. Wiss., Berlin/Reidel, Dordrecht 1986.

[23] R. P. Holzapfel, *The Ball and Some Hilbert Problems*, Lectures in Mathematics, ETH Zürich, Birkhäuser, 1995.

[24] R. P. Holzapfel, On the Nebentypus of Picard modular forms, Math. Nachr. **139** (1988), 115-137.

[25] R. P. Holzapfel, Chern numbers of algebraic surfaces-Hirzebruch's examples are Picard modular surfaces, Math. Nachr. **126** (1986), 255-273.

[26] B. Hunt, *The Geometry of some special Arithmetic Quotients*, Lecture Notes in Math. **1637**, Springer-Verlag, 1996.

[27] D. Husemöller, *Elliptic Curves*, GTM **111**, Springer-Verlag, 1987.

[28] F. Klein, *Lectures on the Icosahedron and the Solution of Equations of the Fifth Degree*, Translated by G. G. Morrice, second and revised edition, Dover Publications, Inc., 1956.

[29] S. Lang, *Elliptic Functions*, GTM **112**, Springer-Verlag, 1987.

[30] R. P. Langlands, Some contemporary problems with origins in the Jugendtraum, In: *Mathematical developments arising from Hilbert problems* (Proc. Sympos. Pure Math., Vol. XXVIII, Northern Illinois Univ., De Kalb, Ill., 1974), 401-418, Amer. Math. Soc., Providence, R. I., 1976.

[31] R. P. Langlands, *Base Change for* $GL(2)$, Ann. of Math. Studies **96**, Princeton University Press, 1980.

[32] Yu. I. Manin, *Cubic Forms, Algebra, Geometry, Arithmetic*, Translated from Russian by M. Hazewinkel, Second Edition, North-Holland, 1986.

[33] Yu. I. Manin and M. A. Tsfasman, Rational varieties: algebra, geometry and arithmetic, Russian Math. Surveys **41**:2 (1986), 51-116.

[34] Yu. I. Manin and A. A. Panchishkin, *Introduction to Modern Number Theory*, Second Edition, Springer-Verlag, 2005.

[35] H. Maschke, Aufstellung des vollen Formensystems einer quaternären Gruppe von 51840 linearen Substitutionen, Math. Ann. **33** (1889), 317-344.

[36] J. McKay and A. Sebbar, Arithmetic semistable elliptic surfaces, In: *Proceedings on Moonshine and Related Topics*, Edited by J. McKay and A. Sebbar, CRM Proceedings and Lecture Notes, Vol. **30**, 119-130, AMS Providence, 2001.

[37] C. Meyer, *Modular Calabi-Yau Threefolds*, Fields Institute Monographs, Vol. **22**, AMS Providence, 2005.

[38] R. Miranda, Smooth models for elliptic threefolds, In: *The Birational Geometry of Degenerations*, Edited by R. Friedman and D. R. Morrison, Progress in Math. **29**, 85-133, Birkhäuser, 1983.

[39] C. Peters, Monodromy and Picard-Fuchs equations for families of K3-surfaces and elliptic curves, Ann. Sci. École Norm. Sup. (4) **19** (1986), 583-607.

[40] E. Picard, Sur les formes quadratiques ternaires indéfinies à indéterminées conjuguées et sur les fonctions hyperfuchsiennes correspondantes, Acta Math. **5** (1884), 121-184.

[41] M. -H. Saito and N. Yui, The modularity conjecture for rigid Calabi-Yau threefolds over $\mathbb{Q}$, J. Math. Kyoto Univ. **41** (2001), 403-419.

[42] C. Schoen, On fiber products of rational elliptic surfaces with section, Math. Z. **197** (1988), 177-199.

[43] A. Sebbar, Classification of torsion-free genus zero congruence groups, Proc. Amer. Math. Soc. **129** (2001), 2517-2527.

[44] B. Segre, *The Nonsingular Cubic Surfaces, A New Method of Investigation with Special Reference to Questions of Reality*, Oxford, 1942.

[45] I. Naruki and J. Sekiguchi, A modification of Cayley's family of cubic surfaces and birational action of $W(E_6)$ over it, Proc. Japan Acad. Ser. A Math. Sci. **56** (1980), 122-125.

[46] J. Sekiguchi and M. Yoshida, $W(E_6)$-action on the configuration space of six lines on the real projective plane, Kyushu J. Math. **51** (1997), 297-354.

[47] J.-P. Serre, *Lectures on the Mordell-Weil Theorem*, Translated from the French and edited by Martin Brown from notes by Michel Waldschmidt, Aspects of Mathematics, E15, Friedr. Vieweg & Sohn, Braunschweig, 1989.

[48] T. Shioda, On elliptic modular surfaces, J. Math. Soc. Japan **24** (1972), 20-59.

[49] T. Shioda, On the Mordell-Weil lattices, Comment. Math. Univ. St. Pauli **39** (1990), 211-240.

[50] T. Shioda, Theory of Mordell-Weil lattices, Proc. ICM Kyoto, Vol. I (1990), 473-489.

[51] J. H. Silverman, *The Arithmetic of Elliptic Curves*, GTM **106**, Springer-Verlag, 1986.

[52] J. Tate, Problem 9: The general reciprocity law, In: *Mathematical developments arising from Hilbert problems* (Proc. Sympos. Pure Math., Vol. XXVIII, Northern Illinois Univ., De Kalb, Ill., 1974), 311-322, Amer. Math. Soc., Providence, R. I., 1976.

[53] C. Vafa, Geometric physics, Proc. ICM Berlin, Vol. I (1998), Doc. Math. 1998, Extra Vol. I, 537-556.

[54] H. A. Verrill, The L-series of certain rigid Calabi-Yau threefolds, J. Number Theory **81** (2000), 310-334.

[55] H. A. Verrill, Picard-Fuchs equations of some families of elliptic curves, In: *Proceedings on Moonshine and Related Topics*, Edited by J. McKay and A. Sebbar, CRM Proceedings and Lecture Notes, Vol. **30**, 253-268, AMS Providence, 2001.

[56] A. Weil, Arithmetic on algebraic varieties, In: *Collected Papers*, Vol. I, 116-126, Springer-Verlag, 1979.

[57] A. Weil, Abstract versus classical algebraic geometry, Proc. ICM Amsterdam, Vol. III (1954), 550-558, In: *Collected Papers*, Vol. II, 180-188, Springer-Verlag, 1979.

[58] A. Weil, Two lectures on number theory, past and present, In: *Collected Papers*, Vol. III, 279-302, Springer-Verlag, 1979.

[59] L. Yang, Product formulas on a unitary group in three variables, arXiv:math.NT/0104259.

[60] L. Yang, Geometry and arithmetic associated with Appell hypergeometric functions, Part I of this monograph.

[61] L. Yang, Hessian polyhedra, invariant theory and Appell hypergeometric functions, Part II of this monograph.

[62] N. Yui, Update on the modularity of Calabi-Yau varieties, In: *Calabi-Yau Varieties and Mirror Symmetry*, Edited by N. Yui and J. D. Lewis, Fields Institute Communications Vol. **38** (2003), 307-362, AMS Providence, 2003.

[63] F. Q. Gouvéa and N. Yui, Rigid Calabi-Yau threefolds over $\mathbb{Q}$ are modular, Expo. Math. **29** (2011), 142-149.

www.ingramcontent.com/pod-product-compliance
Ingram Content Group UK Ltd.
Pitfield, Milton Keynes, MK11 3LW, UK
UKHW052114170926
8624IPUK00026B/397

9 789813 209473